Key Methods in Geography

Key Methods in Geography

Second Edition

Edited by
Nicholas Clifford,
Shaun French and Gill Valentine

Los Angeles | London | New Delhi
Singapore | Washington DC

SAGE Publications Ltd
1 Oliver's Yard
55 City Road
London EC1Y 1SP

SAGE Publications Inc.
2455 Teller Road
Thousand Oaks, California 91320

SAGE Publications India Pvt Ltd
B 1/I 1 Mohan Cooperative Industrial Area
Mathura Road, Post Bag 7
New Delhi 110 044

SAGE Publications Asia-Pacific Pte Ltd
33 Pekin Street #02-01
Far East Square
Singapore 048763

Library of Congress Control Number 2010925556

British Library Cataloguing in Publication data

A catalogue record for this book is available from the British Library

ISBN 978-1-4129-3508-1
ISBN 978-1-4129-3509-8 (pbk)

Typeset by C&M Digitals (P) Ltd, Chennai, India
Printed by MPG Books Group, Bodmin, Cornwall
Printed on paper from sustainable resources

Contents

Notes on Contributors

Paul Aplin is Associate Professor in Geographical Information Science (GIS) in the School of Geography at Nottingham University. He has interests in environmental remote sensing, specializing in land cover investigation, classification, scaling issues and ecological applications. He is Chairman of the Remote Sensing and Photogrammetric Society and Book Series Editor for the International Society for Photogrammetry and Remote Sensing. His current research activities include knowledge transfer in terrestrial Earth Observation technologies (funded by the Natural Environment Research Council) and land cover classification in support of environmental assessment in Panama, Chile and South Africa (funded through various sources).

Rob Bartram is a graduate of University College London (BA Geography) and the University of Nottingham (PhD). He was Lecturer at the University of Sheffield Department of Geography with research and teaching interests in Social and Cultural Geography. He now works in medical general practice.

Michael Batty is Bartlett Professor of Planning and Director of the Centre for Advanced Spatial Analysis (CASA), University College London. His research is in the development of computer-based technologies, specifically graphics-based and mathematical models for cities, and he has worked recently on applications of fractal geometry and cellular automata to urban structure. He was awarded the CBE in the Queen's Birthday Honours List in 2004 and is a Fellow of the Royal Society (FRS). He was formerly Director of the SUNY Buffalo site of NCGIA (1990–1995). His most recent books are *Cities and Complexity* (MIT Press, 2005) and *Virtual Geographic Environments* (edited with Hui Lin, Science Press, 2009). He is also the editor of the journal, *Environment and Planning B*.

Iain S. Black is Fellow in Geography at Sidney Sussex College, Cambridge. He has written widely on the historical geography of money and banking in Britain between 1750 and 1950, with particular reference to the social, economic and architectural transformation of the City of London. Recently, these interests have been extended to work on the role of imperial and colonial banking groups in the diffusion of British banking culture overseas between 1850 and 1950, together with studies of the impact of empire on the landscape and built environment of London in the nineteenth and twentieth centuries.

Michael Bradford is Professor of Geography at the University of Manchester. He was Pro-Vice-Chancellor for Teaching and Learning from 2001 to 2004 and Head of the School of Geography from 1996 to 2000.

Myrna M. Breitbart is Professor of Geography and Urban Studies and Director of the Community Engagement and Collaborative Learning Network at Hampshire College where she has taught since 1977. Her teaching and research interests as well as publications focus on the broad themes of participatory planning and social action (with a special interest in young people), community economic development and struggles over urban public space. She is currently conducting research on the role of the arts and cultural economy in urban redevelopment with a special focus on small, post-industrial cities. Professor Breitbart has a strong commitment to community-based learning and participatory action research. She works closely with a number of community development and housing organizations as well as urban youth and community art organizations in Western Massachusetts.

Joanna Bullard is Reader in Aeolian Geomorphology at Loughborough University. She completed her undergraduate degree at Edinburgh University and her PhD at the University of Sheffield, specializing in the relationships between sand dune geomorphology, vegetation and climate. She has undertaken fieldwork on glaciers, in tropical forests, along various coastlines, and in sandy and rocky deserts around the world. Her current research focuses primarily on aeolian dust emissions in cold and hot deserts. She is the physical geography editor of the RGS-IBG book series and one of the associate editors of the journal *Earth Surface Processes and Landforms*.

Adrian Chappell is Senior Research Scientist with the Commonwealth Scientific and Industrial Research Organisation (CSIRO), Australia's national science agency and one of the largest and most diverse research agencies in the world. Prior to joining CSIRO he was Senior Lecturer at the University of Salford for more than ten years before becoming a Senior Geoscientist for a Geostatistics Consultancy company working on three-dimensional geophysical problems. His interest in geostatistics was initiated during his doctorate at University College London tackling the problem of sampling and mapping the spatial and temporal variation of soil erosion in Niger, West Africa. He developed his knowledge of geostatistics with diverse applications that include the spatial and temporal variation of aeolian sediment transport in Australia and the UK, African and Australian dust, wandering gravel bed-rivers in UK and West African Sahel rainfall. His applications of geostatistics are an attempt to improve across scales, the understanding of spatial variation in geomorphic processes and the implications for land surface formation.

Nick Clifford is Professor of River Science at Nottingham University and Visiting Professor in the School of Geography, Planning and Environmental Policy at University College Dublin. He received his BA and PhD from the University of Cambridge. His principal research interests are in fluvial geomorphology, and the history of ideas and methods in Geography. He is currently Director of the River Science Research Network, which is dedicated to improving the scientific basis for river management and rehabilitation. He is a member of the editorial board of

Catena and *River Research and Applications,* and is Managing Editor of *Progress in Physical Geography.* He has produced diverse publications, such as *Turbulence: Perspectives on Flow and Sediment Transport* (with J. French and J. Hardisty; John Wiley, 1993), and *Incredible Earth* (DK Books, 1996). He teaches courses in river form and processes, river restoration, and the history and philosophy of Geography.

Meghan Cope is Associate Professor and Chair of the Department of Geography at the University of Vermont. Her interests lie in the areas of critical youth geographies, urban race–class–gender intersections and qualitative research. She recently completed a four-year participatory research project on children's conceptualizations of urban space in inner-city Buffalo and she is now exploring teens' independent mobility and access to public space in Vermont and other locations.

Marcus A. Doel is Research Professor of Human Geography, and is the Head of the School of the Environment and Society at Swansea University. He studied Geography, Economics and Accountancy as an undergraduate at the University of Bristol, and pursued his PhD in continental philosophy and human geography there. He currently serves on the editorial boards of *Environment and Planning D: Society and Space* and the *International Journal of Baudrillard Studies,* and is a Fellow of the Royal Geographical Society with the Institute of British Geographers as well as being a member of the Association of American Geographers. In addition to poststructuralist spatial theory, his research interests include photography and film, postmodern fiction and conceptual art, consumer culture and risk society. His most recent (edited) book is entitled *Moving Pictures/Stopping Places: Hotels and Motels on Film* (Lexington Books, 2009). He is currently investigating the spatial theory of Alain Badiou, modernity's optical unconscious, and the work of space in community pharmacy and general practice.

Danny Dorling is Professor of Human Geography, Department of Geography, University of Sheffield. With colleagues, he is author of 25 books and 400 papers. Since 2006, Danny has been working with many others on remapping inequality worldwide (www.worldmapper.org). He is an Academician of the Academy of the Learned Societies in the Social Sciences and, in 2008, became Honorary President of the Society of Cartographers. In 2010, he joined the World Health Organization's Scientific Resource Group on Health Equity Analysis and Research. His most recent book, *Injustice: Why Social Inequality Persists,* is to be published by Policy Press in April 2010.

Richard Field is Associate Professor in Biogeography in the School of Geography at Nottingham University. He gained an MSc from the University of Durham, and a PhD from Imperial College London. His main interests are in global biodiversity patterns (particularly plants), macroecology, conservation ecology (particularly in Cusuco National Park, Honduras) and island biogeography (with a special interest in the ecological dynamics of the Krakatau islands, Indonesia). He is an Associate Editor of both *Global Ecology and Biogeography: A Journal of Macroecology* and

Frontiers of Biogeography. He is a member of six learned societies, including being a Fellow of the Royal Geographical Society and a founder member of the International Biogeography Society. He teaches courses in biogeography, quantitative methods and environmental management.

Shaun French is Lecturer in Economic Geography at the University of Nottingham. He gained his PhD from the University of Bristol and was subsequently seconded to the Bank of England to work on a study of access to finance for small and medium-sized enterprises in deprived areas. He has research interests in the geographies of financial risk technologies, financial subjects and processes of financial exclusion. Recent projects include an investigation of new modalities of life assurance (with James Kneale) and an analysis of the geography of the buy-to-let sector in the UK (with Andrew Leyshon).

Iain Hay is Australian Learning and Teaching Council Discipline Scholar for Arts, Social Sciences and Humanities as well as Professor of Geography at Flinders University, South Australia. He was awarded a LittD in 2009 by the University of Canterbury, New Zealand, for his research work over the past two decades on geographies of oppression and domination. One of his recent books is *Research Ethics for Social Scientists: Between ethical conduct and regulatory compliance* (Sage, 2006), with Mark Israel.

Mick Healey is Professor of Geography and Director of the Centre for Active Learning at the University of Gloucestershire. For many years he researched and taught economic geography. His current interests are in research and development in higher education, particularly the links between research and teaching. He is the Senior Geography Adviser for the National Subject Centre for Geography, Earth and Environmental Sciences. In 2000 he was awarded one of the first National Teaching Fellowships in England and Northern Ireland and in 2007 he was made a Senior Fellow of the Higher Education Academy.

Ruth L. Healey recently completed her PhD in Human Geography at the University of Sheffield. She is now a Lecturer in Human Geography at the University of Chester. Ruth is a social geographer, with particular interests in the settlement experiences of migrants and refugees in the UK. She is a member of the editorial board of the *Journal of Geography in Higher Education*.

Robin A. Kearns completed his PhD at McMaster University, Canada, in 1987 and is Professor of Geography in the School of Environment at the University of Auckland. His research focuses on the geography of health and health care and the cultural politics of place. He co-edits two journals: *Health and Social Care in the Community* and *Health & Place*.

Stuart N. Lane is Professor of Physical Geography and Director of the Institute for Hazard, Risk and Resilience at Durham University. Stuart completed his undergraduate degree at the University of Cambridge, and his PhD at Cambridge and City

University, London. He has wide-ranging interests in geographical methodology and history. He is a fluvial geomorphologist and hydrologist, with major research projects in the process–form relationships in gravel-bedded rivers, the numerical simulation of river flows using complex three-dimensional computer codes, and flood risk and diffuse pollution modelling.

Alan Latham is Lecturer in Geography at University College London. He received a BA from Massey University, and a PhD from the University of Bristol. He is an urban geographer, with interests in sociality and urban life, globalization and the cultural economy of cities. He has recently co-edited *Key Concepts in Urban Geography* (Sage, 2009).

Eric Laurier is Senior Research Fellow at the University of Edinburgh. His research projects have been on cafes and their place in civic life, cars and how we inhabit them and, currently, the workplace skills required for video-editing. Methodology has been an abiding fascination for him, the common methods of everyday life in parallel with the more arcane methods of professions, arts, sciences and social sciences.

Robyn Longhurst is Professor of Geography at the University of Waikato, New Zealand. Her areas of teaching and research include 'the body', feminist geography, the politics knowledge production and qualitative methodologies. She is author of *Maternities: Gender, Bodies and Space* (2008) and *Bodies: Exploring Fluid Boundaries* (2001), and co-author of *Space, Place and Sex: Geographies of Sexualities* (2010) and *Pleasure Zones: Bodies, Cities, Space*s (2001).

Clare Madge is Senior Lecturer in Geography at the University of Leicester. She is a feminist geographer with interests in critical development geographies and cyberspace. Her current research interests span critical development geographies, cybergeographies and feminist geographies. She serves on the editorial advisory board of *ACME: An International E-Journal for Critical Geographies*.

John H. McKendrick is Senior Lecturer in Human Geography and Director of the Scottish Poverty Information Unit at the School of Law and Social Sciences (Glasgow Caledonian University). His research interests span studies of the provision of environments for children, children's use of space, children's play and children and poverty. He has recently completed research for the Scottish Executive ('Life in Low Income Households with Children') and for sportscotland, Play Scotland and Grounds for Learning ('School Grounds in Scotland: a National Survey'). He edited *Children's Playgrounds in the Built Environment* (Alexandrine Press, 1999) and *First Steps*, a collection of 21 'retrospective' autobiographical short notes written by geographers on one of their favourite readings from the geographies of children and youths (2004).

Sara L. McLafferty is Professor of Geography at the University of Illinois at Urbana-Champaign. Her research explores the use of spatial analysis methods and GIS in analysing health and social issues in cities and women's access to social services and employment opportunities.

Miles Ogborn is Professor of Geography at Queen Mary, University of London. He has undertaken archival research on social policy in nineteenth-century Britain, the changing geographies of eighteenth-century London and the global networks of the English East India Company in the seventeenth-century. He has written and edited numerous books, including *Cultural Geography in Practice* (Arnold, 2003), *Georgian Geographies: Essays on Space, Place and Landscape in the Eighteenth Century* (Manchester University Press, 2004), *Indian Ink: Script and Print in the Making of the English East India Company* (University of Chicago Press, 2007), and *Global Lives: Britain and the World, 1550–1800* (Cambridge University Press, 2008).

Chris Perkins is Senior Lecturer in Geography in the School of Environment and Development, University of Manchester and Chair of the International Cartographic Association Maps and Society Commission. He is the author of seven books and numerous academic papers and is currently researching the social lives of mapping.

Stephen Rice is Reader in River Science in the Department of Geography at Loughborough University. He was an undergraduate at Oxford University, and gained his PhD from the University of British Columbia. His research interests fall within the general fields of fluvial geomorphology, fluvial sedimentology and river ecology, and he has active research projects examining the characterization of fluvial sediments using conventional sampling, photogrammetry and image analysis, and the role of tributary sediment inputs in punctuating benthic river ecosystems. He teaches courses in river management, physical geography, earth surface processes and landforms and quantitative methods.

Fiona M. Smith is a Lecturer in Human Geography at the University of Dundee, UK. Her research currently focuses on contemporary political and cultural geographies in Germany, and on geographies of volunteering in the UK. She is also interested in language and cross-cultural research. She has published widely on these issues and is co-author with Nina Laurie, Claire Dwyer and Sarah Holloway of *Geographies of New Femininities* (Longman, 1999).

Catherine Souch gained her PhD from the University of British Columbia. She is Head of Research and Higher Education at the Royal Geographical Society (with IBG) where she oversees the Society's scholarly publishing, grants, academic conferences and workshops, research groups and liaison with the UK Geography Departments in Higher Education. Previously, she was Professor in the Department of Geography at Indiana University, Indianapolis. She has research interests in paleoenvironmental reconstruction using geomorphic evidence in lacustrine/wetland environments, and in human impacts on the environment, and has taught courses in research and field methods in geography, climatic change and physical systems of the environment.

Alice Turkington is Associate Professor of Geography at the University of Kentucky. She is a geomorphologist interested in rock weathering processes in natural and urban landscapes and the relationship between weathering and landscape evolution. She has conducted fieldwork in the USA, Europe, Scandinavia, Australia and New Zealand.

Gill Valentine is Professor of Human Geography and Director of the Leeds Social Science Institute (LSSI) at the University of Leeds. She has an international reputation for theoretically informed empirical work that is methodologically innovative and has popular and policy impacts. Her research interests include social identities, citizenship and belonging, children and parenting, consumption cultures (especially in relation to food and drink) and research methods. Gill has been awarded numerous research grants and contracts from UK Research Councils, charities, government departments and non-governmental organizations.

Bettina van Hoven is a Researcher and Lecturer at the Department of Cultural Geography, University of Groningen, Netherlands. Her research interests are geographies of inclusion and exclusion.

Paul White is Professor of European Urban Geography in the Department of Geography at the University of Sheffield where he is also Pro-Vice Chancellor. His main research interests are in the settlement of international migrant groups in European cities. He has published 10 books, the most recent being *Global Japan: The Experience of Japan's New Immigrant and Overseas Communities* (with R. Goodman, C. Peach and A. Takenaka; RoutledgeCurzon, 2003).

Ellen Wohl is Professor in the School of Geosciences at Colorado State University, and obtained her PhD from the University of Arizona. She is a fluvial geomorphologist, whose current interests lie in the hydraulics, sediment transport and controls on channel morphology, human impacts on bedrock and mountain channels, and the role of floods in shaping channel morphology. In both 2000 and 2003 she was the recipient of the Association of American Geographers, G. K. Gilbert Award for excellence in geomorphological research.

List of Figures

List of Tables

Acknowledgements

We owe a continuing debt of thanks to Robert Rojek at Sage for commissioning this revision and for his editorial advice and support, and to all our contributors for their enthusiasm for the project. The quality of the final product is due to the efforts of the copyeditor Neil Dowden and the Sage Senior Production Editor, Katherine Haw.

We are grateful to the following for kind permission to reproduce their material: Figures 15.1 a,b,c images and data provided by Dr Henri Grissino-Mayer; Figures 15.2 and 15.3.

Figure 19.3 (from Lau, 2000).
Figure 19.5 (after Lau, 2000).

Figures 20.1 – 20.5 original data copyright: 2002 DigitalGlobe.

Figure 21.2 data taken from Frodin (2001).

Figure 21.4 Base map taken from MapInfo Professional version 7.0 in-built map.

Figure 24.2 data provided by J. Leys.

Figure 25.1 Digital Data Types: Raster and Vector Map Representations, adapted from the University of Melbourne's GIS Self-Learning Tool.

Every effort has been made to obtain permission to reproduce copyright material. If any proper acknowledgement has not been made we invite copyright holders to inform us of the oversight.

Getting Started in Geographical Research

Getting Started in Geographical Research: how this book can help

1

Nick Clifford, Shaun French and Gill Valentine

Synopsis

Geography is a very diverse subject that includes studies of human behaviour and the physical environment. It is also a discipline that embraces a very diverse range of philosophical approaches to knowledge (from positivism to post-structuralism). As such, geographers employ *quantitative methods* (statistics and mathematical modelling) and *qualitative methods* (a set of techniques that are used to explore subjective meanings, values and emotions such as interviewing, participant observation and visual imagery) or a combination of the two. These methods can be used in both *extensive research designs* (where the emphasis is on pattern and regularity in large 'representative' data sets, which is assumed to represent the outcome of some underlying (causal) regularity or process) and *intensive research designs* (where the emphasis is on describing a single case study, or small number of case studies, with the maximum amount of detail). Yet, despite this diversity, all geographers, whatever their philosophical or methodological approach, must make common decisions and go through common processes when they are embarking on their research. This means doing preparatory work (a literature review, thinking about health and safety and research ethics); thinking through the practicalities of data collection (whether to do original fieldwork or rely on secondary sources; whether to use quantitative or qualitative methods or a combination of both); planning how to manage and analyse the data generated from these techniques; and thinking about how to present/write up the findings of the research. This chapter aims to guide you through these choices if you are doing research for a project or dissertation. In doing so, it explains the structure and content of this book and points you in the direction of which chapters to turn to for advice on different forms of research techniques and analysis.

The main topics of the chapter are:

- Introduction: the nature of geographical research
- Quantitative and qualitative approaches to geography
- Designing a geographical research project
- The philosophy of research and importance of research design
- Conclusion: how this book can help you get started

INTRODUCTION: THE NATURE OF GEOGRAPHICAL RESEARCH

This book aims to help you prepare for, design and carry out geographical research, and to analyse and present your findings. Geographers have given attention to an enormous range of subject matter. Most aspects of the world, whether physically or environmentally determined, or politically, economically or culturally constructed, have been considered as suitable for geographical research. Moreover, the range of geographical enquiry continues to increase. Traditionally, geographers considered the contemporary human and physical world together with their historical configurations, thus extending geographies to the past as well as to the present. Now, in both physical and human geography, the range is even greater (see, for example, Walford and Haggett, 1995; Gregory, 2000; Thrift, 2002; Gregory 2009). Physical geographers have access to new techniques of absolute and relative environmental dating, and greater ability to gather, analyse and visualize large amounts of data. They can reconstruct palaeoenvironments and landform development, as well as model this into the future, over timescales ranging from years to geological epochs. Physical geography is increasingly conducted under the umbrella of 'Earth System Science' which stresses interconnections between bio-physical atmospheric and earth science, and also includes human activity as a driver or response to earth and environmental change (Pitman, 2005). In human geography, technological advances in areas such as GIS allow more flexible and more creative analysis of data, facilitating 'virtual geographies' which exist only in 'hyperspace'. For the less technically minded, the subject is probing areas traditionally within the domains of psychology and cultural anthropology: there are now, for example, imagined and mystic geographies, whose foundations, or connections with the 'real' world are almost entirely interpretational, rather than empirical. All of these new areas of geographical exploration bring challenges of interpretation, as methods of research associated with them may be radically different (even fundamentally irreconcilable with one another), or so new that they have yet to be formalized into transferable schemes to inform other research programmes.

Until the 1980s, geography was cast largely as either a physical (environmental or geological) science, a social science, or some combination of the two. This implied a commonality of objective, if not entirely of method: there was a shared commitment to the goal of 'general' explanation. More recently, however, some would dispute the use of the term 'science' in any of its forms in certain areas of subject (for an excellent introduction to debates surrounding science – its meaning, construction and application – see Chalmers, 1990). Instead, the 'cultural' turn in human geography (which in part reflects the growing influence of feminist and post-structural approaches) has brought a new emphasis on meanings, representation, emotions and so on that is more readily associated with the arts. (These issues are discussed in more detail by Barnes (2001) and in various chapters in Clifford *et al.*, 2009.)

Given the breadth of geographical enquiry, it is not surprising that the subject is similarly broad with respect to the methods it employs, and the philosophical and ethical stances it adopts. This book reflects the diversity of contemporary geography, both in the number, and the range, of chapters which it contains. In this chapter, we want to briefly introduce you to different approaches to research methods and design, and to offer some guidance on how you might develop your own research design for a geographical project using this book.

QUANTITATIVE AND QUALITATIVE APPROACHES TO GEOGRAPHY

The chapters in the book loosely deal with two forms of data collection/analysis: quantitative and qualitative methods and techniques. Quantitative methods involve the use of physical (science) concepts and reasoning, mathematical modelling and statistical techniques to understand geographical phenomena. These form the basis of most research in physical geography. They first began to be adopted by human geographers in the 1950s, but it was in the 1960s – a period dubbed the 'quantitative revolution' – that their application became both more widespread and more sophisticated in Anglo-American geography. It was at this time that, influenced by the 'scientific' approaches to human behaviour that were being adopted by social sciences such as economics and psychology, some human geographers began to be concerned with scientific rigour in their own research. In particular, they began to use quantitative methods to develop hypotheses and to explain, predict and model human spatial behaviour and decision making (Johnston, 2003). (Collectively, the adoption of 'objective', quantified means of collecting data, hypothesis testing and generalizing explanations is known as positivism.) Much of this work was applied to planning and locational decision making (Abler *et al.*, 1971; Haggett, 1965).

In the 1970s, however, some geographers began to criticize positivist approaches to geography, particularly the application of 'objective' scientific methods that conceptualized people as rational actors (Cloke *et al.*, 1991). Rather, geographers adopting a humanistic approach argued that human behaviour is, in fact, subjective, complex, messy, irrational and contradictory. As such, humanistic geographers began to draw on methods that would allow them to explore the meanings, emotions, intentions and values that make up our taken-for-granted lifeworlds (Ley, 1974; Seamon, 1979). These included methods such as in-depth interviews, participant observation and focus groups. At the same time, Marxist geographers criticized the apolitical nature of positivist approaches, accusing those who adopted them for failing to recognize the way that scientific methods, and the spatial laws and models they produced, might reproduce capitalism (Harvey, 1973). More recently, feminist and post-structuralist approaches to geography have criticized the 'grand theories' of positivism and Marxism, and their failure to recognize people's multiple subjectivities. Instead, the emphasis is on refining qualitative methods to allow the voice of informants to be heard in ways which are non-exploitative or oppressive (WGSG, 1997; Moss, 2001) and to focus on the politics of knowledge production, particularly in terms of the positionality of the researcher and the way 'other' people and places are represented (articles in a special issue of *Environment and Planning: D*, 1992; Moss, 2001).

Humanistic inquiry did not just prompt interest in people's own account of their experiences, but also in how these experiences are represented in texts, literature, art, fiction and so on (Pocock, 1981; Daniels, 1985). Again, such visual methodologies have also been informed and developed by the emergence of post-structuralist approaches to geography which have further stimulated human geographers' concerns with issues of representation.

Despite the evolving nature of geographical thought and practice, both quantitative and qualitative approaches remain important within the discipline of Geography. While taken at face value they appear to be incompatible ways of 'doing' research, it is important not to see these two approaches as binary opposites. Subjective concerns often inform the development and use of quantitative methods. Likewise, it is also possible to work with qualitative material in quite scientific ways. Whatever methods are adopted, some degree of philosophical reflection is required to make sense of the research process. Equally, the two approaches are often combined in research designs in a process known as mixing methods (see below).

DESIGNING A GEOGRAPHICAL RESEARCH PROJECT

Faced with a bewildering array of possibilities, both in what to study and in how to approach this study, it may seem that geographical research is difficult to do well. However, the very range of geographical enquiry is also a source of excitement and encouragement. The key is to harness this variety, rather than to be overwhelmed by it. Essentially, geographical research requires perhaps more thought than any of the other human or physical academic disciplines. Whether this thought is exercised with the assistance of some formal scheme of how to structure the research programme, or whether it is exercised self-critically, or reflexively in a much less formal sense, is less important than the awareness of the opportunities, limitations and context of the research question chosen, the appropriateness of the research methods selected, the range of techniques used to gather, sort and display information, and ultimately, the manner and intent with which the research findings are presented. For student projects, these questions are as much determined by practical considerations, such as the time available for the project, or the funding to undertake the research. These limitations should be built into the project at an early stage, so that the likely quality of the outcomes can be judged in advance. None of the constraints should be used after the research is completed to justify a partial answer or unnecessarily restricted project.

The 'scientific' view

Conventionally, geographical research programmes have been presented as a sequence of steps, or procedures (Haring and Lounsbury, 1983 – see below). These steps were based upon the premise that geography was an essentially scientific activity, that is, a subject identifying research questions, testing hypotheses regarding possible causal relationships, and presenting the results with some sort of more general (normative) statement or context. The aim of separating tasks was to enable time (and money) to be budgeted effectively between each, and to encourage a structuring of the thought processes underpinning the research.

The steps identified in this form of 'scientific geographic research' (Haring and Lounsbury, 1983) are as follows:

- *Formulation of the research problem* – which means asking a question in a precise, testable manner, and which requires consideration of the place and time-scale of the work.
- *Definition of hypotheses* – the generation of one or more assumptions which are used as the basis of investigation, and which are subsequently tested by the research.
- *Determination of the type of data to be collected* – how much, in what manner is sampling or measurement to be done.
- *Collection of data* – either primary from the field or archive, or secondary, from the analysis of published materials.
- *Analysis and processing of the data* – selecting appropriate quantitative and presentational techniques.
- *Stating conclusions* – nowadays, this might also include the presentation of findings verbally or in publication.

Today, there is more recognition that these tasks are not truly independent, and that an element of reflexivity might usefully be incorporated in this process. In some areas of the subject – particularly human geography – the entire notion of a formalized procedure or sequence would be considered unnecessary, and the notion of normative, problem-solving science would, at best, be considered applicable to a restricted range of subjects and methodology. Rather, as outlined above, many human geographers now reject or are sceptical of scientific approaches to human behaviour, preferring to adopt a more subjective approach to their research. Nevertheless, having said this, most qualitative research also involves many of the same steps outlined in the mechanical or scientific formulation above – albeit not conceptualized in quite the same way. For example, qualitative researchers also need to think about what research questions to ask, what data need to be collected and how this material should be analysed and presented. In other words, all research in Geography – whatever its philosophical stance – involves thinking about the relationships between methods, techniques, analysis and interpretation. This important role is filled by *research design*.

The importance of research design

In its broadest sense, research design results from a series of decisions we make as researchers. These decisions flow from our knowledge of the academic literature (see Chapter 2), the research questions we want to ask, our conceptual framework, and our knowledge of the advantage and disadvantages of different techniques (see Chapters 6–10 and 13–18). The research design should be an explicit part of the research: it should show that you have thought about how, what, where, when and why!

There are at least six key things you need to bear in mind to formulate a convincing research design:

I Think about what research questions to ask
On the basis of your own thinking about the topic, the relevant theoretical and empirical literatures (see Chapter 2), and consulting secondary material (see

Chapter 5) – and if possible having discussed it with other students and your tutor – you need to move towards framing your specific research questions. For a human geographer, these might include questions about what discourses you can identify, what patterns of behaviour/activity you can determine, what events, beliefs, and attitudes are shaping people's actions, who is affected by the issue under consideration and in what ways, and how social relations are played out, and so on. For a physical geographer, these might include questions concerning the rate of operation and location of a certain geomorphological process, the morphology of a selected set of landforms, or the abundance and diversity of particular plant or animal species in a given area (many of the chapters in this volume provide examples of research problems).

It is important to have a strong focus to your research questions rather than adopting a scatter-gun approach asking a diverse range of unconnected questions. This also means bearing in mind the time and resource constraints on your research (see below). At the same time as you develop a set of core aims it is also important to remain flexible, and to remember that unanticipated themes can emerge during the course of fieldwork which redefine the relevance of different research questions, likewise, access or other practical problems can prevent some research aims being fulfilled and lead to a shift in the focus of the work. As such, you should be aware that your research questions may evolve during the course of your project.

2 Think about the most appropriate method(s) to employ

There is no set recipe for this: different methods have particular strengths and collect different forms of empirical material. The most appropriate method(s) for your research will therefore depend on the questions you want to ask and the sort of information you want to generate. Chapters 5–22 outline the advantages and disadvantages of core methods used by human and physical geographers. While many projects in human and physical geography involve going out into the field – for example to interview or observe people, or to take samples or measurements – it is also possible to do your research without leaving your computer, living room or the library. For example, research can be based on visual imagery such as films and television programmes (see Chapters 10 and 30); secondary sources including contemporary and historical/archival material (see Chapters 5, 15 and 29); or GIS and remote sensing (Chapters 20 and 25). Some human geographers are also experimenting with conducting interviews and surveys by email or in chat-rooms (see a brief discussion of this in Chapter 13).

In the process of research design it is important not to view each of these methods as an either/or choice. Rather, it is possible (and often desirable) to mix methods. This process of drawing on different sources or perspectives is known as *triangulation*. The term comes from surveying, where it describes using different bearings to give the correct position. Thus, researchers can use multiple methods or different sources of information to try and maximize an understanding of a research question. These might be both qualitative and quantitative (see, for example, Sporton, 1999). Different techniques should each contribute something unique to the project (perhaps addressing a different research question or collecting a new type of data) rather than merely being repetitive of each other.

3 Think about what data you will produce and how to manage it

An intrinsic element of your choice of method should not only involve reflecting on the technique itself, but also how you intend to analyse and interpret the data that you will produce. Chapters 6 and 15–26 all discuss how to analyse different forms of quantitative material, while Chapters 8–14 and 27–30 demonstrate how to bring a rigorous analysis to bear on interview transcripts/diary material, historical and archive sources and cultural material. For example, Chapters 17, 22, 23 and 26 discuss some of the issues you need to think about when deciding which statistical techniques to apply to quantitative data. Chapters 27 and 28 present alternative methods of coding interview transcripts/diary material: one manually and the other using computer software. While qualitative techniques emphasize quality, depth, richness and understanding, instead of the statistical representativeness and scientific rigour which are associated with quantitative techniques, this does not mean that they can be used without any thought. Rather, they should be approached in as rigorous a way as quantitative techniques.

4 Think about the practicalities of doing fieldwork

The nitty-gritty practicalities of who, what, when, where, and for how long inevitably shape the choices we can make about our aims, methods, sample size, and the amount of data we have the time to analyse and manage (see Chapter 17 on sampling and Chapter 6 on handling large amounts of qualitative data using computer software packages). Increasingly, the kind of work which is permissible is constrained by changing attitudes and legislative requirements with respect to safety and risk which ultimately define the range and scale of what you can achieve (see, for example, Chapter 4 on the health and safety limitations of fieldwork). It is important to bear in mind that the research which is written up by academics in journals and books is often conducted over several years and is commonly funded by substantial grants. Thus, the scale this sort of research is conducted on is very different to that at which student research projects must be pitched. It is not possible to replicate or fully develop in a three-month student dissertation or project all of the objectives of a two-year piece of academic research that you may have uncovered in your literature review! Rather, it is often best to begin by identifying the limitations of your proposed study and recognizing what you will and will not be able to say at the end of it. Remember that doing qualitative or quantitative work in human geography, just like fieldwork in physical geography, requires a lot of concentration and mental energy. It is both stressful and tiring, so there is a limit to how much you can achieve in the field in any one day. Other practicalities such as the availability of field equipment, tape recorders, cameras, transcribers or access to transport can also define the parameters of your project.

Drawing up a time-management chart or work schedule at the research-design stage can be an effective way of working out how much you can achieve in your study, and later on can also serve as a useful indicator if you are slipping behind. While planning ahead (and in doing so, drawing on the experience of your tutor and other researchers) is crucial to developing an effective research design it is also important to remember that you should always remain flexible.

5 Think about the ethical issues you need to consider

An awareness of the ethical issues which are embedded in your proposed research questions and possible methodologies must underpin your final decisions about the research design. The most common ethical dilemmas in human geography focus around: participation, consent, confidentiality/safe guarding personal information and giving something back (see Alderson, 1995; Valentine, 1999). In physical geography, ethical issues involve not only questions of consent (for example to access field sites on private land) but also the potential impacts of the research techniques on the environment (e.g. pollution). Thus, while ethical issues may seem routine or moral questions rather than anything which is intrinsic to the design of a research project, in practice they actually underpin what we do. They can shape what questions we can ask, where we make observations, who we talk to, and where, when and in what order. These choices in turn may have consequences for what sort of material we collect, how it can be analysed and used, and what we do with it when the project is at an end. As such ethics are not a politically correct add-on but should always be at the heart of any research design (see Chapter 3 for an overview of ethical issues, Chapters 9 and 11 for the specific ethics of participatory research, Chapter 12 about the specific ethical issues involved working in different cultural contexts and Chapter 13 regarding ethical issues in internet mediated research).

6 Think about the form in which your research is to be presented

The scale and scope of your research design will partially be shaped by your motivations for doing the research and what you intend to use the findings for. If you are presenting your findings in a dissertation it will very different to a piece of work than if it is to be presented as a report or in a verbal seminar. Chapter 31 outlines and illustrates these different forms of presentational style in detail, and Chapter 22 describes how to use maps in your work. Likewise, when you are designing, conducting and writing up your research, it is important to bear in mind the assessment criteria by which your findings will be judged. The final chapter in this volume, 32, explores some of the ways that your work might be assessed.

THE PHILOSOPHY OF RESEARCH AND IMPORTANCE OF RESEARCH DESIGN

The most basic, formal, distinction in research design is that between *extensive* and *intensive* approaches. The important aspects of these contrasting approaches have been explored in some detail for the social sciences by Sayer (1992). This book is an ideal and thought-provoking introduction to the ways in which research seeks to make sense of a complex world. Sayer reviews theories of causation and explanation in which 'events' (what we observe) are thought to reflect the operation of 'mechanisms' which, in turn, are determined by basic, underlying 'structures' in the world. The way in which explanations are obtained reflects differing degrees of 'concrete' and 'abstract' research – that is, how much our generalizations rely

Table 1.1   The essential differences between extensive and intensive research designs

Notes	Intensive	Extensive
Research question	How? What? Why? In a certain case or example	How representative is a feature, pattern, or attribute of a population?
Type of explanation	Causes are elucidated through in-depth examination and interpretation	Representative generalizations are produced from repeated studies or large samples
Typical methods of research	Case study. Ethnography. Qualitative analysis	Questionnaires, large-scale surveys. Statistical analysis
Limitations	Relationships discovered will not be 'representative' or an average/ generalization	Explanation is a generalization – it is difficult to relate to the individual observation. Generalization is specific to the group/population in question
Philosophy	Method and explanation rely on discovering the connection between events, mechanisms and causal properties	Explanation based upon formal relations of similarity and identification of taxonomic groups

Source: Based on Sayer (1992, Figure 13, p. 243)

on observation, and how much they rely on interpretation of the ways in which events, mechanisms and structures are related. In the physical sciences, more attention has been given to the broader topic of scientific explanation of which research design is a part, but recently within physical geography, too, the implications of extensive and intensive research designs and the varying philosophies of research have begun to be explored (e.g. Richards, 1996).

Both extensive and intensive designs are concerned with the relationship between individual observations drawn from measurement programmes or case studies, and the ability to generalize on the basis of these observations. The detailed distinctions are illustrated in Table 1.1 but the essential differences are as follows:

- In an extensive research design, the emphasis is on pattern and regularity in data, which is assumed to represent the outcome of some underlying (causal) regularity or process. Usually, large numbers of observations are taken from many case studies so as to ensure a 'representative' dataset, and this type of design is sometimes referred to as the 'large-n' type of study (see Chapter 17 on sampling).
- In an intensive research design, emphasis is on describing a single, or small number of case studies with the maximum amount of detail. This approach is therefore sometimes known as the 'small-n' type of study. In anthropology, the term 'thick description' has been used (Geertz, 2000). In an intensive design, by thoroughly appreciating the operation of one physical or social system, or by immersion in one culture or social group, elements of a more fundamental, causal nature are sought. 'Explanation' is therefore concerned with disclosing the links

between events, mechanisms and structures. General explanations are derived from identification of the structures underlying observation, and from the possible transferring of the linkages discovered from detailed 'instantiations'.

Importantly, both approaches may be undertaken in quantitative or qualitative fashions – there is no necessary distinction in the *techniques* used. The two approaches are, however, separated to some extent in their philosophical underpinnings, and, more obviously, in the practical, logistical requirements they impose.

Philosophically, the extensive approach relies on the idea that the data pattern necessarily reflects an underlying cause, or process, which is obscured only by measurement error, or 'noise'. However, in the 'real' world, it is rare that one cause would lead directly, or simply, to another 'effect' – the chain of causation is more obscure, and 'noise' may be an essential part of the 'causation', reflecting the presence of some other (unknown or uncontrolled) effect which merely mimics the apparent pattern. There is the related problem of being unable to explain individual occurrences on the basis of 'average' behaviour of entire groups – the so-called ecological fallacy. In an intensive research design, there is a deeper appreciation of the 'layers' which separate observations from an underlying (causal) reality. As such (and at the risk of considerable oversimplification) extensive approaches have often been linked to positivist methodology and philosophy, and intensive approaches to realist methodologies and philosophies.

Practically, the different types of research design have clearly different requirements in both data type and amount, and with respect to cost and time. The extensive design lends itself to situations where large amounts of data are already published, or where large amounts of data can be generated from secondary sources. In many student projects, the need for many observations across comparative or contrasting field sites may be too daunting or logistically impossible if an attempt is made to mount an extensive research design based upon primary data sources. An exception to this is in laboratory-type studies, where a series of experiments may quickly build up a dataset representative of a wider range of conditions. The intensive design is, perhaps more common, but care is needed to 'tease-out' those aspects of the study which might disclose basic, causal processes.

CONCLUSION: HOW THIS BOOK CAN HELP YOU GET STARTED

Geographical research is complicated, because geographical phenomena are many and varied, and because they may transcend multiple scales in space and time. Further, over the last few decades, geographers have adopted a diverse range of philosophic stances, methods and research designs in their efforts to understand and interpret the human and physical worlds. In order to make sense of this variety in a research context, a considerable amount of thought must go in to geographical research at all of its stages. Although the prospect of embarking on your own research might seem daunting this book has been put together to help make it easier for you. The first five chapters in this book all offer advice and guidance about how to prepare your research project. This chapter has explained the process of research design; Chapter 2 describes how to do a literature search to help define your topic and research questions; Chapter 3 raises some of the ethical issues you might need to consider in your research design; Chapter 4 explains the practical and logistical issues you need to plan for in terms of your own health and safety in the field; and Chapter 5 outlines some of the secondary data sources that might inform, or form the basis of your work. In addition, Chapters 11 and 12 locate these issues within the specific situations of working in different cultural contexts and doing participatory research, while Chapter 13 addresses the uses and implications of internet-mediated research.

Other chapters explain how to use a range of quantitative and qualitative methods in human and physical geography, including: questionnaire surveys (Chapter 6), finding historical data (Chapter 7), semi-structured interviews and focus groups (Chapter 8), participant observation (Chapter 9), visual imagery (Chapter 10), diaries (Chapter 14) paleo and historical data sources (Chapter 15), making observations and measurements in the field (Chapter 16), numerical modelling (Chapter 19) and using remotely sensed data (Chapter 20) GIS (Geographical Information Systems) (Chapter 25). In addition, there is a specific chapter on sampling in physical and human geography (Chapter 17). You will obviously not want to use them all, but by reading across the chapters you will get a feel for the different ways that you might approach the same topic, and the advantages and disadvantages of different methods. When you have developed your own research design the chapter(s) appropriate to your chosen method(s) will give you practical advice about how to go about your research.

Further chapters then go on to explain how to analyse the sort of data collected by this diverse range of methods. Specifically, there are chapters on how to analyse qualitative interview and diary materials both manually and using computer software (Chapters 27 and 28), how to analyse historical/archive sources (Chapter 29) and cultural criteria (Chapter 30); and how to use statistics in a variety of different ways (see Chapters 21, 22, 24 and 26).

Finally, Chapter 31 explains how to present your findings in a variety of formats, and Chapter 32 deals with what sort of research criteria might be used to assess your work.

As with this chapter, each chapter in this book contains a synopsis at the beginning which briefly defines the content of the chapter and outlines the way the chapter is structured. At the end of each chapter, is a summary of key points and an annotated list of key further readings. Several of the chapters also contain boxed tips or useful exercises to develop your understanding of the topic in question.

Doing your own research can be one of the most rewarding aspects of a degree. It is your chance to explore something that really interests or motivates you and to contribute to geographical knowledge, so enjoy it. And good luck!

Summary

- Geographers are faced with a vast array of potential subject matter, techniques for data collection, visualization and analysis.
- Research design is crucial to link data collection, methods and techniques together, and to produce convincing, meaningful results.

- The basic choice in research design is between extensive and intensive designs. These have very different implications in terms of data collection, analysis and interpretation, although both quantitative and qualitative methods may be used in either.
- Whichever form of research design is adopted, the ethical dimension of research must be considered.
- Frequently, practical issues of land access, time or financial constraints, or field safety will, together with ethical issues, determine the scope and kind of research adopted. While these are constraints, prior consideration of their likely effects will minimize the loss of intellectual integrity and merit in the project.

Further reading

- The different philosophies underlying human geography research are outlined in numerous volumes including Cloke *et al.* (1991), Graham (1997) and Johnston (1997). Various chapters in Clifford *et al.* (2009) explain the tensions between geography as a physical science, social science or arts subject and show how geographers' understandings of key concepts have evolved as approaches to geographical thought have developed.

- Sayer (1992) is a thorough and extensive treatment of 'methodology' and its connections to the way in which we make sense of the world through observations, experiments, surveys and experiences. It argues that our philosophy of the way in which the world is structured (how things come to be as they are seen) must inform our choice of research design, and our choice of techniques for generalizing on the basis of the information which we collect from putting the research design into practice. The book depends on a particular ('realist') approach, but is an excellent starting point from which to relate philosophy to the practicalities of doing research, and to the strengths and weakness of particular kinds of research methods.

- Chalmers (1990) is a student-centred volume which covers an enormous range of material dealing with more modern 'post-positivist' approaches to the philosophy of science. It examines the nature of scientific explanation (the relations between observations, experiments and generalization), the social and political dimensions of science and scientific research, and the way that scientific explanation and scientists have their own sociology of knowledge.

- The edited book by Limb and Dwyer (2001) provides a critical introduction to qualitative methodologies. Each method is illustrated with examples of the authors' own research experiences and practices.

- Moss (2001) is a collection of essays exploring feminist geography in practice. These essays share a particular concern with the ethics and politics of knowledge production, notably in terms of the positionality of the researcher and the way 'other' people and places are represented.

- Rhoads and Thorn (1996) is an edited volume of papers which were presented at the 27th Annual Binghampton Symposium in Geomorphology, held in 1996. It has some excellent and entertaining chapters dealing with changing ideas concerning science and its methods, the motivation of current geomorphological research and researchers, case studies of contrasting modelling approaches, and chapters dealing with differing approaches to explanation in geomorphology and the earth sciences.

- Gregory (2000) provides an up-to-date account of the way in which physical geography is structured, the research which physical geographers undertake and the methods which they use. It seeks to present physical geography as a changing discipline, but one with strong connections to its past, and with bright prospects for the future.

- The journal *Ethics, Place and Environment* contains articles dealing with all aspects of ethical concerns and practice within Geography.

Note: Full details of the above can be found in the references list below.

References

Abler, R.F., Adams, J.S. and Gould, P.R. (1971) *Spatial Organization: The Geographer's View of the World.* Englewood Cliffs, NJ: Prentice-Hall.

Alderson, P. (1995) *Listening to Children: Children, Ethics and Social Research.* Ilford: Barnardos.

Chalmers, A.F. (1990) *Science and its Fabrication.* Buckingham: Open University Press.

Clifford, N.J., Holloway, S.L., Rice, S.P. and Valentine, G. (2009) *Key Concepts in Geography.* London: Sage. Second edition.

Cloke, P., Philo, C. and Sadler, D. (eds) *Approaching Human Geography.* London: Paul Chapman.

Environment and Planning D: Society & Space (1992) Special issue on the politics of knowledge, vol. 10.

Geertz, C. (2000) *The Interpretation of Cultures* (2nd edn). New York: Basic Books.

Graham, E. (1997) 'Philosophies underlying human geography research', in R. Flowerdew and D. Martin (eds) *Methods in Human Geography.* Harlow: Longman.

Gregory, D.J. (2009) 'Geography', in D.J. Gregory, R.J. Johnson, G. Pratt, M.J. Watts and S. Whatmore (eds) *The Dictionary of Human Geography* (5th edn). Chichester: Wiley-Blackwell, pp. 287–295.

Gregory, K.J. (2000) *The Changing Nature of Physical Geography.* London: Arnold.

Haggett, P. (1965) *Locational Analysis in Human Geography.* London: Edward Arnold.

Haring, L.L. and Lounsbury, J.F. (1983) *Introduction to Scientific Geographical Research.* Dubuque: W.C. Brown.

Harvey, D. (1973) *Social Justice and the City.* London: Edward Arnold.

Holloway, S.L., Rice, S. and Valentine, G. (eds) (2003) *Key Concepts in Geography.* London: Sage.

Johnston, R.J. (1997a) *Geography and Geographers: Anglo-American Human Geography since 1945.* London: Arnold.

Johnston, R.J. (2003) 'Geography and the social science tradition', in S.L. Holloway, S. Rice and G. Valentine (eds) *Key Concepts in Geography.* London: Sage.

Ley, D. (1974) *The Black Inner City as Frontier Outpost: Image and Behaviour of a Philadelphia Neighbourhood.* Monograph Series 7. Washington, DC: Association of American Geographers.

Limb, M. and Dwyer, C. (eds) (2001) *Qualitative Methodologies for Geographers.* London: Arnold.

Moss, P. (ed.) (2001) *Feminist Geography in Practice.* Oxford: Blackwell.

Pitman, A.J. (2005) On the role of Geography in Earth System Science, *Geoforum* 36: 137–148. (See also subsequent discussions.)

Pocock, D.C.D. (ed.) (1981) *Humanistic Geography and Literature: Essays on the Experience of Place.* London: Croom Helm.

Rhoads, B.L. and Thorn, C.E. (eds) (1997) *The Scientific Nature of Geomorphology.* Chichester: John Wiley and Sons.

Richards, K. (1996) 'Samples and cases: generalisation and explanation in geomorphology', in B.L. Rhoads and C.E. Thorn (eds) *The Scientific Nature of Geomorphology.* Chichester: John Wiley and Sons, pp. 171–190.

Sayer, A. (1992) *Method in Social Science.* London: Routledge.

Seamon, D. (1979) *A Geography of a Lifeworld.* London: Croom Helm.

Sporton, D. (1999) 'Mixing methods of fertility research', *The Professional Geographer*, 51: 68–76.

Thrift, N. (2002) 'The future of Geography', *Geoforum*, 33: 291–298.

Valentine, G. (1999) 'Being seen and heard? The ethical complexities of working with children and young people at home and at school', *Ethics, Place and Environment*, 2: 141–155.

Walford, R. and Haggett, P. (1995) 'Geography and geographical education: some speculations for the twenty-first century', *Geography*, 80: 3–13.

WGSG (Women and Geography Study Group) (1997) *Feminist Geographies: Explorations in Diversity and Difference.* London: Longman.

2 How to Conduct a Literature Search

Mick Healey and Ruth L. Healey

Synopsis

Identifying the most relevant, up-to-date and reliable references is a critical stage in the preparation of essays, reports and dissertations, but it is a stage which is often undertaken unsystematically and in a hurry. This chapter is designed to help you improve the quality of your literature search.

The chapter is organized into the following sections:

- The purpose of searching the literature
- Making a start
- A framework for your search
- Managing your search
- Search tools
- Evaluating the literature

THE PURPOSE OF SEARCHING THE LITERATURE

The purpose of this chapter is to support you in developing and using your literature search skills over a range of media, including paper and the web, not just books and journals. It is aimed primarily at undergraduate geography students needing to search the literature for research projects, dissertations and essays in human and physical geography. However, the search methods and principles are applicable to most subjects and, if you are a postgraduate geography student, you should also find it a useful refresher to get you started. Many sources are available worldwide, though details of accessing arrangements may vary. Country-specific sources are illustrated with selected examples of those available in the UK, North America and Australia.

Exercise 2.1   Why read?

Make a list of the reasons why you should read for a research project. Compare your list with those in Box 2.1. Most also apply if you are preparing an essay.

Reading the literature is an important element of academic research. It is a requirement with essays and projects as well as dissertations for you to relate your ideas to the wider literature on the topic. Reading around the subject will also help you broaden and refine your ideas, see examples of different writing styles and generally improve your understanding of the discipline. When undertaking a dissertation or thesis, reading will help you identify gaps, find case studies in other areas which you may replicate and then compare with your findings, and learn more about particular research methods and their application in practice (Box 2.1). Effective reading may, of course, take many different forms depending on your purpose – from skimming, through browsing, to in-depth textual analysis – and will rarely involve just reading from the beginning to the end (Kneale, 2003).

Box 2.1   Ten reasons for reading for research

1 It will give you ideas.
2 You need to understand what other researchers have done in your area.
3 To broaden your perspectives and set your work in context.
4 Direct personal experience can never be enough.
5 To legitimate your arguments.
6 It may cause you to change your mind.
7 Writers (and you will be one) need readers.
8 So that you can effectively criticize what others have done.
9 To learn more about research methods and their application in practice.
10 In order to spot areas which have not been researched.

Source: Blaxter *et al.* (2006: 100)

MAKING A START

Your literature search strategy will vary with your purpose. Sometimes you may want to search for something specific, for example a case study to illustrate an argument. In other situations a more general search may be required; for example, you might wish to identify 15 articles which have been written on a particular topic for an essay. Your search strategy may also vary with the level at which you are in the higher education system and your motivation. Identifying half a dozen up-to-date books on a topic may be appropriate at the beginning stage, while at graduate level you may need to explore whether any PhD theses have been written on the topic you are proposing to undertake for your thesis (e.g. Theses Abstracts: www.theses.com).

Exercise 2.2   Starting your search

You have been set a research project on a topic you know little about (e.g. organic farming). Before reading any further, write down the first three things you would do to find out what has already been written on the topic.

When we have used Exercise 2.2, the most common responses were to look in the subject section of the library catalogue, use a search engine on the internet and ask a lecturer. These are all sensible strategies, though the usefulness of most search engines is exaggerated due to the lack of regulation and quality control on the sites found. However, apart from possibly asking the lecturer/professor who set the assignment for one or two key references to get you going, usually the first things you should do are to identify and define the key terms in the assignment and construct a list of terms to use in your literature search. Only then is it appropriate to turn to the search tools, such as library catalogues, reference books, indexes, databases and websites, and seek help from a librarian.

Making a start is usually the most difficult stage of undertaking a research project or assignment. The issues involved in identifying your own research topic were discussed in Chapter 1. When you have a provisional idea about your topic and the research methods you may use, or when a research project or assignment is given to you, take a little while to plan your literature search. Defining the key terms in the topic or assignment is a good starting point. The dictionaries of human and physical geography are essential references for all geography students (Thomas and Goudie, 2000; Gregory *et al.*, 2009). The indexes of appropriate textbooks will also help. These references will further help you identify search terms, as will a thesaurus, a good English dictionary and a high-quality encyclopedia. For example, the Oxford Reference Online (www.oxfordreference.com) includes subject dictionaries and an encyclopaedia as well as English dictionaries. The Geobase subject classification is another source (see the section below on abstracts and reviews). Remember to allow for American English spellings of words as well as standard English spellings.

In identifying search terms, group them into three categories: broader, related and narrower. The first will be useful in searching for books, which may contain useful sections on your topic, while the second and third will be particularly helpful in identifying journal articles and websites and using indexes to books. Box 2.2 illustrates how to make a start with searching the literature for a research project on organic farming.

Box 2.2   Defining key terms and identifying search terms: an example

Topic: Socioeconomic aspects of the geography of organic farming.

Definition: 'this system uses fewer purchased inputs compared with conventional farming, especially agri-chemicals and fertilizers, and consequently produces less food per hectare of farmland ..., but is compensated by higher output prices' (Atkins and Bowler, 2001: 68–69).

Search terms:

Broader	Related	Narrower
Agricultural geography	Organic farming	Certified organic growers
Farm extensification	Organic agriculture	Organic organizations
Farm diversification	Organic production	Soil Association

Alternative farm systems	Organic growers	Organic food retailers
Food, geography of	Organic food	Organic food markets
Sustainable agriculture	Organic movement	Organic food shops

Note: 'Organic farming' is not listed in *The Dictionary of Human Geography*, although discussion of 'agricultural geography' and 'food, geography of' provides a useful context. The index of Atkins and Bowler (2001), which is on the 'Food and the Environment' course reading list, identifies four mentions of 'organic farming'. These lead to a useful introduction to the topic and to several recent references, and are a source for some of the above search terms.

TIP

If you are collecting data as part of your research project remember to carry out a literature search on research methodologies and techniques as well as on your main topic. Many of the later chapters in this book will help you with this.

A FRAMEWORK FOR YOUR SEARCH

A summary of how to search the literature is given in Figure 2.1 – which also provides a framework for the structure of this chapter. Where you start depends on the purpose of your search. For example, if you are looking to see whether the government has any policy documents on your topic you might begin by searching the Directgov website (www.direct.gov.uk). If you want to check news stories you might go to one of the newspaper database sites given in the other literature sources section, while if you are searching for journal articles to start you on an assignment you might look at a citation index such as Web of Knowledge. Figure 2.1 might suggest that undertaking a literature search and writing an essay or a literature review is a linear process. The reality is much messier. There is frequent interaction between the different stages. As you begin to identify and scan the key references, your knowledge and understanding of the topic will increase, which will lead you to identify particular subtopics that you wish to investigate in more detail. A further search, using new key terms may then be appropriate. Iteration is a key element of the search process. You should not give up after entering the first obvious key word in a search engine. Further thoughts are also likely to arise as you begin to draft your essay or literature review, which may call for additional searches. Saunders *et al.* (2007) capture some of this interactivity by conceptualizing the literature review process as a spiral.

TIP

Identify a few key references, skim read them, then revise your search criteria in the light of your new understanding.

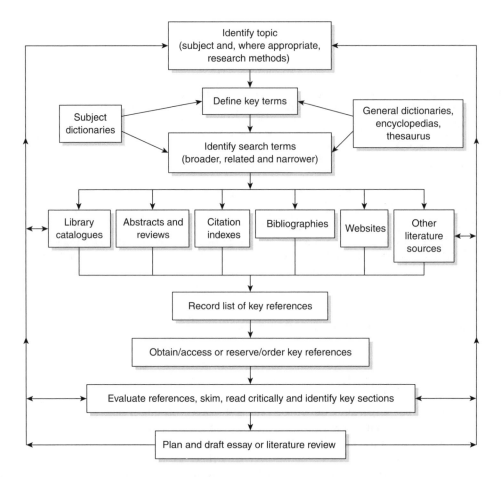

Figure 2.1 A framework for undertaking a literature search

MANAGING YOUR SEARCH

The search process, as just indicated, is one that you will keep coming back to at various stages in your research. It is therefore sensible to keep a search diary, which includes the sources searched, the key words used and brief notes on the relevant references they reveal. This is, perhaps, best done using a word-processing package, which will enable you to list your key words, cut and paste the results from your online searches and keep track of which search engines and sources you have used. Alternatively packages such as EndNote or Reference Manager may be useful in keeping track of your references. However, some programs may require further training as they can be complicated to use. For further advice on working online, Dolowitz *et al.* (2008) provide much useful information on search strategies using the internet, emphasizing the need to develop a search strategy. They suggest that you should begin by asking yourself three main questions: 'What am I looking for?' 'What are the most relevant search terms?' and 'What tool will be the most useful for helping me find it?' (Dolowitz *et al.*, 2008: 52).

TIP

Have a memory stick available when you are searching. Many of the databases enable you to save your searches direct to a memory stick or a PC. Some will also allow you to email them to yourself. Chapter 5 in Ridley (2008) provides useful advice on keeping records and organizing information to help you with your searches.

TIP

When using key words you can easily miss articles, as authors may have chosen different key words than you to describe their work. Look at the key words authors have chosen to use to describe their work in the articles you do find. Making a list of these key words will help to increase your search terms and avoid missing other useful references (Ridley, 2008).

How long to spend on the search process depends on the purpose of your search. For example, are you seeking ten key references for an essay, or 50 or more references for a dissertation? Generally, the broader the topic and the more that has been written on it, the longer the search tends to take. This is because much of the effort is spent in trying to identify the key references from what may be a list of several hundred marginal or irrelevant ones.

TIP

Aim to identify two to three times as many relevant references as you think you will need for your assignment/dissertation. Many may not exactly meet your needs when you obtain them

(Continued)

(Continued)

and/or may not be accessible in the time you have available. If you are finding what appear to be too many relevant references, focus on the most recent ones and the references most frequently cited, and consider narrowing the search by, for example, focusing on a subtopic or restricting the geographical coverage. If you are finding too few relevant references, try some new search terms and consider broadening the topic or the geographical area. Also ask a librarian.

To avoid any possibility of plagiarism (that is, the unacknowledged use of the work of others), be sure to take down the full bibliographic details of the references you find, including, where relevant, the author(s)' name(s) and initials, year of publication (not print date), title of book, edition (if not the first), publisher and town/city of publisher (not printer), title of article/chapter, journal title, volume number and page numbers, and names of editors for edited books. Be sure to put all direct quotations in quotation marks and give the source, including the page number(s) (Mills, 1994). The same applies to material taken from websites. However, avoid direct quoting too much, and cutting and pasting information from websites, instead summarize and paraphrase material. Take particular care in citing websites, giving wherever possible the author/organization responsible for the site, the date the page/site cited was last updated, the title of the page/site and the date you accessed the site, as well as the URL.

TIP

Inconsistent referencing and missing or erroneous information are some of the most common comments made on student writing (see also Chapter 32). Most geographers use the Harvard style of referencing, but departments vary on the format in which they like references to be cited. Guidance on how to provide correct references, including websites, is available from Kneale (2003). It is best to acquire the habit of using one way of formatting references and to apply this consistently whenever you note a reference (down to the last comma, full stop and capital letter!).

TIP

When taking notes remember to put any sentences you copy (or paste from a website) in quotation marks and note down the page number(s) so that if you decide to use the author's direct words later you can acknowledge this properly.

SEARCH TOOLS

Various features are used to search databases and the web. As these vary it is important to check their 'help' facilities. Most allow you to use exact phrases.

Simply place the phrase in double quotes (" ") (e.g. "organic farming"). Most search engines support the use of Boolean operators. The basic ones are represented by the words AND, OR and NOT (e.g. "organic farming" AND "UK"; "organic farming" OR "organic food"; "organic farming" NOT "North America"). The use of wild cards (*) may also be available. For example, 'farm*' will find records containing the word 'farm', including farm, farms, farmer, farmers and farming. It will also identify farmhouse, farmstead and farmyard. Bell (2005), Flowerdew and Martin (2005), and Dolowitz *et al.* (2008) provide further information on search tool techniques.

Library catalogues

In most cases the first place to search for relevant books is the subject index of your library catalogue. You will usually need to use your broader list of search terms. Unfortunately the classification systems used in most libraries put geography books in several different sections of the library. But do not restrict yourself to books with 'geography' in the title. The integrative nature of the subject means that many books written, for example, for sociologists, economists, planners, earth scientists, hydrologists and ecologists may be just as relevant. Once you have found the classification numbers of relevant books check the catalogue for other books with the same number and browse the relevant shelves. Looking at other books on the shelves near to the ones you are looking for often reveals other relevant references. Use your list of related and narrower search terms to explore the book indexes. Do not forget to also check where short loan and oversize books and pamphlets are shelved. Older books may be in store. Furthermore, a search for a topic on a website such as Amazon (www.amazon. co.uk) provides a list of a large number of the books published within that topic area (Ridley, 2008).

TIP

One of the quickest ways to generate a list of references is to find the latest book or article covering your topic and look at its reference list.

For a wider search try the combined catalogues of 29 of the largest UK and Irish research libraries, which are available online through COPAC (copac.ac.uk). The British Library Catalogue provides a national collection (www.bl.uk). The equivalent in the USA is the Library of Congress, which may be searched along with the catalogues of many other libraries in the USA and other countries via the Z39.50 Gateway (www.loc.gov). To check book details try WorldCat (www. worldcat.org). This is a huge database of over 1.2 billion references held in more than 10,000 libraries worldwide.

When you are away from the university it may be worth seeing whether your local university or public library has the book you require in stock. The Library Catalogue of the Royal Geographical Society (www.rgs.org/OurWork/Collections)

holds more than 2 million items tracing 500 years of geographical research and receives over 800 periodical publications. You can use the library free if your department has an Educational Corporate Membership, or otherwise there is a small fee charged per visit.

Abstracts and reviews

Evaluating the relevance of a book or journal article simply from its title is difficult. Generally each journal will articulate the purposes of the journal and its intended audience in each issue published, providing you with some idea of the relevance of the type of articles in that journal (Ridley, 2008). Abstracts give a clearer idea of the contents of articles. One of the most useful set of abstracts for geographers is Geobase, which provides international coverage of the literature (particularly journal articles) on all aspects of geography, geology, ecology and international development. The database provides coverage of over 2000 journals and contains close on 1.4 million records from 1980 onwards. Book reviews which appear in the journals abstracted are also included. These are useful for evaluating the significance of books and finding out what has recently been published. It is available online via Science Direct. There is also a CD version. LexisNexis (www.lexisnexis.com/academic) provides similar coverage, and CSA Sociological Abstracts (www.csa.com) is useful for some human geography topics. You should also check recent issues of review journals, particularly *Progress in Human Geography* and *Progress in Physical Geography*, for appropriate articles and updates on the literature in particular subfields. Environment Complete (www.ebscohost.com), which some libraries are using instead of Geobase, is available through EBSCOhost. It offers deep coverage in applicable areas of agriculture, ecosystems, energy, renewable energy sources, natural resources, geography, pollution and waste management, social impacts, urban planning, and more. This contains more than 1,957,000 records going back to the 1940s.

Citation indexes

Probably the most useful tool that you will find for searching the literature is the ISI Web of Knowledge. So it is well worth investing some time in exploring how to use it effectively. The ISI Web of Knowledge includes proceedings of international conferences, symposia, seminars, colloquia, workshops and conventions as well as the ISI Web of Science. ISI Web of Science consists of three citation indexes covering the social sciences, the arts and humanities, and science. It is the prime source for finding articles published in refereed journals, which is where most research is first published. As well as providing data on the number of times articles published in a wide range of journals are cited by authors of other articles, they also provide a valuable source for identifying journal abstracts and reviews. They can thus be used for generating lists of articles with abstracts on particular topics, as well as identifying influential articles. You can also use the indexes to identify related records that share at least one cited reference with the retrieved item, and hence may be contributing to the discussion of related topics. The ISI Web of Knowledge contains sophisticated ways of restricting searches including by subject area, date of publication and country. However, not all geography or related journals are included in the ISI database.

TIP

Some articles are cited frequently because they are heavily criticized, but they have nevertheless contributed to the debate. In the world of citation analyses the only real sin is largely to be ignored, which is the fate of most published papers. However, a few papers are ahead of their time and are not 'discovered' until several years after they have been published.

The three citation databases include articles published in over 13,000 journals, plus book reviews and editorials. There are over 23 million records extending back to 1981, which are added to weekly. Virtually all higher education institutions in the UK and many elsewhere have taken out a subscription to the ISI Web of Knowledge. In the UK access is through Mimas (www.mimas.ac.uk). Most universities in the UK now offer automatic access to such resources through Intranet services available off campus. Currently access is via Athens but this is gradually being replaced by a new generation of access management service, based on the Shibboleth technology.

TIP

Once you have identified key authors who are writing on your topic it is worth checking abstracts and citation indexes to see what else these authors have written, some of which may be on related topics. Beware when searching that the way authors' first names are cited may differ between publications. For example, the first author of this chapter appears variously as: Michael; Michael J; Mick; M; and M J.

Citation analyses are used to rank the impact that journals have on intellectual debate. They thus provide a crude guide as to which journals to browse through in the library and a possible basis for choosing between which of two, otherwise apparently equally relevant articles, to read first (Table 2.1). Lists of journal ranking may be obtained from ISI Journal Citation Reports (scientific.thomsonreuters. com/products/jcr) if your institution has a subscription. One of their limitations is that many of the key articles used by geographers are not published in mainline geography journals.

Table 2.1   Top geography journals, 2008

Rank	Journal title
1	*Transactions, Institute of British Geographers*
2	*Global Environmental Change*
3	*Progress in Human Geography*
4	*Economic Geography*
5	*Journal of Economic Geography*
6	*Annals of the Association of American Geographers*
7	*Geographical Analysis*
8	*Antipode*
9	*Political Geography*
10	*Environment and Planning D*
11	*Cultural Geography*
12	*Landscape and Urban Planning*
13	*Environment and Planning A*
14	*Area*
15	*Professional Geographer*

Note: Based on ranking journals by their 'impact factor' (a measure of the frequency with which the 'average article' in a journal has been cited in a particular year)
Source: Journal Citations Report, Social Sciences Edition (2008)

Bibliographies

A range of specialized bibliographies are available. The most useful are annotated. Some are in printed form. For example, the Countryside Agency Accession List includes details of new books, reports and pamphlets received at the Countryside Agency Library, and the Countryside Agency Selected Periodical Articles lists articles from recent journals under subject headings. An increasing number are available on the web at no charge and without registration (Table 2.2). Others may be available if your university has taken out a licence.

TIP

As you generate your list of references check whether your library holds the books, and if so whether they are on loan. If they are on loan put in a reservation request. In the case of journals, check whether the library takes them. Also check whether your library has a subscription to the journals identified in your search for accessing the full text of articles online. If so, obtain passwords and check how to access them and whether you can do this off-site. If the library does not hold the book or journal, consider ordering the reference on inter-library loan (ILL). Make sure that the journal article is relevant by reading the abstract first if one is available. To check the relevance of the book it is worth doing a Google Scholar search (scholar. google.co.uk) for the text as you may be able to read a few pages of the introduction or a sample chapter before ordering the book. Journal articles can usually arrive within 24 hours electronically or between a week and ten days for a hard copy. Recalling books, or ordering them through ILL, often takes longer. Check whether your library entitles you to a certain number of free inter-library loans or, if not, what the charge is per loan.

Table 2.2   Examples of web-based geography bibliographies

Bibliography	Comments
Australian Heritage Bibliography (www.environment.gov.au/heritage/ahc)	Full access is provided from this site but professional users will find subscription service on CD-ROM or access through the Informit server at Melbourne which provides greater flexibility in search formulation and output
Bibliography of Aeolian Research (www.csrl.ars.usda.gov/wewc/biblio/bar.htm)	Coverage from 1930; updated monthly
Development Studies – BLDS Bibliographic Database (www.ids.ac.uk/blds)	Web-searchable version of the library catalogue and journal articles database of the British Library for Development Studies at the Institute of Development Studies, UK; contains records of over 91,000 documents
Disability and Geography Resource (DAGIN) (isc.temple.edu/neighbor/research)	Alphabetical list of references compiled for Disability and Geography International Network, last modified in 2003
Gender in Geography (www.emporia.edu/socsci/fembib/index.htm)	Alphabetical list contributed by members of the discussion list for Feminism in Geography
GIS Bibliography (campus.esri.com/library)	ESRI Virtual Campus Library provides a searchable database of over 67,000 references
International Bibliography of the Social Sciences (IBSS)	Produced by the London School of Economics and Political Science, this database includes 2.5 million references dating back to 1951, focusing on four core social science disciplines – anthropology, economics, politics and sociology

Websites

An increasing amount of useful information is being placed on the web. However, identifying this from the huge amount of irrelevant and low-quality information is a time-consuming task. The general search engines, such as Google (www.google.co.uk) and AltaVista (www.altavista.com), and meta search engines, such as Dogpile (www.dogpile.com) and Ixquick (www.ixquick.com), which search other search engines' databases, are indispensable when searching for specific information, such as the URL address of an institution's website. One of the problems with search engines is that 'even the largest of search engines have only indexed a small proportion of the total information available on the internet' (Dolowitz *et al.*, 2008: 62). The 'deep web' consists of databases, non-textual files and content available on sites protected by passwords. Items such as phone directories, dictionary definitions, job postings and news are all part of the deep web. To access these may involve a two-stage process: first searching for database sites (e.g. UK newspapers) and then going to the database itself and searching for the information you want. More advanced search engines, such as Google, are beginning to incorporate access to some parts of the deep web. Google Scholar (scholar.google.co.uk) is a specialized web search database which focuses upon scholarly literature and therefore the quality of the content of results should be higher, although caution is still advised.

This is where internet gateways or portals can be useful because they provide links to sites on particular subjects which have been evaluated for their quality.

Most of the ones mentioned below can be accessed via Pinakes (www.hw.ac.uk/ libwww/irn/pinakes/pinakes.html). Among the more useful internet gateways is BUBL (bubl.ac.uk), which provides a catalogue of selected internet sources for higher education classified by subject and Dewey classification number. Its main interface, BUBL LINK/5:15, offers between 5 and 15 sources for most subjects. It includes many sites relevant to geographers. Intute (www.intute.ac.uk) is another free internet service dedicated to providing effective access to high-quality internet resources for the learning, teaching and research community. There is a separate section for geography but many of the resources used by geographers may be found under other subject headings, such as earth sciences, environment, sociology, or travel and tourism. ELDIS provides a gateway to development information (www.eldis.org). In the USA, Academic Info provides a gateway to college and research-level internet resources for many subjects, including geography (www. academicinfo.net/geogmeta.html).

TIP

Check out the advanced search facilities on search engines and databases. These allow you to focus your searches in a variety of ways and make them much more efficient. For example, you can restrict your search to one domain. So if you want to search for documents published by the US government departments search within '.gov'.

Box 2.3   Searching the literature: an example

Assignment: a 2,000-word essay.

Topic: Socioeconomic aspects of the geography of organic farming.

Library catalogues: 98 books on agricultural geography and 30 books on sustainable agriculture were listed in the University of Sheffield library catalogue; 11 specific ones on 'organic farming'. *COPAC*: key word search found the following number of books: "organic farming" (1,134), "organic agriculture" (181), "organic production" (134), "organic growers" (45), "organic food" (400), "organic movement" (19).

Abstracts and reviews: Environment Complete: key word search resulted in the following number of references: "organic farming" (3,059), "organic agriculture" (307), "organic food" (580), "organic production" (207), "organic growers" (67), "organic movement" (46), but the majority were about scientific aspects. One useful article was found in a recent issue of *Progress in Human Geography*.

Citation indexes: ISI Web of Knowledge: search for "organic AND farming" gave Web of Knowledge (5,873); Web of Science (2,843); Social Science Citation Index (231); Social Science Citation Index (UK OR England OR Wales OR Scotland OR Northern Ireland – alternative key words to cover UK geographical area) (21).

Bibliographies: A search for "organic farming bibliography" on Google Scholar found three listed, but when the search was changed to "organic AND farming AND bibliography" about

5,440 records were found. Several useful references were also found from reference lists of books and articles identified by the above search tools and the economic geography reading list.

Websites: Inputting "organic farming" into general search engines result in a huge number of hits e.g. Google (about 3,300,000; about 178,000, UK). In contrast BUBL identified eight organic farming websites, whereas Intute identified 238 websites. Restricting the Intute search to social science sites gave 22 different websites. Atkins and Bowler (2001) identify five organic food organizations in the UK. A Google search found specific websites for all of them.

Other sources: Proquest Newspapers: key word search found the following number of articles: "organic farming" (150), "organic agriculture" (8), "organic production" (3), "organic growers" (9), "organic food" (87), "organic movement" (0). Direct Government (a 'quick search') revealed 61 matches on "organic farming".

Summary: Given the nature of the assignment (2,000 words, 30 per cent of marks for course) and the number of references identified in early searches, it would be sensible to focus the search (e.g. only on the UK). It would be best to start with the most frequently cited and most up-to-date references, websites and newspaper articles and those that appear to be the most comprehensive. Apart from sources such as ISI Web of Knowledge, which have sophisticated ways of focusing a search, much time will be spent on weeding out non-relevant references which deal with topics such as methods of farming or environmental impacts. Expect to find further references and undertake more specific searches as you become more familiar with the topic. Make a shortlist of references (two to three times as many as you think you are likely to need) to show your lecturer to ask his or her advice on identifying key ones and any major omissions.

Many departments and institutions also provide links to websites relevant to areas in which they teach and research. The University of Colorado (www.colorado.edu/geography), the University of Utrecht (Geosource) (www.library.uu.nl/geosource) and the Universities of Gottingen and Freiberg (Geo-Guide) (www.Geo-Guide.de) all provide extensive lists.

TIP

When doing searches using the internet it may be useful to be more specific with your search terms and select particular fields to search (e.g. search only academic journals or peer-reviewed titles in order to focus your search). QUT (2006) provide a useful tutorial for negotiating the vocabulary to work through when considering your exact search terms.

Other literature sources

For many topics newspapers can be a useful source of information, especially for up-to-date case studies. Proquest Newspapers provides access to library subscribers to UK broadsheets from 1996 or earlier, as does Lexis-Nexis Butterworths (www.lexisnexis.co.uk). Several papers can be searched simultaneously. Dialog@

CARL, is a North American subscription service which provides access to 275 databases, including the full text of most major US newspapers. Many individual newspapers provide free online searchable databases. World Newspapers provides a searchable directory of online world newspapers, magazines and news sites in English (www.world-newspapers.com).

If you are undertaking your own thesis it is important to check whether any-one else has written a thesis on a similar topic by looking at the Index to Theses which has Abstracts Accepted for Higher Degrees by the Universities of Great Britain and Ireland since 1716 (www.theses.com). Dissertation Abstracts Online is a definitive subject, title and author guide to virtually every American disserta-tion accepted at an accredited institution since 1861. An institutional subscription is required to access it.

A wealth of information is available from central and local government via Directgov in the UK (www.direct.gov.uk) and Usgovsearch in the USA (usasearch. gov). The Australian federal Department of the Environment maintains ERIN – Environmental Resources Information Network (www.environment.gov.au). A selection of UK official statistics is available via the Office for National Statis-tics (www.ons.gov.uk). Europa offers centralized access to European Union sites (europa.eu).

The use of videos is becoming increasingly common in higher education. Some universities provide audio and/or video content of many of their lectures (e.g. MIT ocw.mit.edu/OcwWeb). NewsFilm Online provides access to 3,000 hours of foot-age from about 60,000 stories from the ITN/Reuters archives for staff and students at subscribing universities (www.nfo.ac.uk).

TIP

Continue to refine your search as you progress using the cycle illustrated in Figure 2.1 in order to compile a list of useful references. Start with more general sources in the early stages of your search moving towards more specific ones later on (Dolowitz *et al.*, 2008).

EVALUATING THE LITERATURE

Do not be put off undertaking a systematic literature search, such as is illustrated in Box 2.3, because you feel you will not have time to read all the references you find. Indeed, you will not have time to read them all. The purpose of the literature search is to identify the most appropriate references for the task in hand (Table 2.3): 'learning how to determine the relevance and authority of a given resource for your research is one of the core skills of the research process' (Olin and Uris Libraries, Cornell University, 2006). Websites in particular need to be evaluated critically for their origin, purpose, authority and credibility.

If you follow the advice above you should have reduced the list of references several fold before you have even opened a book or journal or read a newspa-per article or website, for example by focusing on the most frequently cited and up-to-date references. The titles and abstracts will also help you to judge those references likely to be most relevant.

> ### TIP
>
> Avoid listing all the references you have found simply to try to impress your lecturer/ professor. You must use some relevant idea or material from each one to justify its inclusion. Ensure that you include a reference to each in the text, usually by following the Harvard system of referencing.

Exercise 2.3

Select four references from different sources on a topic that you are preparing, such as a website, a textbook, a journal article and a newspaper report, and use the criteria in Table 2.3 to evaluate their relevance, provenance and source reliability.

Table 2.3   Reducing your list of references to manageable proportions

Criterion	Possible (Score 4 points)	More doubtful (Score 2 points)	Probably forget it (Score 0 points)
Relevance to my topic – judged by title and/or abstract (double the score for this criterion)	High	Moderate	Tangential
Up-to-date	Last 5 years	6–15 years' old	Over 15 years' old
Authority – the author or paper is cited in the references I have already read	Extensively cited	Recent paper not yet had time to be cited extensively	Older paper cited infrequently or not at all
Respectability and reliability of source publication	Published in major geographical publication or that of sister subject or something very close to my topic	Publication is not in geography or an allied field	Informal publication or unreliable Internet source
Nature of publication	Peer-reviewed academic journal or monograph	Textbook or conference proceedings	Popular magazine
Originality	Primary source of information – the authors generated this information using reliable and recognized methods	The authors take their information from clearly identified and reliable secondary sources	The authors assert facts and produce information without providing appropriate supporting evidence
Accessible	Instant – by download or short walk to library	Obtainable with effort – reserve, interlibrary loan	Unobtainable

Source: Modified from an idea by Martin Haigh (personal communication 29 January 2002)

Exercise 2.4

Take a book or article relevant to your topic. You have five minutes to extract the key points it contains.

Researchers rarely read books from cover to cover and they read relatively few articles in their entirety. Like you, they do not have the time. They are practised at evaluating references in a few minutes by skimming the abstracts, executive summaries, publisher's blurbs, contents pages, indexes, introductions, conclusions and subheadings. This enables them to select the references that deserve more attention. Even then they will usually identify key sections by, for example, reading the first and last paragraphs of sections and the first and last sentences of paragraphs. This is not to suggest that all you need is a superficial knowledge of the literature; rather, that you should read selectively and critically to ensure that you obtain both a broad understanding of the topic and an in-depth knowledge of those parts of the literature that are particularly significant. If you are not familiar with the processes involved in critical and strategic reading, have a look at the relevant chapters in Blaxter *et al.* (2006) and Kneale (2003).

Exercise 2.5

In your next essay or research project try applying the framework (Figure 2.1) for searching the literature outlined in this chapter. Good hunting!

Summary

The aim of this chapter has been to identify effective and efficient ways to systematically search and evaluate the literature:

- The first stage is to define the key terms for your topic and to identify a range of search terms.
- You should then systematically search a range of sources, including library catalogues, abstracts and reviews, citation indexes, bibliographies and websites, being careful to keep a search diary.
- Having made a record of the references you have found you should evaluate each of them for such things as relevance, whether they are up to date, authority, respectability, originality and accessibility.
- Although searching the literature needs to be systematic, it is also iterative and, as your knowledge and understanding of your topic and of the number and quality of the references you are identifying increase, you will inevitably need to make modifications to your search and repeat and refine many of the stages several times.
- For a quick guide, look at the exercises, boxes, tables and tips in this chapter and the framework for undertaking a literature search shown in Figure 2.1.

Further reading

Few books focus only on searching the literature, though Hart (2001) is an exception; most are guides to study skills or how to research, which put the literature search process in a broader context.

- Bell (2005) presents a chapter on finding and searching for literature. This includes a useful introduction to searches using computers and ways to help limit or broaden your criteria. It is part of a guide to doing research.

- Blaxter *et al.* (2006) provide an excellent user-friendly guide on how to research; they include chapters on reading for research and writing up.

- Cohen (2008) has compiled a very useful set of internet tutorials.

- Dolowitz *et al.* (2008) present a thorough guide to searching for information on the internet including advice on how to approach searching and different techniques to use within the search strategy.

- Flowerdew and Martin (2005) provide a guide for human geographers doing research projects; they include a chapter by Flowerdew on finding previous work and a chapter by Clark illustrating the benefits and issues of using secondary data sources which also provides further information on how to use search engines to their best advantage.

- Hart (2001) provides a comprehensive guide for doing a literature search in the social sciences.

- Olin and Uris Libraries, Cornell University (2006) provide a useful guide to searching and evaluating information sources critically.

- Ridley (2008) presents a step-by-step guide to writing a literature review. This book includes a chapter on doing a literature search and a chapter focusing on information management.

Note: Full details of the above can be found in the references list below.

References

Atkins, P. and Bowler, I. (2001) *Food in Society: Economy, Culture and Geography*. London: Arnold.

Bell, J. (2005) 'Literature searching', in *Doing Your Research Project: A Guide for First-Time Researchers in Education and Social Science* (4th edn). Buckingham: Open University Press, pp. 79–98.

Blaxter, L., Hughes, C. and Tight, M. (2006) *How to Research* (3rd edn). Buckingham: Open University Press.

Christchurch Polytechnic Institute of Technology (CPIT) (2008) *Introduction to Research*. Canterbury, NZ: CPIT, pp. 26–44.

Cohen, L. (2008) Internet tutorials: www.internettutorials.net (accessed 30 August 2008).

Dolowitz, D., Buckler, S. and Sweeney, F. (2008) *Researching Online*. Basingstoke: Palgrave Macmillan.

Flowerdew, R. and Martin, D. (eds) (2005) *Methods in Human Geography: A Guide for Students Doing Research Projects* (2nd edn). Harlow: Longman, pp. 48–56.

Hart, C. (2001) *Doing a Literature Search: A Comprehensive Guide for the Social Sciences*. London: Sage.

Gregory, D., Johnston, R.J., Pratt, G., Watts, M. and Whatmore, S. (2009) *A Dictionary of Human Geography* (5th edn). Oxford: Blackwell.

Kneale, P.E. (2003) *Study Skills for Geography Students: A Practical Guide* (2nd edn). London: Arnold.

Mills, C. (1994) 'Acknowledging sources in written assignments', *Journal of Geography in Higher Education*, 18: 263–268.

Olin and Uris Libraries, Cornell University (2006) 'Guide to library research at Cornell: seven steps to effective library research': www.library.cornell.edu/olinuris/ref/research/tutorial.html (accessed 30 August 2008).

QUT (Queensland University of Technology) (2006) Pilot information literacy tutorial: pilot.library.qut.edu.au/module2/2_1/index.jsp (accessed 30 August 2008).

Ridley, D. (2008) *The Literature Review: Step-by-Step Guide for Students.* London: Sage.
Saunders, M.N.K., Lewis, P. and Thornhill, A. (2007) *Research Methods for Business Students* (4th edn). Harlow: Financial Times/Prentice Hall.
Thomas, D. and Goudie, A. (2000) *A Dictionary of Physical Geography* (3rd edn). Oxford: Blackwell.

ACKNOWLEDGEMENTS

The authors are very grateful to Gordon Clark (Lancaster, UK), Ken Foote (Colorado at Boulder, US), Robin Graham (Christchurch Polytechnic Institute of Technology, NZ), Diana Ridley (Sheffield Hallam, UK) and Nicky Williams (Gloucestershire, UK) for their helpful comments in revising this chapter.

3 Ethical Practice in Geographical Research

Iain Hay

Synopsis

Ethical research in geography is characterized by practitioners who behave with integrity and who act in ways that are just, beneficent and respectful. Ethical geographers are sensitive to the diversity of moral communities within which they work and are ultimately responsible for the moral significance of their deeds. This chapter explains the importance of behaving ethically, provides some key advice on the conduct of ethical research and provides some examples of ethical dilemmas.

The chapter is organized into the following sections:

- Introduction
- Why behave ethically?
- Principles of ethical behaviour and common ethical issues
- Truth or consequences? Teleological and deontological approaches to dealing with ethical dilemmas in your research
- Conclusion

INTRODUCTION

To behave ethically in geographical research requires that you and I act in accordance with notions of right and wrong – that we conduct ourselves morally (Mitchell and Draper, 1982).[1] Ethical research is carried out by thoughtful, informed and reflexive geographers who act honourably because it is the 'right' thing to do, not just because someone is making them do it (see, for example, Cloke, 2002; Dowling, 2010).[2]

This chapter seeks to heighten your awareness of the reasons for, and principles underpinning, ethical research. Although I am conscious of the need to avoid ethical prescription, leaving you the opportunity to exercise and act on your own 'moral imagination' (Hay, 1998), I believe it is important to set out a range of specific ethical matters that colleagues and communities will commonly expect you to consider when preparing and conducting research. As a geographer, you must consider carefully the ethical significance of your actions in those contexts within which they have meaning and be prepared to take responsibility for your actions. In some instances (e.g. cross-cultural research), ethically reflexive practice includes acknowledging and working with (negotiating) different groups' ethical expectations in ways that yield satisfactory approaches for all parties involved (see Chapter 12).

The first part of the chapter introduces a few of the reasons why geographers need to behave ethically. I follow this with a discussion of some fundamental principles behind ethical behaviour and offer a few points of guidance that might assist you in your practice as an ethically reflexive geographer. As I note, however, no matter how well you might try to anticipate the ethical issues that might arise in your work, you are still likely to confront ethical dilemmas. In recognition of this, the chapter sets out a strategy that might help you resolve those dilemmas you face. I conclude with some suggestions on ways in which you might continue to develop as an ethical geographer. A number of real ethical cases are included in the chapter to illustrate points made and to provide some material for you to discuss.

In starting, there are two cautions I would like to offer. First, decisions about ethical practice are made in specific contexts. While all people and places deserve to be treated with integrity, justice and respect (see Smith, 2000a), ethical behaviour requires sensitivity to the expectations of people from diverse moral communities and acknowledgement of the webs of physical and social relationships within which the work is conducted (see Chapter 12). For these reasons, rules for ethical practice cannot be prescribed, particularly in work involving people (Hay, 1998; Israel and Hay, 2006). And as you will see from the real quandaries encountered by geographers that are discussed in this chapter, the 'correct' resolution of most ethical dilemmas cannot be dictated. Second, simply because your peers, colleagues and institution (e.g. fellow students, professional geographers, university ethics committee) say that some behaviour or practice is ethical does not necessarily mean it is always so (see Kearns *et al.*, 1998 for a discussion). That determination is ultimately up to you, your conscience and those people with whom you are working. In the end, you must take responsibility for the decisions you make.

WHY BEHAVE ETHICALLY?

Aside from any moral arguments that as human beings we should always act ethically, there are important practical arguments for geographical researchers to behave ethically. These fall into three main categories.

First, ethical behaviour protects the rights of individuals, communities and environments involved in, or affected by, our research. As social and physical scientists interested in helping to 'make the world a better place', we should avoid (or at least minimize) doing harm.

Second, and perhaps a little more self-interestedly, ethical behaviour helps assure a favourable climate for the continued conduct of scientific inquiry. For instance, as Walsh (1992: 86) notes in an early discussion of ethical issues in Pacific Island research, incautious practice and a lack of cultural awareness can lead to community denial of research privileges:

> Not infrequently, the blunderings of an inexperienced and culturally insensitive researcher creates problems for those researched and for [Pacific] Island governments, and some [researchers] have been asked to leave. For these and similar reasons, all Island governments now require intending researchers to apply for research permits, may charge a non-refundable application fee, and it is not uncommon for a permit to be refused.

By behaving ethically, we maintain public trust. From that position of trust we may be able to continue research and to do so without causing suspicion or fear amongst those people who are our hosts (see Case Study 3.1). Moreover, we maintain trust within research communities. Not only is it important that we feel sure we can depend on the integrity of colleagues' work but trustworthy work helps ensure the continuing support of agencies upon whom we depend to fund our research.

Case Study 3.1 Waving, not drowning

While he was conducting surveys of flood hazard perception in the USA, Bob Kates found that in rare cases his questions raised anxieties and fears in those people to whom he was speaking. Even the actions of the research team in measuring street elevations to calculate flood risk created rumours that led some people to believe that their homes were to be taken for highway expansion (Kates, 1994: 2).

For discussion
Is it Kates' responsibility to quash the rumours and suspicions his research engenders? Justify your answer.

Third, growing public demands for accountability and the sentiment that institutions such as universities must protect themselves legally from the unethical or immoral actions of a student or employee mean there is greater emphasis on acting ethically than ever before (see for example, Oakes, 2002).

Clearly, then, there are compelling moral and practical reasons for conducting research ethically. How are these brought to life?

PRINCIPLES OF ETHICAL BEHAVIOUR AND COMMON ETHICAL ISSUES

Around the world a growing number of organizations supporting research have established committees to scrutinize research proposals to ensure that work is conducted ethically (see Israel and Hay, 2006). Because these committees regularly consider the possible ethical implications of research, their workings and the principles behind their operation might provide you with a useful starting point to ensure that your own research is ethical.

In general, the central guiding value for these committees is integrity, meaning a commitment to the honest conduct of research and to the communication of results. However, committees usually emphasize the desirability of three fundamental principles,[3] as set out in Box 3.1. These principles lead to a set of core questions (right-hand column) to guide your personal consideration of ethical matters. You will find it helpful always to consider these simple questions when reflecting on your work. They provide the beginnings of ethical practice. For instance, if you can sustain the argument that your work is just, is doing good and you are demonstrating respect for others, you are probably well on the way to conducting an ethical piece of work. If, by comparison, you believe that your work is just, yet you are not showing respect for others or you are doing harm, you may have an ethical dilemma on your hands. I set out a strategy for dealing with such situations later in this chapter.

Box 3.1   Principles of ethical behaviour

Justice: this gives emphasis to the distribution of benefits and burdens.   Is this just?

Beneficence/non-maleficence: respectively, these mean 'doing good' and 'avoiding harm'. Our work should maximize benefits   Am I doing harm?
and minimize physical, emotional, economic and environmental
harms and discomfort.   Am I doing good?

Respect: individuals should be regarded as autonomous agents   Am I showing respect?
and anyone of diminished autonomy (e.g. intellectually disabled)
should be protected. It is important to have consideration for
the welfare, beliefs, rights, heritage and customs of people
involved in research. Of course, respect should also extend to
consideration of any discomfort, trauma or transformation
affecting organisms or environments involved in the research.

The principles and questions set out in Box 3.1 are a good general framework but if you are just beginning to engage in research, some more specific advice on how to determine whether your research might be understood to be ethical will be helpful (see Case Study 3.2).

Typically, ethics committees give detailed consideration to five major issues when evaluating research proposals. These issues are set out in Box 3.2. I have attached to each of these a set of 'prompts' for moral contemplation. As you can see, the 'prompts' offer no specific direction about the actions you should or should not take in any or all situations. To do so would be virtually impossible given the enormous variability of geographic research, the associated need for flexible research practices (see, for examples and discussion, Bauman, 1993; Valentine et al., 2001) and the recognition that society does not comprise an 'isolable, unitary, internally coherent whole' (Amit-Talai, in Matthews et al., 1998). It would also deny the need to negotiate specific issues with participants.[4]

When you are thinking about the issues in Box 3.2, consider them in terms of the value of integrity and the principles set out earlier – justice, beneficence and respect for others. For example, let us think about the prompt 'time for consideration of the study before consent is provided'. A human geographer might be expected to allow people more time to consider their involvement in a complex, long-term observational study than for a two-minute interview about their grocery shopping behaviour in a retail mall. In another example concerning harm minimization, a physical geographer might think about, and act on, the potential harm caused to nesting birds as a result of fieldwork involving weed clearance. Indeed, there could be no need to conduct the work if similar studies have been conducted previously.[5]

Case Study 3.2 'They did that last week'

University student Ali bin Ahmed bin Saleh Al-Fulani carefully prepares a questionnaire survey for distribution to two groups of 16-year-old students in 'home groups' at two local high schools. The survey is central to the comparative work Ali is conducting as part of his thesis. In compliance with government regulations, Ali secures permission from the students' parents to conduct the survey. He also gets permission for his work from the university's Ethics Committee. The Ethics Committee requires him to include a covering letter to students which states that their participation in the study is voluntary and that no one is obliged to answer any of the questions asked. A few weeks before he intends to administer the questionnaire survey, Ali leaves near-final drafts of it with the students' teachers for comment. The draft copy of the questionnaire does not include the covering letter. It is Ali's intention to revise the questionnaire in the light of each teacher's comments and then return to the schools to administer the questionnaire during 'home group' meeting times. About a week after he leaves the survey forms with the teachers Ali calls them to find out if they have had an opportunity to comment on the questionnaire. The first teacher has just returned the questionnaire – with no amendments – by post. However, Ali finds that the second teacher had already made multiple copies of the forms and had administered the questionnaire to her student 'home group'. She asks Ali to come along to collect the completed forms. Ali scuttles off to the school immediately. He finds that the questionnaires had been completed fully by every student present in the home group. Only one student from the class of 30 had been absent so the response rate was 97 per cent – an extraordinarily high rate. Ali feels he cannot ask the teacher to readminister the survey because she has already indicated several times that she is tired of his requests for assistance and access to the class.

For discussion

It would appear from the circumstances and from the very high response rate that students were not free to refuse to participate in the study. Is it ethical for Ali to use results that have been acquired without free and informed consent? Would your view change if the survey had dealt with some sensitive issue such as sexual assault or if the results had been acquired, without consent, through the use of physical force?

Box 3.2   Prompts for contemplation and action

Before and during your research, have you considered the following?

Consent[6]

- Amount of information provided to participants on matters such as purposes, methods, other participants, sponsors, demands, risks, time involved, discomforts, inconveniences and potential consequences
- Accessibility and comprehensibility to prospective participants of information upon which consent decisions are made
- Time for consideration of the study before consent is provided
- Caution in research requiring deceit[7]
- Caution in obtaining consent from people in dependent relationships

(Continued)

(Continued)

- Recording of consent
- Informed consent issues for others working on the project

Confidentiality

- Disclosing identity of participants in the course of research and in the release of results
- Consent and the collection of private or confidential information
- Relationships between relevant privacy laws and any assurances made by researchers of confidentiality or privacy
- Data storage during and after research (for example, field notes, completed surveys and recorded interviews)

Harm

- Potential physical, psychological, cultural, social, financial, legal and environmentally harmful effects of the study or its results
- Extent to which similar studies have been performed previously
- Issues of harm for dependent populations
- Relationship between the risks involved and the potential advantage of the work
- Opportunities for participants to withdraw from the research after it has commenced
- Competence of researchers and appropriateness of facilities
- Representations of results

Cultural awareness

- Personality, rights, wishes, beliefs and ethical views of the individual subjects and communities of which they are a part

Dissemination of results and feedback to participants

- Availability and comprehensibility of results to participants
- Potential (mis)interpretations of the results
- Potential (mis)uses of results
- 'Ownership' of results
- Sponsorship
- Debriefing
- Authorship[8]

TRUTH OR CONSEQUENCES? TELEOLOGICAL AND DEONTOLOGICAL APPROACHES TO DEALING WITH ETHICAL DILEMMAS IN YOUR RESEARCH

No matter how well prepared you are, no matter how thoroughly you have prepared your research project, and no matter how properly you behave, it is likely that in your geographical research you will have to deal with a variety of ethical dilemmas (see, for example, Case Studies 3.3 and 3.4).

Case Study 3.3   You are being unco-operative

Catriona McDonald has recently completed an undergraduate research project that involved sending out a confidential questionnaire to members of a non-government welfare organization called ANZAC Helpers. Members of this organization are mainly World War II veterans aged over 70 years, and a good deal of their volunteer work requires them to drive around familiar parts of a large metropolitan area. Catriona works part time for the organization and feels that her employment tenure is somewhat precarious. The research project has been completed and assessed formally by Ms McDonald's professor and Catriona is planning to present a modified report to ANZAC Helpers. In the work car-park one afternoon, Catriona meets Mr Montgomery Smythe, one of the organization's members and a man known to Catriona to be something of a troublemaker. After some small talk, Smythe asks what the study results are. Catriona outlines some of the findings, such as the percentage of the membership who are having trouble performing their voluntary duties for the organization due to old age and ill-health. Smythe then asks the student to tell him the names of those members who are having difficulties with their duties. It is possible that he could use the information to help encourage the implementation of strategies to help those members experiencing difficulties. It is also possible that he could campaign to have the same members redirected to less demanding volunteer roles that many of them are likely to find less fulfilling.

For discussion

1   Should Ms McDonald give Mr Smythe the information he wants?

2   Given that some of the members of ANZAC Helpers might actually be putting their lives at risk by driving around the city, should Catriona disclose the names of those people she has discovered to be experiencing sight and hearing problems, for example, to the organization?

Case Study 3.4   The power of maps

Dr Tina Kong has recently commenced work as a post-doctoral fellow with a research organization applying GIS (geographical information systems) to illustrate and resolve significant social problems. This is a position in which full-time research is possible. It is also a position that can allow someone to begin to forge the beginnings of a noteworthy academic career for himself or herself. However, much of that promise depends on producing good results and having them published in reputable journals. Tina decides to conduct work on environmental carcinogens (cancer-producing substances) in a major metropolitan area. She spends about two months of her two-year fellowship conducting some background research to assess the need for, and utility of, the work. After this early research, Tina resolves to use GIS to produce maps which will illustrate clearly those areas in which high levels of carcinogenic materials are likely to be found. At a meeting of interested parties to discuss the proposed research, one of the participants makes the observation that, if broadcast, the results of the study may cause considerable public alarm. For example, there may be widespread individual and institutional concern about public health and welfare; property values in areas with high levels of carcinogenic material may be adversely affected; past and present producers of

(Continued)

(Continued)

carcinogenic pollutants may be exposed to liability suits; and local government authorities might react poorly to claims that there are toxic materials in their areas. Tina is cautioned by senior members of the research organization to proceed cautiously, if at all.

For discussion

Should Tina proceed with the project? Justify your answer. Should Tina 'cut her losses' and move into a less controversial area that might be supported by her colleagues?

How will you deal with such situations? To answer that fully, we will have to make a short excursion into the work of philosophers. But, first, I should point out that to resolve any dilemma you encounter it is likely that you will have to 'violate' one of the three ethical principles – justice, beneficence or respect for others – set out earlier in this chapter. That is why it is a dilemma! But you will probably feel more confident about the difficult decision you will have to make if it is well considered and informed by a basic appreciation of two key normative approaches to behaviour.

Box 3.3   Steps to resolving an ethical dilemma

How do you decide what to do if you are presented with an ethical dilemma? There are two major approaches that you might draw from. One focuses on the practical consequences of your actions (*teleological* or consequentialist approach) and might be summed up brutally in the phrase 'no harm, no foul'. In contrast, the *deontological* approach would lead you to ask whether an action is, in itself, right. For example, does an action uphold a promise or demonstrate loyalty? The essence of deontological approaches is captured by Quinton's (1988: 216) phrase: 'Let justice be done though the heavens fall.' These two positions serve as useful starting points for a strategy for coping with ethical dilemmas.

1 *What are the options?*

List the full range of alternative courses of action available to you.

2 *Consider the consequences*

Think carefully about the range of positive and negative consequences associated with each of the different paths of action before you:

- Who/what will be helped by what you do?
- Who/what will be hurt?
- What kinds of benefits and harms are involved and what are their relative values? Some things (e.g. healthy bodies and beaches) are more valuable than others (e.g. new cars). Some harms (e.g. a violation of trust) are more significant than others (e.g. lying in a public meeting to protect a seal colony).
- What are the short-term and long-term implications?

Now, on the basis of your answers to these questions, which of your options produces the best combination of benefits maximization and harm minimization?

3   *Analyse the options in terms of moral principles*

You now have to consider each of your options from a completely different perspective. Disregard the consequences, concentrating instead on the actions and looking for that option which seems problematic. How do the options measure up against such moral principles as honesty, fairness, equality and recognition of social and environmental vulnerability? In the case you are considering, is there a way to see one principle as more important than the others?

4   *Make your decision and act with commitment*

Now, bring together both parts of your analysis and make *your* informed decision. Act on your decision and assume responsibility for it. Be prepared to justify *your* choice of action. No one else is responsible for this action but you.

5   *Evaluate the system and your own actions*

Think about the circumstances which led to the dilemma with the intention of identifying and removing the conditions that allowed it to arise. Don't forget to think about your own behaviours and any part they may have played.

Source: Adapted from Stark-Adamec and Pettifor (1995); Center for Ethics and Business (2008)

Many philosophers suggest that two categories, teleological and deontological, exhaust the possible range of theories of right action (Davis, 1993). In summary, the former sees acts judged as ethical or otherwise on the basis of the consequences of those acts. The latter suggests, perhaps surprisingly, that what is 'right' is not necessarily good.

In the terms of teleology – also known as consequentialism – an action is morally right if it produces more good than evil (Israel and Hay, 2006). It might therefore be appropriate to violate and make public the secret and sacred places of an indigenous community if doing so prevents the construction of a bridge through those sacred places, because disclosure yields a greater benefit than it 'costs'.

Deontological approaches reject this emphasis on consequences, suggesting that the balance of good over evil is insufficient to determine whether some behaviour is ethical. Instead, certain acts are seen as good in themselves and must be viewed as morally correct because, for example, they keep a promise, show gratitude or demonstrate loyalty to an unconditional command (Kimmel, 1988). It is possible, therefore, for something to be ethically correct even if it does not promote the greatest balance of good over evil. To illustrate the point: if we return to the example of the researcher made aware of the location of an indigenous group's secret, sacred places, a deontological view might require that researcher to maintain the trust with which he or she had been privileged, even if non-disclosure meant that construction of the bridge would destroy those sacred places.

Thus, we have two philosophical approaches that can point to potentially contradictory ethical ways of responding to a particular situation. This is not as debilitating as you might think. As Box 3.3 sets out, you can draw from these approaches to ensure that your response to an ethical dilemma is at least well considered, informed and defensible.

You can really give this scheme a workout by considering the (in)famous work of Laud Humphreys, set out as Case Study 3.5.

Case Study 3.5   The 'watch queen' in the 'tea room'

In 1966–7, and as part of a study of homosexual behaviours in public spaces, Laud Humphreys acted as voyeuristic 'lookout' or 'watch queen' in public toilets ('tea rooms'). As a 'watch queen' he observed homosexual acts, sometimes warning men engaged in those acts of the presence of intruders. In the course of his observations Humphreys recorded the car licence plate numbers of the men who visited the 'tea room'. He subsequently learnt their names and addresses by presenting himself as a market researcher and requesting information from 'friendly policemen' (Humphreys, 1970: 38). One year later, Humphreys had changed his appearance, dress and car and got a job as a member of a public health survey team. In that capacity he interviewed the homosexual men he had observed, pretending that they had been selected randomly for the health study. This latter deception was necessary to avoid the problems associated with the fact that most of the sampled population were married and secretive about their homosexual activity (Humphreys, 1970: 41). After the study, Humphreys destroyed the names and addresses of the men he had interviewed in order to protect their anonymity. His study was subsequently published as a major work on human sexual behaviour (Humphreys, 1970).[9]

For discussion

1 It might be said that Humphreys' research in 'tea rooms' was a form of participant observation, a type of research which is often most successful when 'subjects' do not know they are being observed. Was it unethical for Humphreys to observe men engaged in homosexual acts in the 'tea room'? Does the fact that the behaviour was occurring in a public place make a difference to your argument? Why?

2 Was it ethical for Humphreys to seek and use name and address information – details that appear commonly in telephone books – from police officers who should not have released those details for non-official reasons? Would it have been acceptable if he had been able to acquire that same information without deceit?

3 Upon completion of the research should Humphreys have advised those men who had been observed and interviewed that they had been used for the study? Why? Discuss the significance of your answer. Should only some research results be 'returned' to participants and not others? What criteria might one employ to make that determination? Why are those criteria (more) important (than others)?

4 Should Humphreys have destroyed the name and address information he used? How do we know he was not making the whole story up? How can someone else replicate or corroborate his findings without that information?

5 Humphreys' work offered a major social scientific insight into male homosexual behaviours. It might be argued that his book *Tearoom Trade* contributed to growing public understanding of one group in the broader community. Moreover, no apparent harm was done to those people whose behaviour was observed. Do you think, then, that the ends may have justified the means?

CONCLUSION

Being an ethical geographer is important. It helps to protect those people, places and organisms affected by our research and helps to ensure that we are able to continue to conduct socially and environmentally valuable work. The steps set out in this chapter to help you prepare for research and to deal with the ethical problems you may encounter should go some way to helping achieve these ends. However, your development as an ethically responsible geographer cannot stop with this chapter. It is important that you continue to heighten your awareness of ethical issues and develop your ability to act thoughtfully when confronted with dilemmas like those set out in this chapter. To that end, I shall conclude with some thoughts on ways in which you can continue to become a more ethical geographer.

Good luck!

Summary

What can you do to become a more ethical geographer?

- *Make sure your 'moral imagination' is active and engaged.* There will always be ethical issues in your research. Make ethics as normal a part of your research project discussions as how the stream gauge or questionnaire is working. Discuss ethical issues and possibilities with your colleagues. Read journals like *Ethics, Place and Environment*. Learn to recognize ethical issues in context. Think about the potential moral significance of your own actions and those of other people. Remember that we live in a vast network of moral relationships. The meanings of particular behaviours and moral positions may sometimes be given or understood far from the places they might be expected (Smith, 1998) and interpreted from different ethical standpoints. Look for hidden value biases, moral logic and conflicting moral obligations. Make yourself aware of (local) ethical practices (Mehlinger, 1986).
- *Develop your philosophical and analytical skills.* What is 'right' or 'good'? On what bases are those decisions made? Be prepared to think hard about difficult questions. For example, how can you evaluate prescriptive moral statements such as 'endangered species should (or should not) be protected' or 'research should (or should not) be conducted with the consent of all participants'?
- *Heighten your sense of moral obligation and personal responsibility.* 'Why should I be moral?' or 'Why should I think about ethics?' Embrace ethical thought and action as an element of your professional and social identity as a geographer. Come to terms with the idea that you need to act morally because it is the 'right' thing to do, not just because someone is making you do it (Mehlinger, 1986).
- *Expect – but do not passively accept – disagreement and ambiguity.* Ethical problems are almost inevitably associated with disagreements and ambiguities. However, do not let that expectation of ambiguity and disagreement provide justification for abandoning debate and critical thought. Learn to seek out the core of differences to see if disagreement might be reduced. Be committed enough to follow through on your own decisions.

NOTES

1   Although most of the principles discussed here apply to environmental research ethics, the chapter focuses most heavily on research involving humans. Readers especially interested in environmental research ethics are strongly advised to consult ASTEC (1998), for example.

2   Nevertheless, geographers and geography students increasingly have to take account of the formal codes of ethical practice drawn up by departments, universities and research funding councils (see, for example, Dyer and Demeritt, 2009). When planning to conduct your own research you should always consult your department's or school's own ethical policy and guidelines.

3   These principles place a strong emphasis on individual autonomy. It is important to note that in some societies and situations (e.g. work with some indigenous groups and children) individual autonomy may be limited and influenced by related groups and other individuals who have authority over that individual (NHMRC *et al.*, 2007). It is imperative, therefore, to consider the specific local contexts within which rights are understood.

4   Ethical dimensions of a research project should be negotiated between the researcher and the participant(s) to ensure that the work satisfies the moral and practical needs and expectations of those individuals and communities involved. However, the negotiations will be influenced by the different levels of power between the parties. In some situations a researcher will have more power than the informant (e.g. research work involving children) while in others (e.g. interview with the CEO of a large organization whose comments are critical to your study) informants may have more power. (For a discussion of some of these important issues and ways of dealing with them, see Wilkinson, 1998; Chouinard, 2000; Cloke *et al.*, 2000; D'Cruz, 2000; Dyck, 2000; England, 2002; Kezar, 2003; Dowling, 2005; Smith, 2006; Schoenberger, 2007).

5   For detailed discussions of some of the ethical responsibilities of researchers involved in environmental fieldwork, see ASTEC (1998: esp. pp. 21–4) and Smith (2002).

6   For an interesting discussion on the issue of consent, see Metzel (2000).

7   Wilton (2000) and Routledge (2002) provide fascinating and thought provoking reflections on the ethical dilemmas of geographical work involving deception.

8   Kearns *et al.* (1996) offer a concise and very helpful discussion of the ethics of authorship/ownership of knowledge in geography. In other fields the International Committee of Medical Journal Authors (ICMJE) has formulated a comprehensive statement on writing and editing for biomedical journals. Used with caution, this offers some very helpful suggestions for practice in geography.

9   In a postscript to his book, Humphreys (1970) convincingly addresses issues of misrepresentation, confidentiality, consequentiality and situation ethics associated with his research. His vocation as a religious minister might add, for some, strength to his arguments. For other discussion of this example, Diener and Crandall (1978) is helpful.

Further reading

- A good place to start is the journal *Ethics, Place and Environment.* This publishes research, scholarship and debate on all aspects of geographical and environmental ethics.

- Dowling (2010) provides a helpful introduction to some central ethical issues in geographical research (e.g. harm, consent) and makes a case for critical reflexivity (i.e. ongoing, self-conscious scrutiny of oneself as researcher and of the research process).

- Israel and Hay (2006) elaborate fully and clearly on some of the ideas set out in this chapter.

- Richie Howitt's (2005) paper on ethics and cross-cultural engagement in work with Australian Indigenous people is very helpful. Another challenging but deeply engaging chapter examining issues in cross-cultural research ethics and relationships is that by Howitt and Stevens (2010) – who have vast experience working with Indigenous peoples in Australia and Chomolungma (Mt Everest) respectively.

- Mitchell and Draper (1982) is a classic reference for geographers interested in issues of relevance and research ethics. Though dated, this volume is well worth a read.

- Proctor and Smith (1999) examine the place of geography in ethics and of ethics in geography through the themes of ethics and space, ethics and place, ethics and nature, and ethics and knowledge. It is useful if you have an interest in exploring broader ethical issues than those covered in this 'methods' chapter.

- Scheyvens and Leslie (2000) is a helpful article exploring ethical dimensions of power, gender and representation in overseas fieldwork. It is illustrated by several examples drawn from the authors' recent fieldwork practice.

- Smith (2000b) explores the interface between geography, ethics and morality. At its core is Smith's long-standing concern with the practice of morality.

Note: Full details of the above can be found in the references list below.

References

ASTEC (Australian Science, Technology and Engineering Council) (1998) *Environmental Research Ethics*. Canberra: ASTEC.

Bauman, Z. (1993) *Postmodern Ethics*. Oxford: Blackwell.

Center for Ethics and Business (2008) 'Resolving an ethical dilemma': available at http://www.lmu.edu/Page27945.aspx.

Chouinard, V. (2000) 'Getting ethical; for inclusive and engaged geographies of disability', *Ethics, Place and Environment*, 3: 70–80.

Cloke, P. (2002) 'Deliver us from evil? Prospects for living ethically and acting politically in human geography', *Progress in Human Geography*, 26: 587–604.

Cloke, P., Cooke, P., Cursons, J., Milbourne, P. and Widdowfield, R. (2000) 'Ethics, reflexivity and research: encounters with homeless people', *Ethics, Place and Environment*, 3: 133–154.

Davis, N.A. (1993) 'Contemporary deontology', in P. Singer (ed.) *A Companion to Ethics*. Oxford: Blackwell, pp. 205–218.

D'Cruz, H. (2000) 'Social work research as knowledge/power in practice', *Sociological Research Online*, 5: available at http://www.socresonline.org.uk/5/1/dcruz.html.

Diener, E. and Crandall, R. (1978) *Ethics and Values in Social and Behavioural Research*. Chicago: University of Chicago Press.

Dowling, R. (2010) 'Power, subjectivity and ethics in qualitative research', in I. Hay (ed.) *Qualitative Research Methods in Human Geography* (3rd edn). Toronto: Oxford University Press, pp. 26–39.

Dyck, I. (2000) 'Putting ethical research into practice: issues of context', *Ethics, Place and Environment*, 3: 80–87.

Dyer, S. and Demeritt, D. (2009) 'Un-ethical review? Why it is wrong to apply the medical model of research governance to human geography', *Progress in Human Geography*, 33(1): 46–64.

England, K. (2002) 'Interviewing elites: cautionary tales about researching women managers in Canada's banking industry', in P. Moss (ed.) *Feminist Geography in Practice: Research and Methods*. Oxford: Blackwell, pp. 200–213.

Hay, I. (1998) 'Making moral imaginations: research ethics, pedagogy and professional human geography', *Ethics, Place and Environment*, 1: 55–76.

Howitt, R. (2005) 'The importance of process in Social Impact Assessment: ethics, methods and process for cross-cultural engagement', *Ethics, Place and Environment*, 8: 209–221.

Howitt, R. and Stevens, S. (2010) 'Cross-cultural research: ethics, methods and relationships', in I. Hay (ed.) *Qualitative Research Methods in Human Geography* (3rd edn). Toronto: Oxford University Press, pp. 40–68.

Humphreys, L. (1970) *Tearoom Trade: A Study of Homosexual Encounters in Public Places*. London: Duckworth.

International Committee of Medical Journal Authors (2007) *Uniform Requirements for Manuscripts Submitted to Biomedical Journals: Writing and Editing for Biomedical Journals*: available at: www.icmje.org.

Israel, M. and Hay, I. (2006) *Research Ethics for Social Scientists: Between Ethical Conduct and Regulatory Compliance*. London: Sage.

Kates, B. (1994) 'President's column', *Association of American Geographers' Newsletter*, 29: 1–2.

Kearns, R., Le Heron, R. and Romaniuk, A. (1998) 'Interactive ethics: developing understanding of the social relations of research', *Journal of Geography in Higher Education*, 22: 297–310.

Kearns, R., Arnold, G., Laituri, M. and Le Heron, R. (1996) 'Exploring the politics of geographical authorship', *Area*, 28: 414–420.

Kezar, A. (2003) 'Transformational elite interviews: principles and problems', *Qualitative Inquiry*, 9: 395–415.

Kimmel, A.J. (1988) *Ethics and Values in Applied Social Research*. London: Sage.

Matthews, H., Limb, M. and Taylor, M. (1998) 'The geography of children: some ethical and methodological considerations for project and dissertation work', *Journal of Geography in Higher Education*, 22: 311–324.

Mehlinger, H. (1986) 'The nature of moral education in the contemporary world', in M.J. Frazer and A. Kornhauser (eds) *Ethics and Responsibility in Science Education*. Oxford: ICSU Press, pp. 17–30.

Metzel, D. (2000) 'Research with the mentally incompetent: the dilemma of informed consent', *Ethics, Place and Environment*, 3: 87–90.

Mitchell, B. and Draper, D. (1982) *Relevance and Ethics in Geography*. London: Longman.

NHMRC (Australian National Health and Medical Research Council), Australian Research Council (ARC) and Australian Vice-Chancellors Committee (AVCC) (2007) *National Statement on Ethical Conduct in Human Research*: available at www.nhmrc.gov.au/publications/synopses/e72syn.htm.

Oakes, J.M. (2002) 'Risks and wrongs in social science research: an evaluator's guide to IRB', *Evaluation Review*, 26: 443–479.

Proctor, J. and Smith, D.M. (eds) (1999) *Geography and Ethics: Journeys in a Moral Terrain*. London: Routledge.

Quinton, A. (1988) 'Deontology', in Bullock, A. Stallybrass, O. and Trombley, S. (eds) *The Fontana Dictionary of Modern Thought* (2nd edn). Glasgow: Fontana, p. 216.

Routledge, P. (2002) 'Travelling east as Walter Kurtz: identity, performance and collaboration in Goa, India', *Environment and Planning D: Society and Space*, 20: 477–498.

Scheyvens, R. and Leslie, H. (2000) 'Gender, ethics and empowerment: dilemmas of development fieldwork', *Women's Studies International Forum*, 23: 119–130.

Schoenberger, E. (2007) 'Politics and practice: becoming a geographer', in A. Tickell, E. Sheppard, J. Peck and T. Barnes (eds) *Politics and Practice in Economic Geography*. London, Sage, pp. 27–37.

Smith, D.M. (1998) 'How far should we care? On the spatial scope of beneficence', *Progress in Human Geography*, 22: 15–38.

Smith, D.M. (2000a) 'Moral progress in human geography: transcending the place of good fortune', *Progress in Human Geography*, 24: 1–18.

Smith, D.M. (2000b) *Moral Geographies: Ethics in a World of Difference*. Edinburgh: Edinburgh University Press.

Smith, K.E. (2006) 'Problematising power relations in "elite" interviews', *Geoforum*, 37: 643–653.

Smith, M. (ed.) (2002) *Environmental Responsibilities for Expeditions: A Guide to Good Practice*: available at www.britishecologicalsociety.org.

Stark-Adamec, C. and Pettifor, J. (1995) *Ethical Decision Making for Practising Social Scientists: Putting Values into Practice*. Ottawa: Social Science Federation of Canada.

Valentine, G., Butler, R. and Skelton, T. (2001) 'The ethical and methodological complexities of doing research with "vulnerable" young people', *Ethics, Place and Environment*, 4: 119–125.

Walsh, A.C. (1992) 'Ethical matters in Pacific Island research', *New Zealand Geographer*, 48: 86.

Wilkinson, S. (1998) 'Focus groups in feminist research: power, interaction, and the co-construction of meaning', *Women's Studies International Forum*, 21: 111–125.

Wilton, R.D. (2000) '"Sometimes it's OK to be a spy": ethics and politics in geographies of disability', *Ethics, Place and Environment*, 3: 91–97.

4 Health and Safety in the Field

Joanna Bullard

Synopsis

Health and safety in the field concerns the practical steps that geographers can take to lessen the chances of an incident or accident causing harm to themselves and others during fieldwork. This chapter focuses on minimizing risks, reducing hazards and taking responsibility for health and safety both in group and independent fieldwork. It is structured into the following sections:

- Introduction
- Managing health and safety
- Know your field environment
- Know your limitations
- Know your equipment
- What if something goes wrong?
- Conclusion

INTRODUCTION

Fieldwork is often one of the most rewarding aspects of being a Geography student. Many residential trips and one-day outings are remembered by students long after graduation. Fieldwork can, however, also be time consuming, frustrating, difficult and potentially dangerous. This chapter offers some guidance about minimizing the risks and hazards involved in fieldwork. Following this guidance should reduce the dangers inherent in your chosen field activity and consequently should also make your time in the field more rewarding, more enjoyable and possibly a little easier.

As a Geography student you may undertake fieldwork in a variety of contexts including supervised and unsupervised, group and individual field activities, residential and non-residential field studies, local and overseas environments. These different contexts present a range of health and safety considerations and the roles and responsibilities of both students and staff may change accordingly.

This chapter discusses the assessment and management of risks associated with undertaking fieldwork. It covers evaluating and minimizing risk in a range of common field environments and emphasizes the importance of planning ahead. In particular, the chapter focuses on your responsibility as an individual and as a member of a group undertaking fieldwork to ensure that appropriate health and safety precautions are observed.

In addition to reading this chapter, note that every higher education institution (and frequently school or department within this) has its own health and safety policy, now most commonly available in its website provision. You may have been required to sign a declaration stating that you have read and understood these guidelines. One of your responsibilities as a student is to ensure that you follow any institutional or departmental guidelines when you are out in the field, and often this is a compulsory part of dissertation planning.

MANAGING HEALTH AND SAFETY

In most countries, national legislation provides the legal framework for health and safety management. For example, in the UK, universities abide by the Health and Safety at Work Act 1974, while in Australia, the Workplace Health and Safety Act 1995 has a similar role. These Acts place a duty upon employers to take steps to ensure, as far as possible, the health and safety of their employees and any other people affected by their activities including, in the case of universities, all students and members of the public. Additional or complementary legislation and guidelines may also exist at the local, state or national scale. For example, there is a British standard concerning 'Specification for the provision of visits, fieldwork, expeditions, and adventurous activities, outside the United Kingdom' that your university may have signed up to (BS8848 – see Further Reading). These laws and guidelines typically require that 'risk assessments' are undertaken to identify what should be done in order to manage safety. Assessments of risk are usually focused around 'risk' and 'hazard'. Put simply, hazards result from working in potentially dangerous environments and refer to environmental conditions, agents or substances, that can cause harm. Risk is the chance that someone might be harmed by the hazard. During fieldwork hazards and risks can change rapidly, for example, as a result of changing weather conditions or political actions, and should be continually reassessed.

Many education institutions adopt a five-step approach to risk assessment.

1 *Identify the hazards*: during the fieldwork, what could cause harm – for example, slippery ground, high altitude, weather conditions, civil or political unrest?
2 *Identify who might be harmed and how*: this includes all the field workers and members of the public.
3 *Evaluate and minimize the risks*: once suitable precautions have been taken, for example wearing appropriate protective clothing, how likely is it that someone will be harmed?
4 *Record the findings*: write down the identified hazards and precautions to be taken.
5 *Review the assessment periodically*: situations change and so do the potential hazards and risks, so risk assessment is not an isolated task but an ongoing evaluation to be updated and revised as often as necessary.

A risk assessment should be completed for fieldwork conducted as part of your degree. The extent of your involvement in actually assessing the risks will vary according to the way in which the fieldwork is organized, but you have a responsibility to follow any precautions or safety measures laid down in the risk assessment.

Supervised and unsupervised group fieldwork

Group fieldwork will usually be undertaken as part of a course or module designed by a member of academic staff at your university. During supervised fieldwork, one member of staff will take overall responsibility for field activities and will complete the risk assessment. The leader of the fieldwork will be responsible for ensuring that appropriate health and safety matters have been taken into consideration and are complied with by all accompanying members of staff and students. You are not, however, absolved of responsibility. Be aware that as an individual you have a responsibility for your own safety and also for the safety of others who may be affected by your actions, including staff, other students and members of the public. You should follow any instructions given by the staff members. If you witness any accidents or foresee any unexpected problems these should be reported to the fieldwork leader as soon as possible. Failure to uphold your responsibilities may result in disciplinary measures and, in extreme cases, in a criminal prosecution. It is worth noting that your activities prior to fieldwork, for example having a few alcoholic drinks the evening before, can adversely affect your performance and reliability in the field. Staff may refuse to let you participate in fieldwork if you are considered a danger to yourself or others.

On arrival at university, you may have completed a confidential medical questionnaire. You may be asked to complete a more detailed questionnaire prior to undertaking any fieldwork, especially if it will be conducted overseas. These questionnaires should be used to give details of any diseases, conditions, disabilities or susceptibilities that you may have, such as diabetes, severe allergies, impaired hearing or vertigo. If you are involved in an incident or accident whilst on fieldwork this information may be vital in ensuring that you receive prompt and appropriate medical treatment. If you did not complete a questionnaire or equivalent, and do have medical needs that could compromise the health and safety of yourself or others, you should let staff know well before fieldwork commences so that they can take appropriate action where necessary.

During a field course you may undertake unsupervised group fieldwork. This commonly involves students working in groups of two or more. A member of staff may visit you during the field activities; for example, they may move from group to group along the course of a river or arrange to meet you at a particular location within a city. Alternatively you may be left entirely to your own devices and asked to book in or out of a field centre or to conduct work on campus or in your university city. In these situations you have a responsibility to the group of students with whom you are working. Keep together as a group, discuss any difficulties that you might have and pay attention to anyone who is feeling unwell or is having difficulty keeping up. You should follow any guidelines issued; for example, take a first aid kit, have contact numbers for the university department or individual members of staff and know where you should be working and what time you are expected to check back in.

Individual, unsupervised fieldwork

You may be involved in individual, unsupervised fieldwork at any time during your university career, but this is most likely to be the case when you undertake fieldwork as part of a dissertation. Many universities now require every student to complete an individual risk assessment for their intended project. This can take many forms but will usually be organized around the five-step plan outlined above. Detailed discussion and examples of this procedure are provided by Higgitt and Bullard (1999), templates are now frequently to be found on institutions' websites and most departments will have a nominated Safety Officer who can give advice.

The most important health and safety concerns associated with individual, unsupervised fieldwork arise from lone working which should be avoided wherever possible. If you are conducting an individual project or dissertation a good strategy is to persuade a responsible friend or sibling to act as your field assistant. You might pair up with a friend from university and agree to help them survey a beach if they, in turn, will accompany you delivering questionnaires. If lone working is unavoidable or you are working with just one other person it is vital that you develop a strategy for dealing with emergencies should they arise. At the very least you should leave with a responsible adult written details of where you are going – for example, the map co-ordinates of your field site or name and address of person you intend to interview – what you intend to do and what time you expect to return. You should also write down what you expect the person to do in the event that you do not return on time. If your plans change you must let this person know. In risk assessment, lone working always counts as a hazardous activity. Where the site or conditions are themselves hazardous, then lone working is liable to be dangerous.

Whether you are working in a group or as an individual there are some basic steps that you can always take to minimize any health or safety risks. First know the telephone numbers that can be used to contact the emergency services if necessary. In the UK, 999 or 112 will allow you to contact the fire, police and ambulance services, the coastguard and mine, mountain, cave and fell rescue services. Similar services can be contacted in the USA and Canada by dialling 911, in Australia, 000 and in New Zealand, 111. If you are working overseas, find out local emergency numbers and the contact details of your embassy, high commission or consulate before you start working and keep them written down in a safe place (e.g. on a card inside your first aid kit). Other emergency communication methods include the international Alpine Distress signal – six long whistle blasts or torch flashes or shouts for help in succession repeated at one-minute intervals – and the Morse code signal 'SOS' signalled by three short whistle blasts/torch flashes, three long blasts/flashes, three short blasts/flashes, pause, then repeat as necessary.

Second, have a basic kit that you take into the field with you every day. Never start any fieldwork in any environment without having with you a pen/pencil and paper and any personal medication. If you are working in a rural or remote environment you will also need a first aid kit, mobile phone, whistle, compass, map, additional clothing (hat, gloves, socks, spare sweater), food and drink as a minimum. If you are working in a built-up or urban environment your field kit may include a first aid kit, mobile phone, street map and personal alarm.

Third, make sure that you always know exactly where you are – this is relatively easy in a town centre but can be more problematic in an unfamiliar large city or a remote area with few landmarks. If the worst happens and you need help from the emergency services this will be a vital piece of information that you can provide. Borrowing or buying a GPS (Global Positioning System) unit (see below) is a good idea.

Fourth, consider whether insurance and other forms of cover for travel accidents and breakdowns is necessary. Frequently, cover can be provided for support and assistance before emergency services need to be involved (breakdown assistance, for example, and advice and support for foreign travel, travel incidents and health). Universities may provide this, or you may be expected to make your own provision. There may also be health requirements to meet before travel can be undertaken.

When you carry out the risk assessment for your field project or set off on a field trip with other students there are a number of other things that you need to take into account to ensure the health and safety of everyone involved. These fall broadly into the categories of knowing your field environment, knowing your own limitations and knowing your equipment.

KNOW YOUR FIELD ENVIRONMENT

There are many different locations in which you might find yourself doing fieldwork, from a suburban street to a kibbutz, from a quarry to the top of a mountain. No matter where you intend to go there will always be some information that you can acquire before you start that will provide guidance as to how to act, dress and behave. Gathering this information together is important, because in many cases you will need to complete a risk assessment for your field project *before* you reach the site. If the field site is nearby or very familiar to you then you can draw on your existing experiences when completing the risk assessment, but if the site is not somewhere you have been before or is overseas, you will need some help in anticipating the risks and hazards involved.

A key control on many field activities is the weather or climate of the area in which you are working. As part of your ongoing risk assessment you should consult a daily weather forecast throughout your fieldwork (see Further Reading). Specific hazards associated with weather and climate include hypothermia, frostbite, sunburn, dehydration, heat stroke and heat exhaustion. In most situations being aware of the predicted weather for the local area and having the correct clothing and equipment can reduce the risks associated with these hazards. Specialist information for mountain and coastal environments is available in many countries. When working at the coast make sure you know where the nearest coastguard station is and check local tide timetables.

If you are conducting people-oriented fieldwork at home or overseas you are advised to do some research on the local area and to make yourself familiar with any customs, political issues and religious beliefs that may affect the ways in which you conduct your research and how people react to your project. The aim is to minimize the risk of causing offence, which may lead to personal attack/injury or abuse. Take advice from local contacts about how to respect local customs and about any unrest in the area and dress appropriately for the environment and culture within which you are conducting your research. If possible avoid areas that are known to be 'unpleasant' and try not to enter unfamiliar neighbourhoods alone. A good street map is invaluable, plan your route, and walk purposefully and with confidence.

When taking part in a supervised field trip, the trip leader should check local conditions and is likely to brief you about them. If you are told tide times or other specific pieces of information, such as neighbourhoods to avoid, make sure you write them down. Prior to the field trip you may be given a list of clothing and equipment to bring with you that is suitable for the expected conditions. Depending on the environment, this could include sturdy walking boots, water-proof jacket, gloves, sunglasses, sunhat, water bottle, long sleeves, and so on. In some institutions you will not be allowed to take part in the field activities if you do not have appropriate clothing with you – this is not a minor issue, it is one with potentially profound health and safety implications.

Many government websites provide detailed information for nationals travelling overseas. This can include information on countries or regions to avoid, visa requirements, local customs, pre-travel inoculations and what to do should you be the victim of crime, fall ill or get into trouble of any sort. Make sure you are aware of any political, military or civil unrest in the country – local newspapers can be an invaluable source of information. If you are working alone or as a small independent group you should inform your nearest embassy, high commission or consulate of your presence in the country when you arrive to enable them to keep you informed of any hazardous situations. Nash (2000) provides a discussion of things to consider when planning independent overseas fieldwork and includes a list of essential information sources. Another useful reference both whilst planning and during an overseas field visit will be a guidebook to that country or area. Most popular guidebooks, such as those in the Rough Guide or Lonely Planet series, include sections on health risks, safety precautions and appropriate dress codes.

KNOW YOUR LIMITATIONS

Many people underestimate the time needed to complete fieldwork tasks. The majority of incidents and accidents occur when people are tired. Make sure that the aims of your fieldwork are realistic and that you have allowed enough time to achieve them. Weather conditions can dramatically affect your efficiency in the field – it is hard to survey for six hours in the pouring rain but much easier (and more fun) if the sun is shining, although note again, risks from sun and wind burn, and take appropriate precautions. A useful approach is to make a list of the minimum amount of work that you need do to achieve your basic goals and rank these tasks in order of priority. Then make a list of what would be useful or desirable if you have the time. Work through the list in order, achieve the minimum and then add to it if possible. Do not compromise your safety by trying to work longer hours in unsuitable conditions. If you are doing supervised fieldwork and are uncomfortable about undertaking certain field activities tell the leader and ask whether or not there is an alternative. For example, if you are scared of heights, and are very nervous about walking along a narrow path with a steep drop, find out if there is an alternative route, or whether someone more experienced with the environment can walk with you.

Although you should try to work within your limitations, fieldwork is about facing challenges and doing things you might not normally do, for example interviewing complete strangers, working in a challenging physical environment. There are many things that you can do before you head off in to the field that will extend your skills and abilities and so reduce the risks involved. Some organizations offer courses and workshops and produce publications that may be useful (see Further Reading). If you intend to work in sparsely populated areas then you should consider taking a first aid training course. These are run on a regular basis by bodies such as the St John Ambulance and St Andrew's Ambulance Association and may be available at your college or university as an evening class. Advice on expedition planning and organization is available from a number of sources including the Expedition Advisory Centre (UK) and The Explorers Club (USA) (see Further Reading). More specialist courses, such as winter mountain skills and sea navigation, are run by specific organizations and societies.

KNOW YOUR EQUIPMENT

Whatever equipment you intend to use for your fieldwork, whether a tape recorder, a soil auger or something electronic, make sure that you know how to use it and how to carry it. Some items of equipment are very heavy and need two people to move them safely, even a simple surveying kit comprising staff, level and tripod can be awkward for one person to manage. Take advice from academic and support staff in your institution about the safest ways to collect samples from a river, to dig a deep soil pit, to conduct a house-to-house survey. If your equipment requires batteries and/or cassettes, take spare ones, if you are delivering questionnaires house-to-house make sure you have more copies than you think you need and a good street map. These types of precautions will prevent delays in your field programme and reduce the chance of you being tempted to work at unsuitable times, for example delivering questionnaires after dark or working on a beach with a rapidly rising tide.

Mobile telephones have revolutionized the ways in which we communicate with each other. Although a mobile phone may be a useful way to contact help in a city, reception problems can render a telephone inoperable in isolated areas or in valleys. In addition, technological failure and (more commonly) lack of power can reduce your mobile phone to a useless piece of plastic or metal in any location. The same is true for personal alarms. Carrying a mobile phone or personal alarm is no substitute for a well-thought out system for informing other people of your whereabouts (your lone working strategy).

If you are working in a rural or remote land area or at sea you may decide to take a hand-held GPS with you. This type of device can provide accurate latitude/longitude co-ordinates and can be used for mapping and/or navigation when referenced to national mapping conventions. If you are using a GPS for navigation you must be aware of its limitations. Most hand-held GPSs are only accurate to a few metres latitude/longitude and are notoriously unreliable with regards to altitude readings. In addition they run on batteries, which may expire at an inconvenient moment and, like any technological device, can malfunction or break. If you do take a GPS with you into the field you should still ensure that you have a navigational compass and a map with you (if working overseas your compass should be calibrated for the country in which you are working) and that you know how to use it, for example how to take a bearing and walk along it.

WHAT IF SOMETHING GOES WRONG?

If you take the precautions outlined in this chapter and carry out a full and careful assessment of risks before undertaking fieldwork the chances of any problems should be minimized. However, risks can never be eliminated altogether. Vehicle accidents and breakdowns happen, people trip over, the weather changes. If you do find yourself in an unfortunate situation try to keep calm, assess your options and get help if necessary. If you have insurance cover, consult this, depending on the type and severity of the incident or circumstance. Keeping calm is always easier said than done, but it is important that you try not to panic and not to induce panic in other people. Take some time to think about the situation, use your common sense and never place yourself or others in danger. If an incident or accident does happen that jeopardizes health and safety there are two key things that you, or members of your party, need to do. First, understand what has happened and, second, assess the severity of the situation.

When things start to go wrong try to understand what has happened as fully and quickly as possible. If you are lost on a hillside do you know where you took a wrong turning? If so, can you retrace your steps and get back on the right path? If your vehicle has broken down, do you know why, for example a flat tyre? Can you mend it safely? If someone is injured, do you know how the accident occurred? Was there a rockfall? Did they slide down a steep slope? This will help you to understand not only what their injuries might be, but also whether or not the danger is still present. For example an unstable cliff face may collapse as a series of multiple rockfalls. Look out for any dangers to yourself or to the casualty and never put yourself at risk. If necessary, can you treat the casualty using your first aid kit?

If you cannot resolve your predicament then you need to assess the situation as accurately as possible before you seek help. If you are dealing with a casualty or casualties that you cannot treat, try to determine what their injuries are. The emergency services will need to know how many casualties there are, whether they are conscious or unconscious and whether or not they are breathing and/ or have a pulse. If the person has a known condition, such as asthma or a heart condition, and if you know how the incident occurred, for example immersion in water or traffic collision, inform the emergency services of this.

You also need to know where the incident has occurred. Take a note of your grid reference and landmarks if you are in a remote area, note any road numbers or junction details and have an idea of how far away from the nearest habitation you are. If you have to leave the scene to get help be sure to write this information down before you set off. Emergency services can trace your telephone call to any call box or motorway telephone but if you have to leave the site or are using a mobile phone to make the call you should be extra vigilant about establishing your exact location. If you are lost in hilly terrain on foot and cannot retrace your steps, stop and consider the safest route off the hill or mountain. Use your compass to set a bearing in that direction. Heading towards buildings, a water course or a road are often good options. If your vehicle has broken down in a remote area the best option is usually to stay with the vehicle until help arrives, especially in extreme weather conditions. Finally, if hazardous conditions, such as poor visibility, unstable slopes, flooding, high tides or electricity, have contributed to the accident or may affect the response from emergency services, then inform them of this when you call.

When the crisis is over, try to recall and write down what went wrong and why, including times and dates, if possible. You should give this information to the safety officer in your department as it may be useful for establishing procedures to prevent similar future incidences.

Summary

- Fieldwork can be a risky business, but considering the variety of locations in which Geographers undertake fieldwork and the amount of time they spend in the field, few incidents and accidents actually take place.
- Many of the precautions that are taken to protect health and safety in the field are common sense and you would take the same precautions whilst shopping at the weekend or going walking in the hills with friends – remember to apply this common sense when undertaking fieldwork in all situations (whether as part of a supervised group or as an unsupervised individual).
- Risk-assessment techniques can be used to anticipate likely fieldwork hazards. Identify any precautions that could minimize the risks and ensure that these are fully implemented.
- Occasionally there may be special procedures associated with particular equipment or environments that you need to learn and apply – seek advice from tutors and/or specialist advisory services.
- You have a responsibility to initiate and/or follow appropriate health and safety guidelines and to take all reasonable precautions to ensure the health and safety of yourself and others in the field.
- By following the above advice, you may not only learn new skills and take on new responsibilities, but, more importantly, you may also prevent any serious incidents or accidents occurring.
- Always consult health and safety information for your university, and specific information for your department within this. This should now be easily accessible from the website within the institution.

Further reading

There is not much written specifically for students about health and safety in the field but two useful sources are:

- Higgitt and Bullard (1999) details why and how risk assessments are undertaken for undergraduate dissertations. Using two geography case studies, one human and one physical, the paper illustrates the types of hazards and risks that need to be considered.

- Nash (2000) is the first part of a guide to doing independent overseas fieldwork. It discusses where to go, what to do when you get there and also some of the health, safety and insurance issues that you should consider before setting off.

Many government websites present comprehensive information about travelling overseas including country by country guides to safety and security, local travel, entry requirements and health concerns. Some also include more general sections within the websites, such as travellers' tips, how to get your mobile phone to work overseas and what to do if it all goes wrong. These sites are updated regularly particularly with regard to political disturbances and natural disasters. The sites for UK, USA and Australian citizens are listed below. Other nationals should consult the travel section of their home-government website.

- Foreign and Commonwealth Office website (UK): http://www.fco.gov.uk/en/travel-and-living-abroad/

- Center for Disease Control and Prevention (health advice) (USA): http://wwwnc.cdc.gov/travel/

- United States Department of State (travel advice): http://travel.state.gov/

- Australian Department for Foreign Affairs and Trade: http://www.smartraveller.gov.au/

Up-to-date local weather forecasts are usually available in newspapers, by telephone (check local papers for the number) or from websites, for example:

- The Meteorological Office (UK): http://www.metoffice.gov.uk/

- National Weather Service (USA): http://www.nws.noaa.gov/

- Bureau of Meteorology (Australia): http://www.bom.gov.au/

If you are planning an expedition and need advice or want to develop particular skills before you leave useful contacts include:

- The Royal Geographical Society (with the Institute of British Geographers) (http://www.rgs.org/OurWork/ Fieldwork+and+Expeditions/Fieldwork+Expeditions.htm) runs workshops and seminars on topics such as four-wheel drive training, people-oriented research techniques, risk assessment and crisis management, and produces publications offering advice on logistics and safety in a range of environments from tropical forests to deserts.

- The Explorers Club (USA) (http://www.explorers.org/) can provide expedition planning assistance and has a lecture series which occasionally features sessions on field techniques.

- The International Mountaineering and Climbing Federation (http://www.theuiaa.org/) and British Mountaineering Council (http://www.thebmc.co.uk) provide advice on all aspects of high altitude travel and safety.

Fieldwork opportunities arise from a variety of sources including not only educational institutions but also through working with charities or taking part in expeditions. The British Standards Institution has published 'BS8848 Specification for the provision of visits, fieldwork, expeditions, and adventurous activities, outside the United Kingdom', which sets out the requirements to be met by those organizing adventurous trips in order to comply with good practice. BS8848 (which takes its name from the metric height of Everest) is aimed primarily at 'providers' but also highlights the responsibilities of participants to commit to reasonable behaviour 'as a contribution to keeping themselves and others safe' (p. 39). If you are organizing an expedition or fieldwork, there are useful checklists of things to consider included in the appendices to BS8848, which is available from libraries or from the British Standards Institute (http://www.bsigroup.com/en/).

References

Higgitt, D. and Bullard, J.E. (1999) 'Assessing fieldwork risk for undergraduate projects', *Journal of Geography in Higher Education*, 23: 441–449.

Nash, D.J. (2000) 'Doing independent overseas fieldwork 1: practicalities and pitfalls', *Journal of Geography in Higher Education*, 24: 139–149.

Generating and Working with Data in Human Geography

5   Making Use of Secondary Data

Paul White

Synopsis

Secondary data consist of information that has already been collected for another purpose but which is available for others to use. Such data are an indispensable source of information for many student projects where resource limitations, such as time and money, may preclude data collection for an extensive area, a large population or comparison between places. The chapter first looks at the nature of secondary data: who collects it, what it consists of and why it is of interest to geographers. It then goes on to discuss how to access secondary data: where to obtain it (at scales ranging from the international to the local) and the use of Internet sources. Finally, it looks at the uses of secondary data: to provide a context for a study, for comparisons and as the prime evidence for analysis.

The chapter is organized into the following sections:

- Introduction: the nature of secondary data
- Sources of secondary data
- Utilizing secondary data
- Conclusion

INTRODUCTION: THE NATURE OF SECONDARY DATA

We live in a world where things are counted. Arguably this has always been so. Some of the earliest extant writings consist of counts of people, goods, transactions and possessions. Conquering forces have always been keen to enumerate the attributes of what it is they have conquered. And wherever political power has needed financial funds, tax systems have produced data on wealth indicators of various sorts.

These exercises in data collection are multiplied many times over in the contemporary world, where it often seems that almost every aspect of human life is counted, and where aspects of these counts are stored in databases of some sort – births, examination performances, traffic flows, financial transactions, housing standards, consumption patterns and so on. Complex societies and economies count things more than do other populations with simpler structures and activities. The management, planning and provisioning of complex societies give rise to data needs, while in certain cases government agencies wish to gain insights into the lives of citizens for reasons of control. These exercises all produce secondary data which, once collected for one purpose, may be used for another. Such data can be either qualitative or quantitative (or a combination of the two) in nature, but the focus here is on quantitative data (see Chapter 7 of this book and also Kitchin and Tate, 2000, on qualitative secondary information).

A number of general observations can be made about quantitative secondary data:

1   *The principal source of such data is normally government.* Non-governmental sources of secondary data certainly exist but tend to have only local spatial coverage or be very specific in nature. Commercial data are often not made available to other researchers, for reasons of commercial confidentiality, or can only be acquired at a cost. Government-generated information covering multiple topics, however, is often collected with varied purposes in mind. Population censuses, one of the major government data sources of interest to geographers, normally provide a variety of data, reflecting a belief that it is a 'good thing' for government to have evidence available to support its policy-making, implementation and monitoring functions. A trend in many (but not all) parts of the world is the increasing availability of government data. Where at one time only limited parts of datasets were accessible to the general public (or the academic researcher), or where the costs of access were considerable (in terms of data purchase or the need to travel to inspect original data manuscripts), large amounts of data are now available to all via the internet. Today the desk-bound researcher can call up the figures for the resident population of Mexican municipalities or the number of benefit claimants in the districts of an English city with a few clicks of a mouse.

2   In making use of secondary data, the *researcher needs to be aware that the information has been collected by someone else, for another purpose.* The questions asked might not have been exactly the ones that the researcher would have used, and the release of data to others may have involved the simplification or re-categorization of information in ways that are potentially unhelpful.

3   Nevertheless, because of the massive financial investment governments make in collecting data they are keen to maximize such data's utility. Census questions, data-recording systems and coding schedules are usually subject to intensive pre-testing (Rose and O'Reilly, 1998). *Secondary data (particularly from government) tend to have been collected in ways that are much more robust than the means available to the individual researcher.*

4   *Secondary data may already have been manipulated for particular, possibly political, purposes; hence, secondary data may not be entirely trustworthy.* Local census officers, for example, are sometimes believed to have inflated population sizes within their jurisdictions for reasons connected with local prestige or power. A strong case can be made that all government data inquiries reflect cultural perceptions which influence both what is asked and how the questions are answered (Hoggart *et al.*, 2002).

5   *Available secondary data are often strongly spatially referenced.* Government pertains to spatially bounded units at a hierarchy of scales. Secondary data sources thus often emphasize the variable geography of society. However, the spatial 'containers' for some datasets may not be those the researcher would

choose. Immigration data, to take one example, are generally only available at national levels so that a research project on the role of international migration in a single city would not find relevant data available. Some datasets (e.g. the results of extensive questionnaire surveys) are available at regional or other subnational levels, but it is generally only population census data that are available at the most detailed local level.

6 Although much secondary data is 'factual' in nature (dealing with counts of objects or people), *there is a rapidly growing volume of attitudinal data available for secondary use.*

These general points indicate a series of advantages and disadvantages to the use of secondary data. Although the negative points must be borne in mind, it is possible to overstress them. For many geographical investigations, secondary data are indispensable, since the project could not proceed without such data. This is particularly true for student projects that focus on anything other than a small locality over a short time period. And one reason why the availability of secondary data is growing rapidly lies in the protocols of many research funding bodies (such as the Economic and Social Research Council in the UK) which now require all researchers receiving funding to deposit any data they collect in publicly accessible 'data archives'.

SOURCES OF SECONDARY DATA

Unlike many of the issues dealt with in this book, questions of access to secondary data tend to be specific to the country in which a student is studying. Nevertheless, an attempt is made in what follows to raise a number of general issues concerning the availability of data.

For many student projects, at whatever scale, an obvious starting place is the website of the national statistical office of the country for which information is required. In many countries all available government data can be accessed via this office, even information gathered by disparate government departments and agencies. Table 5.1 provides a list of the web addresses of selected national statistical offices, as well as certain international agencies which either themselves gather or collate from other sources comparative information about major issues.

Amongst the international organizations, UNICEF is particularly useful for comparative development data on countries around the world. An additional source of statistical information, but which requires individual country entries to be compared, is the 'world factbook' produced by the American Central Intelligence Agency (www.cia.gov/library/publications/the-world-factbook/index.html).

Students in England and Wales wishing to make use of detailed local data will find a mass of information available. The results of the 2001 population census (the latest available at the time of writing) are available via the census website: www.census.ac.uk. This site also holds data for 1991 and other censuses back to 1971. It is a first port of call for many studies of recent geographical change in large parts of the UK and is closely linked to a variety of other government

Table 5.1   Selected sources of official statistical data

A. National statistical offices (in each case the web address starts http://www)

Country	Web address
Argentina	indec.mecon.gov.ar
Australia	abs.gov.au
Brazil	ibge.gov.br/english
Canada	statcan.gc.ca
China	stats.gov.cn/english
Colombia	dane.gov.co
Egypt	campus.gov.eg/eng-ver
France	insee.fr/en
Germany	destatis.de/en
Ghana	statsghana.gov.gh
India	censusindia.net
Ireland	cso.ie
Italy	istat.it/english
Japan	stat.go.jp/English
Malaysia	statistics.gov.my
Mexico	inegi.gob.mx
Netherlands	cbs.nl/en
New Zealand	stats.gov.nz
Poland	stat.gov.pl/english
Portugal	ine.pt/en
South Africa	statssa.gov.za
Spain	ine.es/en
Sweden	scb.se/eng
UK	statistics.gov.uk
USA	fedstats.gov

B. International organizations

Organization	Web address
Eurostats	http://epp.eurostat.ec.europa.eu
Food and Agriculture Organization	http://www.fao.org/corp/statistics/en
UNICEF	http://www.unicef.org/statistics
United Nations	http://unstats.un.org/unsd
World Health Organization	http://www.who.int/whosis/en

data. Nevertheless, social change is often reflected in the census itself and in the questions it asks. For example, the inclusion of questions on religion in the most recent censuses reflects new interest in the links between religious affiliation and ethnicity (see Southworth, 2005; Peach, 2006).

The Office for National Statistics collates a wealth of information on individual administrative wards from a range of government sources (www.statistics.gov.uk/hub). Of these the most useful spatially referenced data deal with neighbourhood statistics (neighbourhood.statistics.gov.uk/dissemination) where information is available down to the level of individual census enumeration districts generally containing between 500 and 700 households. The data available include material on population, health, employment, education, housing, crime, the environment and a variety of other topics. In reality few of these are actually reported at neighbourhood level: several indicators show data at local authority level. Clearly the data available, and the scale involved, reflect the administrative interests of government, as well as the ways in which statistical information is used to inform policy – particularly through the indices of deprivation used to identify areas to be targeted for particular initiatives. The use and calculation of such indices will be discussed later in this chapter.

Political devolution in the UK is not new in terms of certain government data services, and students researching in Scotland and Northern Ireland have other statistical services available to them. For Scotland the main site is www.gro-scotland.gov.uk (which includes an online 'data library'), while for Northern Ireland the site is www.nisra.gov.uk (containing impressive data maps as well as other sources). For neighbourhood statistics the Scottish site is at www.sns.gov.uk, whilst for Northern Ireland similar data are available at www.ninis.nisra.gov.uk.

Government statistical databases have developed in recent years to the point where they now cover a vast variety of topics that were once only dealt with by commercial organizations. The student is thus strongly advised to start any search for secondary statistical data with the relevant government site for whatever country is under consideration.

One area where recent advances in data availability have in reality not been of as much use for student projects as might be imagined is that of historical census data. The UK censuses from 1841 to 1901 are now all available online. However, the prime users are family historians who are interested in tracking individuals, and consequently the digitized structures adopted by the providers of such information are most suited to genealogical rather than other forms of research. The costs of digitization are generally also borne by users, in the form of registration or user fees.

Census data, and other forms of spatially referenced information, can give rise to certain problems of interpretation for the unwary user. The most common error is that of the 'ecological fallacy'. This occurs where the aggregate results for an area are (mis)used by inferring that they apply to all individuals within that area. If in a neighbourhood the census shows that there is an above-average proportion of the elderly and an above-average proportion of houses without central heating, the ecological fallacy consists of inferring that the elderly live in unheated houses. It could be that old people in the neighbourhood are in fact wealthy and live in relative luxury, but are surrounded by poorer younger families.

One way of avoiding the ecological fallacy is to use data on individuals rather than on aggregates. Instead of inferring individual characteristics from groups, the researcher can then examine actual people. Inevitably the circumstances under which this is permitted are restrictive, but the 1991 and 2001 UK censuses produced an individual-level sample (the 'Sample of Anonymized Records' or SARs) including 1 per cent of all households in the UK and 2 per cent of all individuals. In addition, a small set of individuals have been followed through and linked up for every census since 1971 to create what is termed a 'longitudinal survey'. However, neither of the anonymized records nor the longitudinal survey are openly accessible to the general public. Such samples are often known as 'microdata' and they are being increasingly made available in a number of countries. Canada has produced them in every census since 1971, while in the USA retrospective sample microdata have been generated from old census returns back to 1850 (see www.ipums.umn.edu). The main US site allows students to download, by file transfer, information on sample populations in any state (see fisher.lib.virginia. edu/pums). In any use of microdata, the user cannot expect a fine level of spatial detail. Because these are samples and contain potentially sensitive information on individuals, the spatial framework for their availability is coarse grained.

One very important further type of secondary data that students should be aware of is derived from sample questionnaire and other surveys. Many of these are carried out by private organizations (sometimes for government and sometimes for other pur-chasers of the data such as trade unions, lobby groups or public corporations). Many private surveys are only available for general use at some cost to the researcher – since the commissioning bodies wish to recoup something of their investment. The major polling organizations, such as MORI or Gallup, thus generally charge for data at any level of significant detail – only the headline figures from their surveys are available free. An increasing number of surveys are now carried out by governments, using sophisticated methodologies to ensure robustness from relatively small samples. Other surveys are carried out by academic researchers as part of major projects. All surveys, whoever they are carried out by, share certain characteristics (some of which also apply to the microdata series already mentioned).

First, they are samples and are thus subject to sampling errors. The size and nature of these sampling errors varies greatly according to the exact manner in which the survey was carried out. In general, surveys undertaken by governmental agencies are likely to have the least variability due to sampling, thus emphasizing the crucial point that the use of secondary data enables the student to achieve levels of confidence that would be unattainable through lone research. Second, surveys commonly focus on opinions as well as facts. They thus bring in a whole range of attitudinal issues as well as the behavioural elements more commonly associated with censuses. Third, surveys, especially when carried out by private polling organizations, sometimes deal with topics that are not of immediate inter-est to government, or which government regards as being too sensitive. Fourth, certain governmental surveys are carried out annually or at a frequency that consid-erably improves on the ten-year periodicity of the census. Finally, even in large-scale government surveys, sample sizes are often relatively small and therefore surveys are not usually spatially referenced to detailed areas. Thus, national or regional results may be published but rarely at a more local scale such as the neighbourhood – unless the survey was itself local in its execution (in which case it applies to only one place and should not be generalized).

For UK researchers, the UK Data Archive (housed at the University of Essex) provides online access (after registration) to a wide variety of survey data, includ-ing such major series as the British Social Attitudes Survey (Park *et al.*, 2001), the Quarterly Labour Force Survey, the Citizenship Survey and the Scottish Household Survey (see www.data-archive.ac.uk). In addition the Data Archive also acts as a depository for qualitative data such as interview transcripts. At an international scale, the University of Amsterdam maintains a useful website listing social-science databases around the world (www.sociosite.net).

The European Union operates a polling organization known as Eurobarometer which conducts surveys across the Union. These principally concern opinions on EU policies and integration, but other themes such as racism and xenophobia have also been investigated in recent years (Eurobarometer's home page is at ec.europa. eu/public_opinion). Apart from a regular survey of standard questions carried out twice a year, Eurobarometer also operates 'flash' surveys on topical issues as required by the European Commission.

These various surveys are of growing importance to researchers, and it is worth considering them in a little more detail. The UK's General Household Survey is a multi-purpose survey carried out every year except two since 1971 (see www.statistics. gov.uk/ssd/surveys/general_household_survey.asp), thus allowing longitudinal comparisons on topics such as the ownership of consumer durables, the use of health services, tobacco and alcohol consumption, and family composition. In addition the survey covers topics also dealt with in the decennial census, but allows finer temporal analysis of change. The General Household Survey therefore complements the census by covering many issues (particularly around household expenditure) that have never been considered suitable for inclusion in the census questionnaire. It is nevertheless at root a factual survey counting attributes. In contrast, the Citizenship Survey (www.communities.gov.uk/communities/racecohesionfaith/research/citizenshipsurvey) poses a variety of attitudinal questions as well as behavioural ones. Thus, it is a source of information on topics such as respondents' feelings about their community, or about social diversity and cohesion. It was set up more recently, in 2001, and takes place every two years with a general sample of 10,000 respondents. It emerged from government anxieties about inter-ethnic-group conflict and thus also targets a further 5,000 respondents from minority groups. Unusually in 2003 there were supplementary booster surveys of young people aged 10 to 15 and of children aged 8 and 9: these are groups that cannot normally be easily surveyed by the lone researcher.

One thing that the General Household Survey and the Citizenship Survey share is that data are normally available only at a national scale – as befits a sample. However, from time to time data are reported at other scales – for example the 2003 Citizenship Survey provided results for 20 local areas.

At an international scale there are now a series of attitudinal surveys that mirror some of those in the UK but which have European or world-wide coverage. At European scale there are the European Values Study (www.europeanvaluestudy. eu) and the European Social Survey (www.europeansocialsurvey.org). Whilst the first of these is commissioned by the European Union and meets the needs of the Commission, the second is an academically driven survey that is now starting its fifth wave of interviews across Europe. The topics covered are wide-ranging, but data are only published at national scales. The European Values Study has now been extended into a World Values Study (www.worldvaluesstudy.org) facilitating interesting international comparisons, although many countries are not yet involved in the survey. Finally, the Gallup polling organization has a world data site which requires registration but is currently free to users: this is a further source of comparative data on attitudes (worldview.gallup.com).

Existing large-scale surveys may be of particular use in a student project because of their rigour and reliability. Questionnaires undertaken by students sometimes run into problems in the field because the wording of questions is not interpreted consistently by respondents, or because the questioner has failed to recognize all the possible answers in his or her coding schedule. In contrast, large-scale publicly funded surveys are, as with the census, rigorously pretested. One possibility therefore is for students to reuse (with suitable acknowledgements) questions that have formed part of existing large-scale surveys. A major 'question bank' that can be utilized in this way relates to British government surveys (surveynet.ac.uk), but the other social surveys discussed above all publish the questions on which their data are based.

It should be clear from this discussion, and from a few minutes spent investigating some of the web addresses given, that there is a huge variety of social scientific data available to a researcher. Much of this information is free to access. However, this raises the question of what are the uses to which secondary data might be put?

UTILIZING SECONDARY DATA

Given the breadth of coverage and detailed nature of much available secondary data, it will be no surprise that the uses to which they can be put are extremely varied. Such data can be used to inform projects in which the primary focus of the research is to collect evidence by other means, but they can also be regarded as the primary form of information and then subjected to analysis of varying degrees of sophistication – from description to causal modelling.

Data as context

The simplest level at which secondary data can be used is to provide a description of the characteristics of the place, space or group that is the focus of a research investigation being carried out by other means. Too many student projects discuss a particular situation in detail without ever providing a wider context.

Secondary data can provide justifications for the choice of topic and location. Consider, for example, the following two projects:

- A study of reasons for the choice of a particular destination by retirement migrants.
- A study of the attitudes of young women to future family building.

In both cases a student could produce a self-contained piece of work in which all the evidence (for example, from questionnaires, in-depth interviews or focus-group discussions) is collected by the student. However, much more rounded studies would result if they were contextualized via (respectively):

- the use of census data to demonstrate the local importance of retirement migration, the origins of the migrants and the numbers of local elderly; and
- the use of population registration data and time series derived therein to demonstrate more general temporal changes in fertility and its characteristics in terms of age of mother, child parities, marital status and so on. This might also be supplemented by national-level attitudinal data derived from some of the social surveys mentioned in the previous section.

It should thus be clear that secondary data can be used to demonstrate that there is a research issue that merits being developed into a project, as well as providing a context for the work carried out. Secondary data are not in these cases being subjected to any further analysis, but quoted as they are (albeit with some recognition of their utility and limitations). Such data may also be used to support a project that is inherently qualitative in nature but for which a justification in terms of importance is derived from statistical information.

The availability of such data for contextual use should also lead to a recognition that, in any assigned work (and not just research projects), secondary data can support an argument. In the twenty-first century it is no longer permissible for student essays to quote outdated sets of figures derived from ten-year-old textbooks in support of their contentions. Ideas derived from published texts should be tested against more recent data, wherever available, to examine their continuing plausibility.

Data as comparison

One of the most significant ways in which all forms of science (social science included) progress is through replication. This involves previous studies being repeated in different circumstances – for example, testing the same ideas but at a different place or with a repeat study in the original location some years after the initial investigation. Many student projects are implicitly based on a comparison of a local situation with a much wider aspect of the phenomenon. In all these endeavours there is a 'benchmark' against which deviation is being evaluated, measured and (possibly) interpreted or explained. This 'benchmark' is best defined (indeed, in many cases, it may only be defined) through the use of secondary data. The following projects are examples of this:

- A study of female employment characteristics in a local labour market (carried out through house-to-house questionnaires), aiming to consider whether or not they mirror national patterns.
- A study of the spatial awareness of school children at a particular age (carried out through exercises undertaken in a school), aiming to consider whether there are differences in such awareness compared with a study carried out ten years earlier.

In the first case national data are needed. Problems may, of course, arise in that the available national data may not relate to exactly the same issues as the student is interested in but, given an awareness from the outset that such data will be used, the local survey can be adjusted to ask comparable questions.

The second case above is rather different because it involves treating the primary evidence produced by an earlier study as secondary evidence (the 'benchmark') for the new study. Questions used in the earlier study can be replicated in the new, and results can be compared, possibly involving statistical testing to evaluate the significance of observed differences.

This second example may involve wider possibilities of the replication of method. As indicated earlier, large-scale surveys have created considerable 'question banks' that are open for perusal and 'mining' for questions to add to student projects. The results obtained by the student can then be compared with the findings, almost certainly at a much larger scale, from the published surveys. To some students such comparisons might seem like 'cheating' – reusing materials that already exist rather than deriving new evidence. But it is partly to enable such comparisons that researchers in the past, including government statistical bodies, have made secondary data available.

In certain projects the scale of the task is one for which secondary data are the only possible source of comparative evidence. Without their use, a literature-based project on, for example, the comparisons between economic or social geographic trends in a number of countries must depend solely on what established journal articles and books say about the phenomena under consideration. By going to statistical and other secondary materials the student can re-examine the conclusions that published researchers have come up with, and can also replicate the research methods they have gone through. Recent examples of desk-based cross-national comparisons based on secondary data include Wolbers (2007), on trends in labour market entry in 11 countries, and Halleröd (2006), on relative deprivation in the UK, Sweden and Finland. Halleröd's study, in fact leads us into another aspect of the use of secondary data – as a source of material for analysis.

Data as the basis for analysis

The third major use of secondary data is as the raw material for analysis. In this way it can play the same role in a project as the original material collected by a student by means of a questionnaire, through a field notebook of participant observation, or through focus group or interview transcripts. A high proportion of research projects use secondary data in such a manner, and without such data – and the analysis undertaken – the project could not be completed. Consider the following three topics:

- Testing the idea that school league-table results are conditional upon the nature of the school's catchment area.
- Comparing the levels of residential segregation of different ethnic groups in a single city.
- Considering what social or economic factors correlate with mortality rates.

Each of these projects requires data that cannot be generated using the resources of an individual student researcher. In two of the projects (the first and third) there is an expectation that government data (school performance in one case, population mortality in the other) are to be matched with further data in an explanatory framework of some kind. Such uses of secondary data are extremely common, both in public-sector research and in the academic community. Indeed, the three projects suggested above have been derived in part from existing studies (Townsend *et al.*, 1987; Peach, 1996; Herbert and Thomas, 1998).

Table 5.2   The characteristics of two composite indices

The Townsend Index	Human Development Index
Current unemployment	Life expectancy
Short-term wealth (car ownership)	Adult literacy
Long-term wealth (house ownership)	GNP per capita
Housing overcrowding	
Measured as standard deviations from the average on each variable, then summed	Measured as differences from the maximum on each variable, then summed

Data derived from secondary sources can often be entered into statistical manipulations in a straightforward manner. For example, population distribution by age groups can be compared between two areas using non-parametric statistics such as chi-square. However, there are also a number of specific techniques that have been developed to utilize secondary data in deeper ways to enhance their analytical value.

There is now a considerable history of geographers constructing overall 'scoring systems' for places – for example, in terms of standards of living (Knox, 1975). These are often known as 'composite indices'. All the methods in use today can be traced back to certain common principles, one of which is the use of secondary data. Two of the best known of the resultant indices are the 'Townsend Index' for the measurement of deprivation (particularly in health research) and the Human Development Index for the international comparison of quality of life. Table 5.2 shows the variables considered in both cases.

The Townsend Index was developed for use in Great Britain and reflects certain cultural specificities of its origin (for example, its emphasis on house ownership as an index of long-term wealth, which would not be appropriate in societies such as Switzerland or Germany where long-term wealth is often held in the form of financial assets). Data are collected on a set of areas (for example, wards within a city or local authority areas within a region). For each of the four variables in the index the mean and standard deviation values of the area datasets are calculated. The variables are defined as follows:

- The percentage of the working population that is currently unemployed.
- The percentage of households without a car.
- The percentage of households that does not own its own home.
- The percentage of households living in overcrowded conditions (sometimes taken as more than one person per room).

For each of these variables a 'high' value represents poor conditions. Each area is analysed to see where its actual value on each variable lies in relation to the mean(s) and standard deviation(s). An area that is much 'better' than the average on all variables will record uniformly low values. The composite index is derived from z scores:

$$z_a = (x_a - \mu) / \sigma$$

where x_a is the value of the given variable for area a; μ is the mean value for the given variable across all areas; and σ is the standard deviation for the given variable across all areas.

For each area the final Townsend Index score is obtained by summing the four z scores. Thus areas of low deprivation have high negative scores. Areas of high deprivation have high positive scores.

The Townsend Index represents a measure where the composite has been argued to encapsulate more than each of the individual variables. A number of variants to it have been made but with the basic structure still visible (Bradford *et al.*, 1995). Townsend Index scores have been shown to be closely correlated with the distribution of various population subgroups in British cities, such as lone-parent families, the long-term sick and the unskilled (White, 2000). The provisos made earlier in this chapter about the ecological fallacy need to be borne in mind in interpreting such analyses. Recent elaborations of the basic Townsend methods have become increasingly sophisticated, sometimes using the composite index methodology in some way on data for individuals that are then referenced to their place of residence. A recent study of childhood obesity in Leeds (Procter *et al.*, 2008) provides an interesting example. Users of the UK neighbourhood statistics referred to earlier quickly become very conversant with composite indices since every neighbourhood in the country is scored and ranked on this basis according to measures such as income, housing or environmental deprivation, with the rankings regularly updated as newer data sets are incorporated in the analysis. It is surprising how infrequently these deprivation scores are used to contextualize students' projects that are themselves neighbourhood based.

As with the Townsend Index, so the Human Development Index (HDI) has been refined – in particular because of problems with international measures of gross domestic product. These are now generally given as 'purchasing power parities', precisely in order to facilitate international comparison. To some extent the HDI as currently used is an 'index of indices' with each of the three separate components resulting from analysis of raw data.

The methods involved in the creation of indices determine that the results should be regarded as relative to one another. The addition of an extra case necessitates the recalculation of the whole index. Similarly, it is not possible to compare the scores for the Townsend Index for neighbourhoods in different cities where these scores have been separately derived: one can say that two neighbourhoods are both affluent, but only in relation to the rest of the neighbourhoods in their respective cities – not in relation to each other.

In a different pattern of analysis based on secondary data there has been a long history of social and urban geographers, particularly those working on class and ethnicity, considering the spatial distributions of population subgroups in comparison to the population as a whole. A series of single-number measures have been constructed to summarize such distributions and to show differences and similarities between them: these measures include the Indices of Dissimilarity and of Segregation. These almost invariably depend on secondary data, almost invariably from census results.

The basic premise is that, in a society characterized by no spatial separation between different subgroups, all subgroups will be similarly distributed. The analysis therefore aims to identify areas where this is not the case. The two principal sets of output measures are as follows:

- *Indices of Dissimilarity*: these measure the differences between population subgroups. Values range from 0 to 100. At 0, two population distributions are identical. At 100, the two distributions are completely dissimilar.
- *Indices of Segregation*: these measure the extent to which a particular population subgroup is segregated within the total population. Again the values run from 0 to 100. At 0, there is no segregation present. At 100, the subgroup is totally segregated and inhabits its own exclusive space.

As a reading of Peach (1996) will show, the number of indices of dissimilarity that can be calculated in any set of subgroups grows rapidly as the number of subgroups is increased. Thus for six subgroups (for example, the six socioeconomic groups commonly utilized in the UK and known as A, B, C1, C2, D, E) there are 15 Indices of Dissimilarity. For ten subgroups there are 45. In contrast, there is only one Index of Segregation for each subgroup.

The calculation of dissimilarity and segregation indices (as with the composite indices referred to earlier) is informed by social theory, but only to the extent of justifying the method used. Whereas in the Townsend Index the individual variables are brought in because of their intuitive relationship with deprivation, in the analysis of spatial distributions there is an assumption that social distance and spatial distance are related. Other types of analyses of spatial distributions may not even start with these assumptions. This particularly applies to GIS-based analysis of spatially referenced secondary data. GIS analysis often operates from an inductive and empirical standpoint – looking for patterns and possible relationships rather than testing their theoretical existence.

A further type of analysis that a student might want to perform on secondary data involves the consideration of relationships between datasets and the possibility of producing statistical explanations of causality. Such explanations may not necessarily be backed by social theory: it is often easier to show that there is an empirical relationship between two datasets or variables than to provide convincing theoretical arguments as to why these relationships come about. Nevertheless, there are many areas of geographical research that depend heavily on the analysis of such relationships: the field of electoral geography, for example, would be very much reduced if research of this kind was excluded.

Many projects seek ultimately to suggest reasons for the appearance of a particular pattern (spatial or otherwise) in a single variable. To the example quoted at the start of this section (variations in school league table performance) can be added many others – such as spatial variations in the electoral success of a particular party or (at an international scale) in medal performance in the Olympic Games. In each of these the generalized form of the project is to calibrate a regression model based on the assumption that variations in the phenomenon being investigated result from the influences (singly or collectively) of a set of other measured factors. Certain specific issues arise in the use of secondary data for such analyses.

First, there may be no other way of examining the hypothesized relationships other than through the use of secondary data. No researcher can expect to collect school performance measures as part of a primary data-gathering exercise, nor is there any need to do so. Similarly, there are issues of scale in most of the problems just outlined: the datasets needed to operationalize the project generally cover a lot of spatial ground.

A second issue, however, is the temporal coincidence (or lack of it) of particular datasets. Consider the use of population census data in certain of the projects just outlined. In the UK (and in many other countries) the census is a decennial exercise. Elections take place at more frequent intervals, while league tables of school or hospital performance are produced annually. In a study in which a student aims to explain voting for the Scottish National Party by reference to the social and economic characteristics of constituencies there is an immediate dilemma. If the bulk of the constituency data are drawn from the 2001 census, should the study be limited to the 2001 general election (in the year of the census), or can the study use more recent elections even though the temporal coincidence of the dependent variable (voting patterns) and independent (census-based) variables is becoming ever more remote? The purist argument would privilege the former solution while the pragmatic view would be to mix datasets from different years on the basis that there is no plausible alternative. Most census users in government, commerce and academia accept that the data they use is often somewhat dated. Some attempts are made to update such datasets by various estimation and correction techniques, but only for major variables (such as the population sizes of administrative areas).

A third issue concerning the use of secondary data in analysis may be the lack of spatial coincidence of datasets. School catchments, for example, are drawn for admission purposes and do not necessarily coincide with postcode sectors or administrative wards. Certain spatial units in one dataset may need amalgamation to fit the spatial framework of another before correlation and regression analysis can take place. This is often known as the 'modifiable areal unit problem' (Morphet, 1993). The relevant procedures can best be accomplished via the use of a GIS, although 'traditional' methods of 'eyeballing' areas and allocating them manually to coincide with others can still be effective, if time consuming.

Fourth, even with the abundant nature of secondary data today, there are often still gaps in availability (Dorling and Simpson, 1999). The researcher may hypothesize that household income is an important explanatory factor for the problem at hand, but data on this are not available in most countries at anything less than regional level. Income questions are regarded as too sensitive (and too liable to result in false answers) to be posed in most government surveys, although they are used (generally in the form of predefined bands) in commercial questionnaires. Other issues that are seen as sensitive (such as sexual orientation or certain disease histories) are similarly rarely raised. Finally, the interpreter of the results of correlation and regression analysis on secondary data has to be alert to the dangers of inferring cause from correlation.

CONCLUSION

There is room for a consideration of secondary data in almost any student project on contemporary aspects of human geography. The uses of such data vary from the contextual at one end of the scale to the fundamental basis for analysis at the other. Consideration of what is already known about a topic should play a role alongside the literature review that identifies what is already thought about it. Early examination of available data can help in the formulation of a research project, help to identify gaps in understanding, provide a justification for the choice of areas, groups or case studies for research and demonstrate the importance of the field in which a project is set. Such uses of secondary data do not necessarily mean that the resulting project need follow a positivist-empirical methodology. The data may be seen as providing the 'extensive' basis and context for a more 'intensive' investigation that takes a very different approach.

Public opinion, political discourses, policy discussions and commercial actions are all today increasingly based on a reading (or sometimes a misreading) of secondary data. Information is gathered for a multitude of purposes and used by many different sections of society. In contributing to the well-being of society, geographers need to be able to investigate issues at scales for which only secondary data provide plausible sources of evidence; they need to be able to present carefully supported conclusions to the policy community; and they need to be able to advance aspects of their own discipline through the understanding of variability at a variety of scales. For these endeavours secondary data analysis is an essential tool.

Summary

- Secondary data are primarily, but not exclusively, collected and made available by governments.
- Researchers have no control over what or how information is collected.
- Many secondary datasets contain information that is strongly spatially referenced.
- Increasing amounts of secondary data are becoming publicly available, particularly via the internet.
- Secondary data can be used to provide a context for a wide range of geographical studies, they can be used in comparisons and they can provide the basis for analysis.
- Problems over the inference of cause from outcome must be borne in mind, as in the use of many other geographical methods.

Further reading

- Hakim (1982) is a wide-ranging classic text dealing with a number of different ways in which secondary data can be used, including the re-analysis of statistical data, the use of existing questionnaire results and the use and abuse of official data.

- Peach (1981) provides a good illustration of the debates about causality that arise from the discerning of patterns or outcomes in secondary data and the need to theorize process as a supplementary stage in secondary analysis.

- Townsend *et al.* (1987). Townsend's name is associated with the deprivation index he suggested. This study demonstrates the richness of the social geographical argument that can be developed from the analysis of official data, in this case in the sphere of health in the UK.

- Walford (2002) provides a recent compendium of secondary source locations, as well as further consideration of the issues involved in secondary analysis.

Note: Full details of the above can be found in the references list below.

References

Bradford, M.G., Robson, B.T. and Tye, R. (1995) 'Constructing an urban deprivation index: a way of meeting the need for flexibility', *Environment and Planning A*, 27: 519–534.

Dorling, D.F.L. and Simpson, S. (1999) 'Introduction to statistics in society', in D.F.L. Dorling and S. Simpson (eds) *Statistics in Society: The Arithmetic of Politics*. London: Arnold, pp. 1–5.

Hakim, C. (1982) *Secondary Analysis in Social Research*. London: Allen & Unwin.

Halleröd, B., Larsson, D., Gordon, D. and Ritakallio, V-M. and Tye, R. (2006) 'Relative deprivation: a comparative analysis of Britain, Finland and Sweden', *Journal of European Social Policy*, 16: 328–345.

Herbert, D.T. and Thomas, C.J. (1998) 'School performance league tables and social geography', *Applied Geography*, 18: 199–210.

Hoggart, K., Lees, L. and Davies, A. (2002) *Researching Human Geography*. London: Arnold.

Kitchin, R. and Tate, N.J. (2000) *Conducting Research into Human Geography: Theory, Methodology and Practice*. Harlow: Prentice Hall.

Knox, P.L. (1975) *Social Well-Being: A Spatial Perspective*. Oxford: Oxford University Press.

Martin, D. (1991) *Geographic Information Systems and their Socioeconomic Applications*. London: Routledge.

Morphet, C. (1993) 'The mapping of small-area census data – a consideration of the effects of enumeration district boundaries', *Environment and Planning A*, 25: 1267–1277.

Park, A. *et al.* (eds) (2001) *British Social Attitudes: Public Policy, Social Ties: The 18th Report*. London: Sage.

Peach, C. (1981) 'Conflicting interpretations of segregation', in P. Jackson and S. Smith (eds) *Social Interaction and Ethnic Segregation*. London: Academic Press, pp. 19–33.

Peach, C. (1996) 'Does Britain have ghettos?', *Transactions, Institute of British Geographers*, 21: 216–235.

Peach, C. (2006) 'Islam, ethnicity and South Asian religions in the London 2001 census', *Transactions, Institute of British Geographers*, 31: 353–370.

Procter, K.L., Clarke, G.P., Ransley, J.K. and Cade, J. (2008) 'Micro-level analysis of childhood obesity, diet, physical activity, residential socioeconomic and social capital variables: where are the obesogenic environments in Leeds?', *Area*: 40: 323–340.

Rose, D. and O'Reilly, K. (1998) *ESRC Review of Government Social Classifications*. London: Office for National Statistics/Swindon: Economic and Social Research Council.

Southworth, J.R. (2005) '"Religion" in the 2001 census for England and Wales', *Population, Space and Place*, 11: 75–88.

Townsend, P., Phillimore, P. and Beattie, A. (1987) *Health and Deprivation: Inequality and the North*. London: Croom Helm.

Walford, N. (2002) *Geographical Data: Characteristics and Sources*. Chichester: Wiley.

White, P. (2000) 'Who lives in deprived areas in British cities?', *Geocarrefour (Revue de Geographie de Lyon)*, 75: 107–116.

Wolbers, M.H.J. (2007) 'Patterns of labour market entry: a comparative perspective on school-to-work transitions in 11 European countries', *Acta Sociologica*, 50: 189–210.

6 Conducting Questionnaire Surveys

Sara L. McLafferty

Synopsis

Questionnaire survey research is a research method for gathering information about the characteristics, behaviours and/or attitudes of a population by administering a standardized set of questions, or questionnaire, to a sample of individuals. In geography, questionnaire surveys have been used to explore people's perceptions, attitudes, experiences, behaviours and spatial interactions in diverse geographical contexts. This chapter explains the basics of why and how to carry out survey research.

The chapter is organized into the following sections:

- Introduction
- Questionnaire design
- Strategies for conducting questionnaire surveys
- Sampling
- Conclusion

INTRODUCTION

Survey research has been an important tool in geography for several decades. The goal of survey research is to acquire information about the characteristics, behaviours and attitudes of a population by administering a standardized questionnaire, or survey, to a sample of individuals. Surveys have been used to address a wide range of geographical issues, including perceptions of risk from natural hazards; social networks and coping behaviours among people with HIV/AIDS; environmental attitudes; travel patterns and behaviours; mental maps; power relations in industrial firms; gender differences in household responsibilities; and access to employment. In geography, questionnaire surveys were first used in the field of behavioural geography to examine people's environmental perceptions, travel behaviour and consumer choices (Rushton, 1969; Gould and White, 1974). Survey research methods quickly spread to other branches of human geography, and today they are an essential component of the human geographer's toolkit.

Questionnaire survey research is just one method for collecting information about people or institutions. When does it make sense to conduct a questionnaire, rather than relying on secondary data (see Chapter 5) or information collected by observational methods (see Chapter 9), for example? Survey research is particularly useful for eliciting people's attitudes and opinions about social, political and environmental issues such as neighbourhood quality of life, or environmental problems and risks. This style of research is also valuable for finding out about complex behaviours and social interactions. Finally, survey research is a tool for gathering information about people's lives that is not available from published sources (e.g. data on diet, health and employment characteristics). In developing countries where government data sources are often out of date and of poor quality, questionnaire surveys are a primary means of collecting data on people and their characteristics.

Before embarking on survey research, it is critically important to have a clear understanding of the research problem of interest. What are the objectives of the research? What key questions or issues are to be addressed? What people or institutions make up the target population? What are the geographical area and time period of interest? These issues underpin how the survey is designed and administered. Surveys can be expensive and time-consuming to conduct, so the quality and type of information gathered are all important.

Although each survey deals with a unique topic, in a unique population, the process of conducting survey research involves a common set of issues. The first step is *survey design*. Researchers must develop questions and create a survey instrument that both achieves the goals of the research and is clear and easy to understand for respondents. Second, we need to decide how the survey will be administered. Postal (or mail-back) questionnaires and telephone interviews are just a few of the many *strategies for conducting surveys*. Third, survey research involves *sampling* – identifying a sample of people to receive and respond to the questionnaire. This chapter provides a brief introduction to each of these issues, drawing upon examples from geographic research.

QUESTIONNAIRE DESIGN

Questionnaires are at the heart of survey research. Each questionnaire is tailor made to fit a research project, including a series of questions that address the topic of interest. Decades of survey research have shown that the design and wording of questions can have significant effects on the answers obtained. There are well established procedures for developing a 'good' questionnaire that includes clear and effective questions (Fowler, 2002).

Good questions are ones that provide useful information about what the researcher is trying to measure. Although this may appear to be simple, straightforward advice, it is often challenging to implement. Questions can range from factual questions that ask people to provide information, to opinion questions that assess attitudes and preferences. Writing good questions requires not only thinking about what information we are trying to obtain but also anticipating how the study population will interpret particular questions. Let's examine the following question: 'Are you concerned about environmental degradation in your neighbourhood?' This item raises more questions than it answers. What does a person mean when he or she says that he or she is concerned? What does environmental degradation mean? Do people understand it? How does each respondent define his or her neighbourhood? Questions should be clear and easy to understand for survey respondents, and they should provide useful, consistent information for research purposes.

One of the most important rules in preparing survey questions is keep it simple. Avoid complex phrases and long words that might confuse respondents. Do not ask two questions in one. The question 'Did you choose your home because it is close to work and inexpensive?' creates confusion because there is no obvious response if only one characteristic is important. Jargon and specialized technical terms cause problems in survey questions. Terms like 'accessibility' or 'power' or 'GIS' are well known among geographers, but ambiguous and confusing for most respondents. Don't assume that respondents are familiar with geographic concepts! Define terms as clearly as possible and avoid vague, all-encompassing concepts. Asking people about their involvement in community activities, without specifying what kinds of activities and what level of involvement, is unlikely to produce useful responses. It is better to ask a series of questions about specific types of involvement rather than a single vague question. Finally, one should avoid negative words in questions. Words like 'no' and 'not' tend to confuse respondents (Babbie, 2003) (see Box 6.1).

Responses to survey questions are as important as the questions themselves. *Open-ended* questions allow participants to craft their own responses, whereas *fixed-response* questions offer a limited set of responses. Open-ended questions have several advantages. Respondents are not constrained in answering questions. They can express in their own words the fullest possible range of attitudes, preferences and emotions. Respondents' 'true' viewpoints may be better represented.

Box 6.1   Guidelines for designing survey questions

Basic principles:

- Keep it simple.
- Define terms clearly.
- Use the simplest possible wording.

Things to avoid:

- Long, complex questions.
- Two or more questions in one.
- Jargon.
- Biased or emotionally charged terms.
- Negative words like 'not' or 'none'.

Open-ended questions provide qualitative information that can be analysed with qualitative methodologies, as discussed elsewhere in this volume. Increasingly, geographers are using open-ended responses in questionnaire surveys as part of the broader shift towards qualitative methodologies. Relying on a mix of open-ended and fixed-response questions, Gilbert (1998) analysed survival strategies among working, poor women and their use of place-based social networks. The fixed-response questions provided data on the demographic and household characteristics of the women and their social interaction patterns, while the open-ended questions offered detailed insights about women's coping strategies and life circumstances.

Fixed-response questions are commonly used in survey research, and the principles for designing such questions have been in place for decades. There are several advantages to fixed responses. First, the fixed alternatives act as a guide for respondents, making it easier for them to answer questions. Second, the responses are easier to analyse and interpret because they fall into a limited set of categories (Fink and Kosecoff, 1998). The downside is that such responses lack the detail, richness and personal viewpoints that can be gained from open-ended questions.

A simple type of fixed-response question is the factual question that asks about, say, age, income, time budgets or activity patterns. Responses may be numerical or involve checklists, categories or yes/no answers. The key in framing these types of questions is to anticipate all possible responses. As in all phases of survey design, it is important to think about the kind of information needed for research as well as characteristics of the study population that might influence their responses. A 'don't know' or 'other' option is generally included to allow for the fullest range of responses. For numerical information (age, income), one must decide between creating categorical responses (i.e. age < 15, 15–25), or recording the actual numerical value. Creating categories involves a loss of information – a shift from interval to ordinal data[1] – but the categorical information may be easier to analyse. Also, for sensitive topics such as age, respondents are more likely to answer if the choice involves a broad category rather than a specific number.

Although factual questions appear straightforward, they often reflect an uneasy balance between the needs and views of survey administrators and those of respondents. Questions about 'race' are a good example. Race is a social construction that does not fit easily into the discrete response categories used in questionnaire surveys. In the census of the United States, the categories and options used to elicit information about race have changed over time reflecting changes in social understandings of race. The 1850 Census presented three options – 'white', 'black' and 'mulatto'[2] – for respondents to choose in identifying their race. In contrast, the 2000 Census provided six racial categories and permitted respondents to check multiple categories to identify themselves as mixed race. Even with this wider range of alternatives, in completing the recent census many people did not select a racial category, responding instead by writing in an ethnic identification.

Finding out about attitudes and opinions involves more complex kinds of fixed-response formats. In general, respondents are asked to provide a rating on an ordinal scale that represents a wide range of possible responses. The Likert scale presents a range of responses anchored by two extreme, opposing positions (Robinson, 1998). For example, residents may be asked to rate the quality of the schools in their neighbourhood from 'excellent' to 'satisfactory' to 'poor'. The two extreme positions, 'excellent' and 'poor', serve as anchor points for the scale, and any number of alternative responses can be included in between (Box 6.2). It is best to use an odd number of responses – 3, 5 and 7 are common – so that the middle value represents a neutral opinion. Respondents often want the option of giving a neutral answer when they do not have strong feelings one way or the other. Odd-numbered scales give such an option, whereas even-numbered scales force the response to one side. Another approach is to present the scale as a continuous line connecting the two anchors. Respondents are asked to draw a tick mark on the line at the location representing their opinion, and the distance along the line shows the strength of opinion. This gives maximum flexibility, but respondents are often confused about the process and it is difficult to compare results among respondents. Consequently most researchers work with fixed Likert scales.

Box 6.2 Examples of Likert-type responses

Please rate the quality of schools in your neighbourhood:

Excellent				Poor	(continuous)
Excellent		Satisfactory		Poor	(three-point scale)
Excellent	Good	Satisfactory	Fair	Poor	(five-point scale)

Attitudinal scales can be difficult to evaluate because there is no 'objective' standard for knowing whether or not a response is accurate. However, researchers can take several steps to improve validity. In general, it is better to offer respondents more possible answers than fewer – i.e. a five-point scale provides more information than a three-point scale. But as the number of categories increases, respondents lose their ability to discriminate among categories and the responses lose meaning. An intermediate number of categories (five or seven are commonly used) works best. Because responses often vary depending on how a question is worded, another good practice is to use multiple questions, with different wording and formats, to measure the same concept. By comparing responses across questions one can check if people give consistent responses. If so, the responses can be averaged or combined statistically to represent the underlying concept or attitude. This strategy was used in a study of the links between residents' perceptions of neighbourhood quality in Bristol and objective indicators of social deprivation (Haynes *et al.*, 2007). To measure perceived neighbourhood quality, the authors asked respondents to evaluate levels of noise, pollution, friendliness, crime and social interaction in the local neighbourhood. Responses to these diverse questions were combined statistically to create composite measures that were correlated with area-based indicators of housing quality and social deprivation.

Questionnaires should also include a clear set of instructions to guide individual responses. For self-administered surveys – those that do not involve an interviewer – the questionnaire has to be self-explanatory. The instructions for respondents must be written in simple, direct language and be as clear and explicit as possible so that the questionnaire can be filled out without assistance.

Fixed-response questions work best in self-administered questionnaires, and the design and layout of the questionnaire are critically important. For questionnaires involving interviewers, the key is to have a clear and consistent set of instructions for interviewers to follow. There are well-tested guidelines for designing and formatting interviewer-administered questionnaires (Fowler, 2008).

The final and critically important step in questionnaire construction is pre-testing (pilot-testing). In this phase, we test the questionnaire on a small group of people to check the questions, responses, layout and instructions. Are the questions understandable? Does the questionnaire allow all possible responses? Are the instructions clear and easy to follow? Is the questionnaire too long? Do any questions make respondents uncomfortable? Pretesting often reveals flaws in the questionnaire that were not obvious to researchers. The questionnaire is then modified, and it may be pretested again before going to the full sample. Several pretests may be needed to achieve a well-designed questionnaire. For interview-based surveys, pretesting has other benefits. It builds interviewing skills and helps interviewers develop confidence and rapport with respondents. In sum, pretesting is an essential step in ensuring a successful questionnaire survey.

STRATEGIES FOR CONDUCTING QUESTIONNAIRE SURVEYS

There are many strategies for conducting questionnaire surveys. Among the traditional methods are telephone surveys, face-to-face interviews and postal surveys. Advances in computer technology have stimulated the growth of internet and email-based survey research. Survey strategies differ along many dimensions – from practical issues like cost and time, to issues affecting the quality and quantity of information that can be collected. Some survey strategies require the use of interviewers whereas others utilize self-administered questionnaires.

Face-to-face interviews

Face-to-face interviews are one of the most flexible survey strategies. They can accommodate virtually any type of question and questionnaire. The interviewer can ask questions in complex sequences, administer long questionnaires, clarify vague responses and, with open-ended questions, probe to reveal hidden meanings. The personal contact between interviewer and respondent often results in more meaningful answers and generates a higher rate of response. Interviews require careful planning. Interviewers need training and preparation to ensure that the process is consistent across interviewers. Thus, face-to-face interviews are generally the most expensive and time-consuming survey strategies. Another drawback is the potential for interviewer-induced bias. The unequal relationship between interviewer and respondent, embedded in issues of gender, 'race', ethnicity and power, can influence responses (Kobayashi, 1994).

Telephone interviews

Telephone interviews are widely used in market research and are becoming a more common strategy in social-science research. They combine the personal touch of interviews with the more efficient and lower-cost format of the telephone. In many places, firms can be hired to conduct telephone surveys, saving researchers the time and expense of training interviewers and setting up phone banks. Phone surveys, however, are generally limited to short questionnaires with fixed-response questions. Such surveys miss people who do not have telephones or who are frequently away from home. Finally, although the interviewer and interviewee are only connected remotely, issues of power and bias can creep into phone surveys.

Postal surveys

Postal (or mail) surveys are self-administered questionnaires distributed in a post-out, post-back format. A stamped, addressed envelope is included for returning the completed survey, and reminder notes may be sent later to encourage people to respond. For interviewees, there is no time pressure to respond; forms can be completed at a convenient time. The main weakness of postal surveys is the low response rate. Typically, less than 30 per cent of questionnaires will be completed and returned, and those who respond may not be representative of the target survey population. People with low levels of education or busy lives are less likely to respond. The unevenness of responses often violates random or stratified sampling plans and makes it difficult to estimate sample sizes. Finally, low response means that more surveys must be sent out, increasing the cost of the survey effort.

Drop and pick-up questionnaires

A related strategy is the drop and pick-up questionnaire. This involves leaving self-administered questionnaires at people's homes and picking the surveys up at a later date. The person dropping off the surveys can give simple instructions and a brief description of the survey effort. The personal contact in dropping off the survey gives response rates close to those for face-to-face interviews, but with much less time and interviewer training. Thus, the method combines the strengths of interview and self-administered strategies. This comes at a cost – the costs are substantially higher than are those for postal or telephone surveys, though still less than those for personal interviews.

Internet surveys

A new approach is the internet survey, which is similar to a postal survey but conducted via email or the internet. The questionnaire can have the same format as a standard postal questionnaire or it may be an 'intelligent', computer-assisted questionnaire that checks and directs people's responses (Couper et al., 1998). Geographers Claire Madge and Henrietta O'Connor (2002) used an internet questionnaire to find out how new parents use the internet in acquiring information on parenting and in developing social-support networks. Every phase of their research methodology relied on the internet. Respondents were solicited online: they volunteered for the project by clicking on a 'cyberparents' hotlink on a prominent parenting website. The web-based questionnaire included a series of fixed format questions and hyperlinks to related websites (caspian.geog.le.ac.uk/baby/babyworldform. asp). Online interviews and discussion groups were conducted. Because their study emphasized internet usage, it made sense to gather information online.

Internet surveys like the one used by Madge and O'Connor have several advantages. They are inexpensive to administer; they provide access to geographically dispersed populations and they can be used to reach physically immobile groups (Madge and O'Connor, 2004). Another important advantage for geography research is that web-based questionnaires can include detailed colour graphics, such as maps, photographs, video clips and animations. On the negative side, distributing questionnaires via the internet raises a host of sampling issues. Who are the respondents? Where are the respondents? Do they represent the target population? What types of people respond and don't respond to internet surveys? Clearly people without access to email and the internet will be left out of the sample. Although many questions remain, internet surveys represent a significant innovation whose use will continue to expand in the years to come (see Chapter 13).

Each survey strategy has distinct advantages and disadvantages and the 'best' choice varies from one research project to another. Choosing a survey strategy involves weighing practical considerations, such as time and cost constraints, with research considerations, such as response rate, types of questions and the need (or lack of it) for interviewer skills. Frequently, the research context limits one's choice. Surveys in developing countries often rely on personal interviews (Awanyo, 2001); map-perception surveys often use computer-assisted questionnaires and the internet. Regardless, researchers should note the limitations of the chosen method and attempt to minimize their effects.

SAMPLING

Sampling is a key issue in survey research because who responds to a survey can have a significant impact on the results. The sample is the subset of people to whom the questionnaire will be administered. Typically the sample is selected to represent some larger population of interest – the group of people or institutions that are the subject of the research. Populations can be very broad – e.g. 'all people in the UK' – or they can be quite specific, for example 'married women with children who work outside the home and live in Chicago'. Populations are bounded in time and space, representing a group of people or institutions in a particular geographical area over a particular time period. Effective sampling requires that this population of interest be clearly defined.

The first step in sampling is to identify the sampling frame – those individuals who have a chance to be included in the sample (Fowler, 2008). The sampling frame may include the entire population or a subset of the population. Sometimes the design of the survey limits the sampling frame. For instance, in a telephone survey drawn from a telephone directory, the sampling frame only includes households that have telephones and whose telephone numbers are listed in the directory. Similarly, an internet survey excludes people who do not have access to the internet or who do not use it. The resulting sample will be biased if those excluded from the sampling frame differ significantly from those included.

Sampling also involves decisions about how to choose the sample and sample size. Commonly used sampling procedures include random sampling, in which individuals are selected at random, and systematic sampling, which involves choosing individuals at regular intervals – i.e. every tenth name in a telephone directory (Robinson, 1998). The former ensures that each individual has the same chance of being selected, whereas the latter provides even coverage of the population within the sampling frame. Sometimes the population consists of subgroups that are of particular interest – for example, different neighbourhoods in a city or ethnic groups in a population. If these subgroups differ in size, random sampling will result in the smaller subgroups being under-represented in the sample.

Stratified sampling procedures ensure that the sample adequately represents various subgroups. In stratified sampling, we first divide the population into subgroups and then choose samples randomly or systematically from each subgroup. Surveys that explore differences among groups or geographical areas often rely on stratified sampling. A recent study by Fan (2002) utilized stratified sampling to examine differences in labour-market experiences among three groups – temporary migrants, permanent migrants and non-migrants – in Guangzhou City in China. The sample consisted of more than 1500 respondents, stratified to represent not only the three migrant groups but also various occupational groups and districts within the city (Fan, 2002). Respondents were chosen randomly within each occupational and geographical group, with adjustments to ensure that the three migrant groups were appropriately represented.

Another important issue is how large a sample should be. Large sample sizes give more precise estimates of population characteristics and they provide more information for addressing the research problem. However, large samples also mean more questionnaires and more time and effort spent in interviewing and analysis. The cost of survey research increases proportionately with sample size. In choosing a sample size, analysts must trade off the benefits of added information and better estimates with the costs of administering and analysing the surveys.

One way to decide on a sample size is to focus on subgroups rather than the population as a whole. The sample must be large enough to provide reasonably accurate estimates for each of the subgroups that are being compared and analysed. Large sample sizes shrink quickly when divided into subgroups. For example, in analysing travel patterns by gender, urban/rural residence and three categories of ethnic origin, there will be 12 ($2 \times 2 \times 3$) subgroups. An overall sample size of 100 yields just eight responses on average for each subgroup, which is too small to produce reliable subgroup estimates. To avoid this problem, the researcher should first identify the various subgroups and choose an adequate sample size for each. This is easy to do in a stratified sampling design. With other sampling procedures the issue is trickier. Small subgroups will be missed unless the overall sample size is large. Researchers should carefully assess the various subgroups of interest in choosing a sample size.

Sample size decisions also involve thinking about how much precision or confidence is needed in various estimates. Precision always increases with sample size, but the improvements in precision decrease at larger sample sizes. The benefits of larger samples begin to level off at sample sizes of 150–200 (Fowler, 2008).

It is also important to think about how the survey data will be analysed. Typically researchers use statistical procedures such as chisquare and analysis of variance (ANOVA) in analysing survey responses. These procedures require sample sizes of approximately 25 or more. Larger sample sizes are needed for multivariate statistical procedures such as multiple regression analysis and logistic regression. If separate statistical analyses will be performed for different subgroups in the sample, each subgroup must have a sample size that is sufficient for statistical analysis.

Finally, sample size decisions also involve very real budgetary and time constraints and these may well be beyond the researcher's control. In sum, there is no single answer to the sample size decision. The decision involves anticipating what the data will be used for and how it will be analysed, and balancing those considerations with the realities of time and money.

Sampling procedures are important because they can introduce various sources of bias into a research project. Sampling bias arises when the sample size is not large enough accurately to represent the study population or subgroups within it. More importantly, the sampling frame may be biased, as occurs in telephone or Internet surveys. Many survey procedures under-represent disadvantaged populations, including poor and homeless people and ethnic and racial minorities. Special efforts are needed to ensure that these groups are not excluded from the research project. Finally, non-response bias occurs when those who refuse to respond differ significantly from those who do respond. Non-response often correlates with age, social class, education and political beliefs, resulting in a sample that is not representative of the study population. Although non-response bias cannot be eliminated, its effects can be minimized with a good sampling design. Because survey results are often highly dependent on the characteristics of the sample, bias is a crucial issue in sampling and survey design.

CONCLUSION

Conducting questionnaire surveys involves a series of steps, including designing and pretesting the questionnaire; choosing a survey strategy; identifying a sample of potential respondents; and administering the survey. These complex decisions are closely interconnected. The design of the questionnaire affects whether or not face-to-face interviews are needed. For many projects, financial constraints dictate the use of postal or telephone surveys and relatively small sample sizes. Thus, in any survey project there is a continual give and take between different factors, framed by the goals and constraints of the research endeavour.

Questionnaire surveys have well-known limitations, as discussed at various points in this chapter. For geographical research, poorly worded questions, ambiguous responses and non-response bias are all issues that raise major concerns. Some geographers contend that survey information is of limited value, especially when compared to the rich and detailed information that can be gleaned from depth interviews and participant observation (Winchester, 1999). A more balanced view recognizes the strengths of questionnaire surveys – their ability to gather information from large samples, about large and diverse populations; their ability to incorporate both open and fixed questions and their use of trained interviewers to elicit information; and, finally, in the internet era, their ability to reach widely dispersed populations with innovative, computer-assisted, graphically based questionnaires. Despite their limitations, surveys remain the most efficient and effective tool for collecting population-based information.

Questionnaire surveys have a long history in geographic research, a history that continues to evolve as the discipline of geography changes. During the 1970s, survey methods facilitated the shift away from statistical analysis of secondary data towards behavioural and environmental perception research. During the 1980s and 1990s, survey methods became less popular as a result of a 'qualitative turn' in human geography. Today, as geographers search for a common ground between quantitative and qualitative methods, questionnaire surveys are playing an important role in innovative 'mixed' methodologies (Sporton, 1999). As these developments unfold, questionnaire surveys will continue to provide a rich array of information about people's lives and well-being in their diverse geographical contexts.

Summary

- Questionnaire surveys are useful for gathering information about people's characteristics, perceptions, attitudes and behaviours.
- Before embarking on survey research, clearly identify the goals and objectives of the research project. Decide what information you are trying to gather via the questionnaire survey.
- Survey research involves three key steps: designing the questionnaire, choosing a survey strategy and choosing the survey respondents (sampling).
- The questionnaire should be designed to acquire useful information about the research problem of interest. Questionnaires can include both open-ended and fixed-response questions. In either case, the questions should be clearly and simply worded and should avoid jargon.
- The types of survey strategies include face-to-face interviews, telephone surveys, postal surveys, drop and pick-up surveys and internet surveys. Each has distinct advantages and disadvantages, and the choice among them depends on the type of questionnaire, desired response rate, and time and budgetary constraints.
- Sampling involves identifying the group of people to whom the questionnaire will be administered. The sample should be selected to represent well the target population and to minimize non-response bias. The sample size should be large enough to represent various subgroups in the population and to allow effective statistical analysis of results.

NOTES

1   That is, a shift from simply seeking to collect numerical data that is comparable as it is based on the same interval (i.e. the measurement of age in years, or income in US dollars), to data that is more readily ordered, ranked or rated (i.e. age categories, or income bands).
2   The 'mulatto' category being used at the time to refer to a person who has one black and one white parent; a term that is now widely considered offensive.

Further reading

A great deal has been written about questionnaire surveys in the social sciences. The following are just a few of the sources I have found useful:

- Babbie (1990) is a comprehensive, well-written book that covers both the theory and practice of survey research and its role in the social sciences.

- Fink and Kosecoff (1998) provide a very clear primer on how to conduct questionnaire surveys, focusing on 'how to' and practical advice.

- Fowler (2008) is an excellent, detailed, up-to-date discussion of survey research methods with an extensive bibliography. This book emphasizes methodological topics such as non-response bias, validity, questionnaire evaluation and ethical issues.

- Survey Research Laboratory, University of Illinois at Chicago – Sites Related to Survey Research (www. srl.uic.edu/srllink/srllink.htm). This is a comprehensive internet site for survey research, including links to journals, organizations, sample questionnaires, software packages for analysing survey data, and sampling-related software and websites.

Note: Full details of the above can be found in the references list below.

References

Awanyo, L. (2001) 'Labor, ecology and a failed agenda of market incentives: the political ecology of agrarian reforms in Ghana', *Annals of the Association of American Geographers*, 91: 92–121.

Babbie, E. (1990) *Survey Research Methods* (2nd edn). Belmont, CA: Wadsworth.

Babbie, E. (2003) *The Practice of Social Research* (10th edn). Belmont, CA: Wadsworth.

Couper, M., Barkin, R., Bethlemhem, J., Clark, C., Martin, J., Nicholls, W. and O'Reilly, J. (eds) (1998) *Computer-Assisted Survey Information Collection*. New York: Wiley.

Fan, C. (2002) 'The elite, the natives, and the outsiders: migration and labor market segmentation in urban China', *Annals of the Association of American Geographers*, 92: 103–124.

Fink, A. and Kosecoff, J. (1998) *How to Conduct Surveys: A Step-by-Step Guide* (2nd edn). Thousand Oaks, CA: Sage.

Fowler, F. (2008) *Survey Research Methods* (4th edn). Thousand Oaks, CA: Sage.

Gilbert, M. (1998) '"Race", space and power: the survival strategies of working poor women', *Annals of the Association of American Geographers*, 88: 595–621.

Gould, P. and White, R. (1974) *Mental Maps*. Baltimore, MD: Penguin Books.

Haynes, R., Daras, K., Reading, R. and Jones, A. (2007) 'Modifiable neighbourhood units, zone design and residents' perceptions', *Health and Place*, 13: 812–825.

Kobayashi, A. (1994) 'Colouring the field: gender, "race" and the politics of fieldwork', *The Professional Geographer*, 46: 73–79.

Madge, C. and O'Connor, H. (2002) 'On-line with e-mums: exploring the Internet as a medium for research', *Area*, 34: 92–102.

Madge, C. and O'Connor, H. (2004) 'Online methods in geography educational research', *Journal of Geography in Higher Education*, 28: 143–152.

Robinson, G. (1998) *Methods and Techniques in Human Geography*. New York: Wiley.

Rushton, G. (1969) 'Analysis of spatial behavior by revealed space preference', *Annals of the Association of American Geographers*, 59: 391–406.

Sporton, D. (1999) 'Mixing methods in fertility research', *The Professional Geographer*, 51: 68–75.

Winchester, H.P.M. (1999) 'Interviews and questionnaires as mixed methods in population geography: the case of lone fathers in Newcastle, Australia', *The Professional Geographer*, 51: 60–67.

7 Finding Historical Sources

Miles Ogborn

Synopsis

Historical sources in geography are materials that can be used to provide interpretations and analyses of the geographies of past periods. Finding such sources is a matter of their survival, and of discovering where they are now kept and how they can be accessed. Historical material is not just something found in libraries and archives but also includes letters, personal diaries, photographs, maps, objects and works of art.

This chapter is organized into the following sections:

- Introduction
- Sources for questions and questions for sources
- Survival and archives
- Finding historical sources
- Using the archives

INTRODUCTION

There is a huge variety of historical material that can be used within human geography. As well as all sorts of written and numerical material which can tell us about social and economic history – diaries, perhaps, or the census – historical geographers have also used methods as diverse as oral history (interviewing people about what they and others did and thought in the past – see Rose, 1990 on ideas of community in east London); dendrohistory (determining the types of the wood used in building to understand regional economies, societies and cultures – see Biger and Liphschitz, 1995); and the cultural interpretation of visual images and maps (see, for example, Heffernan, 1991, for an analysis of French representations of desert landscapes in the nineteenth century, and Harley, 2001), or of material objects (see Hill, 2006 on the movement of artifacts from Henry Wellcome's historical medical collection). This range suggests that many of the issues raised by the use of historical sources are the same as for work on contemporary human geographies (see Chapters 10 and 30). However, there is a particular set of questions which always need to be considered in relation to historical work: what sources of information have been kept, where and how are they kept, who can get access to them and what you should do when you get access to them? While this chapter cannot give you a guide as to how to find exactly the sources you are looking for, it can give you an indication of how to begin to look and what sorts of issues you will need to consider. To do so it will deal first with the relationship between *research questions* and *historical sources*; second, with the question of the *survival* of material from the past; third, with the practicalities of *finding historical sources* in libraries and archives; and, finally, with the issues of *access* to archives and the *collection* of historical data. The place to start is deciding what you are looking for.

SOURCES FOR QUESTIONS AND QUESTIONS FOR SOURCES

In any historical work there are always limitations on the sources you can use. As one historical geographer has put it, 'the dead don't answer questionnaires' (Baker, 1997: 231), so you are restricted by the 'survival' of sources: whether that is those people who have survived to tell you about the past or, more usually, which maps, documents, pictures, sound recordings, physical objects or landscape features have survived for you to interpret. This means that anyone considering historical work needs to think both about the research questions they want to answer and about whether the sources they will need to use to provide those answers exist and are accessible. Indeed, it is sensible to think about your research questions in terms of the available sources as well as thinking about the availability of sources for particular research questions which may have been devised through more theoretical work or through reviewing the existing literature.

You can come at this problem from either end. It is possible to derive a set of research questions about the past from just a few interests and ideas. For example, following feminist debates and contemporary concerns it would be interesting to ask about women's use of public space in the nineteenth-century city (see, for example, Domosh, 1998; Walkowitz, 1998; Rappaport, 2001; Koven, 2004). You might want to know what sorts of things women did in public space (and what sorts of public space they did those things in): was it leisure, shopping, paid work, education, political activism or charity work? You might need to ask whether this differed by social class or age. You might also want to know what people thought of the women who did those things, and what the women who did them thought of them too. All these possibilities can be worked through without considering the sources of information you might use. However, beginning to do so means making some choices and ruling lines of inquiry in or out on the basis of the sources that can be found to answer the questions. For example, you would have to decide on which city or cities to look at. This is a theoretical question: should it be an industrial city (Manchester, Lyons, Chicago), a capital city (London, Paris, Washington), an imperial, colonial or neo-colonial city (Calcutta, Melbourne, Buenos Aires), or some other sort of city? It is also a practical question: can you read the necessary languages and get to where the sources are to be found? You would also have to consider which sources would tell you about the activities you want to focus on. What are the sources for the history of women shopping in nineteenth-century Calcutta? What are the sources for women's charitable work in nineteenth-century

Melbourne (see Gleeson, 1995)? In this way you can generate a sense of what historical material you are looking for and can begin to think about where you might look for it. In doing so you will realize that different sources will answer different sorts of questions, and that some questions are more easily answered than others. For example, due to differences in work, leisure and education in the nineteenth century it is much more likely that you would find reflections by middle- and upper-class women about their lives and activities than for working-class women, and much more likely that you would find European women writing about life in the colonial city than African or Asian women (Blunt and Rose, 1994; Bondi and Domosh, 1998). This does not mean that the material does not exist at all, that it is not important to look for it (you never know what you might find) or that it is not possible to expand the definition of what counts as a source when asking new sorts of questions (Gagen *et al.*, 2007). It does mean that you have to keep formulating your research questions and considering how you can find material that will help you answer them.

This means thinking about the other end of the problem too. What are the sources that are accessible to you, and what sorts of questions will they provide answers to? In the absence of any really clearly defined research questions – but with some interests in particular periods, places, events or activities – you can begin by looking at a source or set of sources and thinking about what sorts of questions they could answer. This does not have to involve anything more elaborate than a trip to the nearest library, but it is a way of tying the material available to you into wider research questions and broader literatures. So, for example, if you picked up Elizabeth Gaskell's novel *Mary Barton* (1848) about life in early nineteenth-century Manchester (Gaskell, 1970), or Edwin Chadwick's (1842) *Report on the Sanitary Condition of the Labouring Population of Great Britain* (Chadwick, 1965), or a set of letters written from or to your great, great grandmother who lived in Liverpool in the 1880s, you could begin to ask questions about women's lives in nineteenth-century British cities. You could also begin to think about the research questions to which these sources might begin to provide answers, what other sources you would need and what wider theoretical and substantive literatures you would need to cover in order to make sense of them.

In fact, it soon becomes apparent that you need to do both at the same time: working out what sort of sources will answer the questions you want to ask and deciding what sorts of questions are appropriate for the sources you have available. This means that there is a need to think about ways of finding sources that allow both general and specific searches to be carried out, and ways of quickly evaluating the sources that you have found prior to your full analysis or interpretation of them (see Chapter 29).

SURVIVAL AND ARCHIVES

It is obvious that not everything that happens leaves a record (in writing, sound or image), and equally obvious that not every record that is made survives and is stored away for later use. If any place where such records are kept so that they can be used as sources of information is thought of as an 'archive', we must include our own personal archives (of emails, letters, photographs, perhaps even diaries) as well as the official archives of companies, organizations and public bodies, as well as the libraries and galleries (local and national libraries and art galleries, university libraries, sound libraries, map libraries and picture libraries) where books, recordings and images are kept for later enjoyment and use. What is created and what survives in these archives is a social and political process which can tell us much about the conditions under which information of different sorts is produced, used and evaluated (Ogborn, 2003). As was pointed out above, a middle-class woman in the nineteenth century was more likely to have kept a diary than a working-class woman, and that diary was more likely to have survived in a family or public archive, or even in a later published form, than that of her working-class counterpart. One woman had more time, space and power than the other to construct this sort of version of her life for herself, and possibly others. Yet it does not always work this way. For example, there is no record in the Indian Mughal empire's archival record of many of the earliest encounters with English traders in India in the seventeenth century. We only have the letters and journals of English merchants and diplomats. This is not because they were more powerful than the Mughal emperors; far from it. They do not figure in the Mughal record because they were deemed to be an insignificant force in India compared with the other Asian rulers and traders with whom the Mughal leaders had to deal. Their activities in India were not worth recording. Although these cases work out differently, in each one the creation and survival of a record are at least in part a matter of power. For a fragile manuscript, magnetic tape or photograph to survive, someone has to think it is worth keeping and have the ability to keep it secure and legible (Ogborn, 2004).

Yet it is also more than just a matter of power. Archives, understood in this broad sense, are the sites of memory. They are the places – whether a cardboard box in the attic or an imposing public building – where people can begin to construct accounts of the past. This means that they are also full of emotion because they are the places where people's lives are remembered, and where we have a responsibility to think carefully about how we reconstruct those lives in the present and for the future. What is held within these archives and how we can use that information are shaped by the commitments of many people to maintaining a record of the past (Ogborn, forthcoming). This can take many forms (Samuel, 1994), but each one contains within the selection of material that is kept, stored and catalogued – the word 'archived' serving to cover all this – a commitment to remembering the past, a valuing of certain sorts of relationships and representations, and a sense of how that material might be used. Trying to understand each archive and each source within it in terms of both power relations and emotional investment in the past can help us to understand better the historical material we are using.

In the remainder of the chapter I want to deal with some very broad categories of archive and archived historical sources, how to find them and what questions to ask about using them.

FINDING HISTORICAL SOURCES

Libraries

As was pointed out above, the search for historical sources can certainly begin with the libraries you have access to – whether these are local libraries, university libraries or specialist libraries. I have made a distinction here between libraries and archival collections to deal with the difference between printed sources which can be available in many places and those, usually manuscript (meaning hand-written), sources which are by their nature only available in one place (although see the later discussion of online archives). This does not mean that libraries do not also hold archival collections; many of them do. What it does mean is that they are certainly places where printed sources of various kinds (including printed editions of manuscript sources) are available to you.

Books

It may be an obvious point but printed books from the periods and places, and on the subjects, that you are interested in are a crucial source. There are, for example, many different ways in which geographers have begun to make use of fictional literary representations, particularly novels, to explore the geographies of the past (Sharp, 2000). For example, both Mandy Morris (1996) and Richard Phillips (1995) have worked on ideas of gender, childhood and nature in two quite differ-ent children's books: Frances Hodgson Burnett's *The Secret Garden* (1911) and R.M. Ballantyne's *The Fur Trappers* (1856). In a quite different context, David Schmid (1995) and Philip Howell (1998) have explored the representations of the city in detective fiction. There are also many examples from literary studies in all languages of interpretations that are attentive to questions of space, place and landscape. Beyond the fiction section, it is worth considering whether the books which you might otherwise pass over as 'out of date' could become the sources for a historical study. The resurgence of interest in the history of geography and of geographical knowledge has been based upon using old geography books and periodicals to try to understand and explain the sorts of geographical ideas – and the representations of people, places and environments – that were part of understandings of geography in different periods and places (for example, Livingstone, 1992; Matless, 1998; Mayhew, 2000; Driver, 2001; Withers, 2007). Indeed, the same sorts of ideas and methods can be applied to any set of books or periodicals from the past on any subject you are interested in. One example is Chris Philo's (1987) examination of the changing ideas of madness and its treat-ment through the location of institutions in the now discontinued *Asylum Journal*, a periodical for those involved in what he calls the 'mad business' of nineteenth-century England and Wales. Another example is Mona Domosh's (2001) use of urban exposés of New York in the 1860s to interpret the sorts of ideas about women and public space outlined earlier in this chapter (see also Howell, 2001; and Dennis, 2008). Finding these books is simply a matter of using the library catalogue and scanning along the shelves in the sections you are interested in.

However, you also need to be aware that any one library collection may not have all the books which you will require: perhaps all those by one author, or those which are referred to by the authors you are interested in as influences on them or that put arguments they want to challenge. Finding these texts involves some detective work, and one useful resource is the integrated catalogue of the British Library (available at www.bl.uk/catalogue) since you can use this to search for books by author and subject whether or not you have access to that particular London library. You should also be aware that there are many specialist libraries for all sorts of subjects which may be able to provide access to collections of books that will answer your research questions. Some libraries specialize in terms of their subject matter – for example, in Britain, the Wellcome Library for the History of Medicine, the Cornish Studies Library or the Manx National Heritage Library and, in the USA, the Library for Caribbean Research in New York or the library of the Black Film Center in Bloomington, Indiana. Other libraries are defined by the sort of material they hold – for example, map libraries like the collections at the British Library, the National Library of Wales, the Royal Geographical Society with the Institute of British Geographers and the Newberry Library in Chicago, or newspaper libraries such as the British Library Newspaper Library at Colindale in north London (http://newspapers.bl.ukblcs). These libraries can be found in the same way as archives (see below).

Printed sources

Many libraries, especially university libraries, also hold printed sources of various sorts. We might certainly include printed maps under this heading. Another good example, for Britain, are the nineteenth-century parliamentary papers. These are the record of the inquiries and reports made by contemporary politicians, civil servants and reformers on a whole range of subjects of concern: poverty, prisons, prostitution, the conditions of work in factories, public health and so on. Through the collection of opinion and statistics they tried to reach conclusions about what could or should be done. As a result, they provide a very rich source of information on the subjects they were dealing with and on contemporary ideas about those issues and how they could be addressed (see, for example, Kearns, 1984, on public health; Driver, 1993, on the poor law; and Ogborn, 1995, on prisons). Your library may have them in their original series, in the facsimile editions produced by the Irish University Press, or have access to them online (see http:// parlipapers.chadwyck.co.uk/home.do). It should also be noted that other countries have comparable forms of official publication on all sorts of issues of concern to them (see the *Checklist of United States Public Documents, 1789–1909* (US Government, 1911) and, for an example that uses US government inquiries into early twentieth-century immigration, see King, 2000).

Parliamentary papers were printed in the nineteenth century to make them available to as many people as possible, and other institutions have also used printing to do the same for manuscript sources. For example, there are printed versions of seventeenth- to nineteenth-century diaries, memoirs and journals (for example, those of Samuel Pepys (1970–1983), Giacomo Casanova (1997) and Fanny Burney (1972–1984)). There are also series like the Hakluyt Society's publications of travellers' accounts or the Chetham Society which, since 1843, has published material relating to the counties of Lancashire and Cheshire (see: www.chethams.org.uk). The former series, which has produced over 350 volumes since the mid-nineteenth century, provides annotated (and sometimes translated) transcripts of ships' logs, journals and letters for a huge range of voyagers to and from a wide range of places, giving easy access to material which would otherwise be only available to a few in forms that are difficult to read and understand (for an example of work based on these and other printed versions of manuscript sources, see Ogborn, 2002).

Archival collections

There is no cast-iron distinction that can be made between libraries and archives. What I want to stress here is that archival collections can be thought of as holding material that is unique to them. Because of that, and also due to the increasing importance of visual sources in historical geography, I include art galleries in this section as well. This is not simply because of the particular set of things that archival collections and galleries have, but also because of the nature of much of the material they hold which – in forms like handwriting or oil painting – are one-offs which have to change their form to be reproduced as printed sources or photographs, or to be digitized. Having said that, archival collections can be full of all sorts of material, both rare and commonplace, and another aspect of them is that their collections have often been put together in relation to a particular individual, family, institution or theme. People have their own archives, as do charities, businesses and public bodies (even university geography departments; see Johnston and Withers, 2008). The largest archives are those of governments or states (and those of Europe and North America the largest of those) which have been established as gatherers and collectors of material for some considerable time, and they are often organized according to the government departments that collected and archived the material.

Finding whether there is archival material on a historical subject you are interested in means thinking carefully about who – in terms of individuals or organizations – would have produced information about it at the time and where that might now be stored (if it has survived at all), whether it is still with the person or organization concerned or deposited in a public archive. As with finding artistic works it means searching for those who produced the sources and for thematic collections within which they are now held (for examples of work in historical geography that use visual images, see Stephen Daniels' (1999) study of the landscape garden designs of a particular individual, Humphry Repton, and James Ryan's (1997) study of photography in the British Empire, which uses lots of thematic collections of photographs including the Royal Geographical Society with the Institute of British Geographers picture library: images.rgs.org). In both cases this means careful attention to individuals and to the institutional contexts within which they operated in order to track down appropriate archives.

There is, therefore, a huge number and variety of archives. ARCHON (the UK National Archives' principal information gateway for users of manuscript sources: www.nationalarchives.gov.uk/archon) lists over two thousand archives and libraries of all sorts in the UK and the Republic of Ireland from the Abbey House Museum in Leeds to the archives of Zurich Financial Services Ltd. There are many thousands of others all over the world. Finding out which ones hold sources that might be useful for your project is made easier by reference guides both online and in print. A few of these are listed below:

- ARCHON (www.nationalarchives.gov.uk/archon) itself gives contact details (including addresses, online maps, contact names, telephone and fax numbers, email and website addresses) and access information for the archives and museums listed, as well as cross-referencing them to their NRA listings which tell you more about what sources they contain (see below). Similar access for Australia is made available through the Directory of Archives in Australia (www.archivists.org.au/directory-of-archives). There are also sites such as Libdex (www.libdex.com) that provide access to many thousands of library home pages and online catalogues worldwide.
- Access to Archives (A2A) (www.a2a.org.uk) is a database of catalogues of UK archives. It allows you to search the titles that appear in those catalogues. This means that you can do subject keyword searches. However, it should be noted that searches for the names of places, people and organizations are likely to be more productive as they are more specific and are likely to appear in the titles of archival records. You should also be aware that although this site contains the catalogues of over four hundred archives in the UK, this means that there is a great deal of archival material that is still not included. Again, a similar service is provided for Australian archives by the National Library of Australia's 'Trove' service (http://trove.nla.gov.au) and for New Zealand by the National Register of Archives and Manuscripts (http://thecommunityarchive.org.nz).
- The National Register of Archives (www.nationalarchives.gov.uk/nra): the NRA's indexes – available through this website – contain references to the papers of about 150,000 corporate bodies, persons and families which are held in archives and libraries across the UK. The indexes can be searched by corporate name, personal name, family name or place name, or they can be browsed alphabetically. This means that you need to know the names of the people and organizations you want to trace since there is no subject keyword search to allow searching by topic. References in NRA are linked directly to ARCHON.
- Janet Foster and Julia Sheppard's (2002) British Archives is a single-volume guide with entries on over a thousand archives giving contact details, information on opening times, access, finding aids and facilities, as well as brief outlines of major holdings. The entries are organized alphabetically by town, with an index by archive name, an Index to Collections which is predominantly made up of personal and organizational names, and a Guide to Key Subjects which offers broader categories. There are other similar guides for other countries, including the National Historical Publications and Records Commission (1988) Directory of Archives and Manuscript Repositories in the United States, which is organized by state, gives full contact details along with brief descriptions of the main holdings and has a full subject index.

In many cases, where the project is based to any extent upon material gathered or generated by government departments, the sources will be held by the appropriate national archives. These have very extensive holdings of material and, in many cases (including that of Britain), their catalogues can be searched online: see, for example, the British National Archives (in Kew in London) at www.nationalarchives.gov.uk; the French Archives Nationales (in Paris) at www.archivesnationales.culture.gouv.fr; the US National Archives and Records Administration (in Washington, DC) at www.archives.gov; the National Archives of Canada (in Ottawa) at www.collectionscanada.gc.ca; and the Jamaica Archives and Records Department (in Spanish Town) at www.jard.gov.jm.

It is also useful to be aware that some archives are dedicated to specific forms of material. In Britain, much oral history material and other audio sources are held at the National Sound Archive at the British Museum (sounds.bl.uk). Moving images can be found at the National Film and Television Archive at the British Film Institute (www.bfi.org.uk/nftva), and the Mass Observation archive (www.massobs.org.uk) at Sussex University holds a collection, gathered since 1937, of material based on observations of everyday life.

Online archives

As well as being able to gain increasing amounts of information about archival collections online it is also increasingly possible to get access to digitized versions of archival sources themselves on the internet and in other electronic forms (for example, the CD-ROMs of the British Calendar of Colonial State Papers for North America and the West Indies for 1574–1739, also available online via www.proquest.com, and the Trans-Atlantic Slave Trade Database which contains details of nearly 35,000 slaving voyages, also available online at www.slavevoyages.org). This is making historical materials that were previously only available, often in single versions, as manuscripts, printed books, or paintings or photographs, or extensive collections of data gathered by particular researchers, accessible to many more people. Sometimes payment (usually by an institution) is required, but these sites often have open access policies. In terms of sources already mentioned, the British House of Commons Parliamentary Papers (parlipapers.chadwyck.co.uk), Samuel Pepys's diaries (www.pepysdiary.com) and some of the books published by the Hakluyt Society in the nineteenth century (www.hakluyt.com) have been made available online.

It is impossible to summarize here the variety of sources that are being made accessible in these ways. (For a digital library of primary and secondary sources for the history of Britain, see British History Online (www.britishhistory.ac.uk), and for a broader listing of digital versions of primary sources, see History Online (www.history.ac.uk). However, it is significant that as well as manuscripts and printed books, sources available include statistical material (for example, the Great Britain Historical GIS project (www.visionofbritain.org.uk) which makes available both geographically located social, economic and demographic data (from the period 1801–2001) and the GIS through which they can be mapped; and the History Data Service which holds digital resources deposited by previous research projects: hds.essex.ac.uk); maps (for example, the University of Edinburgh's Charting the Nation: Maps of Scotland and Associated Archives, 1550–1740 project (www.chartingthenation.lib.ed.ac.uk) which has over 3,500 cartographic images and descriptions on its website); and other visual images and objects (for example, the City of London Libraries and Guildhall Art Gallery (collage.cityoflondon.gov.uk/collage) and the John Carter Brown Library in Rhode Island (which specializes in the history of the Americas, see: www.brown.edu/Facilities/John_Carter_Brown_Library) have made their picture collections available online, and the National Maritime Museum has done that and included images of many of the objects it holds too (www.nmm.ac.uk/collections)). The great benefit of these resources is not only that they make these sources available to a wider community of researchers, albeit with the requirement that those researchers have the necessary computing facilities to access the internet, but that they make these sources and collections searchable (by keywords, etc.) and, sometimes, open to new forms of analysis. For example, the Old Bailey Online project (www.oldbaileyonline.org) makes available the full proceedings of London's Central Criminal Court for the period 1674–1913 (nearly 200,000 trials) in a fully searchable format. This allows searches by names, time periods, keywords (for example, types of crimes, objects or places), and it also allows the user to compile statistical tables, diagrams and maps based on this data (for published research based on this resource, see Hitchcock and Shoemaker, 2005). However, when using such online and electronic resources it is crucial that you understand what the underlying database holds (and what it does not) and what has been done in order to facilitate searches. You need to think about the limitations of keyword searches as well as their benefits, especially how much of the whole set of sources you need to read to understand the contexts within which your keywords appear. In particular, you have to ask yourself for each particular online source whether you need to get access to the original. The need to see the original will be more significant for some sources than others (e.g. perhaps more for oil paintings than for printed books) and for some forms of analysis than others.

USING THE ARCHIVES

Since the vast majority of historical sources are not yet available online in digital form it is likely that you will have to go to the material rather than having it come to you. This means thinking about some practicalities. First, is the material available to you? Different libraries and archives have different rules on who is allowed to access the material. Private archives of families, companies or charities are not obliged to allow you access. You will have to find out who is in charge and explain your project and the reasons why you want to see the material they have. They may restrict certain items or restrict the use you can make of them – for example, not using people's names or not publishing work based on their collection without permission. Public archives often have rules that restrict material for time periods from 30 to 100 years for reasons of confidentiality and sensitivity, but they are also committed to making their collections accessible to public use. Second, when is the archive open? Opening times vary, and you need to make sure that the archive is open when you get there and not going to close shortly after! Finally, is it worth the trip? Often items in lists and catalogues sound more interesting and useful than they are. You need to find out as much as possible about them before you commit your time and resources to a research visit. You should certainly get as much information as you can from the archive or library's website, particularly if you can search their collections online. You can also write to, telephone or email the archive and ask about the material you want to see. You will want to ask for a description of what is there in qualitative and quantitative terms. The archivists know their collections better than anyone so they can be a lot of help to you. If you have to travel a long way and stay away from home for an extended period, archival research can be a costly business. Make sure you have thought about the resource implications of your research before you devise a project on, for example, 'race' and urban change in the 1940s which can only be undertaken in the municipal archives of the City of Los Angeles.

Once you have located the material and decided that an archive needs to be visited, there are a few hints about working methods that might come in handy. First, find out (from the website if possible) as much as you can about how that particular archive works – what are the opening hours, what are the security arrangements (what can you bring in, and do you need to present ID or references?), do you need to book an appointment or a seat, how do the catalogues work, how do you order material to read, can you order material online in advance, how long does material take to be delivered when ordered on site, how much can you have on your desk at any one time? You will need to ensure that you have a steady flow of material coming to you. If you find yourself waiting around with nothing to read, use the time productively by searching the catalogues or the book shelves for other relevant material, or by talking to other researchers. Second, think about how you are going to collect the material that you need. What suits your requirements and those of the project being undertaken: taking notes on paper or on a laptop; inputting into a database; photocopying; or taking digital photographs of the material? What are the archive's policies and procedures on data collection? Some allow digital photography (with some charging to do so); others do not (and you should certainly experiment in advance at home to ensure that your camera will take usable images, and to work out how many images you can store and how long your battery will last). Photocopying charges vary enormously, where it is permitted at all. And some archives have limited accommodation for laptop users. Third, talk to the archivists early on and ask them about the material you are interested in. They may also be able to point you towards other useful material that you have not located in advance. Fourth, try to work out how long the archival research is going to take you. You should, perhaps even before you get to the archive, have a full list of the material you want to look at. After you have looked at roughly a tenth of that material, you should assess how long the job will take if you keep working at that rate. On the basis of those back-of-the-envelope calculations and the time (and resources) you have available for the research, make a judgement about whether you need to begin sampling or collecting the material differently or prioritizing what to read in a different way. It is very unlikely that you will be able to speed up significantly unless you change how you are doing the research, since the limits are usually set by the material you are using. You need to come away from the archive with what is most useful to you in the time available. Fifth, keep a clear record of what you have read and what you still have to cover. You must also always ensure that both the notes you take on the material and your lists of what to read use the reference system which the archive uses, whereby each item will have an individual reference number (the one you use to order it for consultation). If you have to recheck a quotation or some figures then you have to know, at any point in the future, where they came from so you can go back to it. And when you use those sources in your work you will also have to tell others where the material came from using this reference number. Finally, always take a pencil, a pencil sharpener, an eraser and some paper. Most archives only let you use pencils and, whether you are equipped with a laptop or not, you will certainly need one for jotting down research notes and so on. There is nothing worse than spending the first precious hours of archival research trying to find a pencil in a place you don't know.

Summary

- Work out what sort of sources will answer the questions you want to ask, and decide what sorts of questions are appropriate for the sources you have available.
- Remember that the nature of the material that survives in 'archives' is a matter of both power relations and an emotional investment in the past.
- Historical sources can come in many different forms – literary, visual, printed official publications, manuscripts – and there are many different libraries, galleries, archives and museums that contain this material.
- Historical sources are increasingly being made available online so check to see if you can get access to what you need that way.
- Find out about access to the archives and plan your collection strategy carefully.

Further reading

- Discussions of archival research methods in historical geography can be found in Baker (1997), Ogborn (2003) and Gagen *et al.* (2007).

- Useful surveys of the sorts of work that has been done and is being done in historical geography can be found in Dodgshon and Butlin (1990), Graham and Nash (2000) and in articles published in the *Journal of Historical Geography*.

- Details of British archives and their collections can be browsed through online in the resources listed above or in Foster and Sheppard (2002) (this is its fourth edition). Previous editions (written by Janet Foster) are still useful if in danger of having some out-of-date details.

Note: Full details of the above can be found in the references list below.

References

Baker, A.R.H. (1997) '"The dead don't answer questionnaires": researching and writing historical geography', *Journal of Geography in Higher Education*, 21: 231–243.
Biger, G. and Liphschitz, N. (1995) 'Foreign tree species as construction timber in nineteenth-century Palestine', *Journal of Historical Geography*, 21: 262–277.

Blunt, A. and Rose, G. (1994) *Writing Women and Space: Colonial and Postcolonial Geographies.* New York: Guilford Press.

Bondi, L. and Domosh, M. (1998) 'On the contours of public space: a tale of three women', *Antipode*, 30: 270–289.

Burney, F. (1972–84) *The Journals and Letters of Fanny Burney (Madame D'Arblay).* Oxford: Clarendon Press (12 vols).

Casanova, G. (1997) *History of My Life.* Baltimore (trans. W.R. Trask). MD: Johns Hopkins University Press (6 vols).

Chadwick, E. (1965, originally published in 1842) *Report on the Sanitary Condition of the Labouring Population of Great Britain*, 1842 (edited by M.W. Flinn). Edinburgh: Edinburgh University Press.

Daniels, S. (1999) *Humphry Repton: Landscape Gardening and the Geography of Georgian England.* New Haven, CT: Yale University Press.

Dennis, R. (2008) *Cities in Modernity: Representations and Productions of Metropolitan Space, 1840–1930.* Cambridge: Cambridge University Press.

Dodgshon, R.A. and Butlin, R.A. (1990) *An Historical Geography of England and Wales* (2nd edn). London: Academic Press.

Domosh, M. (1998) 'Those "gorgeous incongruities": polite politics and public space on the streets of nineteenth-century New York City', *Annals of the Association of American Geographers*, 88: 209–226.

Domosh, M. (2001) 'The "women of New York": a fashionable moral geography', *Environment and Planning D: Society and Space*, 19: 573–592.

Driver, F. (1993) *Power and Pauperism: The Workhouse System, 1834–1884.* Cambridge: Cambridge University Press.

Driver, F. (2001) *Geography Militant: Cultures of Exploration and Empire.* Oxford: Blackwell.

Foster, J. and Sheppard, J. (2002) *British Archives: A Guide to Archival Resources in the United Kingdom* (4th edn). Basingstoke: Macmillan.

Gagen, E., Lorimer, H. and Vasudevan, A. (eds) (2007) *Practising the Archive: Reflections on Method and Practice in Historical Geography.* Historical Geography Research Series 40. London: Historical Geography Research Group.

Gaskell, E. (1970, originally published in 1848) *Mary Barton: A Tale of Manchester Life.* Harmondsworth: Penguin Books.

Gleeson, B.J. (1995) 'A public space for women: the case of charity in colonial Melbourne', *Area*, 27: 193–207.

Graham, B. and Nash, C. (2000) *Modern Historical Geographies.* London: Prentice Hall.

Harley, J.B. (2001) *The New Nature of Maps: Essays in the History of Cartography.* Baltimore, MD: Johns Hopkins University Press.

Heffernan, M.J. (1991) 'The desert in French orientalist painting during the nineteenth century', *Landscape Research*, 16: 37–42.

Hill, J. (2006) 'Travelling objects: the Wellcome collection in Los Angeles, London and beyond,' *Cultural Geographies*, 13: 340–366.

Hitchcock, T. and Shoemaker, R. (eds) (2005) 'Special Issue: Tales from the Old Bailey', *The London Journal*, 30: 6–70.

Howell, P.M.R. (1998) 'Crime and the city solution: crime fiction, urban knowledge, and radical geography', *Antipode*, 30: 357–378.

Howell, P.M.R. (2001) 'Sex and the city of bachelors: sporting guidebooks and urban knowledge in nineteenth-century Britain and America', *Ecumene*, 8: 20–50.

Johnston, R. and Withers, C.W.J. (2008) 'Knowing our own history? Geography department archives in the UK', *Area*, 40: 3–11.

Kearns, G. (1984) 'Cholera and public health reform: the significance of the geographical patterns', *Bulletin of the Society for the Social History of Medicine*, 35: 30–32.

King, D. (2000) *Making Americans: Immigration, Race, and the Origins of the Diverse Democracy.* Cambridge, MA: Harvard University Press.

Koven, S. (2004) *Slumming: Sexual and Social Politics in Victorian London.* Princeton, NJ: Princeton University Press.

Livingstone, D.N. (1992) *The Geographical Tradition.* Oxford: Blackwell.

Matless, D. (1998) *Landscape and Englishness.* London: Reaktion.

Mayhew, R.J. (2000) *Enlightenment Geography: The Political Languages of British Geography*, 1650–1850. Basingstoke: Macmillan.

Morris, M.S. (1996) '"Tha'lt be like a blush-rose when tha' grows up, my little lass": English cultural and gendered identity in *The Secret Garden*', *Environment and Planning D: Society and Space*, 14: 59–78.

National Historical Publications and Records Commission (1988) *Directory of Archives and Manuscript Repositories in the United States* (2nd edn). Phoenix, AZ: Oryx Press.

Ogborn, M. (1995) 'Discipline, government and law: separate confinement in the prisons of England and Wales, 1830–1877', *Transactions, Institute of British Geographers*, 20: 295–311.

Ogborn, M. (2002) 'Writing travels: power, knowledge and ritual on the English East India Company's early voyages', *Transactions, Institute of British Geographers*, 27: 155–171.

Ogborn, M. (2003) 'Knowledge is power: using archival research to interpret state formation', in A. Blunt *et al.* (eds) *Cultural Geography in Practice*. London: Arnold, pp. 9–20.

Ogborn, M. (2004) 'Archives', in S. Harrison, S. Pile and N. Thrift (eds) *Patterned Ground: Ecologies of Nature and Culture*. London: Reaktion, pp. 240–242.

Ogborn, M. (forthcoming) 'Archive', in J. Agnew and David Livingstone (eds) *The Sage Handbook of Geographical Knowledge*. London: Sage.

Pepys, S. (1970–83) *The Diary of Samuel Pepys* (ed. R. Latham and W. Matthews). London: Bell (11 vols).

Phillips, R.S. (1995) 'Spaces of adventure and the cultural politics of masculinity: R.M. Ballantyne and *The Young Fur Traders*', *Environment and Planning D: Society and Space*, 13: 591–608.

Philo, C.P. (1987) 'Fit localities for an asylum: the historical geography of the nineteenth-century "mad business" in England as viewed through the pages of the Asylum Journal', *Journal of Historical Geography*, 13: 398–415.

Rappaport, E.D. (2001) *Shopping for Pleasure: Women in the Making of London's West End*. Princeton, NJ: Princeton University Press.

Rose, G. (1990) 'Imagining Poplar in the 1920s: contested concepts of community', *Journal of Historical Geography*, 16: 415–437.

Ryan, J.R. (1997) *Picturing Empire: Photography and the Visualisation of the British Empire*. London: Reaktion.

Samuel, R. (1994) *Theatres of Memory. Volume I. Past and Present in Contemporary Culture*. London: Verso.

Schmid, D. (1995) 'Imagining safe urban space: the contribution of detective fiction to radical geography', *Antipode*, 27: 242–269.

Sharp, J.P. (2000) 'Towards a critical analysis of fictive geographies', *Area*, 32: 327–334.

US Government (1911) *Checklist of United States Public Documents, 1789–1909*. Washington, DC: US Government Printing Office.

Walkowitz, J.R. (1998) 'Going public: shopping, street harassment, and streetwalking in late Victorian London', *Representations*, 62: 1–30.

Withers, C.W.J. (2007) *Placing the Enlightenment: Thinking Geographically about the Age of Reason*. Chicago: Chicago University Press.

8 Semi-structured Interviews and Focus Groups

Robyn Longhurst

Synopsis

A semi-structured interview is a verbal interchange where one person, the interviewer, attempts to elicit information from another person by asking questions. Although the interviewer prepares a list of predetermined questions, semi-structured interviews unfold in a conversational manner offering participants the chance to explore issues they feel are important. A focus group is a group of people, usually between 6 and 12, who meet in an informal setting to talk about a particular topic that has been set by the researcher. The facilitator keeps the group on the topic but is otherwise non-directive, allowing the group to explore the subject from as many angles as they please. This chapter explains how to go about conducting both interviews and focus groups.

The chapter is organized into the following sections:

- Introduction
- What are semi-structured interviews and focus groups?
- Formulating questions
- Selecting and recruiting participants
- Where to meet
- Recording and transcribing discussions
- Ethical issues
- Conclusion

INTRODUCTION

Talking with people is an excellent way of gathering information. Sometimes in our everyday lives, however, we tend to talk too quickly, not listen carefully enough and interrupt others. Semi-structured interviews (sometimes referred to as informal, conversational or 'soft' interviews) and focus groups (sometimes referred to as focus-group interviews) are about talking with people but in ways that are self-conscious, orderly and partially structured. Krueger and Casey (2000: xi) explain that focus-group interviewing (and we could add here, semi-structured interviewing) is about talking but it is also

> ... about listening. It is about paying attention. It is about being open to hear what people have to say. It is about being nonjudgmental. It is about creating a comfortable environment for people to share. It is about being careful and systematic with the things people tell you.

Over the last few decades there has emerged in geography interesting debates (especially amongst feminist geographers) about the utility and validity of qualitative methods, including semi-structured interviews and focus groups (Eyles, 1988; Pile, 1991; Schoenberger, 1991, 2007; McDowell, 1992a, 2007; Nast, 1994; Crang, 2002, 2003, 2004; Davies and Dwyer, 2007). Many geographers have moved towards what Sayer and Morgan (1985) call 'intensive methods' to examine the power relations and social processes constituted in geographical patterns.

Geographers employ a range of intensive or qualitative methods. Some included in this book are participant observation, keeping a research diary and visual methodologies. Semi-structured interviews, however, are probably one of the most commonly used qualitative methods (Kitchin and Tate, 2000: 213). Focus groups are not as commonly used but they have become increasingly popular since the mid-1990s (see *Area*, 1996: Vol. 28, which contains an introduction and five articles on focus groups).

Geographers have used focus groups to collect data on a diverse range of subjects. As early as 1988 Burgess, Limb and Harrison used focus groups (which they called 'small groups') to explore people's environmental values (Burgess *et al.*, 1988a, 1988b). A decade later Miller *et al.* (1998) conducted focus groups (as well as surveys and ethnographic research) on shopping in northern London to explore links between shopping and identity. Myers and Macnaghten (1998) conducted focus groups to investigate 'rhetorics of environmental sustainability'. Wolch *et al.* (2000) ran a series of focus groups in Los Angeles with an aim of exploring the role played by cultural difference in shaping attitudes towards animals in the city. Skop (2006) examined the methodological potential of focus groups for population geography.

Geographers have also used semi-structured interviews to collect data on an equally diverse range of subjects. Winchester (1999: 61) conducted interviews (and questionnaires) to gather a range of information about the characteristics of 'lone fathers' and the causes of marital breakdown and post-marital conflict in Newcastle, Australia. Valentine (1999) interviewed couples, some together, some apart, in order to understand gender relations in households. Johnston (2001) conducted interviews (and focus groups) with participants (or subjects – see McDowell, 1992b: footnote 4, on the contested nature of the terms 'participant' and 'subject') and organizers at a gay pride parade in Auckland, New Zealand. Johnston was interested in the relationship between people taking part in the parade (hosts) and people watching the parade (guests). Punch (2000) conducted interviews (and participant observation) with children and their families in Churquiales, a rural community in the south of Bolivia. She wanted to 'document the ways in which children devise ways to contest adult's [sic] power and control in their lives' (Punch, 2000: 48). Yantzi and Roseberg (2008) interviewed 11 women in Ontario, Canada who identified themselves as the main caregiver for their child with a disability.

In this chapter I define briefly what I mean by semi-structured interviews and focus groups. These two methods share some characteristics in common; in other ways they are dissimilar. I also discuss how to plan and conduct semi-structured interviews and focus groups. This discussion includes formulating a schedule of questions, selecting and recruiting participants, choosing a location, transcribing data and thinking through some of the ethical issues and power relations involved in conducting semi-structured interviews and focus groups. Throughout the chapter empirical examples are used in an attempt to illustrate key arguments.

WHAT ARE SEMI-STRUCTURED INTERVIEWS AND FOCUS GROUPS?

Interviews, explains Dunn (2005: 79), are verbal interchanges where one person, the interviewer, attempts to elicit information from another person. Basically there are three types of interviews: structured, unstructured and semi-structured, which can be placed along a continuum. Dunn (2005: 80) explains:

> Structured interviews follow a predetermined and standardised list of questions. The questions are always asked in almost the same way and in the same order. At the other end of the continuum are unstructured forms of interviewing such as oral histories ... The conversation in these interviews is actually directed by the informant rather than by the set questions. In the middle of this continuum are semi-structured interviews. This form of interviewing has some degree of predetermined order but still ensures flexibility in the way issues are addressed by the informant.

Semi-structured interviews and focus groups are similar in that they are conversational and informal in tone. Both allow for an open response in the participants' own words rather than a 'yes or no' type answer.

A focus group is a group of people, usually between 6 and 12, who meet in an informal setting to talk about a particular topic that has been set by the researcher (for other definitions see Krueger, 1988; Morgan, 1997; Merton and Kendall, 1990; Swenson *et al.*, 1992; Greenbaum, 1993; Stewart *et al.*, 2006; Gregory *et al.*, 2009). The method has its roots in market research. The facilitator or moderator of focus groups keeps the group on the topic but is otherwise non-directive, allowing the group to explore the subject from as many angles as they please. Often researchers attempt to construct as homogeneous a group as possible (but not always – see Goss and Leinback, 1996). The idea is to attempt to simulate a group of friends or people who have things in common and feel relaxed talking to each other. When Honeyfield (1997; also see Campbell *et al.*, 1999) conducted research on representations of place and masculinity in television advertising for beer he carried out two focus groups: one with five women, one with seven men. In both groups the participants had either met before, were friends or lived together as 'flatmates'.

Focus groups tend to last between one and two hours. A key characteristic is the interaction between members of the group (Morgan, 1997: 12; Cameron, 2005). This makes them different from semistructured interviews which rely on the interaction between interviewer and interviewee. Focus groups are also different from interviews in that it is possible to gather the opinions of a large number of people for comparatively little time and expense.

Focus groups are often recommended to researchers wishing to orientate themselves to a new field (Morgan, 1997; Greenbaum, 1993). For example, in 1992 I began some research on pregnant women's experiences of public spaces in Hamilton, New Zealand. There was no existing research on this topic so I wanted to establish some of the parameters of the project before using other methods. I did not know what words pregnant women in Hamilton used to refer to their pregnant bodies – tummies? stomachs? breasts? boobs? – therefore, it would have been difficult to conduct interviews. Focus groups provided an excellent opportunity to gather preliminary information about the topic (see Longhurst, 1996, for an account of these focus groups).

Both semi-structured interviews and focus groups can be used as 'stand-alone methods', as a supplement to other methods or as a means for triangulation in multi-methods research. Researchers often draw on a range of methods and theories. Valentine (2005: 112) explains:

> Often researchers draw on many different perspectives or sources in the course of their work. This is known as triangulation. The term comes from surveying, where it describes using different bearings to give the correct position. In the same way researchers can use multiple methods or different sources to try and maximize their understanding of a research question.

To sum up thus far, semi-structured interviews and focus groups can be used for a range of research, are reasonably informal or conversational in nature and are flexible in that they can be used in conjunction with a variety of other methods and theories. It is also evident that semi-structured interviews and focus groups are more than just 'chats'. The researcher needs to formulate questions, select and recruit participants, choose a location and transcribe data while at the same time remaining cognizant of the ethical issues and power relations involved in qualitative research. In the section that follows I address these topics.

FORMULATING QUESTIONS

Dunn (2005: 81) explains: 'It is not possible to formulate a strict guide to good practice for every interview [and focus group] context.' Every interview and focus group requires its own preparation, thought and practice. It is a social interaction and there are no hard and fast rules one can follow (Valentine, 2005). Nevertheless there are certain procedures that researchers are well advised to heed.

To begin, researchers need to brief themselves fully on the topic. Having done this it is important to work out a list of themes or questions to ask participants. People who are very confident at interviewing or running focus groups often equip themselves with just a list of themes. Personally, I like to be prepared with actual questions in case the conversation dries up. Questions may be designed to elicit information that is 'factual', descriptive, thoughtful or emotional. A combination of different types of questions can be effective depending on the research topic. Researchers often start with a question that participants are likely to feel comfortable answering. More difficult, sensitive or thought-provoking questions are best left to the second half of the interview or focus group when participants are feeling more comfortable. In Box 8.1 is a list of questions I drew up in order to examine large/fat/overweight people's experiences of place. This schedule could be used for semi-structured interviews or focus groups. Follow-up questions are in parentheses.

I would not necessarily ask these questions in the order listed. Allowing the discussion to unfold in a conversational manner offers participants the chance to explore issues they feel are important. At the end of the interview or focus group, however, I would check my schedule to make sure that all the questions had been covered at some stage during the interview or focus group.

It is important to remember that it can take time for participants to 'warm up' to semi-structured interviews and focus groups. If possible, therefore, it is worth offering drinks and food as a way of relaxing people. It is also useful at the beginning of a focus group to engage participants in some kind of activity that focuses their attention on the discussion topic. For example, participants might be asked to draw a picture, respond to a photograph or imagine a particular situation. This technique tends to be used more by market researchers but it can also prove effective for social scientists. Kitzinger (1994) presented focus group members with a pack of cards bearing statements about who might be 'at risk' from AIDS. She asked the group to sort the cards into different piles indicating the degree of 'risk' attached to each 'type of person'. Kitzinger (1994: 107) explains that '[s]uch exercises involve people in working together with minimal input from the facilitator and encourage participants to concentrate on one another (rather than on the group facilitator) during the subsequent discussion'.

Box 8.1   Semi-structured interview and focus-group schedule

- Can you remember a time in your life when you were *not* large/fat/overweight? (Tell me about that. How did people respond to you then?)
- Are there places that you avoid on account of being large? (Why? How do you feel if you do visit these places?)
- Are there places where you feel comfortable or a sense of belonging on account of your size? (Tell me about these places and how you feel in them.)
- In New Zealand there is a strong tradition of spending time at the beach. Do you go to the beach? (Explain. What is it like for you at the beach?)
- Describe your experience of clothes shopping. (Where do you shop? Are shop assistants helpful? Are the changing rooms comfortable? Do you ever feel that other shoppers judge you on account of your size?)
- When you shop for groceries or eat out in a public space, how do you feel? (Why?)
- Are there any issues concerning your size that arise at work? (What are these issues?)

(Continued)

(Continued)

- Do you feel cramped in some spaces? (For example, movie-theatre seats, small cars, planes?)
- Do you exercise? (If so, what do you do and where do you do it?)
- Have you made any modifications to your home to suit your size? (For example, altered doorways, selected particular furniture, arranged furniture in specific ways, modified bathroom/toileting facilities. Explain.)
- Do you imagine that your life would be different if you were smaller? (Explain.)

Are there any issues that you would like to raise that you feel are important but that you haven't had a chance to explore in this interview/focus group?

SELECTING AND RECRUITING PARTICIPANTS

Selecting participants for semi-structured interviews and focus groups is vitally important. Usually people are chosen on the basis of their experience related to the research topic (Cameron, 2005). Burgess' (1996, cited in Cameron, 2005: 121) study of fear in the countryside is a useful example of this 'purposive sampling' technique. When using quantitative methods the aim is often to choose a random or representative sample, to be 'objective' and to be able to replicate the data. This is not the case when using qualitative methods. Valentine (2005: 111, emphasis in original) explains that, unlike with most questionnaires, 'the aim of an interview [and a focus group] is *not* to be representative (a common but mistaken criticism of this technique) but to understand how individual people experience and make sense of their own lives'.

For example, if you were studying 'racial violence' you might anticipate interviewing and/or running focus groups with people from different ethnic groups, especially those thought to be involved in the violence. However, you might also want to examine the ways in which people's ethnic or racial identities intersect with other identities such as gender, sexuality, 'migrant status' and age in order to explore more fully the processes shaping racial violence. It is not only participants' identities that need to be considered, however, when conducting research. Valentine (2005: 113) makes the important point that 'When you are thinking about who you want to interview it is important to reflect on who you are and how your own identity will shape the interactions that you have with others.' She explains this is what academics describe as being *reflexive* or recognizing your own *positionality* (see England, 1994: 82; Moss, 2002; and Bondi, 2003 on 'empathy and identification' in the research process).

There are many strategies for recruiting participants for semi-structured interviews and focus groups. Some strategies work for both methods while others are more appropriate for one or the other. If you are recruiting participants for interviews it is common practice 'to carry out a simple questionnaire survey to gather basic factual information and to include a request at the end of the questionnaire asking respondents who are willing to take part in a follow-up interview to give their address and telephone number' (Valentine, 2005: 115). Asking for an e-mail contact is another useful way to follow up. It is also possible to advertise for participants in local newspapers or on radio stations requesting interested parties to contact you.

Another method for recruiting participants for interviews is 'cold calling' – that is, calling on people (usually strangers) to ask if they would be prepared to be interviewed. When I was studying the ways in which managers in the central business districts of Auckland and Edinburgh present themselves at work (their dress, comportment and grooming) I called into retail stores and businesses, introduced myself and asked to speak with the manager. I then explained the research and requested an interview. This can be a nerve-racking process because interviewers often get a high refusal rate. In my research on managers, however, approximately 70 per cent of those approached agreed to take part, resulting in 26 interviews (see Longhurst, 2001).

As mentioned already, focus groups are often made up of people who share something in common or know each other. Group membership lists, therefore, can be a useful tool for recruiting. People who already know each other through sports clubs, social clubs, community activities, church groups or work can make an ideal focus group. When I conducted focus groups on men's experiences of domestic bathrooms (a private space rarely discussed by geographers) I succeeded in enlisting (with the help of friends) four groups of men. The first group belonged to the same rugby club, the second were colleagues in a government department, the third were 'job-seekers' and the fourth were family/friends.

Another route useful for securing participants for focus groups is what Krueger (1988: 94) refers to as 'recruiting on location' or 'on-site recruiting'. I used this strategy to recruit first-time pregnant women to talk about their experiences of public places. Pregnant women were approached at antenatal classes, midwives' clinics and doctors' surgeries. These women 'opened doors' to me speaking with other pregnant women. Social scientists refer to this as 'snowballing': 'This term describes using one contact to help you recruit another contact, who in turn can put you in touch with someone else' (Valentine, 2005: 117).

WHERE TO MEET

Not only is it necessary to decide how to select and recruit participants but also to decide where to conduct the interview or focus-group meeting. It comes as no surprise to most geographers that where an interview or focus group is held can make a difference (Denzin, 1970). Ideally, the setting should be relatively neutral. I once made the mistake of helping to facilitate a focus group about the quality of service offered by a local council at the council offices. The discussion did not flow freely and it soon became apparent that the participants felt hesitant (understandably) about criticizing the council while in one of their rooms. However, it is worth noting that 'In most cases if you are talking to business people or officials from institutions and organizations you will have no choice but to interview them in their own offices' (Valentine, 2005: 118; but also see McDowell, 1997, on interviewing investment bankers in the City of London. Being in the environs you are studying can also prove useful).

It is not always possible to conduct interviews and focus groups in 'the perfect setting' but if at all possible aim to find a place that is neutral, informal (but not noisy) and easily accessible. For example, if you are conducting a reasonably small focus group it is possible to sit comfortably around a dining-room table (see Fine and Macpherson, 1992, for an account of a focus group that took place 'over dinner'). Needless to say, if it is a larger focus group a larger space will be required, perhaps a room at a school, church or club. The main consideration for both semi-structured interviews and focus groups is that interviewees feel comfortable in the space. It is important that the interviewer also feels comfortable (see also Chapter 4). Valentine (2005: 118) warns: 'For your own safety never arrange interviews with people you do not feel comfortable with or agree to meet strangers in places where you feel vulnerable.'

RECORDING AND TRANSCRIBING DISCUSSIONS

When conducting semi-structured interviews or focus groups it is possible to take notes or to audio/video record the discussion. I usually audio(tape) the proceedings. This allows me to focus fully on the interaction instead of feeling pressure to get the participants' words recorded in my notebook (see Valentine, 2005). Directly after the interview I document the general tone of the conversation, the key themes that emerged and anything that particularly impressed or surprised me in the conversation. Taking these notes, in a sense, is a form of data analysis (for information on qualitative data analysis, see Miles and Huberman, 1994; Kitchin and Tate, 2000).

It is advantageous to transcribe interviews and focus groups as soon as possible after conducting them (for how to code a transcript, see Chapter 27). Hearing the taped conversation when it is still fresh in your mind makes transcription much easier. Focus groups, especially large groups, can be difficult to transcribe because each speaker, including the facilitator, needs to be identified. In Box 8.2 is an example of a transcript from a focus group of men who met to discuss their experiences of domestic bathrooms and toilets. Note the 'dynamism and energy as people respond to the contributions of others' (Cameron, 2005: 117). In this focus group excerpt one of the participants puts a question to other group members. Wayne's question about bidets (small, low basins for washing the genitals and anal areas) spearheads a discussion on cultural difference. Note the various transcription codes: the starts of overlap in talk are marked by a double oblique //; pauses are marked with a dot in parenthesis (.); non-verbal actions, gestures and facial expressions are noted in square brackets; and loud exclamations are in **bold** typeface (for more detailed transcription codes, see Dunn, 2005: 98).

Box 8.2 Transcription of a focus group

Wayne: The other question you didn't ask is: has anyone got a bidet? And if they have, then the other question is: does anybody know how bidets are supposed to be used? (.)

Brent: I've never understood how it works.

Robert: No, nor have I.

Christopher: Well, I've got an idea about how it works but it just doesn't seem very efficient.

Robert: Crocodile Dundee just uses a water fountain [laughter].

Brent: Exactly.

Facilitator: That's an interesting question. What do you think about using, I mean, how do you use a bidet?

Robert: It's supposed to wash your bloody bottom [sic] out isn't it?

Christopher: Mm, that's the whole object. And then you use a towel?

Robert: I don't know, I think so, instead of wiping //

Wayne: // In Asia for instance all you get is a bloody hole in the ground and a dipper and you dip the water out and go 'woof' and over and it washes it clean.

Robert: **Oh my god! That's terrible!**

Christopher: It's another culture.

Wayne: And that is probably why Asians are so good at squatting on their heels 'cause they are used to it. You look at a kid, a tiny tot, and they can squat on their heels all day long.

Source: Audio-tape excerpt from a focus group conducted by David Vincent in 1999 (see Longhurst, 2001)

As this transcript illustrates, sometimes data can be 'sensitive'. It is not surprising, therefore, that there are numerous ethical issues to consider when conducting semi-structured interviews and focus groups (see Chapter 3).

ETHICAL ISSUES

Two important ethical issues are confidentiality and anonymity. Participants need to be assured that all the data collected will remain secure under lock and key or on a computer database accessible by password only; that information supplied will remain confidential and participants will remain anonymous (unless they desire otherwise); and that participants have the right to withdraw from the research at any time without explanation. It is also sound research practice to offer to provide participants with a summary of the research results at the completion of the project and to follow through on this commitment. This summary might take the form of a hard copy or an electronic copy posted on a website (for example, the Department of Geography, Durham University provides reports on various research projects conducted by staff, see: www.dur.ac.uk/geography/research/researchprojects).

Focus groups pose a further complication in relation to confidentiality because not only is the researcher privy to information but also members of the group.

Therefore, participants need to be asked to treat discussions as confidential. Cameron (2005: 122) explains:

> As this [confidentiality] cannot be guaranteed, it is appropriate to remind people to disclose only those things they would feel comfortable about being repeated outside the group. Of course, you should always weigh up whether a topic is too controversial or sensitive for discussion in a focus group and is better handled through another technique, like individual in-depth interviews.

Another ethical issue is that participants in the course of an interview or focus group may express sexist, racist or other offensive views. In an earlier quotation, Krueger and Casey (2000: xi) claim that researchers ought to listen, pay attention and be non-judgemental. Sometimes, however, being non-judgemental might simply reproduce and even legitimize interviewees' discrimination through complicity (see Valentine, 2005). Researchers need to think carefully about how to deal with such situations because there are no easy solutions.

Researchers also need to think carefully about how to interview or run focus groups in different cultural contexts (see Chapter 12). For example, 'First World' researchers investigating 'Third World' 'subjects' need to be highly sensitive to local codes of conduct (Valentine, 2005). In short, there is a web of ethical issues and power relations that need to be teased out when conducting semi-structured interviews and focus groups (see Law, 2004 on 'mess' in social science research). Feminist geographers in particular have made a useful contribution in this area (for example, see Katz, 1992, 1994; McDowell, 1992b; Dyck, 1993; England, 1994; Gibson-Graham, 1994; Kobayashi, 1994; Moss, 2002).

CONCLUSION

In this chapter I have outlined two qualitative methods – semi-structured interviews and focus groups – and how they can be employed in geographical research. Both methods involve talking with people in a semi-structured manner. However, whereas semi-structured interviews rely on the interaction between interviewee and interviewer, focus groups rely on interactions amongst interviewees. Both methods make a significant contribution to geographic research, especially now that discussions about meaning, identity, subjectivity, politics, knowledge, power and representation are high on many geographers' agendas. Critically examining the construction of knowledge and discourse in geography (see Rose, 1993) has led to an interest in developing alternative methodological strategies coupled with greater reflexivity about the process of research. Semi-structured interviews and focus groups are useful for investigating complex behaviours, opinions and emotions and for collecting a diversity of experiences. These methods do not offer researchers a route to 'the truth' but they do offer a route to partial insights into what people do and think.

Summary

- Semi-structured interviews and focus groups are about talking with people but in ways that are self-conscious, orderly and partially structured.
- These methods are useful for investigating complex behaviours, opinions and emotions and for collecting a diversity of experiences.
- Every interview and focus group requires its own preparation, thought and practice.
- There are a range of methods that can be used for recruiting participants, including advertising for participants, accessing membership lists, on-site recruiting and 'cold calling'.
- Interviews/focus groups ought to be conducted in a place where participants and the interviewer feel comfortable.
- When conducting semi-structured interviews or focus groups take notes and/or audio/video record the discussion.
- There is a web of ethical issues and power relations that need to be teased out when using these methods.
- Semi-structured interviews and focus groups make a significant contribution to geographic research, especially now that discussions about meaning, identity, subjectivity, politics, knowledge, power and representation are high on many geographers' agendas.

Further reading

There are numerous excellent books, book chapters and articles on semi-structured interviews (and interviewing more generally) and focus groups written by geographers and other social scientists. I have listed below some of the more recently published titles:

- Krueger and Casey (2000). This popular book, first published in 1988, has been reprinted three times. The third edition is easy to read, well illustrated and offers numerous examples of how to use focus groups. It is one of the most comprehensive guides on focus groups available.

- Cameron (2005) offers a geographer's perspective on focus groups, explaining the various ways they have been used, how to plan and conduct them and how to analyse and present results.

- *Area* (1996: Vol. 28) contains an introduction by Goss and five articles on focus groups (by Burgess; Zeigler, Brunn and Johnston; Holbrook and Jackson; Longhurst; and Goss and Leinback). The collection illustrates effectively the range of research carried out by geographers using focus groups.

- Valentine's (2005) chapter on 'conversational interviews' is highly readable and provides advice on whom to talk to, how to recruit participants and where to hold interviews. Valentine raises interesting questions about the ethics and politics of interviewing and alerts readers to some of the potential pitfalls that can occur in research.

- Dunn (2005) discusses structured, semi-structured and unstructured interviewing in geography, critically assessing the relative strengths and weaknesses of each method. His chapter provides advice on interview design, practice, transcription, data analysis and presentation. Like Valentine, Dunn has a useful guide at the end of the chapter to further reading.

Note: Full details of the above can be found in the references list below.

References

Area (1996) 28(2) 'Introduction to focus groups' by J.D. Goss and five papers on using focus groups in human geography by Burgess; Zeigler, Brunn and Johnston; Holbrook and Jackson; Longhurst; and Goss and Leinback.

Bondi, L. (2003) 'Empathy and identification: conceptual resources for feminist fieldwork', *ACME: International Journal of Critical Geography*, 2: 64–76 (http://www.acme-journal.org/vol2/ Bondi.pdf).

Burgess, J. (1996) 'Focusing on fear: the use of focus groups in a project for the Community Forest Unit, Countryside Commission', *Area*, 28: 130–135.

Burgess, J., Limb, M. and Harrison, C.M. (1988a) 'Exploring environmental values through the medium of small groups. 1. Theory and practice', *Environment and Planning A*, 20: 309–326.

Burgess, J., Limb, M. and Harrison C.M. (1988b) 'Exploring environmental values through the medium of small groups. 2. Illustrations of a group at work', *Environment and Planning A*, 20: 457–476.

Cameron, J. (2005) 'Focusing on the focus group', in I. Hay (ed.) *Qualitative Research Methods in Human Geography* (2nd edn). Melbourne: Oxford University Press, pp. 116–132.

Campbell, H., Law, R. and Honeyfield, J. (1999) '"What it means to be a man": hegemonic masculinity and the reinvention of beer', in R. Law, H. Campbell and J. Dolan (eds) *Masculinities in Aotearoa/New Zealand*. Palmerston North: Dunmore Press, pp. 166–186.

Crang, M. (2002) 'Qualitative methods: the new orthodoxy?', *Progress in Human Geography*, 26: 647–655.

Crang, M. (2003) 'Qualitative methods: touchy, feely, look-see?', *Progress in Human Geography*, 27: 494–504.

Crang, M. (2005) 'Qualitative methods: there is nothing outside the text?', *Progress in Human Geography*, 29: 225–233.

Davies, G. and Dwyer, C. (2007) 'Qualitative methods: are you enchanted or are you alienated?', *Progress in Human Geography*, 31: 257–266.

Denzin, N. (1970) *The Research Act: A Theoretical Introduction to Social Research*. Chicago, IL: Aldine.

Department of Geography, Durham University, Projects: http://www.dur.ac.uk/geography/research/research projects/ (accessed 3 September 2008).

Dunn, K. (2005) 'Interviewing', in I. Hay (ed.) *Qualitative Research Methods in Human Geography* (2nd edn). Melbourne: Oxford University Press, pp. 79–105.

Dyck, I. (1993) 'Ethnography: a feminist method?', *The Canadian Geographer*, 37: 52–57.

England, K. (1994) 'Getting personal: reflexivity, positionality and feminist research', *The Professional Geographer*, 46: 80–89.

Eyles, J. (1988) 'Interpreting the geographical world: qualitative approaches in geographical research', in J. Eyles and D.M. Smith (eds) *Qualitative Methods in Human Geography*. Cambridge: Polity Press, pp. 1–16.

Fine, M. and Macpherson, P. (1992) 'Over dinner: feminism and adolescent female bodies', in M. Fine (ed.) *Disruptive Voices: The Possibilities of Feminist Research*. East Lansing, MI: University of Michigan Press, pp. 175–203.

Gibson-Graham J.K. (1994) '"Stuffed if I know!": reflections on post-modern feminist social research', *Gender, Place and Culture*, 1: 205–224.

Goss, J.D. and Leinback, T.R. (1996) 'Focus groups as alternative research practice: experience with transmigrants in Indonesia', *Area*, 28: 115–123.

Greenbaum, T. (1993) *The Handbook for Focus Group Research*. Lexington, MA: Lexington Books.

Gregory, D., Johnston, R.J., Pratt, G., Watts, M. and Whatmore, S. (2009) *A Dictionary of Human Geography* (5th edn). Oxford: Blackwell.

Honeyfield, J. (1997) 'Red blooded blood brothers: representations of place and hard man masculinity in television advertisements for beer.' Master's thesis, University of Waikato, New Zealand.

Johnston, L. (2001) '(Other) bodies and tourism studies', *Annals of Tourism Research*, 28: 180–201.

Katz, C. (1992) 'All the world is staged: intellectuals and the projects of ethnography', *Environment and Planning D: Society and Space*, 10: 495–510.

Katz, C. (1994) 'Playing the field: questions of fieldwork in geography', *The Professional Geographer*, 46: 67–72.

Kitchen, R. and Tate, N.J. (2000) *Conducting Research into Human Geography*. Edinburgh Gate: Pearson.

Kitzinger, J. (1994) 'The methodology of focus groups: the importance of interaction between research participants', *Sociology of Health and Illness*, 16: 103–121.

Kobayashi, A. (1994) 'Coloring the field: gender, "race", and the politics of fieldwork', *The Professional Geographer*, 46: 73–79.

Krueger, R.A. (1988) *Focus Groups: A Practical Guide for Applied Research*. Thousand Oaks, CA: Sage.

Krueger, R.A. and Casey, M.A. (2000) *Focus Groups. A Practical Guide for Applied Research* (3rd edn). Thousand Oaks, CA: Sage.

Law, J. (2004) *After Method: Mess in Social Science Research*. London: Routledge.

Longhurst, R. (1996) 'Refocusing groups: pregnant women's geographical experiences of Hamilton, New Zealand/ Aotearoa', *Area*, 28: 143–149.

Longhurst, R. (2001) *Bodies: Exploring Fluid Boundaries*. London: Routledge.

McDowell, L. (1992a) 'Valid games?', *The Professional Geographer*, 44: 219–222.

McDowell, L. (1992b) 'Doing gender: feminism, feminists and research methods in human geography', *Transactions, Institute of British Geographers*, 17: 399–416.

McDowell, L. (1997) *Capital Culture. Gender at Work in the City*. Oxford: Blackwell.

McDowell, L. (2007) 'Sexing the economy, Theorizing Bodies', in A. Tickell, E. Sheppard, J. Peck and T. Barnes (eds) *Politics and Practice in Economic Geography*. London: Sage, pp. 60–70.

Merton, R.K. and Kendall, P.L. (1990) *The Focused Interview: A Manual of Problems and Procedures* (2nd edn). New York: Free Press.

Miles, M.B. and Huberman, A.M. (1994) *Qualitative Data Analysis: An Expanded Sourcebook*. Thousand Oaks, CA: Sage.

Miller, D., Jackson, P., Thrift, N., Holbrook, B. and Rowlands, N. (1998) *Shopping, Place and Identity*. London: Routledge.

Morgan, D.L. (1997) *Focus Groups as Qualitative Research*. Thousand Oaks, CA: Sage, pp. 75–79.

Morgan, D.L. (1988) 'Focus groups as qualitative research', *Qualitative Research Methods*, 16.

Moss, P. (ed.) (2002) *Feminist Geography in Practice: Research and Methods*. Oxford: Blackwell.

Myers, G. and Macnaghten, P. (1998) 'Rhetorics of environmental sustainability: commonplaces and places', *Environment and Planning A*, 30: 335–353.

Nast, H. (1994) 'Opening remarks on "women in the field"', *The Professional Geographer*, 46: 54–55.

Pile, S. (1991) 'Practising interpretative geography', *Transactions of the Institute of British Geographers*, 16: 458–469.

Punch, S. (2000) 'Children's strategies for creating playspaces', in S.L. Holloway and G. Valentine (eds) *Children's Geographies. Playing, Living, Learning*. London and New York: Routledge, pp. 48–62.

Rose, G. (1993) *Feminism and Geography: The Limits of Geographical Knowledge*. Cambridge: Polity Press.

Sayer, A. and Morgan, K. (1985) 'A modern industry in a reclining region: links between method, theory and policy', in D. Massey and R. Meegan (eds) *Politics and Method*. London: Methuen, pp. 147–168.

Schoenberger, E. (1991) 'The corporate interview as a research method in economic geography', *The Professional Geographer*, 43: 180–189.

Schoenberger, E. (2007) 'Politics and practice: Becoming a geographer', in A. Tickell, E. Sheppard, J. Peck and T. Barnes (eds) *Politics and Practice in Economic Geography*. London: Sage, pp. 27–37.

Skop, E. (2006) 'The methodological potential of focus groups in population geography', *Population, Space and Place*, 12: 113–124.

Stewart, D.W., Shamdasani, P.N. and Rook, D.W. (2006) *Focus Groups: Theory and Practice* (2nd edn). Newbury Park, CA: Sage.

Swenson, J.D., Griswold, W.F. and Kleiber, P.B. (1992) 'Focus groups: method of inquiry/intervention', *Small Group Research*, 24: 459–474.

Valentine, G. (1999) 'Doing household research: interviewing couples together and apart', *Area*, 31: 67–74.

Valentine, G. (2005) 'Tell me about . . . using interviews as a research methodology', in R. Flowerdew and D. Martin (eds) *Methods in Human Geography: A Guide for Students Doing a Research Project* (2nd edn). Edinburgh Gate: Addison Wesley Longman, pp. 110–127.

Winchester, H.P.M. (1999) 'Interviews and questionnaires as mixed methods in population geography: the case of lone fathers in Newcastle, Australia', *The Professional Geographer*, 51: 60–67.

Wolch, J., Brownlow, A. and Lassiter, U. (2000) 'Constructing the animal worlds of inner-city Los Angeles', in C. Philo and C. Wilbert (eds) *Animal Spaces, Beastly Places*. London and New York: Routledge, pp. 71–97.

9   Participant Observation

Eric Laurier

Synopsis

Participant observation involves spending time being, living or working with people or communities in order to understand them. In other words, it is, as the name implies, a method based on participating and observing in which field-notes, sketches, photographs or video recordings are used as a method of data collection. The basis of this approach is to become, or stay, as close to the spatial phenomenon being studied as possible and it is thereby quite distinct from methodologies that emphasize distance and objectivity.

The chapter is organized into the following sections:

- What is participant observation?
- Commentators and players
- Doing participant observation/becoming the phenomenon
- Adequate commentaries on culture and society
- Results: respecifying the generalities of social science
- Final words of advice

WHAT IS PARTICIPANT OBSERVATION?

> I have no great quickness of apprehension or wit ... my power to follow a long and purely abstract train of thought is very limited ... [but] I am superior to the common run of men in noticing things which easily escape attention, and in observing them carefully. (Charles Darwin, Preface to *The Expression of the Emotions in Man and Animals*)

Participant observation is perhaps the easiest method in the world to use since it is ubiquitous and we can all already do it. From the moment we are born we are, in various ways, observing the world around us and trying to participate in it. Children, acquiring language for the first time, listen and watch what, when and how their parents are doing what they are doing. They observe greetings and have greetings directed at them, and attempt to participate by, at first, looking and, later, waving and making sounds that approximate, and eventually are, hellos and goodbyes. It is, of course, not just children who use this method to acquire skills: surgical students spend a great deal of time observing surgery and are gradually entered into the practical demands of operating on patients as fully participating surgeons. International migrants, finding themselves in foreign countries, have the massive task of observing a multitude of activities and exactly how they are done in order to participate in new cultures. Amongst other background knowledge they have to acquire the locals' taken-for-granted ways of getting everyday things done, such as greetings, ordering coffee, queuing for buses, making small talk, paying their taxes and so on.

So far so good: participant observation is easy – it does not require the mastering of arcane skills or technical lexicons. Before you rush to adopt this method of fieldwork on the basis of its minimal social science requirements there are some other aspects of successful participant-observation that we will have to return to. You may have guessed already from the mention of surgeons and migrants that participant observation as a way of engaging with familiar and unfamiliar life-worlds will present you with a different curve to your learning, the steep climbing is elsewhere than on campus. Also, for better and for worse, because it is not an *external* method administered on research subjects, such as a questionnaire or a lab test, participant observation has *no preset formal steps*. Or rather the stages that anyone doing participant observation must go through are the stages *which arise out of the phenomenon and settings* you are investigating.

If you do not know how to do, or be something, then learning how to do or be that thing will be as hard for you as for anyone attempting to participate in it. Think of the effort and time required to do informative participant observation studies of air-traffic control (Harper and Hughes, 1993), mathematical problem-solving (Livingston, 1987) or playing jazz music (and these expert cultures have been studied through participant observation.) Before I give the impression that this method ought only to be used for the study of highly skilled groups and their settings, participant observation can be turned to such seemingly lowly skilled spatial phenomena as shopping in the supermarket or going clubbing (Malbon, 1999) or walking in the city (Watson, 1993). In these latter examples you may already be able to do them and the demands on you will then be to provide a commentary that describes them in revealing and interesting ways. Key to your success in doing participant observation is, as Charles Darwin says of his own powers, that you notice things that otherwise escape attention and that you observe *carefully and patiently*.

COMMENTATORS AND PLAYERS

A common mistake, made as often by well-qualified social researchers as by those new to participant observation, is to take observation as the dominant element of participant observation. This is to some extent a legacy of both actual behaviourist research and endless comedy routines about white-coated observers who watch their research subjects in a detached, emotionless manner and are thereby able to provide objective descriptions of what was occurring. Even though most of us who use participant observation no longer pursue this kind of objective observation, and gave up the white coats for t-shirts, we still underestimate the importance of participation, assuming that it is sufficient to pay attention to what is going on and then write down our observations.

To give you a sense of why it may not be enough to be a diligent observer, let us move on to what your participant observation should produce: *commentary*. Being able to comment on the culture, society and geography of various spaces and places is indeed the major requirement of doing geography. In that sense all geographers are commentators and many of them exceptionally good ones. If we think about sports commentators for a moment, as against social and cultural commentators, we can see that they are seldom the ones playing the game: they are sitting to one side observing it. Some sports commentators provide exasperatingly bad and irrelevant commentary because they have never played the game they are commenting on. One ingredient of a decent sports commentator is that he or she should be, or have been, even a bad or half-hearted player to offer any kind of insight into the game. Knowing *how* to play the game in no way guarantees insightful remarks since many of the outstanding players and competitors have very little to say. Shifting back to social and cultural research, it is the case that far too many of its researchers are only commentators and have never played. The point that is being reiterated here is that the best participant observation is generally done by those who have been involved in and tried to do and/or be a part of the things they are observing.

DOING PARTICIPANT OBSERVATION/BECOMING THE PHENOMENON

Despite my having suggested that there is no template for doing participant observation there are features in its course which, while not specifying what is to be done, will give you a sense of whether you are making any progress or not. If you have never been involved in the event, activity, phenomenon, group or whatever you are investigating then, on your arrival, you will find yourself cast into some category. You will be called, if not a 'greenhorn', 'beginner', then something worse such as an 'incompetent', 'tourist' or 'outsider'. This is not simply a pejorative term for you, it assigns you to a type which shapes up expectations of what you will be able to do, the perspective you will have on events and what you will need to be taught. It is also not such a bad place to start because people's assumptions about you will not be too high and you will be expected to be observing in order to learn how to become one of the group/company/band/players. Recording your observations at this point is vital since, if all goes well, they should have changed by the time you are finished (see Box 9.1). They are also, at this point, the perspective of 'any person' who may well be the audience you will wish to write your report for at the end of your fieldwork. Without keeping a record of your own struggles to get 'the knowledge' you are likely to forget the lay member's perspective once you no longer have it. Consequently you will no longer appreciate what it is that may seem odd, irrational or otherwise mysterious to those, like yourself, now in the know.

Box 9.1   Recording observations

During an ESRC research project on the cultural role and social organization of cafés in UK cities I spent hours with my notebook trying to bring to notice taken-for-granted features of cafés. These features I would place in square brackets, a notational device adopted from phenomenology to help suspend our familiar understandings of various words. One of the things that struck me was that particular times of days in the café showed themselves in different ways: *Staff can be present in the café but without customers the café is [empty]. What more is there to this easily recognizable state of affairs? There is: how the customer recognizes [empty] which is bound up with the typical interior architectural construction of this café, and many others like it, which allows those entering to look around as they enter and note at a glance how busy the café is. It is bound up also with the expectations of 'this early hour' of 7am in this café known for its 'appearances as usual' on a weekday at this sort of time or as an environment of expectations. By contrast, at the same time of day, an airport or flower market café is likely to be full. What the customer makes of [empty] is related to their orientation to the 'awakening' of the day. That is, the reasons for empty-ness are temporally located – it is 'just opened'. A customer is not put off or curious about this observable empty-ness during the opening time, the way they would be were it to be observably empty at 1pm ('Why is it empty? Is the food bad? Are the staff rude? Is it expensive?').* (For more details see Laurier, 2008.)

Just how long it takes to become competent in what you choose to study, and indeed whether it is possible to reach that state, will vary according to what you choose to study using this method. Should you choose to study a supermarket as a shopper or stacking the shelves as a member of the staff then it is not so demanding to become competent in these activities. However, other communities of practice, such as mathematicians, dairy farmers or jazz piano players, may take considerable time and effort before you will be recognized as an accepted member of their groups. For whatever activity you decide to participate in there will be different ordinary and expert ways in which you are instructed in how to do it, from the more formal (i.e. lessons, workshops, courses, rulebooks, etc.) to the informal (tips, jokes, brief chats). For some rural villagers you will always be an 'incomer' even if you live in the village until your dying day. A successful participant observation does not turn entirely on becoming *excellent* in the activity (becoming an Olympic athlete or leading cardiac surgeon) or passing as, say, a homeless Italian-American or even as a member of the opposite sex. Yet by the end you should possess a degree of the particular know-how, appropriate conduct and common knowledge of the place and/or people you have chosen to study.

Here is a brief extract where David Sudnow (1978: 82) offers a sense of what he was doing in what he called 'going for the jazz'. He had already acquired a basic competency in piano but was trying to *become* a jazz piano player, so he was, among other methods, starting to look very carefully at what the experienced players did with their body at the piano:

> At live performances I had watched the very rapid improvisation of players whose records had served as my models, but their body idioms in no way seemed connected in details to the nature of their melodies,

and my occasional attempts to emulate the former had no appreciable bearing on my success with the latter. This, for example, had a little shoulder tic, but mimicking that (which I found myself doing after a night of watching him) did not make his sorts of melodies happen. Another sat tightly hunched over the piano, playing furiously fast, but assuming that posture seemed to have no intrinsic relationship to getting my jazz to happen as his did.

Sudnow spent a *decade* learning, first, how to play a piano; second, how to play jazz that was recognized by other jazz musicians as jazz; and, finally, how to instruct students in jazz piano. The above quotation gives only a hint of the experiences involved in Sudnow's odyssey. However long you devote to your study, and it's more likely to be days or weeks rather than years, what is important is to keep notes, audio-visual records and, ideally, a journal along the way. Basically, record as much as you possibly can, and even more importantly, try to write straightforward and detailed descriptions of the phenomena you are interested in. Record what you learn and can make sense of and what puzzles or upsets you in a journal and supplement those notes with photographs and film where possible. Sudnow video-taped himself playing the piano so that he could then repeatedly watch what his hands were doing on the piano keyboard in relation to the production of the music. These records will form the base materials for your commentary at the end of your time doing fieldwork. It is not impossible to write a delightful report on your participant observation without having kept a record of it at the time but it is considerably more difficult and the details will inevitably slip away. In my own research practice I try to take photographs and shoot video wherever and whenever possible. Video, in particular, provides you with a rich set of records to work with once you are writing up your research. It is a 'retrievable dataset' and reviewable to find unanticipated details that you could not formulate in words at the time, nor may even have noticed, since you were too busy being engaged in the situation. Photos and video clips are also very helpful for presenting your results to lay and expert audiences. With video, audio and still photographs you can share your data and I would strongly encourage you to do data-review sessions in a group to see what different people are able to find in the visual material you are presenting them with. Note down what you all see in common and try to consider how you do so. Materials that may initially seem quite uninteresting should, by the end of their close viewing and description, provide you with surprises. Drawing conclusions and developing ideas from the materials you have gathered remains reliant on your insight.

Fieldnotes

The notes in Box 9.2 were taken in the passenger seat of a company car after a conversation with a regional manager about her boss and some of her co-workers.

Box 9.2   Fieldnotes

Transcript

. . . She feels equally that she is not competent in that 'talk', not as 'real' saleswoman – so it's a question of identity, there's also a lingering suspicion of the salesman and the real sales-man's attitudes. Somehow they aren't trustworthy – once all other comparative morals have disappeared into doing sales. They're described as 'smooth', 'able to talk themselves out of anything'. M describes herself as 'all over the place' and I guess by contrast, rough around the edges. These very misfootings in her performance reinforce her authenticity however? She's also fairly hierarchical – at lunch suggesting she'd never tell tales to her employers that were personal or about losing face – being incompetent. Since her boss's story about going out on her first date with her current boyfriend – with a hangover and then having to nip away half-way through for a camouflaged barf. For M there are further reasons why she's doing so much just now and why she's so exhausted. It's the learning cycle of the job – at first blissfully ignorant of all that was involved and fluking her way through and then as she learns about the job realising there's a whole lot more she should be doing – symbolised by the territorial strategy. And I guess further . . .

Fieldnotes or, rather, certainly *my* fieldnotes are often badly written, dull, cryp-tic and not the kind of thing I would want to show anyone. However, I have included a 'cheat' sample of a not too dire section to give you at least a flavour of how roughly notes tend to be written at the time. What is important to bear in mind is that notes should be taken because you will not be able to recall suf-ficient details of what happens or what people say during a lengthy engagement with them. As it happens, getting your notebook out is also quite useful in show-ing your researcher status in the field or setting so that you will be seen to be at work doing your research and (may) be taken seriously as a result. You should be aware that there are sensitive settings where you will want to keep your status as a researcher more low key. In such an event you may end up scribbling notes in the toilet or in the bus afterwards, or some other hidden place, and there are plenty of amusing stories from experienced ethnographers about doing so. The notes reproduced in Box 9.2 were written up while the regional manager was out of the car dropping in to visit one of her clients.

As importantly, note-taking can help you concentrate during situations which may otherwise allow your mind to wander. The uses of note-taking to help you pay attention will, I am sure, be familiar from your use of them to do so during university lectures. To repeat myself, the most significant use of your fieldnotes will be *after the event*, in helping you recall the details of the situation which you wish to describe, analyse or reflect on. As you can see from the example of my notes they generally need to be tidied up, expanded upon and details frequently need to be filled in which will allow your jottings to be decrypted.

Video transcripts

The video footage drawn on in Box 9.3 was taken later during another research study on pedestrian tourists' wayfinding practices. The extended transcript uses stills from the video clips to bring in elements of the visual and interactional organization of wayfinding.

Box 9.3   Video notes

Fragment 1 – Gladstone's Land

As the episode begins the group of four tourists are all standing together in a group at a crossroads trying to recall and make sense of a recommendation they have been given. The two women on the right walk away to look down a street close to where Barry is standing with the camcorder.

1	Tourist1:	Maybe it's down there. It could be down there Fran ----------->

[pause]

2	Barry:	(to the two women standing beside him looking along the street) Are you looking for a street?
3	Tourist1:	Nooo it's a [pause] a very old house, is it Gladstone or Livingstone. [pause] Very old place. I think it's to the left of Deacon Brodie's ehh

The group of tourists begin, closely packed, all looking at the Edinburgh Old Town map in their 'insight' guidebook to Scotland. In their close-knit standing arrangement they have arranged themselves behind the guidebook with the map (Mondada, forthcoming). As their inspection of the map provides no immediate solution, two of the members of the group split off, walking away to scout. Standing at a distance they can gather additional perspectives up and down

---

the street orthogonal to where the remaining group members are standing (see Brown, 2006 for more discussion of tourist collaboration). From this we can see what tourists can do when they are a group rather than a lone traveller – they can distribute and co-ordinate the activities involved in orienting their unit.

The mapreader has rotated the guidebook so that the map is held in one of the possible matches with the cross of the junction – one aspect of a junction is that it limits you to four possible alignments with the map. Further adjustment is required of the book, the available horizon and the group. To produce a shared orientation of left and right, up and down, in front and behind etc. the group has squeezed together in order to stand behind the book facing the same way. In other words, the alignment between the map and what they can see comes after their alignment with each other as pedestrians. With this side-by-side arrangement any one in the unit, on pointing, can have the potential building or sign or street they are picking out easily found by the others in the group. The alignment here then is a group alignment – standing together close, but polite, so as to see in common both the map and the city. A hand held on the map can be raised to point toward features found to correspond with it.

Source: Laurier and Brown, 2008

Digital video records of events during fieldwork also require writing up afterwards so, even though they save you making notes at the time, you cannot escape making some notes eventually. The clips you are interested in can be imported from compact cameras (though the video quickly fills their memory), tape-based camcorders, hard-drive and solid state memory camcorders to computers that have the appropriate connectivity (USB2 or firewire). Once video is imported then, using various software packages (i.e. Quicktime, Moviemaker, Adobe Premiere, Final Cut Pro or AVID), stills can be extracted to be added to transcripts as in Box 9.3. If you are working towards multimedia documents of your research, the sequences can be kept as moving 'real time' audio-visual data and added to whatever you are constructing your media in (e.g. html, flash). There are plenty of technical guides to this available online and, given the rapid changes in video technology, I will not go into further technical detail on dealing with digital video. Broadly it is worth pointing out that it has been getting easier and cheaper to use and geographers have barely scratched the surface of what might be done with it for describing social practices or presenting their results.

As with using a notebook, using a camcorder makes your status as a researcher highly visible, only more so. When you switch the camera on people frequently feel obliged to make faces or talk to the camera as if they were starring in a docu-soap. For instance, during the café research customers frequently made funny faces as they passed the camera (see Laurier and Philo, 2006). Serious documentary makers emphasize getting the groups or individuals you wish to video familiar with the presence of the camera to the point of ignoring it and urging them to do what they would normally do were the camera not there. In fact many anthropologists hand over the camcorder and allow their communities to video themselves in an attempt to hand over control of the video-making to those being represented. In my own practice I have done both, while filming the tourists below it was Barry Brown holding the camera but in other research on social interaction in cars the cameras were handed over to the drivers and passengers of the car (Laurier *et al.*,

2008). In the clips and their accompanying transcripts I try to capture something of the interaction as it happens 'naturalistically' (in contrast to experimentally or by questionnaire or interview).

Video clips or transcriptions with stills, as I have noted already, assist in sharing your original data. In watching video clips during data sessions we try to describe what is happening without jumping to any particular conclusions. As you can see from the example in Box 9.3, they seem in some ways obvious, yet it is the obvious that we tend to overlook precisely because it is what we take for granted. Just providing a patient attentive description of what is happening is also a first stage towards your analysis, much like tabulating the descriptive statistics from a numerical study. Unlike statistics they do not give you 'results' at this stage but are reliant on your further description and analysis of what is occurring. In the example with the tourists, Barry Brown and myself were investigating how maps and guidebooks are used in real-world situations when a group of people, rather than a lone individual, are finding their way around unfamiliar places. Our aim was, through close examination of a limited set of instances, to explain what is going on when people rotate themselves or their maps while trying to work out where they are.

ADEQUATE COMMENTARIES ON CULTURE AND SOCIETY

First, an oft-quoted remark from Harvey Sacks (1992: 83) on reading his students' reports after he had sent them to do participant observation work on people exchanging glances:

> Let me make a couple of remarks about the problem of 'feigning ignorance'. I found in these papers that people [i.e. the students in his class] will occasionally say things like, 'I didn't really know what was going on, but I made the inference that he was looking at her because she's an attractive girl.' So one claims to not really know. And here's a first thought I have. I can fully well understand how you come to say that. It's part of the way in which what's called your education here gets in the way of your doing what you in fact know how to do. And you begin to call things 'concepts' and acts 'inferences', when nothing of the sort is involved. And that nothing of the sort is involved is perfectly clear in that if it were the case that you didn't know what was going on – if you were the usual made up observer, the man from Mars – then the question of what you would see would be a far more obscure matter than that she was an attractive girl, perhaps. How would you go about seeing in the first place that one was looking at the other, seeing what they were looking at, and locating those features which are perhaps relevant?

The warning Sacks is making to his students is not to exercise a kind of professional scepticism ('feigning ignorance') which subverts the intelligibility of the things and actions they are able to observe – in this case a guy checking out an attractive girl. Sacks is warning of the dangers of acting like a Martian who has landed in a city and is without the myriad methods and experiences we have for making sense of our local environment. Moreover, Sacks is saying that what we see in any setting is tied to the fact that we are participants and that there are classifications that we *are* able to make almost instantly and definitely pre-theoretically as part of our natural attitude to the world. In doing participant observation of places we are already competent inhabitants of and can take their appearance for granted, the solution to doing adequate descriptions of them is not to import strange labels for the things we see or hear or otherwise sense almost instantly. It is the categories that the locals do and would use to describe their observations that we are interested in.

If you are a 'local' already you have huge advantages in providing adequate descriptions of how and why things get done in the way they get done. Yet you are also at the disadvantage of no longer noticing how such things get done because they are so familiar as to be *seen but unnoticed* and you may never have attempted to make them into any kind of formal description (Garfinkel and Sacks, 1986). The exercise Sacks set his students is one that you might also 'try at home without supervision' (see Exercise 9.1).

Exercise 9.1

In a public place such as a busy street, university library or park, set yourself down with a notebook, camera or camcorder. Watch the people there and look for exchanges of glances between people who are not otherwise interacting with one another. Write notes at the time and afterwards if you have video to re-view and check your observations. The notes are to be on what kind of persons exchange glances with what kinds of second (or third) persons. Try to consider what kinds of actions take place during exchanges: recognition, snubs, reprimands, warnings, challenges, thanks. Can exchanged glances be hostile? Friendly? Flirty? Defensive? And if so, how so?

After gathering your observations they can be further described and analysed in order for you to consider the social categories you have used. For instance, that a 'well-dressed woman' exchanged 'friendly' glances with a 'mother and baby', or that a 'teenage girl avoided the glance of another teenage boy'. What you will start to show in your analysis is something of what you know and use already to make sense of your everyday life. Moreover, if you carry out this exercise you should start to get a grasp on how it is that we use glances and how you are able to see what someone else sees. That is, you can see what it is they are looking at, not so much from working out the exact focus of their look but by seeing what it is in the scene that they would glance at or who they would exchange glances with (i.e. a beggar lying in a doorway, two dog owners, a kid on a scooter). This exercise is done with a minimum level of participation by you as a researcher and a minimum level of disruption to the place you are investigating. While easy to do, the test is to get a really good description done that makes available how glancing works as a social and cultural activity. Your description will be adequate when you show some parts of how glancing works in the particular observations you have in hand. This exercise in doing a participant observation is illustrative of a broad type of participant observations, which are those of our common or everyday life-world.

Things will be slightly different if you have pursued participant observations of new and uncommon sets of skills such as mathematical proving, walking a police officer's beat (Sacks, 1972), learning mountaineering, nursing the dying, living among the homeless on the streets (Rose *et al.*, 1965), dropping in at a drop-in centre for the mentally ill (Parr, 2000) or concierges maintaining high-rise buildings (Strebel, forthcoming). From these more practically ambitious projects, as I suggested earlier, your adequacy as a commentator turns on you having learnt things which lay members and indeed geographers cannot be expected to know. This certainly makes delivering 'news' easier since, unlike 'exchanging glances' or 'answering the telephone' or 'buying a newspaper', not everyone knows how these more obscure, expert, secret or exotic activities are done. Not everyone could tell whether what you were saying about these activities indicated that you really know what you were talking about or whether it was sense or nonsense. For that reason you should consider testing the veracity of your descriptions on the people they were purported to represent.

Competence in the particular field you have selected will be one way in which your comments will attain a reasonable degree of adequacy. You yourself and other competent members like you will be able to see your comments as closely tied to the activities they are describing. Ideally you will be able to show some things that are in many ways known already but have simply never been closely described and analysed before. Doing adequate descriptions is already a challenge but in this second case is certainly no easy matter. It will be additionally hard because you will be trying to write or speak to two different audiences: towards the more abstract concerns of human geographers and other researchers and, just as importantly, writing for and as a member of the group you are describing.

RESULTS: RESPECIFYING THE GENERALITIES OF SOCIAL SCIENCE

Human geography, like most other social sciences, has a host of classic topics such as power, class, race, identity and landscape. If, like Charles Darwin, your 'power to follow a long and purely abstract train of thought is very limited' then these classic topics may not best be pursued as purely theoretical matters. As big-picture issues they sit uncomfortably with the more modest and small-picture concerns of actually doing a participant observation. Are you really going to resolve disputes that have dogged the social sciences for a century from your study of a drop-in centre? Perhaps not, but you may be able to respecify what appear to be overpowering abstract problems into worldly, ordinary practical problems.

And what might your results look like? Your results should be ones that you could *not* have guessed from the big topics. Nor should you have been able to imagine them before doing your study, or else why bother going to have a look if you can work it out without ever consulting anything in the world? Darwin's methods were to pay close attention to animals in all kinds of places, including his pets at home, and observe them in extended detail in ways and to ends that had not been pursued before. Despite their quite ordinary non-technical provenance, Darwin's patient observations of animal life overturned our view of ourselves, our origins and our relation to nature and God. Now *that* is a result. But what has been made of his work should not be confused with the actual lowly and lengthy observations he documented that ended up being drawn into these larger conflicts. His 'results' were mostly descriptions, alongside sketches and still photographs, interspersed with reiterations of longstanding problems in biology and zoology and occasional bursts of inspiration and insight.

The strengths of participant observation are, hopefully, quite clear by now in that it is easy to do and it provides a more direct access to phenomena than some of the more complex methodologies of social science. It allows you to build detailed descriptions from the ground up that should be recognizable to the groups whose lives you have entered into. Its limits are that it does not have a handle that you can turn to make results pop out. Nor can it be shoe-horned in as a replacement for statistical methods since it will provide only very weak answers to the kinds of questions that could be hypothesis tested.

The kind of evidence that arises out of a one-off description allows your study to bring into view certain types of phenomena that are too complex for methodologies that seek and detect general features. Good data from a participant observation can be, and usually are, a particular instance of some practice or event or feature that elicits your interest. Sometimes the instance is an exception to the rule that teaches you the rule such as when someone does a foot fault at Wimbledon. At that moment you discover a rule in the game of tennis that you never knew before. It is not a well-known nor commonly broken rule but it is there in the game. Sometimes an instance is simply one that you find recurring all the time, such as in the video notes example from earlier. Groups of tourists rotate themselves and their maps all the time to find their way. Looking at one instance of it revealed methods that tourists commonly use when they are in unfamiliar environments. If you consider that in each and every place the locals have at hand just what they need, then and there, to produce locally the spatial phenomena and interactional events you are observing, then it becomes apparent that wherever you start there is material for your analysis.

There is no need for you to climb a ladder to get a view nobody else has. As Wittgenstein (1980: 257) remarks:

> I might say: if the place I want to get to could only be reached by a ladder, I would give up trying to get there. For the place I really have to get to is a place I must already be at now. Anything I might reach by climbing a ladder does not interest me.

FINAL WORDS OF ADVICE

It is with a mild hint of self-irony that my finishing words are: avoid reading books which claim to describe 'how to do' participant observation. If you must, read just the one and then throw it away afterwards. A preparatory way to learn how this kind of fieldwork is done, if you cannot get hold of someone else who has done participant observation already, is to read an actual study which has been based in participant observation (see the further reading at the end of the chapter). None of these is a substitute for doing a study yourself. Much like learning to play the piano or work out a mathematical proof or describe what a strawberry tastes like, you have to take a bite. Being told how playing the piano or mathematical proving or tasting fruit is done, in a book, does not and cannot provide what you need to know. Participant observation is not difficult, nor obscure, though the topics, places, people, subjects and more to which you apply it may be. Since it acquires the shape and scale of its phenomena, in your first studies choose things you reckon you can handle.

Summary

- Participant observation is a simple skill of doing and watching that we all do as part of our everyday lives without realizing it.
- It is important to participate as well as just observe.
- This approach can be applied to new places and practices but can also be used to make visible familiar places and practices.
- As a participant observer you need to keep written fieldnotes and/or video notes of your research.
- The data collected need to be analysed like other empirical material and can be used to understand and make sense of more abstract problems.

Further reading

The following books and articles provide good examples of participant observation in practice:

- Crang (1994) is an example of participant observation used to examine how a waiter's work gets done and how looking closely at this work teaches us about surveillance and display in workplaces.

- Goffman (1961) is a classic from a very readable author, by turns thought-provoking, amusing and disturbing. It is not just geographers who have found this book of value; it is a key text on many courses in sociology, social psychology, psychiatric nursing and doctoring.

- Harper *et al.* (2000) is based on two of the authors spending time working alongside the employees of new telephone-banking facilities and traditional banks and building societies. Similarly, Zaloom (2006) charts the everyday practices of traders in the City of London and Chicago, and in so doing illustrates how practice is integral to the production of global financial markets. Both studies show how participant observation can be carried out in a business environment.

- Livingston (1987) is a very accessible and practical introduction to what is a sometimes puzzling and certainly distinctive approach in the social sciences which heavily utilizes participant observation.

- Sudnow (1978). Mentioned earlier in the main text, this book was a best-seller when released and is read by both academics and musicians. In addition, if you can play the piano then it can be used as a tutorial in the basics of jazz. For those unfamiliar with playing music, some sections will be hard to grasp since it relies on a basic knowledge of notation, chords, etc.

- Wieder (1974). Based on the author's residence as a researcher in a halfway house for ex-convicts, this study illuminates how the 'convict's code' is used as a device for making sense of and producing events at the halfway house. It provides a good basis for seeing how a particular place and its inhabitants organize their everyday lives.

Note: Full details of the above can be found in the references list below.

References

Crang, P. (1994) 'It's showtime: on the workplace geographies of display in a restaurant in South East England', *Environment and Planning D: Society and Space*, 12: 675–704.

Garfinkel, H. and Sacks, H. (1986) 'On formal structures of practical actions', in H. Garfinkel (ed.) *Ethnomethodological Studies of Work*. London: Routledge & Kegan Paul, pp. 160–193.

Goffman, E. (1961) *Asylums*. Harmondsworth: Penguin Books.

Harper, R. and Hughes, J.A. (1993) '"What a f-ing system! Send 'em all to the same place and then expect us to stop 'em hitting": making technology work in air traffic control', in G. Button (ed.) *Working Order: Studies of Work, Interaction and Technology*. London: Routledge, pp. 127–144.

Harper, R., Randall, D. and Rouncefield, M. (2000) *Organisational Change and Retail Finance: An Ethnographic Perspective*. London: Routledge.

Laurier, E. (2008) 'How breakfast happens in the café', *Time and Society*, 17: 119–143.

Laurier, E. and Brown, B. (2008) 'Rotating maps and readers: praxiological aspects of alignment and orientation', *Transactions of the Institute of British Geographers*, 33: 201–221.

Laurier, E., Brown, B.A.T., Lorimer, H., Jones, O., Juhlin, O., Noble, A., *et al.* (2008) 'Driving and passengering: notes on the natural organization of ordinary car travel and talk', *Mobilities*, 3: 1–23.

Laurier, E. and Philo, C. (2006) 'Natural problems of naturalistic video data', in H. Knoblauch, J. Raab, H.-G. Soeffner and B. Schnettler (eds) *Video-analysis: Methodology and Methods, Qualitative Audiovisual Data Analysis in Sociology*. Oxford: Peter Lang, pp. 183–192.

Laurier, E., Whyte, A. and Buckner, K. (2001) 'An ethnography of a cafe: informality, table arrangements and background noise', *Journal of Mundane Behaviour*, 2 (http://mundanebehavior.org/issues/v2n2/laurier.htm).

Liberman, K. (1985) *Understanding Interaction in Central Australia: An Ethnomethodological Study of Australian Aboriginal People*. London: Routledge and Kegan Paul.

Livingston, E. (1987) *Making Sense of Ethnomethodology*. London: Routledge and Kegan Paul.

Lynch, M. (1985) *Art and Artifact in Laboratory Science: A Study of Shop Work and Shop Talk in a Research Laboratory*. London: Routledge.

Malbon, B. (1999) *Clubbing: Dancing, Ecstasy and Vitality*. London: Routledge.

Parr, H. (2000) 'Interpreting the hidden social geographies of mental health: a selective discussion of inclusion and exclusion in semi-institutional places', *Health and Place*, 6: 225–238.

Rose, E., Gorman, A., Leuthold, F., Singer, I.J., Barnett, G., Bittner, E. and O'Leary, J.C. (1965) *The Unattached Society: An Account of the Life on Larimer Street among Homeless Men*. Denver, CO: Institute of Behavioral Science, University of Colorado.

Sacks, H. (1972) 'Notes on police assessment of moral character', in D. Sudnow (ed.) *Studies in Social Interaction*. Glencoe, IL: Free Press, pp. 280–293.

Sacks, H. (1992) *Lectures on Conversation. Vol. 1*. Oxford: Blackwell.

Strebel, I. (forthcoming) 'Formulating keys: order and disorder of access in Red Road, Glasgow', *Journal of Material Culture*.

Sudnow, D. (1978) *Ways of the Hand: The Organization of Improvised Conduct*. London: MIT Press.

Watson, D.R. (1993) 'Fear and loathing on West 42nd Street: a response to William Kornblum's account of Times Square', in J.R.E. Lee and D.R. Watson (eds) *Final Report: 'Plan Urbain' Interaction in Public Space (Appendix II)*. Manchester: Manchester University, Department of Sociology.

Wieder, D.L. (1974) *Language and Social Reality: The Case of Telling the Convict Code*. The Hague: Mouton.

Wittggenstein, L. (1980) *Culture and Value* (trans. P. Winch). Blackwell: Oxford.

Zaloom, C. (2006) *Out of the Pits: Traders and Technology from Chicago to London*. Chicago, Chicago University Press.

ACKNOWLEDGEMENT

The author is very grateful to Barry Brown for permission to use the video still in Box 9.3.

10 Geography and the Interpretation of Visual Imagery

Rob Bartram

Synopsis

'Visual imagery' is a broad term that embraces cinematic film, photography, promotional materials, computer games, etc. Geographers are interested in the way key concepts like space, place and landscape are used and created in visual imagery. The visual methodologies they adopt are based on philosophical and theoretical ideas about the interpretation of visual imagery that have been developed in other disciplines. Visual methodologies involve thinking not only about the choice and production of images and image aesthetics but also about audiences and the sensory affects of visual imagery.

The chapter is organized into the following sections:

- Introduction
- Visual imagery: some key philosophical and theoretical ideas
- Visual imagery and cultural meaning
- Interpreting visual imagery
- Additional thoughts – image affects

INTRODUCTION

Geography is a highly visual subject. Geographers place great emphasis on the importance of visual imagery to substantiate claims about landscape, place and process – in lectures, in research papers and in coursework. At a populist level, through journals such as the *National Geographic*, geography is as much about the quality of visual representation as textual or cartographic detail. Indeed, my early interest in geography stemmed precisely from an interest in the photography that accompanied such publications. Yet while visual imagery has formed a dominant way of expressing geographical knowledge, our interpretation of visual imagery has sometimes lacked a critical awareness: all too often, visual imagery has been used as a straightforward reflection of reality, with no sense of how, when and by whom the image has been produced. I think this can be partly explained by the complexity of philosophical and theoretical debates that are involved with the interpretation of visual imagery. I am sure this has acted as a deterrent to geographers who have wanted to know more. But I am also fairly certain that the lack of critical engagement with visual imagery can be attributed to an unfortunate misconception about geography – that only cultural geographers should be concerned with the interpretation of visual imagery. The main purpose of this chapter, then, is to put forward some rudimentary ideas about how geographers can go about interpreting visual imagery, irrespective of their interests in the discipline. I will begin this chapter by unravelling some of the key philosophical and theoretical ideas relating to the interpretation of visual imagery. I will then explore the relationship between visual imagery and cultural meaning and provide some practical advice about choosing and using appropriate visual materials. For the interpretative method itself, I will highlight three areas of concern: the production of the image, image aesthetics and interpreting audiences. Finally, I will offer some additional thoughts on the sensory affects of visual imagery.

It should be noted from the outset that the ideas expressed in this chapter are not put forward as a rigid framework for the interpretation of visual imagery. Rather, they provide some useful pointers as to how you might work with visual materials, and how you might allow ideas and issues to emerge through the interpretation process itself. Some of you might be concerned about some of the implicit ambiguities of this 'non-prescriptive' methodology. Don't be. One of the most stimulating things about interpreting visual imagery is that it is never predictable. You never quite know what you will end up seeing.

VISUAL IMAGERY: SOME KEY PHILOSOPHICAL AND THEORETICAL IDEAS

Throughout this chapter I use the phrase 'visual imagery'. It is a broad term of reference that embraces diverse visual forms, from cinematic film and photography to promotional materials, art and computer games. In other relevant literature, you will sometimes find reference to 'visual culture' or, simply, 'the visual'. Although I prefer to use the term 'visual imagery' to describe myriad visual forms, this does not suppose that all types of visual imagery can be lumped together and approached and interpreted in the same way. Indeed, you need to be sensitive to the codes, conventions and genres associated with each particular visual image.

Geographers have shown a keen interest in visual imagery in the last 20 years with some notable early work being carried out on cinematic geographies (Aitken and Zonn, 1994; Clarke, 1997), landscape art (Daniels, 1993) and landscape photography (Crang, 1997). In becoming attuned to the dominance of visual culture, critical frames of reference subsequently emerged in the discipline. In much of the early work, geographers were preoccupied with understanding how power relations were integral to the production and reception of visual imagery. For example, Gillian Rose (1995) argued that Gainsborough's portrait of *Mr and Mrs Andrews*, painted in 1748, came to symbolize quintessential English landscape, despite its partial, 'southern' perspective. This mode of interpretation was characterized by the forging of interdependent links between the aesthetics of the image (for example, symbolic content, colour, atmosphere, perspective) and wider social meaning. So, in the case of Gainsborough's portrait of *Mr and Mrs Andrews*, it was argued that the portrait is not just a likeness of the two landowners, or indeed their land, but a portrait that we can interpret as symbolic of English national identity.

Geography's interest in visual imagery can be related to the broader contours of twentieth-century philosophy that have challenged the dominance of visual culture and its implicit separation of viewer and viewed, subject and object, individual and world. To put this simply, there is a range of moral and intellectual dilemmas posed by how we view the world, and how this relates to individual and collective empowerment. What makes this debate all the more intriguing is that for some commentators such as Jean Baudrillard or Paul Virilio the contemporary world is becoming 'hypervisible' – modern technological developments in media and telecommunications have made much of the world present to us in real time (see, for example, Zurbrugg, 1997; Virilio, 2007). With such an array of visual images, how do we know what is real or authentic? More importantly, how are these visual images manipulated to exercise power, and how might they induce a form of detrimental 'social inertia' (see Bartram, 2004).

We don't need to get too bogged down in this debate, but it is important to bear in mind that interpreting visual imagery carries with it the weight of extensive philosophical debate. I will leave it up to you to decide how far you want to pursue your interest in the philosophical debate, although there are several references at the end of this chapter that should help if you want to take your interests further. I will take just two ideas from this preliminary discussion and expand on them in further detail: What is the relationship between visual imagery and cultural meaning? How do we go about interpreting the meaning of visual imagery?

VISUAL IMAGERY AND CULTURAL MEANING

It should already be apparent that there is a strong link between visual imagery and cultural meaning. To be more precise, visual imagery *always* produces cultural meaning, whether it involves passing a photograph around family and friends in your sitting room or watching a blockbuster film at the cinema. We can understand this relationship in terms of *sign* and *signification* – the visual form that has been 'encoded' with meaning, and its 'decoding', or interpretation. Cultural signs have systems of reference points – referents – that allow us to interpret their complex and interconnected meaning despite their often minimalist representation. Put simply, it allows certain signs to be symbolic of additional or associated ideas and images. Think of how corporate logos, such as those used by Shell or Nike, have become reduced to very basic signs. The Nike 'swoosh' says it all; it symbolizes all that we need to know as consumers when we purchase Nike goods. Modern art and even advertising work on very similar principles – we can reduce complex meaning to a single colour, texture or form.

Cultural signs are also seldom static. They connect, and are continually connected, to other signs, and we would be foolish to think that we can capture them, pin them down to a 'true', essential meaning. As you will come to appreciate if you interpret visual imagery, it is easier to suggest *how* cultural meaning is manipulated than discover its essential truth. Furthermore, we should not assume that cultural signs have an original condition or form that we should privilege with greater meaning: reproductions of well-known visual images can be powerful in their own right. For example, John Constable's *The Hay Wain*, painted in 1820–1821, has been reproduced on tea towels, calendars, biscuit tins and other consumer goods. The reproductions of *The Hay Wain* are just as important in terms of cultural meaning as the original painting.

Finally, more recent developments in the field of visual theory have turned debates about cultural meaning on their head – it has become more fashionable to think of visual imagery as not so much representing the world that we live in, as intervening in it and perhaps destabilizing it. For example, recent interpretations of Damien Hirst's conceptual art highlight how art can become a new 'capacity' for nature, rather than simply a representation of it (Bartram, 2005). For Hirst, placing dead animals in tanks of formaldehyde is not a matter of trying to represent the 'death' of nature, as encouraging us to think about how nature is endlessly perfected and duplicated.

INTERPRETING VISUAL IMAGERY

Students interpreting visual imagery frequently ask me to provide them with a specific method they can use. Although we can describe the interpretation of visual imagery as a form of 'textual analysis' (see Chapter 30), there is no singular or formal visual methodology that can be referred to. Besides, it is far more important to think about the pragmatics of interpreting visual imagery than to think of interpretation as a strictly applied science. So on this basis, here are some practical things to think about.

Choosing and using an appropriate visual image to interpret

If you are fortunate enough to be given the freedom to choose a visual image to work with, my advice is to select an image that you have either produced yourself (typically, from your own photography) or an image that particularly inspires or interests you. This will ensure that you will overcome a hurdle often identified by geography students – that without necessary training in media or arts-based courses, they lack the necessary skills and knowledge to interpret visual imagery. If you are already familiar with the visual image, this can sometimes generate sufficient confidence for a detailed interpretation to unfold. While students increasingly plum for cinema film and occasionally landscape art and photography to conduct their interpretation, you could also think about documentary film, modern art, computer games, Internet websites or 'non-commercial' photography and video as potential sources of interesting visual imagery.

You should then think about doing some background research. Basic information can be tracked down by using a search engine on the Internet. You will find that most contemporary films, artists and photographers have their own websites, although there are some generic websites such as the Internet Movie Database (IMDb). Look out for critical reviews, perhaps through newspaper websites or through e-journals. Of course, there is the university library – be prepared to explore the journals that do not form part of your regular geography reading list such as *Frieze, Modern Painters* and *Sight and Sound*. Finally, you will find exhibition and gallery catalogues a fruitful source of information.

You need to be careful with your use of visual imagery. For example, if you intend to use visual imagery for a seminar presentation, you need to spend time thinking about how your choice of visual image will relate to the oral part of the presentation. If you are interpreting art or photography, it is better to choose just a few, indicative images and explain them carefully and in depth, than try to present too many and provide only a superficial interpretation. Equally, if you intend to interpret a film, do not attempt to show clips of longer than, say, ten minutes. Quite often, students feel compelled to show an extended clip in an attempt to explain the whole film. A brief synopsis of the film at the beginning of your presentation will suffice. If you have time and editing facilities to hand, you can of course put together the most important scenes from the film on one video.

Interpreting the visual image

Once you have chosen an image, or have been presented with an image to work with, there are a few pragmatic issues to bear in mind before moving on to the more detailed aspects of interpretation. You need to be familiar with the visual image before you interpret it. This involves spending time watching and viewing, making a note of your initial impressions and taking time away from the image for further reflection. As much as anything, this sharpens the eye and makes you attentive to minute detail. It is, after all, the minute detail that provides the most interesting and intriguing ideas about visual imagery (see Exercise 10.1).

Exercise 10.1   Interpreting historical photographs

This activity is designed to hone your interpretative skills and involves the use of historical photographs that depict a local scene that is familiar to you. Historical photographs can be sourced from local libraries, the university library or from local publications. Take just one photograph of a street, park, building or agricultural land, and try to locate the scene on a detailed map. Once the scene has been located, try to replicate the same shot by taking your own photograph. When you have got the two photographs side by side, begin to compare and contrast them. Don't just concentrate on the obvious differences between the two images but look at the more subtle changes as well. For example, look at the differences in terms of architectural design, building materials, landscaping and boundaries. This exercise will accustom you to the depth of interpretation that can be achieved and the most appropriate method of recording this information. If the exercise is performed well, you should be able to express ideas about how and when the photograph was taken, and how this might influence the representation of the scene chosen.

We can now turn to the detailed part of the interpretation. This will involve three areas of concern:

1   Production of the image
2   Image aesthetics
3   Interpreting audiences

It might well be that you focus on only one of the areas of concern. If this is the case, you should still bear in mind that the three areas are interlinked.

Production of the image

Taking the production of the image as the first area of concern, we can begin to ask some interpretative questions:

- Who produced the image?
- What do we know about the producer of this image in terms of background education and training, and social identities (age, ethnicity, gender, sexuality)? Do these influences inspire the production of the image?
- Who commissioned the image, when was it made and when was it exhibited/ shown/broadcast?

Image aesthetics

Next we can consider image aesthetics:

- Describe the image (easier than it sounds!): try to convey a sense of how the image works in terms of colour, composition, atmosphere, angle of view, perspective, speed and tempo, sound and narrative.
- Try to identify the symbolic elements of the visual image – how does the image relate to other cultural images and ideas?
- How does the visual image relate to specific cultural genres? Is it indicative of a particular moment in cultural history? Does it represent a departure from the conventions of its genre?
- How does the exhibition/display/broadcast of the visual imagery affect its critical reception?

Interpreting audiences

While the first two areas of concern are relatively straightforward, the interpretation of audience engagement with visual imagery is far more complicated and can require extensive, painstaking research. Indeed, geographers have preferred to limit their interest to the production of the image and image aesthetics. A compromise can be reached. We can *infer* how, why and where audiences engage with visual imagery. We can also make inferences about the intended audiences by asking questions about the availability of the image. For a more detailed interpretation of audience engagement with visual imagery, we would have to use focus groups, in-depth interviews or questionnaires involving participating audiences (see Chapters 8 and 6, respectively) (Exercise 10.2).

Exercise 10.2   Interpreting David Lean's film adaptation of Oliver Twist (1948)

Let me explain how I initiate the interpretation of visual imagery by using cinematic film. I use an exercise that involves a basic interpretation of *Oliver Twist*, a well-known film. The exercise is designed to inspire confidence in the interpretative procedures already discussed in this chapter, and to provide a sense of the depth of interpretation that can be achieved. At the beginning of the exercise, I point out that it takes time to develop a sophisticated level of interpretation, to know what to look for and what to do when you hit the inevitable 'dead-end'. After explaining in detail the key areas of concern, I then show my students a series of short clips from David Lean's 1948 film adaptation of *Oliver Twist*. The film should be fairly familiar to the students because it is shown quite frequently on terrestrial TV and, if it is not familiar to them, they should at least be familiar with the plot and main characters from reading the novel. I make sure the students get to see a variety of landscapes, scenes and settings used in the film, from the opening sequence that shows the bleak, wide panoramas of the moorland that Oliver's mother has to struggle across to get to the workhouse, through to the closing scene that reveals the labyrinth of dark and dingy alleyways that make up the East End of London. I show the scenes sequentially so that I can provide a brief overview of the story at the same time. I then get the students to work in twos or threes to answer the following questions about the film's aesthetics:

- How does David Lean create different moods and atmospheres in this film? (The kinds of prompts I use relate to dialogue, soundtrack, camera angle, lighting, film continuity and visual symbols.)
- How does David Lean make use of landscape to create these moods and atmospheres? (The prompts I use to address this question relate to scale of the scene or setting and how this might be enhanced through cinematic effects by the use of camera lenses, lighting, etc.)

This film is particularly useful to inspire confidence in interpreting visual imagery, not least because symbolic landscapes can be clearly identified. It can also be used to have a wider discussion about the importance of relating the production of visual imagery to the appropriate moment in cultural history: some students are initially tempted to make the links between this film and the Victorian period in which it is set. However, they soon realize that the film is more appropriately discussed in the context of British postwar culture. The film clearly relates to a sense of postwar optimism and pride. For example, David Lean's dramatic use of St Paul's Cathedral as a backdrop can be related to the symbolic status of the cathedral during the war. St Paul's Cathedral became emblematic of the national war effort, and specifically the 'Blitz spirit' of London's East End communities.

This exercise requires little background preparation and even with a basic understanding of British postwar culture, some of the important links between the image aesthetics and the context of production can be recognized.

ADDITIONAL THOUGHTS – IMAGE AFFECTS

So far I have emphasized how we can interpret visual imagery by using the concepts of sign and signification. There are problems with using these concepts, not least because they rely on the assumption that the viewer is detached from the image and able to *extract* meaning from it at will. As Jean Baudrillard and other commentators have recognized, when we try to extract meaning from visual imagery, sometimes the visual image becomes just *too* good to be true because it yields meaning all too easily. This is what can happen when we turn to a visual image already knowing the meanings that we want to find. Interpreting visual imagery then becomes a ritual of 'discovering' what we already know.

There are alternative ways of interpreting visual imagery and we can turn to these if we reconceptualize vision itself. If we accept that viewing the world is also about being in the world and not detached from it, visualizing imagery becomes meaningful in its own right. Let me explain this carefully. To view the image is to experience a visual sensation. We can 'enter' into a film, a photograph, a painting and 'feel' movement, tempo, touch, smell, taste, noise and atmosphere. The image is a sensory affect, whether it causes us to laugh or cry, experience fear or engenders compassion. It doesn't matter that these images are invariably not real, they still stimulate the senses. How many times have you watched a hospital-based soap opera on TV and had to turn away at the sight of a gory operation? The operation is realistic but it is not real, and yet we experience real feelings when we watch it.

Visualizing the image is just one sensory affect. We can also acknowledge the importance of bodily senses in the making of the image itself. For example, artists such as Stanley Spencer wrote extensively about the significance of painting to his personal wellbeing. The act of painting, and the skills and techniques that it entailed, was Stanley Spencer's way of sensing and making sense of the world. It allowed him to be immersed in community life, took him to remote and idyllic spots in the countryside and allowed him to contemplate the rhythm and routines of his own life.

So we should not be dismissive of the sensory affects of visual imagery. On the contrary, we should recognize that they determine how we think, feel and act, and that we produce distinctive geographies accordingly. Anthropologists, psychologists and therapists have long understood the sensory affects of visual imagery. In geography too, John Wylie has begun the process of explaining the importance of this concept in relation to landscape (Wylie, 2007) (see Exercise 10.3).

Exercise 10.3   The moving image

This activity requires you to think about visual 'affects'. Choose a piece of art, photography or a film clip that you find emotionally powerful. For example, I find Edvard Munch's *The Scream* a particularly harrowing image – you can almost feel yourself being sucked into the vortex of pain in this image.

When you have found an appropriate image, write about the thoughts and feelings that it inspires in the form of expressive prose, preferably over different time-frames (this will get you to think about visual 'affects' being transformative and emergent over time). Write specifically about the impact of colour, texture, 'framing', scale and some of the more abstract feelings that you might have about movement, tempo, touch, smell, taste, noise and atmosphere. If you give enough time to this part of the exercise, you find that you have a rich, descriptive account.

In small discussion groups, try to share some of your ideas and perhaps reflect on why your chosen image engendered the particular feelings it did. As a concluding part of the discussion, you might want to discuss the problem of 'translating' your thoughts and feelings to the written word, and the more general problem of writing expressively.

The exercise lends itself to the interpretation of modern and mostly abstract art, although appropriate images could be taken from contemporary cinema or even the internet.

Summary

- Geographers need to engage with the interpretation of visual imagery in a critical fashion.
- The interpretation of visual imagery is intricately linked to philosophical and theoretical debates about the production and experience of culture.
- Visual imagery can be explored through an understanding of the relationship between the 'sign' and 'signification'.
- We can interpret visual imagery by asking detailed, pragmatic questions about the production of images, image aesthetics and audiences.
- Visual images affect. They inspire us to think, feel and act, and we produce distinctive geographies accordingly.

Further reading

There are a growing number of books and articles written on the interpretation of visual imagery. I have highlighted some of the most important contributions written by geographers below:

- Rose (2007) is by far the most accessible explanation of visual methodologies. The book takes the explanation of visual methodologies further by making necessary connections to theories of the visual.

- On photography, see Crang (1997) and Vasudevan (2007).

- On film, see Aitken and Zonn (1994), Fish (2007), Clarke (1997) and Clarke *et al.* (2009).

- On art, Daniels (1993) and Bartram (2005).

Note: Full details of the above can be found in the references list below.

References

Aitken, S.C. and Zonn, L.E. (1994) *Place, Power, Situation and Spectacle: A Geography of Film*. London: Rowman & Littlefield.

Bartram, R. (2004) 'Visuality, dromology and time compression: Paul Virilio's new ocularcentrism', *Time and Society*, 13(2/3): 285–300.

Bartram, R. (2005) 'Nature, art and indifference', *Cultural Geographies*, 12: 1–17.

Clarke, D. (ed.) (1997) *The Cinematic City*. London: Routledge.

Clarke, D., Crawford Pfannhauser, V. and Doel, M. (eds) (2009) *Moving Pictures/Stopping Places: Hotels and Motels on Film*. Lanham, MD: Lexington Books.

Crang, M. (1997) 'Picturing practices: research through the tourist gaze', *Progress in Human Geography*, 21: 359–373.

Daniels, S. (1993) *Fields of Vision: Landscape Imagery and National Identity in England and the United States*. Cambridge: Polity Press.

Fish, R. (ed.) (2007) *Cinematic Countrysides (Inside Popular Film)*. Manchester: Manchester University Press.

Rose, G. (1995) 'Place and identity: a sense of place', in D. Massey and P. Jess (eds) *A Place in the World? Places, Cultures and Globalization*, pp. 87–132. Oxford: Oxford University Press.

Rose, G. (2007) *Visual Methodologies: An Introduction to the Interpretation of Visual Materials* (2nd edn). London: Sage.

Vasudevan, A. (2007) '"The photographer of modern life": Jeff Wall's photographic materialism', *Cultural Geographies*, 14(4): 563–589.

Virilio, P. (2007) *Art as far as the Eye can See*. Oxford: Berg.

Wylie, J. (2007) *Landscape*. Oxford: Routledge.

Zurbrugg, N. (1997) *Jean Baudrillard: Art and Artefact*. London: Sage.

Myrna M. Breitbart

Synopsis

Participatory research seeks to democratize research design by studying an issue or phenomenon with the full engagement of those affected by it. It involves working collaboratively to develop a research agenda, collect data, engage in critical analysis, and design actions to improve people's lives or effect social change.

The chapter is structured into the following sections:

- What is participatory research and why do it?
- Participatory action research principles
- Formulation of the research questions
- Research design and data-collection methods
- Some areas of challenge and concern
- With all the constraints, why do it?

WHAT IS PARTICIPATORY RESEARCH AND WHY DO IT?

In 1990, Kenneth Reardon, a professor of urban planning, discussed the prospect of doing research in East St Louis with Ms Ceola Davis, a 30-year community outreach worker. She greeted him with the following: 'The last thing we need is another university person coming to East St Louis to tell us what any sixth grader here already knows' (Reardon, 2002: 17). At the second meeting, according to Reardon, Ms Davis indicated on what terms she and others *might* be willing to work in partnership with university faculty and students. In what came to be known as the 'Ceola Accords', she laid down the following terms:

- Local residents, not folks from the university, would decide what issues would be addressed through the partnership.
- Local residents would be actively involved in all stages of the planning process.
- The university would have to agree to a six-month probationary period and, if passed, a minimum five-year commitment.
- The resident group wanted help incorporating as a non-profit organization (Reardon, 2002: 18).

The suspicion with which Ms Davis viewed outsiders coming to do research in her community is not unusual. Communities are often treated as laboratories, provided no role in the research process and benefit little from the results of studies conducted within their borders. In the case of East St Louis, by the time Reardon arrived, there were already 60 reports sitting on the shelf that had not resulted in any improvements to the neighbourhood. The partnership that has since developed between the city and the University has nevertheless become one of the best examples of participatory research (see Box 11.1).

Box 11.1   The East St Louis Action Research Project (ESLARP)

The East St Louis Action Research Project began in 1987 when a State Representative in Illinois challenged the president of the University in Champaign-Urbana (UICU) to serve the research and education needs of this low-income neighbourhood that is more than 175 miles away from campus. What began as a series of research projects within the University's Architecture, Landscape Architecture, and Urban and Regional Planning programmes, eventually evolved into an on-going collaboration between many disciplinary departments at UICU and several community-based organizations in East St Louis (1997, 2002).

When the collaboration began, community leaders set the ground rules. All research was to be driven by resident-defined needs and the skills of analysis were to be transferred to community participants by academics. The project began with the collection of data that was used to produce a map of the untapped resources and problems in the neighbourhood. Based on early discussions of this preliminary information, decisions were made about what additional data to collect and how. Utilizing photographs, landownership records, more GIS-produced maps of physical conditions, and census data, the city of East St Louis was compared with its suburbs. Patterns of uneven development, documented in a very real and accessible format, became the basis for further research and actions to produce change. School performance was evaluated, street clean-ups were begun and direct actions were organized to move the city to address its problems. Additional research and planning was undertaken to create recreational space for children, affordable housing and re-direct proposed light rail lines through the neighbourhood. Community partners have also made significant demands on the university, to the point of requiring them to create a satellite Neighborhood College for adult learners. East St Louis residents choose the curriculum, which includes courses on developing affordable housing, political economy and the 'ABCs of Organizing' (Reardon, 2001). Still under way with more than 30 community partners, this participatory action research combines technical and capacity-building with attention to addressing real problems through a commitment to sustaining long-term projects and the constitution of interdisciplinary teams.

Participatory research or, as it is more commonly known, participatory action research (PAR) involves the study of a particular issue or phenomenon with the full engagement of those affected by it. Its most distinguishing features are a commitment to the democratization and demystification of research, and the utilization of results to improve the lives of community collaborators. Most definitions combine data collection, critical inquiry and action. Where PAR departs most from other forms of social research that also focus on effecting change is in the *means* used to achieve this end. In PAR, the *means* by which data is co-generated and interpretations debated are a key part of the change process.

PAR was developed in part as a response to exploitative research and as a component of a much larger radical social agenda. There is a vast collection of writing on its history, theory and practice (see, for example, Fals-Borda and Rahman, 1991; Park *et al.*, 1993; Greenwood and Levin, 1998; Reason and Bradbury, 2001; Bloomgarten *et al.*, 2006; Kindon *et al.*, 2007). Many participatory researchers situate their practice within a broader tradition of liberationist movements and early critiques of international development work that questioned the purposes, ethics and outcomes of social research conducted *on behalf* of other people. Fals-Borda and Paulo Freire, in particular, stress the role of participatory research in recovering knowledge 'from below' (Fals-Borda, 1982; Paulo Freire, 1970; Fals-Borda and Rahman, 1991). Feminists have also been active in promoting participatory research (see Maguire, 1987; Gluck and Patai, 1991; McDowell, 1992; Stacey, 1998). They believe that research 'has an obligation to create social spaces in which people can make meaningful contributions to their own well-being and not serve as objects of investigation' (Benmayer, 1991: 160). PAR has also sought to address the root causes of social injustice and was a practice embedded in the Civil Rights and Black Nationalist movements (Bell, 2001: 49).

William Bunge's 'Geographical Expeditions' in Detroit at the end of the 1960s and early 1970s provide early examples of the application of participatory data collection within the field of geography. Bunge sought to elicit the perspectives of people living in poverty through creative data collection and analysis, creating 'folk geographers' of neighbourhood residents (Bunge, 1977). Bunge's work embraced some key concepts of PAR:

- The importance of field research as an educational tool.
- The power of knowledge when used to design social actions to make a difference in people's lives.
- The acceptance of professionals as both teachers and students.

One notable product, the 'Geography of the Children of Detroit', creatively maps the landscape from the perspective of children (Field Notes, 1971). While children were not active participants in the research, adult residents were involved in all aspects of data collection.

In the years since Bunge's ground-breaking work, a number of geographers and environmental educators have employed participatory research techniques. Colin Ward pioneered participatory research with young people and the use of the built environment as a resource for critical learning through his work for the Town and Country Planning Association (Ward and Fyson, 1973; Ward, 1978; Breitbart, 1992). Roger Hart took up this work in the 1970s. He and several additional researchers have continued to develop highly effective models for engaging children from multiple cultural backgrounds in genuine PAR projects around the world (Hart, 1997; Chawla, 2002; Driskell, 2002). The field of participatory research in geography with adults has also accelerated in recent years (e.g. Cahill, 2007b; Kindon *et al.*, 2007).

PARTICIPATORY ACTION RESEARCH PRINCIPLES

PAR is not a specific methodology with exact procedures, nor is it about data collection alone. In a participatory data collection process, however, great value is placed on the knowledge of those conventionally researched (Park, 1993: 3). For this reason, participatory research often relies on less formal data collection methods and seeks to foster a community's capacity to problem solve and design actions without having to rely solely on outside experts.

The most basic and distinguishing principle of participatory research is *sustained dialogue* between external and community researchers. It is through *dialogue* that formal academic knowledge is meant to work 'in a dialectical tension' with popular insider knowledge to produce a more complete understanding of a situation or environment. This form of dialogue involves not only the exchange of information and ideas but also of feelings and values (Park, 1993: 12). Rachel Pain (2004) highlights the importance of uncovering people's relationships with space and place in the PAR process, while Caitlin Cahill points to the ways in which active participatory research on neighbourhood change that builds upon personal observation and feeling can open up many new avenues of inquiry and understanding. Cahill's work with a racially diverse group of young women on the Lower East Side of New York produced greater clarity about the causes and complicated personal effects of gentrification as part of a global process of uneven development. It also challenged social exclusion by initiating a series of environmental actions and media projects designed to challenge stereotypes and representations of the young women held by many of the people moving into the neighbourhood (2007b).

PAR is also about *sharing power* and involves a commitment to see that the outcomes benefit the community in measurable ways by building on assets. In practice, this means that community members should be hired and trained, whenever possible, as research partners. They should have the power to define research objectives and offer opinions about data collection methods, utilizing their lived experience as a beginning basis for the investigation. The application of these ideas is described in rich detail by Cahill in an article that examines how new knowledge was produced by the young women she worked with on the Lower East Side of New York. Included here is a discussion of how to build the collective research capacity of the participants and enable them to learn particular skills by beginning with their lived experience and continuing with a process of 'learning by doing'. Cahill also describes how she and the participants worked to create a 'community of researchers' with a real collective investment in the outcomes of the project (Cahill, 2007a).

Participatory research is often initiated *outside* the university, with academics assuming the role of consultant or collaborator (supplying information and technical skills). Personal skills (e.g. the ability to facilitate conversation without dictating choice, and an awareness of one's own biases and their impact on research) are more important than technical skills. The research is clearly *not* participatory if you *do everything* – frame questions, conduct the research, and write up and disseminate the results. This is true even if you spend a lot of time with community members asking their opinions or having them review your findings.

It is important to remember too that participatory inquiry can be used as a tool to build knowledge of community assets and strengths as well as to identify or address problems. The Holyoke Community Arts Inventory that I am currently involved in is part of a broader study of new urban revitalization strategies that use the arts to promote economic and community development. As a participatory research project, we are collaborating with Holyoke artists in an exploration of their own cultural assets. This involves a complicated search for the diverse meanings that residents give to the creative pursuits in their lives. The data collected, as well as the process of generating and making it public, are both meant to contribute to the social change and planning process.

In general, the literature on participatory research provides considerably more detail about the ideology and politics of the approach than about the research process (Reason, 1994: 329). Yet it is the application of the above principles within stages of inquiry that may otherwise be quite similar to traditional social research that best distinguish its practice. A great challenge in writing this chapter has been the recognition that every participatory research project differs markedly in the methods that it employs.

FORMULATION OF THE RESEARCH QUESTIONS

In participatory research, the process of formulating a research topic originates with the people who are affected most by it. In this early stage, community members may form partnerships with faculty and students. It is not uncommon for community-based organizations to want to interview prospective collaborators. These initial meetings prior to data collection can be an important time when knowledge is shared and trust is built.

Data collection in PAR begins with the lived experiences of the collaborators and the acquisition by outside researchers of basic knowledge about the neighbourhood or geographic entity under study. It is also common for co-researchers to try to create a shared base of knowledge. Many techniques can be employed to do this. For the first year of a three-year participatory research project on the labour-managed work cooperatives of Mondragón, Spain, Davydd Greenwood and his team read and discussed existing literature on cooperatives together. They also shared their own individual expertise on particular subjects related to their research (Greenwood, 1991: 95).

Deciding what data to collect and how to collect it depends on whether or not a specific question or topic has been targeted. If the research question is undecided, certain methods may provide an informational basis upon which to determine this question. Some common tools include brainstorming, focus groups, neighbourhood walks and photography story boards. The Holyoke Community Arts Inventory began with a focus group attended by city residents who discussed how their involvement in arts-related activities currently impacts their lives. They also shared future visions of the city in the event that 'arts' and 'culture' were to play a larger role. Ideas generated during the discussion helped to define a research project, which focused on producing a community arts map and learning more about indigenous talent so that this information could be taken into account in current urban revitalization plans.

In working with adults or young people, brainstorming workshops, social mapping and model construction provide other effective ways for drawing out participants' existing knowledge. Whenever the young people who work with YouthPower begin a new project (Box 11.2), their initial brainstorming sessions are designed to generate opinions about which aspects of their environment they feel are most in need of improvement. One project that resulted in the Youth Vision Map for the Future of Holyoke began with a youth-led walk around the city. Participants recorded on film those aspects of the environment that they felt had potential and those in need of immediate attention. These images were then placed on a base map for later assessment and analysis. Similar data-gathering techniques and resource inventories were employed in the East St Louis Action Research Project (ESLARP), where residents used disposable cameras to identify important issues and sites in the neighbourhood with 'untapped' resources. In both projects, a preliminary gathering of information and opinions was used to help participants prioritize the issues they would like to study further.

The *YouthPower Guide* (2000), a PAR project involving Holyoke youth and university faculty and students, provides a step-by-step guide for engaging youth in community planning. With respect to data collection, it includes over 20 activities that can be completed in a one-to-two-hour session and that are all youth led. Some of the basic rules it sets out for brainstorming apply to work with adults as well. They include:

- Begin with ice breakers (strategies to get people talking to each other).
- Make sure the goal is to generate ideas not make decisions.
- Encourage everyone to participate.
- Write down ideas without debating them.
- Do not stop the process because of occasional silences, as these are often the precursor of more good ideas.

To this list, I would add that any events run with young people or children also need to include refreshments and lots of breaks!

**Box 11.2   Youth-driven Community Research
and Planning in Holyoke, Massachusetts**

This participatory action research project is an on-going effort on the part of youth and collaborating adults in the city of Holyoke, Massachusetts, to acquire more information about the city's strengths and weaknesses and plan for its future. As one of the oldest planned industrial cities in the US, Holyoke now has many assets but also suffers the effects of a global economy and decades of de-industrialization. It is also a city divided socially between an older white immigrant and growing Puerto Rican population.

While many organizations are involved in on-going planning, one after-school programme, El Arco Iris [the Rainbow], has been especially important in the promotion of a participative process. El Arco Iris offers young people from 10 to 19 an opportunity to take art classes and develop skills as youth leaders. TMAD/YMAD (Teens Making a Difference and Youth Making a Difference), along with YouthPower, are adjunct programmes begun by co-directors Jose Colon and Imre Kepes to initiate community-based research and design urban improvements. Since they began, young people have conducted extensive research on the city of Holyoke, identified a range of problems and opportunities, and brainstormed possible solutions. My work with this programme began over ten years ago, when I initiated a banner design project that combined neighbourhood exploration and critical assessment with the design of visual and public representations of young people's visions for the future of the city (Breitbart, 1995; Breitbart, 1998). I have continued to work with Hampshire College students and the staff of our community-based learning programme to support subsequent youth-initiated planning efforts.

What is most inspiring about this work is the extent to which young people have provided a role model for adult decision makers in the city around participatory research and alternative planning practices. They have provided the spark for revitalization initiatives that might otherwise have been overlooked. Since 1995, YouthPower has contributed to the development of a Master Plan for Holyoke and directly enhanced the quality of life through the reclamation of vacant lots for parks and community gardens, as well as the production of murals and banners. YouthPower has also become the most active citizen group in the city to help plan a large-scale urban initiative to build a walkway and arts corridor along the historic canals, as one way to tie downtown to its neighbourhoods. It has also been a driving force behind a variety of collaborative research and planning projects with other youth organizations. This has led to the creation of periodic Youth Summits and a Youth Commission to advise the city. YouthPower has collaborated with faculty and students from the University of Massachusetts in the production of an engaging book for other youth, *The YouthPower Guide: How to Make Your Community Better* (Urban Places Project, 2000). The book provides many techniques for doing participatory, community-based research to improve the environment and quality of life.

A new youth-initiated Community Visioning process is currently under way to revisit an earlier Youth Vision Map, prioritize action projects and develop more effective ways of incorporating adult participation to promote the chances of implementation (Breitbart and Kepes, 2007).

Some of the 'dos' and 'don'ts' of brainstorming are illustrated in a meeting convened to design a walkway along Holyoke's historic canals that involved architects, adult residents and youth. Everyone was given colour-coded stickers used to divide themselves into brainstorming teams. Participants were also given red and black stickers to mark their most favourite and next favourite design ideas. Following introductions, the session began with one of the architects showing slides and talking about what factors generate a 'sense of place' in other cities that have redesigned waterfront areas, such as San Antonio, Texas. Youth peer leaders then led a slide presentation of the prior data they had collected about key problem areas in Holyoke. Groups divided up for further discussion. When called back together to share ideas, interesting patterns emerged. Adults, it appears, had come up with many ideas that bore a striking similarity to those shared in the slide presentation by the architect – coffee carts with umbrellas, a microbrewery, wild metal sculptures floating in the canals. Obviously, the presentation of earlier images, though well intentioned, served to limit rather than expand adults' imaginations, thus defeating the purpose of brainstorming.

RESEARCH DESIGN AND DATA-COLLECTION METHODS

Once some preliminary sharing of knowledge takes place, participatory research requires decisions about what further information to collect and how to collect it. This can lead to discussions about how to train individuals who may be unfamiliar with the chosen research techniques. The choice of research methods in all stages of data collection is driven, however, by more than the likelihood that they will produce desired information. Data collection methodologies are also chosen to ignite a process of personal and social change, liberating old knowledge, and nourishing the critical and creative capacities of participants.

Participatory research requires consideration of how the methods chosen further these goals, demystify the research process, encourage maximum involvement and contribute to the increased equalization of power among participants. Full participation in these decisions requires discussion about everything from the wording of questions to their sequencing. It also necessitates an acceptance of democratic process and full understanding of the decision-making structure and rules of discourse, no matter how young or old the participants (Hart, 1997). Some important things to consider in these deliberations are the range of skills that participants have to offer and the diversity of the group, including the mix of age, gender and cultural or ethnic background. Issues of safety and access to equipment or transportation may also come into play. In almost all cases, more than one investigative approach is chosen so as to draw upon the widest range of skills.

Both more quantitative (see Chapter 6) and qualitative data collection methods (see Chapters 8 and 9) are used in participatory research. Common qualitative data collection methods that go beyond brainstorming or photographic documentation, and are used with children and adults, include interviewing, oral history, drawing, social mapping and local surveys or environmental inventories. Participatory research with adults may also include the use of story telling and focus groups. Kimbombo, a theatre group in Holyoke, Massachusetts created within a community organization that provides GED (General Education Development high-school equivalency) and English as a second language training to Spanish-speaking residents, currently involves students in active participatory research. This culminates in the writing and performance of plays on issues that the community identifies as important (e.g. domestic violence, diabetes awareness, and the acceptance of residents with HIV/AIDS). The resulting scripts and performances provide an opportunity for the community to control its own representation while simultaneously addressing critical health issues.

Since many of these data collection techniques are described elsewhere in this book, I will focus here on their specific use in participatory research. Let's begin with *interviews* (see Chapter 8). In the Roofless Women's Action Research Mobilization, a collaborative research project involving the University of Massachusetts-Boston, homeless women, the city and other community-based women's organizations, interviews and focus groups were used to collect data on the causes of women's homelessness (Williams, 1997: 14–17). Based on the experiences of those in the research team who had been homeless, it was decided that they would do individual anonymous interviews with currently homeless women because their prior experiences would elicit trust. Other members of the research team collaborated in the conduct of focus groups with formerly homeless women. Group discussions allowed the women to reflect on the experience of *having been* homeless. Some thought here was obviously given to a purposeful division of labour. It is also common in PAR for academics and community researchers, adults and youth, to do paired interviews. These not only draw on the diverse experiences of those working together but allow for differing interpretations of the results later on.

Oral histories and *group story telling* are common data collection vehicles in PAR that enable partners to share perspectives and generate empirical data. In discussing her collaboration as an academic on a participatory research project with adult women in the El Barrio Popular Education Program in New York City, Benmayer describes how she and the other women on the research team discussed and wrote about what they were learning collectively (1991: 159–160). The organic process of introducing questions to the group, examining them and then reformulating the questions and beginning again, contrasts with the more typical linear progression of much social research. It reflects the fact that PAR can be a 'continuous educational process' that does not necessarily begin and end with one project (Park, 1993: 15). On-going evaluation and critique can generate new questions, issues and strategies that build upon a deepening understanding of an issue or topic.

There are many examples in PAR of the use of *qualitative surveys*. In the Immigration Law Enforcement Monitoring Program, researchers with the American Friends Service Committee and a local university collaborated with community-based organizations to collect data on the geography of human rights violations along the Mexico–US border. Mexican immigrants had found that their anecdotal information was insufficient to prompt a response and that a graduate student-designed computer database was too difficult to analyse. The solution was to design their own documentation form based on their experiences – one that allowed the people surveyed to categorize their abuse experiences by type, location, immigrant status, and so on. The group then hired a computer programmer to put this form online at four sites along the border accessed by immigrant workers and their families. The result was a highly detailed human-rights survey that could be used to analyse and address the abuses (Williams, 1997: 8–13).

There are many ways to generate spatial representations, from crude drawings to annotated photographic essays, and various forms of mapping. *Social mapping* is a particularly useful low-technology data-collection tool that has been widely employed by geographers engaged in participatory research as a repository for the raw data resulting from *environmental inventories*. Social maps encourage critical reflection and policy analysis, and help to set research agendas. They can also suggest plans for action. All social mapping begins with the acquisition of a good base map of the area you wish to survey. Planning departments are usually accommodating in sharing large-format maps. In Holyoke, youth and adults, divided into small groups, enhance the information on base maps by designing their own icons or using self-stick dots of varying colours to designate aspects of the environment that they consider important (for example, red dots for areas you like; green dots for locations you hang out in and enjoy; blue dots for places you avoid, etc.).

Analysis of the geographical information on social maps compiled by multiple research partners often involves discussion of the relationship between the 'real' and perceived characteristics of an environment. In the early stages of participatory research, these contrasting spatial representations can be useful in identifying the problems or assets of an area from different perspectives based on participants' gender, age, social class or ethnicity. This information can provoke debate and help decide on foci for further research. In doing participatory research with residents of a neighbourhood in the Bronx, New York, Roger Hart used templates representing such things as dangerous places, places used by teenagers alone, and so on to map the differing perceptions of the neighbourhood by diverse groups of residents. The resulting composite map used tapes of the conversations and analysis of spatial patterns to make recommendations for new types of recreational spaces (Hart *et al.*, 1991).

It is helpful for researchers to walk or drive around the study area with cameras prior to social mapping. Annotated photographs can then be used in place of dots or icons to represent and categorize environmental data. In a collaboration between Hampshire College interns and youth from Nuestras Raices, a community gardening and development organization, co-researchers combed the city with papers, pads and cameras to collect data on sites they considered ecologically sound and those they felt were potentially dangerous. Expanding on a set of symbols already devised for the international Green Map project (www.greenmap.com), they produced a map of their findings that included Puerto Rican restaurants and locally owned businesses in their definition of 'green sites'. Health information compiled through neighbourhood interviews was also mapped, suggesting geographic patterns of illness such as asthma that necessitated further examination. The 'Holyoke Youth Green Map' was also used to begin to acquire vacant lots for future community gardens.

The data collection process in PAR has its own educative value, in part because it necessitates the pooling of information and perspectives among diverse research partners. The commonalities and differences in perspective among researchers that visual displays such as social maps present can generate fruitful discussions that imply avenues for further research or action. However, the meaning of information collected through social mapping is not always readily apparent, nor is it always possible to decide upon a focus for research based solely on a visual or graphic representation. Additional discussion among participants is necessary to uncover varied interpretations and help define priorities.

The Youth Vision Map for the Future of Holyoke resulted from an amalgamation of ideas drawn from many brainstorming sessions. It contains bright visual symbols, placed at various sites, of what modifications young people would like to introduce into the urban landscape. Examples include murals, a Store for Safety (to discreetly provide youth with safe-sex products), a Teen Café (to hang out in, access cheap food and listen to music) and a Youth Van (staffed by adults and youth, to bring youth information about health issues and youth programmes). In this case, because the data suggested several possibilities for action, a summit for all of the youth organizations involved in the data collection provided an opportunity for each to discuss what *its* priorities were. The results were shared and each group went home with its own project to work on.

GIS (Geographic Information Systems) is another form of mapping that can be used by geographers as a tool for data analysis, especially with adults (see Chapter 25). However, the time involved in training and the access required to expensive technology, may limit its use. One possibility is for a PAR team with access to a geography department, to ask for technical assistance with GIS to analyse their field-generated data. The ESLARP provides many GIS-generated maps for the use of community organizations on its website. It also furnishes limited instruction on how to employ GIS to analyse environmental, economic and social data. Sarah Elwood has also developed many effective ways to generate a critical cartography and produce a powerful tool for community empowerment by combining GIS and PAR (Elwood, 2006).

In general, while PAR draws heavily upon a variety of social-research methods, it minimizes approaches that are beyond the material or technical resources of the individuals involved. With quantitative data, this may mean the use of descriptive statistics that help to tell a story, allow comparisons to be made, or identify causal relationships without requiring a significant level of quantitative expertise. In the ESLARP, census and other statistical data are put online by students and faculty, along with the software that allows community residents to pose questions, plug in the data and produce a result in a number of different formats, including charts, diagrams and maps. Another strategy is to divide the labour so that, while some researchers are collecting information, others are taking that data and putting it in a graphic or statistical form that allows it to be analysed.

Students who want to undertake participatory research should have some working knowledge of basic social-research techniques, as described in this book. A community-based learning course, where the reading and analysis of geographic theory and case studies are combined with work on a hands-on field project, is one effective way to gather experience using such techniques. Impressive examples of courses linked to the East St Louis Action Research Project can be found at www.eslarp.uiuc.edu. If data-collection methods are also unfamiliar to community researchers, they too should be trained. This transfer of skills of geographic analysis was the core of Bunge's work in Detroit, and is currently under way in Holyoke, as young people are taught the techniques of social mapping, and in the ESLARP, where adult residents interested in acquiring social-research skills are offered free courses.

While PAR does not require everyone to be equally involved in all phases of research, there is a strong commitment to encouraging active participation. Utilizing a variety of research methods and a division of labour that consciously seek to make use of the particular strengths of each collaborator is one way to assure widespread participation in the collection of information and its exchange.

SOME AREAS OF CHALLENGE AND CONCERN

Participatory data collection presents a number of challenges. One is coordination of the different skills and levels of participation that each partner brings to the process. These differences may be due to 'cultures of silence' inculcated in those whose ideas have never been seriously valued or who lack experience expressing their views. Academics used to a culture of expression may also tend to silence others by jumping in to fill the gaps. Good facilitation skills are therefore essential.

One of the strengths of the on-going youth projects in Holyoke is the commitment of its facilitators to providing young people with the tools of oral presentation (e.g. making outlines of main notes, using visual aids such as slides), and then allowing the young people to 'run the show'. This often involves the exercise of some serious restraint on the part of adults if a young person occasionally misses a point, speaks too quickly or otherwise flounders.

The fact that participatory research seeks everyone's full involvement can present further challenges, given differing levels of personal commitment. Community partners are often over-extended or lack money and resources. Students and faculty also face competing demands on their time, and work within a time frame that differs markedly from the 'real world'. The fact that most colleges and universities operate on what I would call the 'Brigadoon' principle, disappearing at crucial times of the year, only to reappear and try to pick up where they left off after seasonal breaks, can make PAR projects very difficult to sustain. Since PAR is premised on a commitment to greater social justice and social change, it is especially important to consider how the projects we become involved in move from research alone to social action. Academic calendars may preclude students from involvement past one semester, but faculty have the capacity, and many would say the obligation, to sustain projects over the long term. This may require a renegotiation of teaching commitments and it can also mean networking within and outside one's discipline to mentor new students and faculty into the process to insure its continuation. Participatory research is time-consuming and predicated on trusting relationships and a commitment to the project's duration. While it may be possible to incorporate this into a semester-long class, it is more commonly used for long-term projects.

Since a project will be evaluated by all participants, it must also simultaneously try to meet divergent goals. These can include the building of community and the development of a critical understanding of the assets and problems of a place. The writing you may be asked to do as part of a participatory research project may be quite different from that which would be done for a course paper or a professional geography journal. Tensions can develop between 'practically oriented' community partners and 'theoretically oriented' students or faculty (Perkins and Wandersman, 1997).

Just as Ceola Davis made demands on Ken Reardon and his students, community collaborators can also set down the criteria for investigation that academic partners must agree to. These demands can constrain the parameters of a project or redirect its focus. Additionally, since one project is not likely to address all the needs that stakeholders articulate, participatory research almost always begets more projects, soliciting the continued involvement of outside research partners. This issue of sustainability of the partnership presents its own challenges.

Among many practical obstacles to engaging in PAR are the unpredictable aspects of all projects that attempt to address community-defined priorities. Priorities frequently change mid-stream due to political, personal, or socio-economic circumstances that necessitate a shift in focus. When my colleagues and I collaborated with our community partners to write a Housing and Urban Development Community Outreach Partnership Center grant, many project ideas emerged from discussions. Once the grant was funded a year later to develop projects to address these priorities, massive cuts in social services required us to re-order original priorities. Given such situations, it is incumbent upon us to prepare ourselves and our students for unexpected events and situations when key participants may not agree among themselves or fulfil obligations (Pain, 2009). The social relationships that faculty build over time with individual community collaborators and a deepening of trust to support frank communication and problem-solving, are key to adjusting success-fully to inevitable unanticipated events. They also contribute in valuable ways to keeping community/college partnerships alive.

Restrictions placed on 'human subjects' research connected to Institutional Review Board (IRB) policies set out by institutions of higher education in the United States, and university and research council ethical guidelines and frameworks in the UK present further challenges to students who seek to engage in PAR (Pain, 2004; Cahill *et al.*, 2007; Dyer and Demeritt, 2009) (see also Chapter 3). These policies often require an articulation of research questions, methods and goals before these have been negotiated with community partners. Academic critics of IRB policies and the growing bureaucracy of ethical procedures more generally point out that the principles and practices of PAR, informed by a long history of grass-roots organiz-ing and critical race and feminist theory, seek to address unequal power relationships within society and between the university and the community. Human-subjects regu-lations can reinforce those unequal relationships and inhibit the exercise of agency on the part of community participants. An important on-going agenda results from these tensions and contradictions, directing greater attention to the need to generate more effective ethical practices and a more active challenging and reformulation of IRB policies on college campuses (Cahill *et al.*, 2007: 309). Balanced against these chal-lenges is a tendency to hold colleges and universities more accountable as citizens of the cities, towns and region in which they reside (Bloomgarten *et al.*, 2006).

The problematization of these and other ethical dimensions of PAR, as well as increasingly frank discussions of projects that have either failed or met serious chal-lenges, emphasize the importance and the inherent complexity of doing this work. They represent an important commitment to improving transformative practices while also countering tendencies to over-romanticize participatory practice as the only type of research that can promote social justice.

WITH ALL THE CONSTRAINTS, WHY DO IT?

For those of us who do participatory research, it is a humbling experience. After years of working with local neighbourhoods on collaborative projects, Randy Stoeker (1997) still confronts what he refers to as the 'haunting question' of how he can conduct this type of research so that it is both 'empowering and libera-tory'. He nevertheless urges students and faculty not to be so concerned about 'doing the right thing' that they become 'paralysed'. Community collaborators of the likes of Ceola Davis or the youth at El Arco Iris will always tell you when you have erred, and will always give you the benefit of the doubt if you are honest, respectful and follow through. Furthermore, there are no 'pure' forms of PAR; only degrees of participative practice.

The primary reasons why I and other proponents continue to advocate participatory research as a method rest with our commitment to its political aims. We also believe that the data it produces are more likely to be useful, accurate and lead to actions that address people's real needs and desires. When adults and young people in Holyoke came together to consider design elements to incorporate into a proposed walkway along the industrial canals, adults in the room came up with many fanciful ideas for metal sculptures, umbrella food carts, and so forth. The young people, in contrast, voiced a range of practical concerns that had emerged from their research prior to the brainstorming session. This included the need for bike racks so they could access the area from neighbourhoods far away, garbage cans to collect the litter, a police presence so their parents would allow them to come and gardens to enhance the beauty of the space. While the architects may have categorized these ideas as 'details to be thought about later', to the young people, they represented key design features crucial to a successful plan.

Nadinne Cruz, consultant and former Director of the Haas Canter for Public Service at Stanford University, posed a thought-provoking question at a recent service learning conference: what if our institutions' exercise of social responsibility is to assess how adequate the knowledge of our faculty and students is for addressing critical issues in the world? In spite of the complexities, experience with participatory research suggests that we are more likely to address this challenge if we collaborate in methodology design, data collection, analysis and action with those individuals and groups most likely to gain from the investigative process.

Summary

- In PAR the collaborative means by which data is co-generated, interpreted and used to design actions play a key role in social transformation.
- In PAR university researchers and their collaborators share knowledge, power and a decision-making role.
- Participatory research begins with the lived experiences of the collaborators and the acquisition by outside researchers of basic knowledge about the geographic entity under study.
- In addition to producing information, PAR data-collection methods should draw on the experiences, and nourish the critical and creative capacities of all participants.
- Utilizing a variety of data collection methods and a division of labour that consciously builds upon the strengths of each member of the team is one way to assure widespread participation.

Further reading

The following books provide a range of examples of different forms of PAR:

- Hart (1997) provides an introduction to children's participation in community-based environmental planning and identifies certain principles and practices that have worked in diverse cultural settings around the world. It includes detailed case studies and information about concrete methods of participative practice such as mapping, interviews and surveys. The book is also visually rich with photographs and diagrams.

- Kindon *et al.* (2007) is an inventive and comprehensive edited collection on participatory action research that includes a mixture of honest and critical reflection along with case studies from around the world. Chapters are relatively short and they are written in a highly accessible manner. There is also a clear emphasis on space and place, and a range of practical advice on the application of PAR in geography.

- Park *et al.* (1993) begins with a foreword by Paulo Freire, one of the world's experts on participative research as a vehicle for personal and political transformation. Cases referenced here are drawn from the North American experience and address the relationships between power and knowledge, research methods and social action. An appendix identifies key organizations that promote participatory research.

- Reason and Bradbury's (2001) comprehensive collection of articles on action approaches to social science is directed at an academic audience. It is divided into four sections that address theories and methods of participatory research as well as the application of these approaches and the skills necessary for implementation. It also explores the role of universities in action research.

Note: Full details of the above can be found in the references list below.

References

Bell, E. (2001) 'Infusing race into the US discourse on action research', in P. Reason and H. Bradbury (eds) *Handbook of Action Research: Participative Inquiry and Practice.* London: Sage, pp. 48–58.

Benmayer, R. (1991) 'Testimony, action research, and empowerment: Puerto Rican women and popular education', in S. Gluck and D. Patai (eds) *Women's Words: The Feminist Practice of Oral History.* New York: Routledge, pp. 159–174.

Breitbart, M. (1992) '"Calling up the community": exploring the subversive terrain of urban environmental education', in J. Miller and P. Glazer (eds) *Words that Ring Like Trumpets.* Amherst, MA: Hampshire College, pp. 78–94.

Breitbart, M. (1995) 'Banners for the street: reclaiming space and designing change with urban youth', *Journal of Education and Planning Research,* 15: 101–114.

Breitbart, M. (1998) '"Dana's mystical tunnel": young people's designs for survival and change in the city', in T. Skelton and G. Valentine (eds) *Cool Places: Geographies of Youth Cultures.* London: Routledge, pp. 305–327.

Breitbart, M. and Kepes, I. (2007) 'The YouthPower story: how adults can better support young people's sustained participation in community-based planning', *Children, Youth and Environments,* 17: 226–253.

Bloomgarten, A., Bombardier, M., Breitbart, M., Nagel, K. and Smith, P. (2006) 'The Holyoke planning network: building a sustainable college/community partnership in a metropolitan setting', in R. Forrant and L. Silka (eds) *Inside and Out: Universities and Education for Sustainable Development.* Amityville, NY: Baywood, pp. 105–118.

Bunge, W. (1977) 'The first years of the Detroit Geographical Expedition: personal report', in R. Peet (ed.) *Radical Geography.* London: Methuen, pp. 31–39.

Cahill, C., Sultana, F. and Pain, R. (2007) 'Participatory ethics: politics, practices, institutions', *ACME: An International E-journal for Critical Geographies,* 6: 304–318.

Cahill, C. (2007a) 'The personal is political: developing new subjectivities in a participatory action research process', *Gender, Place, and Culture,* 14: 267–292.

Cahill, C. (2007b) 'Negotiating grit and glamour: young women of color and the gentrification of the Lower East Side', *City and Society,* 19: 202–231.

Chawla, L. (ed.) (2002) *Growing Up in an Urbanising World.* London: Earthscan.

Driskell, D. (2002) *Creating Better Cities with Children and Youth: A Manual for Participation.* London: Earthscan.

Dyer, S. and Demeritt, D. (2009) 'Un-ethical review? Why it is wrong to apply the medical model of research governance to human geography', *Progress in Human Geography,* 33(1): 46–64.

Elwood, S. (2006) 'Critical issues in participatory GIS: deconstruction, reconstructions and new research directions', *Transactions in GIS,* 10: 693–708.

Fals-Borda, O. (1982) 'Participation research and rural social change', *Journal of Rural Cooperation*, 10: 25–40.

Fals-Borda, O. and Rahman, M. (1991) *Action and Knowledge: Breaking the Monopoly with Participatory Research.* New York: Apex.

Field Notes (1971) *The Geography of the Children of Detroit.* Discussion Paper # 3. East Lansing, Michigan: Detroit Geographical Expedition & Institute.

Freire, P. (1970) *Pedagogy of the Oppressed.* New York: Seabury.

Gluck, S. and Patai, D. (eds) (1991) *Women's Words: The Feminist Practice of Oral History.* New York: Routledge.

Greeenwood, D. (1991) 'Collective reflective practice through participatory action research: a case study from the Fagor Cooperatives of Mondragón', in D. Schön (ed.) *The Reflective Turn: Case Studies in and on Educational Practice.* New York: Teachers' College Press, pp. 84–107.

Greenwood, D. and Levin, M. (1998) *Introduction to Action Research: Social Research for Social Change.* Thousand Oaks, CA: Sage.

Hart, R. (1997) *Children's Participation.* London: Earthscan Publications.

Hart, R., Iltus, S. and Mora, R. (1991) 'Safe Play for West Farms'. Play and Recreation Proposals for the West Farms Area of the Bronx Based Upon the Residents Perceptions and Preferences. Children's Environments Research Group, City University of New York.

Kindon, S., Pain, R. and Kesby, M. (eds) (2007) *Connecting People, Participation and Place: Participatory Action Research Approaches and Methods.* London: Routledge.

McDowell, L. (1992) 'Doing gender: feminism, feminists and research methods in human geography', *Transactions: Institute of British Geographers*, 17: 399–416.

Maquire, P. (1998) *Doing Participatory Research: a Feminist Approach.* Amherst, MA: Center for International Education, University of Massachusetts.

Pain, R. (2004) 'Social geography: participatory research', *Progress in Human Geography*, 28: 652–663.

Pain, R. (2007) 'Guest editorial: participatory geographies', *Environment and Planning A*, 39: 2807–2812.

Pain, R. (2009) 'Working across distant spaces: connecting participatory action research and teaching', *Journal of Geography in Higher Education*, 33: 81–87.

Park, P. (1993) 'What is participatory research? A theoretical and methodological perspective', in P. Park and M. Brydon-Miller, B. Hall and T. Jackson (eds) *Voices for Change: Participatory Research in the U.S. and Canada.* Westport, CT: Greenwood Press, pp. 1–20.

Park, P. and Brydon-Miller, M., Hall, B. and Jackson, T. (eds) (1993) *Voices for Change: Participatory Research in the U.S. and Canada.* Westport, CT: Greenwood Press.

Perkins, D. and Wandersman, A. (1997) 'You'll have to work to overcome our suspicions', in D. Murphy, M. Scammel and R. Sclove (eds) *Doing Community-based Research: A Reader.* Amherst, MA: LOKA Institute, pp. 93–102.

Participatory Geographies Working Group (2006) http://www.pygywg.org.

Reardon, K. (1997) 'Institutionalizing community service learning at a major research university: the case of the East St Louis Action Research Project', *Michigan Journal of Community Service Learning*, pp. 130–136.

Reardon, K. (2002) 'Making waves along the Mississippi: the East St. Louis Action', *New Village: Building Sustainable Cultures*, 3: 16–23.

Reason, P. (1994) 'Three approaches to participatory inquiry', in N. Denzin and Y. Lincoln (eds) *Handbook of Qualitative Research.* Thousand Oaks, CA: Sage, pp. 324–339.

Reason, P. and Bradbury, H. (eds) (2001) *Handbook of Action Research: Participative Inquiry and Practice.* London: Sage.

Stacey, J. (1998) 'Can there be a feminist ethnography?', *Women's Studies*, 11: 21–27.

Stoeker, R. (1997) 'Are Academics Irrelevant? Roles for Scholars in Participatory Research': http://uac.rdp.utoledo.edu/comm-org/papers98/pr.htm (downloaded 30 June 1999).

Urban Places Project (eds) (2000) *The YouthPower Guide: How to Make Your Community Better.* Amherst, MA: UMass Extension.

Ward, C. (1978) *The Child in the City.* New York: Pantheon.

Ward, C. and Fyson, A. (1973) *Streetwork.* London: Routledge & Kegan Paul.

Williams, L. (1997) *Grassroots Participatory Research: A Working Report from a Gathering of Practitioners.* Knoxville, TN: Community Partnership Center.

12 Working in Different Cultures

Fiona M. Smith

Synopsis

Cross-cultural research is the term used to describe researching 'other' cultures using other languages. This may often involve working in distant places but can also include working with 'other' communities closer to home. It requires a sensitivity to cultural similarities and differences, unequal power relations, fieldwork ethics, the practicalities and politics of language use, the position of the researcher, consideration of collaborative or participatory research, and care in writing up the research. All these issues are explored in this chapter. The chapter is organized into the following sections:

- Fieldwork in 'different' cultures: understanding difference and sameness
- Working with different languages
- Working with difference: power relations, positionality and beyond
- (Re)presenting 'other' cultures
- Conclusion

FIELDWORK IN 'DIFFERENT' CULTURES: UNDERSTANDING DIFFERENCE AND SAMENESS

The first encounters many Geography undergraduates have with fieldwork in different cultural contexts are on field courses, during dissertation fieldwork, or participating in an expedition (Nash, 2000a and 2000b; Smith, 2006). At the University of Dundee, for example, we take students to southeast Spain. It is somewhere 'different' from Scotland for students to experience, providing 'an intense learning curve of different environments and cultures', as one student recently put it. Such a 'classic' approach to fieldwork shapes many geography undergraduate careers. However, encounters with 'different' cultures need not always involve travel to distant places. Nairn *et al.* (2000) explain how New Zealand students participated in a field trip which involved meeting migrants to New Zealand from different cultures and thereby seeing New Zealand from a 'different' perspective.

To the popular imagination, travel to 'other' places and cultures defines 'Geography' as a subject, particularly as it is presented in publications such as the *National Geographic* or in television programmes about grand adventures.[1] These ideas contribute to the romance of fieldwork (Nairn, 1996; Madge and Bee, 1999). Fieldwork, particularly if abroad or in a difficult setting, is often seen as 'character building' (Sparke, 1996: 212), a rite of passage to becoming a 'proper' geographer (Rose, 1992). As attractive as these images are, however, geographers have a responsibility to think critically about how we undertake fieldwork in different cultures, whether distant from home or just down the road (Smith, 2006). Nash (2000a: 146) argues 'all geographers undertaking fieldwork overseas [or in more local 'different' cultures] need to be sensitive to local attitudes and customs', in a manner that 'respects the cultural as well as the physical environments you encounter'. Nash suggests a code of ethics applicable to such fieldwork (Box 12.1) which highlights the need to think about the 'difference' or 'similarity' of other cultures, to consider uneven and unequal power relations, and to move away from 'ethnocentric' approaches to fieldwork (see also Chapter 3).[2]

Box 12.1   A code of ethics for tourists which is equally applicable to fieldwork in different cultures

- Travel in a spirit of humility and with a genuine desire to learn more about the people of your host country. Be sensitively aware of the feelings of other people, thus preventing what might be offensive behaviour on your part. This applies very much to photography.
- Cultivate the habit of listening and observing, rather than merely hearing and seeing.
- Realize that often the people in the country you visit have time concepts and thought patterns different to your own. This does not make them inferior, only different.
- Instead of looking for the 'beach paradise', discover the enrichment of seeing a different way of life, through other eyes.
- Acquaint yourself with local customs. What is courteous in one country may be quite the reverse in another – people will be happy to help you.
- Instead of the western practice of 'knowing all the answers', cultivate the habit of asking questions.
- Remember that you are only one of thousands of visitors to this country and do not expect special privileges. If you really want your experience to be a 'home away from home', it is foolish to waste money on travelling.
- When you are shopping, remember that the 'bargain' you obtained was possible only because of the low wages paid to the maker.
- Do not make promises to people in your host country unless you can carry them through.
- Spend time reflecting on your daily experience in an attempt to deepen your understanding. It has been said that 'what enriches you may rob and violate others'.

Source: Nash (2000a), after Weaver (1998).

In the heading for this section the word 'different' is deliberately placed in inverted commas to raise the question of how we think of cultures as 'other' or 'different'. Cultures and their relations to each other are understood in a variety of ways (Hall, 1995; Skelton and Allen, 1999; McEwan, 2008). 'Other' cultures are sometimes regarded as unchanging, even 'traditional' or 'primitive', with change caused by outside forces (usually from the West) (Cloke, 2005). Such approaches emphasize the 'exotic appeal' of other places or their difference and danger. This was particularly evident in the representations of other cultures produced by Europeans during the period of colonialism and imperial rule which emphasized their strangeness and exotic qualities, the apparent danger posed by 'savage' peoples, and the 'need' for the 'civilizing' and 'modernizing' influence of European colonial cultures. This mix of approaches is often summarized under the term 'orientalism', from the work of Edward Said (1978; see also Driver, 2005; Phillips, 2005).[3] Contemporary accounts where westerners assume their own experiences are the model to which other cultures aspire or see other cultures as a backdrop for their own enrichment can also be 'orientalist'. Think about how western travellers often focus on the excitement and adventure provided by other countries with little understanding of the cultures of the countries themselves. Such approaches produce 'self-centred' or 'ethnocentric' geographies, where one's own culture is set as the measure for all others (Cloke, 2005).

Other approaches regard cultures as more dynamic and interconnected (McEwan, 2008; Potter *et al.*, 2008). Some argue globalization erases differences between cultures in the technologically and culturally interconnected 'global village', or highlight the emergence of one homogenized, global consumer culture built around brands such as McDonald's, Coca-Cola or Nike. These approaches see 'other' cultures as 'just like us'. However, this rather overlooks the diverse experiences of cultural, economic and technological globalization in which geographers are fundamentally interested. One event may be watched simultaneously on television by many people across the world, but its significance for each of them will vary, as will the ability to influence responses to the event. Taking diversity into account provides a third set of ideas about culture where, instead of thinking of other cultures as either 'strange' and 'unchanging', or as 'just the same as us', we seek to understand other cultures in and of themselves while also understanding how local places and cultures are connected to national and global processes in uneven and unequal ways (Massey and Jess, 1995). For example, cities are important centres in the new economic geographies of globalization, but they are unevenly connected to these processes – New York is a hub of global finance, Mexico City is less central. Urban residents in these two cities vary in their ability to connect to these processes, depending on their position in the labour market, access to education, class, gender, ethnicity and so on. Researchers should consider how research participants may be situated in these uneven social relations so that the research does not exacerbate or perpetuate inequalities or stereotypes. And the researcher should be aware of how their research is often made possible by their own relatively privileged position in these wider processes (Laurie *et al.*, 1999; Skelton, 2001; Nagar *et al.*, 2003).

A further approach to culture considers how the encounters, relocations and flows of globalization may stretch out cultural relations across space to produce cultural forms which are 'hybrid' or 'synchretic', reflecting the multiple forms of belongings in and across different places particularly associated with experiences of transnational communities, migration and diaspora populations.[4] Claire Dwyer (2005: 504) argues that it is often in fields such as music, fashion, writing, food and the media that diaspora cultures, characterized by 'interconnections, fusions and links that cut across or transform geographical boundaries', are often found. An example in the British context is Bhangra music, characterized by fusions of Punjabi folk music, hip-hop, soul and house, and later developing in post-Bhangra and 'Asian Kool' musical forms in diverse South Asian-British youth cultures (Dudrah, 2002). Another example is the emergence of hyphenated identities such as 'Black-British', 'Afro-American' or 'Japanese-Canadian' (Hall, 1995). These examples suggest 'different' cultures and transnational communities are often found in close proximity to each other, perhaps most obviously in the diverse ethnic geographies of major cities (Banglatown/Brick Lane in London: Dwyer, 2005; Chinatown in San Francisco: McEwan, 2008). However, these need to be seen not just as exoticized spaces of the 'other', but as spaces of encounter and contestation between diverse peoples, and where versions of hybrid cultures are often commodified and promoted for economic gain. Finally, we should be wary of focusing on what appear to be 'obviously different' cultures. In the North American or European context, for example, a diversity of 'white' cultures should also be analysed, rather than seeing them as the 'norm' and 'other' cultures as those that are novel or 'different'.

Rather than cultures being slow-to-change, fixed sets of beliefs, values and behaviours, with permanent connection to places, Stuart Hall (1995: 187) argues culture is 'a meeting point where different influences, traditions and forces intersect', formed by 'the juxtaposition and co-presence of different cultural forces and discourses and their effects' and consisting of 'changing cultural practices and meanings.' This suggests our 'own' cultures are as caught up in change as other cultures, and that connections between cultures and places can be highly dynamic. However, sometimes people or states re-emphasize fixed, homogeneous and bounded cultures in the face of such fluidity, often claiming landscapes or territories. Nationalist movements are one such form (Grundy-Warr and Sidaway, 2008).

Using more fluid conceptualizations of culture means we must question what ideas about similarity and difference we bring to our cross-cultural studies. Nevertheless, it is important not to overplay the changing and dynamic nature of cultures. Instead we need to pay attention to how cultures are embedded in, and part of, ongoing global inequalities. Likewise it is important to explore how different people's experiences of culture 'will be affected by the multiple aspects of their identity – 'race', gender, age, sexuality, class, caste position, religion, geography and so forth – and [are] likely to alter in various circumstances' (Skelton and Allen, 1999: 4). The remainder of the chapter looks at the ways geographers grapple with the challenges of cross-cultural research, starting with work in different languages.

WORKING WITH DIFFERENT LANGUAGES

Translation (dealing with written texts) and interpretation (dealing with the spoken word) are both skilled professional activities and common everyday practices for people around the globe in business, education, tourism, among migrants and diaspora communities, in diplomacy and politics. Indeed some writers suggest that working between languages and cultural references increasingly characterizes the very nature of our lives, whether in the 'translation nation' of bilingual Hispanic communities in the USA (Tobar, 2005), the multilingual, multinational European

Union (Eco, 1993), or the lives of Somali young people in Sheffield (Valentine *et al.*, 2008). Many people around the globe speak English, either as their first language or as a second or third language, so students who speak English are often in a fortunate position but reliance on this can make native English-speakers lazy in learning other languages. There are many contexts where cross-cultural fieldwork will be hampered if you do not make some effort to communicate in the relevant language. While this can seem daunting, it is possible to make a big difference to fieldwork with some practical strategies and, although this will vary between projects, by paying attention to some key questions about the use of language in research (Box 12.2). Here we address issues about questionnaires, working with interpreters and interviews, and working as a bilingual researcher.

Box 12.2   Key questions for working in different languages

Data collection and fieldwork

- What languages are being used?
- Are you using written translation or working with interpreters?
- Have you considered the translation strategy for your project?
- How might this affect the research and data collected?
- Does language use itself become part of the focus of the research?

Data analysis

- Are you assuming the data will mean the same in the other language as in your own? Do you have some way to check the meaning of the data?
- If you are analysing interviews, are you analysing the original speech or the translation of this material? Does this affect how you can analyse the data?

Dissemination and publication

- Do you mention issues of translation, with ambiguities or difficulties explained?
- Will the language of the report affect who can access it?

Ethics and political issues

- Does your use of particular languages mean some social groups are excluded, or others privileged in the research?
- Have you been careful when translating sensitive topics and concepts between different languages and cultural contexts?

Consider how the questions in Box 12.2 might be dealt with in a survey using questionnaires. One common strategy is to aim to have the questionnaire translated in such a way that there is 'concept equivalence' between the original questionnaire and the translated one (something even more important if the results of questionnaires administered in different languages are to be compared directly). A key technique for this is to translate first into the target language and then to have someone else 'back translate' into the original language, allowing ambiguities and mistakes to be observed and ironed out. While this may seem relatively straightforward, it is important to check if the terms being translated are meaningful in the target language, and David Simon (2006) suggests being sensitive to different calendars, units of measurement and local frames of reference or cultural practices might help make the questionnaire meaningful to research participants. Back translation also helps to weed out ambiguous, insensitive or offensive terminology.

Sometimes, however, relatively simple strategies can be very useful. For example, students investigating the local labour market of the tourist resort of Benidorm on the University of Dundee field course produced a bilingual questionnaire before going to Spain, getting the English translated into Spanish by a teacher they knew. The questionnaire explained in both languages who they were and what they were doing and then gave dual language versions of each question. Using this they were able to gather data on the employment experiences not only of expatriate English speakers and of Spaniards with high-level English skills, but also of people with limited English who found it difficult to access better paid jobs in a resort dominated by British tourists. The questionnaire also stimulated further conversations, with friends or relatives of those completing it who spoke more English joining in to translate parts of the discussions between the students and the person completing the questionnaire. Making this effort allowed the students to understand how language abilities were one of the factors structuring differential access to the labour market in this internationalized tourist resort, as well as to gather data about labour-market processes which had been the original purpose of the research.

Of course, it is possible to spend time learning the relevant language. Indeed it is almost essential if fieldwork is to last for a longer period, not least because even everyday life can be difficult without basic language skills. However, while some writers suggest that being able to speak the language of the research participants is the ideal situation, most researchers will at some time find it necessary or even desirable to work with an interpreter.

In the context of undergraduate research, finding an interpreter with suitable language skills and who you can afford to pay may be a real challenge and you might have to negotiate carefully about the expectations you have of the interpreter and that they have of you. This is worth doing even if the interpreter is a fellow student. Interpreting is a challenging process for the person doing it, but it can also lead to frustrations between different members of a research team. Discussing beforehand such issues as how much they will translate there and then for others in the team or whether they will just get on with the interview and translate key findings later can help immensely in smoothing the process. It also avoids leaving some team members feeling marginalized while others feel they are doing all the work. Having found an interpreter, however, the act of translation/interpretation is not merely a straightforward process of simply transmitting meaning from one language to another. Here two issues are worth considering: the approach to translation appropriate for the fieldwork; and the social relations between the interpreter, researcher and research participants, as well as the relation of each to the research.

Reporting on her research on domestic labour in Kenya and Tanzania, Janet Bujra (2006) outlines how she discussed carefully with the interpreters what kinds of terms were appropriate for asking people about domestic labour and people's social roles in order to understand the processes involved and, crucially, the meanings people ascribed to these practices. She also negotiated with the interpreter about the purpose of the research and what kind of translation was most suitable, agreeing that a rougher translation which preserved local terminology and slang was more appropriate than translating interviews and discussions into standard English or standard Swahili. This was a choice about the function of translation in the research. Often translation is assumed to be a direct transmission of meaning from one language to the other, and in many cases the aim (in professional technical translation, for example) would be to make the equivalent meaning clear in the target language. This is known as 'domesticating translation' (Venuti, 2000). However, for Bujra, such an approach would have meant details of how people understood their own lives being lost. Thus she chose what can be termed a 'foreignizing translation' (Venuti, 2000), where the aim instead is to get a sense of people's meanings in their own context, even if that does not produce a highly polished translated text. Neither of these strategies is necessarily right or wrong. Rather these choices indicate both the possibilities of meaning-translation and the impossibility of fully articulating all cultural references and meanings across languages and cultures (Smith, 1996).

Rather than assuming interpreters are neutral, almost invisible transmitters of meaning between the researcher and research participants, it is worth considering their active role in the research. Would an outside interpreter bring less bias, or lack local knowledge? Does a local interpreter understand the issues better? Might they introduce bias by tending to guide you to interview people they already know? Does their gender, class, ethnic or age position mean access to some participants is easier or more problematic? Remember that in many situations those who have the skills to act as interpreters may also be the more educated or affluent people in a society. Thus it is useful to consider how interpreters are involved in meaning-making in the research.

Twyman *et al.* (1999) discuss the use of interpreters in their study of society–environment links in rangelands management in Botswana. The British fieldworkers found the range of local languages too great and depended on high-level language skills supplied by their interpreters. Taking translation seriously within the research process, they argue that 'translation is a practice of intercultural communication [...] in which we understand other cultures as far as possible in their own terms but in our language' (p. 320). This neat formulation suggests translation involves 'mapping ideas and practices' between cultures (Twyman *et al.*, 1999: 320). Rather than only recording answers to questions once they were translated into English, they taped and transcribed the whole process of communication back and forwards between the various research participants, including the translators. Analysing this revealed how the person interpreting often had to summarize roughly the meaning of what the interviewee said in order to let the interview proceed, but after the event the interviewer and the translator could discuss in more detail the ideas articulated. This revealed it was actually very difficult to 'map' the ideas and meanings of the research participants on to English language terms. In fact many interviewees were already talking in what was not their first language and used a variety of different language terms to communicate their attitudes and practices in land and livestock management. In writing up their research, Twyman *et al.* explored how meaning was mapped between and across cultures, noting particularly where this was problematic or provided new insights.

Awareness of language use and translation remains important even when you know and speak a language fluently. While being bilingual might make you more directly aware of the details of what is being discussed or alert you to how challenging it can be to translate meaning between different contexts, thinking consciously about the translation strategies used and the ethics and politics of translation remains important. Kathrin Hörschelmann (2002) describes choices made in her research on the cultural and social consequences of German unification. She interviewed in her native German, transcribed the interviews in German and did the analysis on the original transcripts. Only when writing the final article in English did she translate quotations, but she tried to keep as much of the meaning and style of the original speech in her translated, written quotations, even leaving some terms untranslated and explained in footnotes (a 'foreignizing' approach rather than a 'domesticating' one).

Strategies that take translation seriously as a meaning-making process and as a political one are discussed by Martin Müller (2007). He calls for geographers to adopt critical approaches to translation and to understand the political and cultural impact of the loss of meanings for key terms when they come to be translated between languages, particularly when they deal with key terms of geographical research such as home, nation, family, environment, power, inequality, community or citizenship. Too often apparently simple translations (typically into internationally standard English terms) become the dominant interpretations of situations which in fact require more nuanced understanding.

These examples suggest it is sometimes possible to work in relatively simple ways with different languages, even in contexts where we are not fluent in the language, and through this to gain an insight into the cultures studied. However, if we want detailed understandings of the processes involved and of the meanings people attach to these processes, we need to pay attention to the possibilities and problems of translation and interpretation between languages. Furthermore, we must be aware of the politics of language use, especially where abilities to communicate in one language or another confer status or access to benefits or privileges, and of the possibility that translation itself can be a political act (Smith, 2009).

WORKING WITH DIFFERENCE: POWER RELATIONS, POSITIONALITY AND BEYOND

Moving on to more general questions of difference, unequal power relations and the position of the researcher, we now consider a range of practical responses by geographers which illustrate the challenges and possibilities of cross-cultural research.

During the 1990s, whilst researching urban change in eastern Germany, I became aware of the unease many local residents felt about comparing their experiences to those of western societies. Claims about the 'victory' of the West after the collapse of communism and the practical need to adapt to western administrative and legal frameworks had served to devalue their own cultural and political experiences. In my research I tried to be open to the diverse experiences on which people were drawing to develop new community politics. At one meeting with a group which was developing a local heritage museum, members discussed the problems of industrial decline and unemployment facing their city. Then one person asked why I was interested in their city and what my home city was like. However, after my attempt to explain some similarities in the effects of deindustrialization and labour-market restructuring between Glasgow, in Scotland, and their city, several people rebuffed my arguments, claiming the situation in Glasgow was in no way like that in their own city.

Initially, I was embarrassed and felt I had lapsed into the problematic stance of what they saw as a 'typical westerner', comparing everything outside western society (usually unfavourably) to the 'normal' West. On further reflection it seemed the group members were partially correct, since the pace of change in their city was greater than in Glasgow. However, what was more important was that I had set up a framework where I claimed to be interested in their experiences in and of themselves, without any claims to 'know best'. I had deliberately tried not to essentialize differences between 'East' and 'West', or between capitalism and communism, instead seeking connections as well as differences. Into this framework, I had then introduced a note suggesting what was new, difficult and often very painful for them, with many experiencing redundancy, was, on a global scale, not so unusual, special, or particular. My comparisons inadvertently belittled the severity of people's experiences, denying their individual significance and the particular social and political relations affecting their city. As I developed my analysis, it seemed to me that this moment was not just one where I made a 'mistake' in the fieldwork. The fact some group members found my comparisons unacceptable highlighted precisely the politics of naming processes, cultures and experiences as 'similar' or 'different' which were tied up with the negotiation of these post-communist transitions.

Our research can never escape from the power relations shaping the situations in which we research. We must address these carefully and take account of them in the choices we make in our research practices as well as in the interpretations we develop. One strategy for addressing such inequalities is to work through the complex positionality[5] of the researcher, subjecting the research process itself to scrutiny and not assuming the researcher is a disembodied presence, removed from the research process. This does not mean adopting a self-centred view of research where 'other cultures and other people' become merely the 'exotic backdrops of authorial self-discovery' (Lancaster, 1996: 131). Nor does it mean assuming the researcher can know exactly how they are viewed by the research participants or can account for the significance of every element of their own identity in the research (Rose, 1997). Rather 'we must recognize and take account of our own position, as well as that of our research participants, and write this in to our research practice' (McDowell, 1992: 409) in ways that are sensitive to the difference our presence makes in the research, and how the process of research itself can shape social relations.

Many researchers provide detailed discussions of questions of positionality in cross-cultural research (Madge *et al.*, 1997; Laurie *et al.*, 1999; Limb and Dwyer, 2001; Cloke *et al.*, 2004). Tracey Skelton (2001: 89) provides a definition:

> By positionality I mean things like our 'race' and gender [...] but also our class experiences, our levels of education, our sexuality, our age, our able-ness, whether we are a parent or not. All of these have a bearing upon who we are, how our identities are formed and how we do our research. We are not neutral, scientific observers, untouched by the emotional and political contexts of places where we do our research.

This means being aware of how aspects of our own identities are significant, or might change as we 'travel' (spatially or culturally) to different contexts. For Skelton, not having children as a young researcher marked her out as different from the women she interviewed about gender relations in Monserrat. Rather than avoiding their questions about this, she used this as a point of discussion, recognizing that some of the women felt luckier, or more mature than her. 'I found this a healthy way of letting the power I had in the interview context – I was the one asking questions – dissipate and shift into complex positions within the interviews' (Skelton, 2001: 91).

At times, thinking about positionality leads researchers to question whether it is even appropriate to undertake studies where they are 'outsiders', especially where their outsider status might perpetuate the ways less powerful groups and cultures have been represented by those in more powerful positions, such as westerners representing the experiences of people in developing countries (Madge, 1994; Radcliffe, 1994), or where those being researched might prefer their experiences are not opened up to public gaze (Barnett, 1997). Kim England (1994) withdrew from researching the lesbian community in Toronto, Canada because she felt she was too much of an 'outsider' and the context of homophobia in Canada meant her trying to 'speak for' or 'give voice to' the lesbian community led to the danger of her colonizing the experiences of this group of women. She felt it more appropriate to leave research of the community to other lesbian women. However, Robina Mohammad's (2001) discussion of research on Pakistani women in Britain suggests that the need to consider positionality does not disappear where we appear to be 'insiders', since we are also partly 'outsiders' by the very fact that we are engaged in research. Likewise, other aspects of our own identities (dress, accent or education) can be markers of our difference as well as our similarities. Rather than her apparent 'insider' status allowing access to 'the truth', Mohammad suggests that within her study participants presented 'multiple truths'. The challenge was to understand 'which truth' was being told and whose interests were served by particular representations.

However, a number of geographers engaged in cross-cultural research have argued that addressing positionality, while important, may not be enough to tackle the fundamentally uneven and unequal social relations which structure the research process and which may indeed be the focus of the research. These researchers have explored the role that collaborative approaches (see also Chapter 11), working with individuals and groups to establish a shared set of goals and practices, can play in helping to move 'beyond positionality' and to develop research practices which do not simply involve reflecting on, but also acting in ways which might diminish such inequalities. Examples include collaboration in the study of women's oral histories in India between Richa Nagar, Farah Ali and Sangatin Women's Collective (2003), Geraldine Pratt's long-term engagement with migrant communities in Vancouver, Canada (Pratt, 2007), and the collaboration between Red Thread (2000) and Linda Peake on issues of reproductive and sexual health among women in Guyana. These examples demonstrate a commitment not just to 'give voice' to those involved (thereby preserving the existing unequal social relations between the privileged researcher and what appear to be the otherwise silent research 'subjects'), but to work with people, to listen to their priorities, to engage with the politics and practices of social change, and to take seriously the challenges of not just noting difference but to work with it and across it. Often adopting participatory methods, these collaborative approaches do not remove the need to address the challenges outlined in this chapter. Instead they are 'partnerships in which the questions around how power and authority would be shared cannot be answered beforehand, but are imagined, struggled over and resolved through the collaborative process itself' (Nagar *et al.*, 2003: 369).

To some extent, then, we are always involved in working with 'different' cultures and must negotiate the power relations of similarity and difference in our research whether these cultures are 'remote' or close at hand. As Heidi Nast (1994: 57) argues, 'we can never *not* work with "others" who are separate and different from ourselves; difference is an essential aspect of all social interactions that requires that we are always everywhere in between or negotiating the worlds of me and not-me'. The challenge is to address this in the research strategies we adopt and in the social relations surrounding our research practice. As Saraswati Raju (2002) asks, 'We are different, but can we talk?'

(RE)PRESENTING 'OTHER' CULTURES

Finally, it is important to consider how to represent the people and places studied in the field report or dissertation. Writing to thank those who have helped and, where appropriate, sending a copy of the report is a good start. If you worked in another language or your academic findings would be inappropriate (people might be interested in what you discovered, but not necessarily in your latest theoretical insight), a revised feedback report might be more appropriate, as might be giving a presentation, writing an article for the local newspaper, or setting up a project website where people can access your findings. You may ask for comments from participants on your analysis or may even develop the analysis in collaboration with them. These are often complex situations and it does not automatically follow that the research collaborators are right and you are wrong. You can instead work with that difference of interpretation to decentre your position as the apparently all-knowing researcher, compared to the research participants. You may need to discuss whether different kinds of analysis, interpretation or publications may be more important or appropriate for particular audiences (Nagar *et al.*, 2003).

It is important to consider the language in which your research is to be written, though this may also be constrained by the requirements for work to be read and assessed in particular languages. More widely in geographical scholarship this question is important because of the global dominance of English-language publishing in academic writing. This affects which languages and concepts are more central to the development of geographical knowledges and which are marginalized (Desbiens, 2002; Garcia-Ramon, 2003). Your writing strategy might engage with this, for example by writing a bilingual text (Cravey, 2003). Such consideration applies to visual representations as well as to written texts, as the following example illustrates.

American anthropologist Kathleen Kuehnast (2000) had been researching the economic burdens faced by women in regions affected by farm privatization in Kyrgyzstan, a central Asian post-Soviet state. For the cover of her report she chose a photograph of 'a Kyrgyz elderly woman and her daughter-in-law, holding a baby, each dressed in the warm clothing of semi-nomadic herders' (Kuehnast, 2000: 105). She was surprised when some women in government jobs, whom she knew, were offended by this, feeling it presented their country in a poor light. After initially dismissing this as the unwillingness of higher-status women to address problems facing nomadic women in their own country, Kuehnast wondered why else the photograph was problematic. Perhaps it failed to represent women's achievements in education and employment, buying in too strongly to the notion of women as victims of communist repression? Did it contradict important Soviet-era ideas of women as strong and competent workers? Alternatively, many of the women interviewed by Kuehnast saw in the western media images of glamorous women, which flooded the country after independence, an ideal of 'western', 'modern' or 'American' womanhood (with the 'right' clothes, make-up and leisured lifestyle). In this context, the portrayal of Kyrgyz women in traditional clothing and as working women may have seemed to illustrate too sharply the apparent 'failings' of Kyrgyz women in adopting suitably 'westernized' or 'modernized' gender identities.

Representation is fundamentally problematic. The demands of 'analysis, writing up and dissemination of information often force us to detach ourselves, switch back to "western mode" to produce texts and develop "distance" to use information' (Madge, 1994: 95). As Kuehnast found, even analysis and representation which is meant to be helpful is not immune to being problematic for the research participants. Each researcher therefore must make his or her own choices, often resulting in pragmatic responses to his or her situation. Clare Madge (1994: 96), in her research on the Gambia, decided that some of the information she gathered talking to people who became her friends could not be included in her analysis, since to use that information would 'betray the trust of my friends'. Audrey Kobayashi (2001) opted to stop working on 'other' cultures outside Canada and instead to focus on working with Japanese-Canadians within Canada in collaborative activist research. Tracey Skelton at times felt it was impossible to write about her research on Montserrat without reproducing the unequal colonial relations she was trying to combat, so for some time she decided not to write on these topics. However, in the end she decided that 'as part of the politics of reflective and politically conscious feminist and/or cross-cultural research, we have to continue our research projects, we must publish and disseminate our research. If we do not, others without political anxieties and sensitivities about their fieldwork processes take the space' (2001: 95).

CONCLUSION

There are many challenges in working in different languages and different cultures and even experienced geographers do not always get it right. However, the key issue is to pay attention to the issues raised in this chapter as we plan and under-take fieldwork, and analyse and represent what we find, keeping in mind 'why we are doing it and what the research we do means to other people' (Skelton, 2001: 96). When done well, research in different languages and different cultures can be incredibly enriching and challenges us to think about difference and diversity in productive and sensitive ways.

Summary

- Cross-cultural research is challenging, enriching and rewarding.
- Concepts of cultural difference should inform such research, taking into account the fluidity of cultures, unequal social relations, the need to avoid ethnocentrism and to be open to hybrid cultural forms.
- Simple strategies may address language differences. Attention to language use and the articulation of meanings provides insight into the 'other' culture. Translation requires careful consideration of how meanings 'map' between cultures.
- Addressing the power relations surrounding research and writing the positionality of the researcher into our research accounts, or adopting collaborative strategies can help avoid ethnocentrism or even begin to challenge unequal social relations. 'Outsiders' and 'insiders' should consider carefully their relation to the research. This may provide a variety of accounts without necessarily producing the single 'correct' answer to interpreting a particular situation.
- Choices about representation in written, verbal, or visual formats should avoid reinforcing unequal power relations or stereotypes and be informed by the ethics and politics of the research.

NOTES

1 'Other' is often used in cultural analysis with inverted commas, and sometimes an initial capital letter, e.g. the 'Other', or 'other' cultures, to imply that such cultures, social groups or societies may not be as different as is implied by cultural and social norms.

2 Ethnocentric or ethnocentricism relates to the prioritizing of one's own world-view and experiences, of one's own culture as the 'norm' against which others are measured for their strangeness, lack of development, difference, or exoticism. It usually implies taking western experiences as the norm.

3   Drawing on the work of Edward Said (1978), 'orientalism' describes and critiques the set of ideas common among Europeans in their depictions of 'the Orient' during the era of imperialism. Such ideas emphasized both the apparent attractions and exotic nature of such people, places and cultures, and their supposed danger and lack of civilization, helping to justify European colonial endeavours. The terms have subsequently been applied more widely to all colonial situations and also to contemporary representations of 'other' cultures, peoples and places.

4   Diaspora describes the dispersal or scattering of a population. It can also refer as a noun to dispersed or scattered populations such as the Black diaspora or Jewish diaspora. In more theoretical terms, the idea of diaspora, or diasporic communities challenges notions of fixed connections between cultures, identity and place. The term transnational community is used to describe social and cultural relations that transcend and escape the bounded spaces of the nation-state, possibly as a result of migration and diaspora or because the social group does not fit existing national boundaries (such as the Kurdish population).

5   Positionality can be understood as consideration of how the relative position of the researcher may affect the process of the research. For example, information given by participants may depend on how the researcher is viewed in that particular context (threatening, insignificant, powerful). The researcher may also be relatively privileged by their positionality.

Further reading

This guide to further reading identifies references for the key themes of the practicalities of fieldwork abroad, debates about cultural change and negotiating power relations:

- Two articles by Nash (2000a, 2000b) discuss practical issues in undertaking independent fieldwork abroad. The first addresses establishing contacts, legal requirements for visas, collecting and exporting samples, health and safety issues, and training. The second considers budgeting and fundraising.

- Hall (1995), Skelton and Allen (1999) and McEwan (2008) explore current debates about contemporary cultural change.

- Smith (1996, 2009), Twyman *et al.* (1999), Bujra (2006) and Müller (2007) all consider the issue of translation between different languages and of working with a translator. Each discusses how translation itself can become part of the focus for analysis.

- Pratt (2007), Nagar (2003), Madge *et al.* (1997) and Laurie *et al.* (1999) provide detailed examples of negotiating power relations, positionality and representation in cross-cultural research.

Note: Full details of the above can be found in the references list below.

References

Barnett, C. (1997) '"Sing along with the common people": politics, postcolonialism and other figures', *Environment and Planning D: Society and Space*, 15: 137–154.

Bujra, J. (2006) 'Lost in translation? The use of interpreters in fieldwork', in V. Desai and R. Potter (eds) *Doing Development Research*. London: Sage, pp. 172–179.

Cloke, P. (2005) 'Self-Other', in P. Cloke, P. Crang and M. Goodwin (eds) *Introducing Human Geographies* (2nd edn). London: Arnold. pp. 61–77.

Cloke, P., Cook, I., Crang, P., Goodwin, M., Painter, J. and Philo, C. (2004) *Practising Human Geography*. London: Sage.

Cravey, A. (2003) 'Toque una ranchera, por favour', *Antipode*, 35: 603–621.

Desbiens, C. (2002) 'Speaking in tongues, making geographies', *Environment and Planning D: Society and Space*, 20: 1–25.

Driver, F. (2005) 'Imaginative geographies', in P. Cloke, P. Crang and M. Goodwin (eds) *Introducing Human Geographies* (2nd edn). London: Arnold. pp. 144–155.

Dudrah, R.K. (2002) 'Drum n dhol: British Bhangra music and diasporic South Asian identity formation', *European Journal of Cultural Studies*, 5(3): 363–383.

Dwyer, C. (2005) 'Diasporas', in P. Cloke, P. Crang and M. Goodwin (eds) *Introducing Human Geographies* (2nd edn). London: Arnold, pp. 495–508.

Eco, U. (1993) *La ricerca della lingua perfetta nella cultura europea*. Rome: Laterza.

England, K. (1994) 'Getting personal: reflexivity, positionality and feminist research', *Professional Geographer*, 46(1): 80–89.

Garcia-Ramon, M.D. (2003) 'Globalization and international geography: the questions of languages and scholarly traditions', *Progress in Human Geography*, 27(1): 1–5.

Grundy-Warr, C. and Sidaway, J. (2008) 'The place of the nation-state', in P. Daniels, M. Bradshaw, D. Shaw and J. Sidaway (eds) *An Introduction to Human Geography*. Harlow: Prentice Hall. pp. 417–437.

Hall, S. (1995) 'New cultures for old', in D. Massey and P. Jess (eds) *A Place in the World? Places, Cultures and Globalization*. Oxford: Oxford University Press, Open University, Press, pp. 175–213.

Hörschelmann, K. (2002) 'History after the end: post-socialist difference in a (post)modern world', *Transactions of the Institute of British Geographers*, 27: 52–66.

Kobayashi, A. (2001) 'Negotiating the personal and the political in critical qualitative research', in M. Limb and C. Dwyer (eds) *Qualitative Methodologies for Geographers*. London: Arnold. pp. 55–70.

Kuehnast, K. (2000) 'Ethnographic encounters in post-Soviet Kyrgyzstan: dilemmas of gender, poverty and the Cold War', in H. de Soto and N. Dudwick (eds) *Fieldwork Dilemmas: Anthropologists in Postsocialist States*. Madison, WI: University of Wisconsin Press, pp. 100–118.

Lancaster, R.N. (1996) 'The use and abuse of reflexivity', *American Ethnologist*, 23(1): pp. 130–132.

Laurie, N., Dwyer, C., Holloway, S.L. and Smith, F.M. (1999) *Geographies of New Femininities*. Harlow: Longman.

Limb, M. and Dwyer, C. (eds) (2001) *Qualitative Methodologies for Geographers*. London: Arnold.

Madge, C. (1994) 'The ethics of research in the "Third World"', in E. Robson and K. Willis (eds) *DARG Monograph No. 8: Postgraduate Fieldwork in Developing Areas*. Developing Areas Research Group of the Institute of British Geographers, pp. 91–102.

Madge, C. and Bee, A. (1999) 'Women, science and identity: interviews with female physical geographers', *Area*, 31(4): 335–348.

Madge, C., Raghuram, P., Skelton, T., Willis, K. and Williams, J. (1997) 'Methods and methodologies in feminist geographies: politics, practice and power', in Women and Geography Study Group, *Feminist Geographies: Explorations in Diversity and Difference*. Harlow: Longman. pp. 86–111.

Massey, D. and Jess, P. (1995) 'Places and cultures in an uneven world', in D. Massey and P. Jess (eds) *A Place in the World? Places, Cultures and Globalization*. Oxford: Oxford University Press, Open University Press, pp. 215–240.

McDowell, L. (1992) 'Doing gender: feminism, feminists and research methods in human geography', *Transactions of the Institute of British Geographers*, 16: 400–419.

McEwan, C. (2008) 'Geography, culture and global change', in P. Daniels, M. Bradshaw, D. Shaw and J. Sidaway (eds) *An Introduction to Human Geography*. Harlow: Prentice Hall. pp. 273–289.

Mohammad, R. (2001) '"Insiders" and/or "outsiders": positionality, theory and praxis', in M. Limb and C. Dwyer (eds) *Qualitative Methodologies for Geographers*. London: Arnold, pp. 101–117.

Müller, M. (2007) 'What's in a word? Problematizing translation between languages', *Area*, 39(2): 206–213.

Nagar, R. in consultation with F. Ali and Sangatin Women's Collective, Sitapur, Uttar Pradesh, India (2003) 'Collaboration across borders: moving beyond positionality', *Singapore Journal of Tropical Geography*, 24(3): 356–372.

Nairn, K. (1996) 'Parties of geography fieldtrips: embodied fieldwork', *New Zealand Women's Studies Journal*, 12: 88–97.

Nairn, K., Higgitt, D. and Vanneste, D. (2000) 'International perspectives on fieldcourses', *Journal of Geography in Higher Education*, 24(2): 246–254.

Nash, D.J. (2000a) 'Doing independent overseas fieldwork 1: practicalities and pitfalls', *Journal of Geography in Higher Education*, 24(1): 139–149.

Nash, D.J. (2000b) 'Doing independent overseas fieldwork 2: getting funded', *Journal of Geography in Higher Education*, 24(3): 425–433.

Nast, H. (1994) 'Opening remarks on "Women in the Field"', *Professional Geographer*, 46: 54–66.

Phillips, R. (2005) 'Colonialism and postcolonialism', in P. Cloke, P. Crang and M. Goodwin (eds) *Introducing Human Geographies* (2nd edn). London: Arnold, pp. 326–338.

Potter, R., Binns, T., Elliott, J. and Smith, D. (2008) *Geographies of Development: An Introduction to Development Studies* (3rd edn). Harlow: Prentice Hall.

Pratt, G. (2007) in collaboration with the Phillippine Women Centre of B.C. and *Ugnayan ng Kabatang Pilipino sa Canada*/Filipio-Canadian Youth Alliance 'Working with migrant communities: collaborating with the Kalayan Centre in Vancouver, Canada', in S. Kindon, R. Pain and M. Kesby (eds) *Participatory Action Research Approaches and Methods: Connecting People, Participation and Place*. London: Routledge, pp. 95–103.

Radcliffe, S. (1994) '(Representing) post-colonial women: authority, difference and feminisms', *Area*, 26(1): 25–32.

Raju, S. (2002) 'We are different, but can we talk?', *Gender, Place and Culture*, 9(2): 173–177.

Red Thread (2000) *Women Researching Women: Study on Issues of Reproductive and Sex Health and of Domestic Violence against Women in Guyana*. Report of the Inter-American Development Bank (IDB) Project TC-97-07-40-9-GY conducted by Red Thread Women's Development Programme, Georgetown, Guyana, in conjunction with Dr. Linda Peake, available at http://www.sdnp.org.gy/hands/wom_surv.htm.

Rose, G. (1992) 'Geography as a science of observation: the landscape, the gaze and masculinity', in F. Driver and G. Rose (eds) *Nature and Science: Essays in the History of Geographical Knowledge*. London: IBG Historical Geography Research Group, pp. 8–18.

Rose, G. (1997) 'Situating knowledges: positionality, reflexivity and other tactics', *Progress in Human Geography*, 21(3): 305–320.

Said, E. (1978) *Orientalism*. Harlow: Penguin.

Simon, D. (2006) 'Your questions answered? Conducting questionnaire surveys' in V. Desai and R. Potter (eds) *Doing Development Research*. London: Sage, pp. 163–171.

Skelton, T. (2001) 'Cross-cultural research: issues of power, positionality and "race"', in M. Limb and C. Dwyer (eds) *Qualitative Methodologies for Geographers*. London: Arnold, pp. 87–100.

Skelton, T. and Allen, T. (eds) (1999) *Culture and Global Change*. London: Routledge.

Smith, F.M. (1996) 'Problematizing language: limitations and possibilities in "foreign language" research', *Area*, 28(2): 160–166.

Smith, F.M. (2006) 'Encountering Europe through fieldwork', *European Urban and Regional Studies*, 13(1): 77–82.

Smith, F.M. (2009) 'Translation', in R. Kitchin and N. Thrift (eds) *International Encyclopedia of Human Geography*. London: Elsevier, pp. 361–367.

Sparke, M. (1996) 'Displacing the field in fieldwork: masculinity, metaphor and space', in N. Duncan (ed.) *BodySpace*. London: Routledge, pp. 212–233.

Tobar, H. (2005) *Translation Nation: Defining a New American Identity in the Spanish-speaking United States*. New York: Riverhead Books.

Twyman, C., Morrisson, J. and Sporton, D. (1999) 'The final fifth: autobiography, reflexivity and interpretation in cross-cultural research', *Area*, 31(4): 313–326.

Valentine, G., Sporton, D. and Nielsen, K.B. (2008) 'Language use on the move: sites of encounter, identities and belonging', *Transactions of the Institute of British Geographers*, 33(3): 376–387.

Venuti, L. (ed.) (2000) *The Translation Studies Reader*. London: Routledge.

Weaver, D.B. (1998) *Ecotourism in the Less Developed World*. Wallingford: CAB International.

13 Internet Mediated Research

Clare Madge

Synopsis

Internet mediated research is research that is conducted through the medium of the internet. It is one of the newer methodological approaches adopted by geographers and holds great potential for collecting data in an innovative manner. This chapter first looks at three main types of internet mediated research: online questionnaires, virtual interviews and virtual ethnographies, outlining some of the potentials and limitations of each of these online research approaches. It then discusses some of the ethical issues involved in internet mediated research. Finally, the chapter concludes by stressing the importance of the online researcher conducting their research with rigour and reflexivity, and highlighting the potential of web 2.0 technologies for developing more novel online research approaches. The chapter is organized into the following sections:

- Introduction: the nature of internet mediated research
- Types of internet mediated research
- Ethics of internet mediated research
- Conclusions

INTRODUCTION: THE NATURE OF INTERNET MEDIATED RESEARCH

Recent years have seen the proliferation of new online research methods (ORM) conducted via the medium of the internet: these have been collectively termed internet mediated research (IMR) or online research practice (ORP). Internet mediated research includes online questionnaires, virtual interviews of various types, virtual ethnographies and online experiments. Internet mediated research has been heralded as a new 'methodology frontier' in some quarters and there is excitement about the potential that the internet offers in collecting data. Some even suggest it has added much to the potentials of survey research generally (Vehovar and Lozar Manfreda, 2008). Within the discipline of geography, there have been a small but growing number of projects utilizing internet mediated research (O'Lear, 1996; Parr, 2002; Kagawa, 2007; Madge et al., 2009).

However, despite this excitement, the use of internet mediated research must be carefully considered. As Best and Krueger (2004: 85) suggest, attempting to undertake online data collection 'is far easier than successfully accomplishing it. For those who choose to perform it, they must do so deliberately and cautiously.'

Thus, we should avoid the use of the internet as an 'easy option' and 'encourage a more developed focus on the justification, applicability and benefits of internet research' to a particular project (Illingworth, 2001: 18.4). The effectiveness of internet mediated research depends on whom is being researched, what is being researched and why. Moreover, the potential of internet mediated research should not be exaggerated: many of the issues and problems of conventional research methods still apply in virtual space. As Smith (1997: 4) concludes: 'The new technology offers a spate of problems layered over the old.' Thus, from the outset, some caution must be expressed about these new internet mediated methodologies: faster, cheaper and more technologically advanced is not necessarily synonymous with 'better' quality research. As Dodd (1998: 60) argues, we must ensure that 'cheap entry costs and glowing attractiveness of internet fieldwork do not result in shoddy "cowboy" research'. There is, therefore, a need for online researchers to tread with caution and practise their 'craft' with reflexivity (Madge and O'Connor, 2002).

It is very unlikely that internet mediated research is going to replace more conventional face-to-face research; rather, it represents another option in the geographers' 'methodological toolkit'. Indeed, internet mediated research should not be seen simplistically in opposition to onsite research; online research methods may be best used in conjunction with face-to-face methods. This 'triangulation' of online and face-to-face methods may produce a more nuanced analysis and suggests that a mixed (online/onsite) methodology might be the most fruitful approach for geography researchers. That said, it is my hope that this chapter will move geography students towards the careful and critical use of online research methods because the long-term success of internet mediated research will ultimately depend on the quality, rigour and credibility of the data generated. Below, the three most popular types of internet mediated research are outlined.

TYPES OF INTERNET MEDIATED RESEARCH

There are three main types of internet mediated research currently used that are relevant to geographers. The choice of research method will depend on the specific aims of the research. A brief summary of each type is given, with indications of further reading.

Online questionnaires

Online questionnaires can provide fast and cheap alternatives to postal, face-to-face and telephone surveys. Online questionnaires are designed as web pages and located on a host site. They can include an inherently flexible design, with a wide variety of question types, a superior questionnaire interface compared to paper-based surveys, and can enable multilingual formats and inclusion of audiovisual stimuli. Data can also be loaded automatically into a spreadsheet or database increasing the speed and accuracy of data collection. According to Dillman *et al.* (2009), online questionnaires can enhance the effectiveness of research and increase response rates when used carefully and appropriately as part of a mixed-mode methodology. Numerous examples of online questionnaires exist in the literature (see Best and Krueger, 2004; Dillman, 2007; Lefever *et al.*, 2007; Sue and Ritter, 2007; Witte, 2009). Box 13.1 gives a summary of the principles of good design for online questionnaires.

Box 13.1   Principles of good design for online questionnaires

Online questionnaire design should follow principles for good design of onsite questionnaires but may also include elements only associated with the online environment. Design issues are an extremely important consideration for online questionnaires because of the highly visual nature of the internet environment and the variety of technical skills of respondents. The massive range of purposes of online questionnaires and diversity of the populations to be studied mean that there is no single design approach that is appropriate for all online questionnaires. However, some general design principles are noted below:

- Allocation of *adequate time* for producing and piloting initial draft, undertaking usability testing (see explanation below), redesigning and producing revised version and pursuing non-respondents.
- *Inclusion of a welcome screen with background information* including the purpose of the research, estimated time it will take to complete the questionnaire, a consent form, reassurances of confidentiality and anonymity (if appropriate), details of how to contact the researcher.
- *Clear and precise instructions* for completing each question, with examples if necessary, and precise details of how to submit the online questionnaire.
- Use of plain, *easy to understand language* and avoidance of technical terms, jargon, acronyms, 'netlingo'.
- *Care with questions* to avoid duplicate or unnecessary questions; avoidance of questions that respondents will not have the knowledge or experience to answer; avoidance of leading, biased or ambiguous questions; use of non-sexist, non-racist, non-homophobic language.
- Use of a *variety of question types* appropriate to the research question and type of respondent; these might include open and closed questions, ranking and rating questions, open-ended or multiple select responses (but bear in mind for online questionnaires open-ended questions and grid questions are less likely to be completed by respondents).
- *Organization of questions* in logical groups with important questions asked first and questions sub-divided into clearly labelled sections which may relate to separate pages of the online questionnaire.
- Use of a *simple questionnaire format and avoidance of unnecessary graphics*. If colour is essential, use of opponent-colours, such as yellow and blue; use of 12- or 14-point font Times Roman, Arial or Verdana font; only use multi-media stimuli when essential to the research project.
- Keep the *questionnaire length and time as short as possible* to get the information you require but 15 minutes maximum is recommended: a shorter questionnaire will increase the response rate. The best response rate is achieved from questionnaires with 15 questions or less.

(Continued)

(Continued)

- Include a *progress indicator* to reduce drop-outs; include 'social presence' or missing data messages to reduce item non-response.
- *Finish the questionnaire* with a place for respondents to add comments, with a thank-you for assistance and with details of how and when participants can get a copy of the completed survey results.
- *Ensure consistency of appearance and accessibility* of online questionnaire with different browsers and computers; ensure your questionnaire is compliant with the latest standards of web design and will be possible to use with a screen reader or text only version.
- Send one or two *follow-up reminders* if online questionnaire is not completed; more than this is considered spamming.

Usability testing refers to checking that the online questionnaire performs the function for which it is designed, with the minimum amount of user frustration, time and effort. It is a type of piloting that occurs prior to the release of the online questionnaire. Usability testing usually involves observation of potential participants, who are asked to verbalize their thoughts about using the questionnaire as they complete it. Any glitches, problems and design issues are identified through the process and remedied before the online questionnaire is finally hosted.

However, although online questionnaires may be of possible use in student projects, especially through their potential to access large and geographically distributed populations and to achieve quick returns, many problems arise in their use. For example, it must be noted that the time taken to prepare an online questionnaire can be substantial and will outweigh some of the other time savings. Response rates may also be lower than with face-to-face questionnaires due to online respondent fatigue, and the technological expertise and facilities of both the researcher and participant may limit the quality of the online questionnaire. Indeed, Lefever *et al.* (2007) warn that the success of online questionnaires is becoming limited through 'over surveying' and caution that they may no longer be as universally appealing as was once believed, particularly because reaching the population sample remains a problem. Braunsberger *et al.* (2007) also argue that, although online questionnaires can produce more reliable data estimates and are cheaper and less time consuming than telephone surveys, they may not be appropriate for all survey research endeavours. Care must be taken to dispel superficial ideas that online questionnaires can quickly, efficiently and unproblematically replace face-to-face surveys. This is especially the case for questionnaires of a sensitive nature, where the personal touch of face-to-face interaction is important to encourage response rates, or where the target population might not have easy and cheap online access or adequate computer literacy. So, despite the growing use of online questionnaires, it is equally clear that some researchers are taking an increasingly critical outlook on their use for the production of good-quality research: their use must be appropriate and justified for each particular research project, especially given increasing levels of 'survey fatigue' (Witte, 2009).

Virtual interviews

There are two main types of virtual interviews: asynchronous and synchronous. Asynchronous or 'non real time' exchanges are most common and can be facilitated by email (Murray and Sixsmith, 1998; Selwyn and Robson, 1998; Bampton and Cowton, 2002; Meho, 2006; James, 2007), discussion board services (also known as bulletin boards, discussion groups or web/internet forums; Stieger and Goritz, 2006) or in a mailing list or listserv environment (Gaiser, 1997, 2008). With asynchronous interviews, it is not necessary for the researcher and respondent to be online at the same time which eliminates the need to set up mutually convenient interview times. However, respondents can post their interview reply at anytime and so the researcher cannot play an active role in moderating the interview or intervening in its progress. Greater spontaneity, group interaction and higher levels of immediacy and engagement in the interview process can only occur during 'real time' synchronous online interviews. Synchronous online interviews are those which most closely resemble a traditional research interview in that they take place in 'real time' in an environment such as an internet chat room or through the use of conferencing software (for examples, see Chen and Hinton, 1999; Rezabek, 2000; Sweet, 2001; Davis *et al.*, 2004; Hughes and Lang, 2004; Madge and O'Connor, 2006; Klein *et al.*, 2007; Fox *et al.*, 2008; Ayling and Mewse, 2009). The researcher and respondents must be online simultaneously and questions and answers are posted in a way which mimics a traditional interview.

Research comparing face-to-face and virtual interviews (see Curasi, 2001 and Schneider *et al.*, 2002) suggests that advantages of the virtual interview include:

- The ability to carry out interviews with a geographically dispersed population;
- The opportunity to interview individuals or groups who are often difficult to reach, such as the less physically mobile, the socially isolated or those living in dangerous places;
- The savings in costs provided to the researcher (for example, costs associated with travel and venue hire);
- The qualitative data that are gained from virtual interviews are already in a text format and can be saved directly to file, reducing transcription time and costs;
- And the reduction in 'interviewer effect' as participants and interviewer cannot 'see' each other.

Although in many respects online and face-to-face interviews are similar processes, there are also important differences to be considered. These include the following:

- There are a number of issues surrounding interview design. For example, the online interviewer must consider ways in which to build rapport in the absence of visual and non-verbal cues (O'Connor *et al.*, 2008).
- It can be difficult to access a relevant sample so sampling strategies and access concerns must be given full consideration (Hamilton and Bowers, 2006).

- Issues of online interview interaction and computer mediated communication must be reflected upon (Watson *et al.*, 2006; Beusch, 2007; Ayling and Mewse, 2009).
- The technology involved in online interviews can potentially lead to the systematic denial of participation to particular groups, either through the need for particular equipment or expertise, or through the accessibility needs of groups with particular disabilities (Schmidt, 2007).
- Technical problems can occur related to access from a work computer or from behind a firewall, or owing to lost connections which interrupt the flow of an interview or make it necessary to reschedule (Fox *et al.*, 2008; Gaiser, 2008).
- Questions surrounding the ethical rights of respondents, including informed consent, confidentiality and privacy, are significant (James and Busher, 2006, 2007).

Before conducting any virtual interviews, these issues must be considered. O'Connor *et al.* (2008) provide more details of how to conduct a virtual interview, and for details of how to set up a virtual interview, including the software available to do so, see http://www.geog.le.ac.uk/orm/. Box 13.2 summarizes some of the design issues involved with virtual interviewing.

Box 13.2 Some issues associated with the design of virtual interviews

Designing the interview script

- Before commencing the interview, decide how to inform participants about the interview procedure, e.g. a brief introduction to the aims of the interview, the estimated length of the interview, the types of question and informed consent, confidentiality and anonymity issues.
- At the start of the interview remind participants how best to contribute to an online discussion and set ground rules for the interview, e.g. respondents should try to stick to questions asked by the interviewer, 'flaming' or aggressive interactions will not be tolerated.
- At the start of the interview introduce yourself to the participants. In the absence of visual clues you may want to create a text-based picture of yourself to facilitate rapport and to elicit profile data from the respondents which would have been visually apparent in a face-to-face interview.

Rapport

- In the virtual online interview, both the interviewer and interviewee are relying on the written word as a means of building rapport.
- One way of doing this is to consciously share personal information. Kivits (2005) recommends developing an online relationship through sharing information on holidays, family life and work.
- Another way of building rapport is through the use of photographs. Madge and O'Connor (2006) directed respondents to a project website, which included information about the project and photographs of the researchers, so respondents had the opportunity to 'see' the researchers prior to the interview.
- Finally, building rapport through personal emails and telephone conversations to arrange interview times and dates can also be useful.

Language use, netlingo, netspeak and emoticons

- Experienced users of online technology often use a specific shorthand or internet language which Thurlow *et al.* (2004) refer to as 'netlingo' and 'netspeak' to communicate. This includes 'emoticons' such as *:-)* (representing smiles) or electronic paralinguistic expressions such as *lol* (laugh(ing) out loud) or *rofl* (rolling on the floor laughing).
- Amongst many online communities the use of such visual and paralinguistic expressions is central to the style of online communication. Indeed, the use of emoticons such as the smiley face can serve to enhance users' sense of connection with other online participants. In the absence of visual pointers such as body language and facial expressions, emoticons play an important role in conveying the 'tone' of an interview response.
- The online interviewer must make themselves aware of the most common netlingo and emoticons used by their respondents. A good source is http://www.netlingo.com/index.cfm. However, you should be cautious about over using netlingo as it can be misunderstood and can suggest naivety.

Source: Adapted from Henrietta O'Connor (http://www.geog.le.ac.uk/orm/interviews/intdesign.htm).

Virtual ethnographies

Interest in online cultures and communities has resulted in virtual ethnographies (Hine, 2000) of cyberspace cultures. Providing a 'thick' description through the 'immersion' of the researcher in the online lives of their subjects, a virtual ethnography is created, and the type of data gained is qualitative, in-depth and contextual. Virtual ethnographies may range from an observational study, in which the researcher is 'passive', to a more participative approach, in which the researcher fully participates and engages as a member of an online community. Virtual ethnography has also variously been termed online ethnography, webnography, digital ethnography, connective ethnography, networked ethnography and cyberethnography, each of which has its own specific methodology. There is great debate about whether conducting a virtual ethnography involves a distinctive methodological approach or whether it mirrors the more conventional onsite ethnographic approach (Domínguez *et al.*, 2007). However, broadly speaking, according to Hine (2005), virtual ethnography extends the traditional notions of field and ethnographic study from the observation of co-located, face-to-face interactions, to technologically mediated interactions in online networks and communities.

Virtual ethnographies have included detailed studies on particular online communities such as feminist online networks (Fay, 2007), computer-mediated fan clubs (Baym, 1997) and the Paraguan diaspora (Greschke, 2007). More general ethnographies have also been conduced on virtual reality technologies (Green, 1999), MUDS (Simona, 2007), types of computer mediated communications (Soukup, 1999) and online identities (Rybas and Gajalla, 2007). Methods used in virtual ethnographies range from systematic observation of online communities, laboratory experiments in computer mediated communication and production of webographies (see Hine, 2000 for a detailed discussion). Numerous ethical issues arise from using unobtrusive techniques as detailed by Paccagnella (1997) and Estalella and Ardèvol (2007). Virtual ethnographies are often undertaken in conjunction with place-based onsite ethnographies to explore the intersection of offline and online communities (Leander and McKim, 2003). For details of how to conduct a virtual ethnography, see Hine (2000) or Bird and Barber (2002).

ETHICS OF INTERNET MEDIATED RESEARCH

The use of internet mediated research using online research methods raises the issue of online research ethics. But, as Mann and Stewart (2000: 8) note, 'Because online research practice is still in its infancy, the critical researcher will be confronted by quandaries at almost every point in the research process'. Thus, the debate surrounding online research ethics is a 'work in progress' and the ethical challenges are not simple. Indeed, it is clear that many nuances to this debate will evolve as internet mediated research becomes a more mainstream and sophisticated methodology. As Jones (2004: 179) suggests: 'At present for most internet researchers it is likely that gaining access is the least difficult aspect of the research process...What has become more difficult is determining how to ensure ethical use is made of texts, sounds and pictures that are accessed for study'. This raises the question: is there anything unique to internet mediated research that requires a new set of ethical guidelines? Obviously, many ethical issues of onsite research are directly translatable to the online context, but there is also a need for existing ethical principles to be examined in the light of new virtual online research strategies. Indeed, according to the Association of Internet Researchers (AoIR) Ethics Working Committee (quoted by Ess, 2002: 180), internet mediated research can entail greater risk to individual privacy and confidentiality, greater challenges to a researcher in gaining informed consent and greater difficulty in ascertaining participants' identities. Hewson *et al.* (2003: 51) concur, proposing that ensuring pre-existing ethical standards are properly met online can be more difficult, due to the novel features of the internet environment.

Three key issues of ethical conduct are commonly identified in the literature pertaining to online research ethics: confidentiality, privacy and informed consent (see Madge, 2007 for more). With respect to confidentiality, clearly internet mediated research should aim to ensure the confidentiality of participants, as with onsite research. However, online research can add additional issues of concern with respect to confidentiality/anonymity. These issues revolve around whether participants' identities should be protected. Prior to the start of the online project, the researcher must decide whether the subject's identity is to be disguised, and to what degree. According to Bruckman (2002), subject confidentiality can range from no disguise, light disguise, moderate disguise to complete disguise. In no disguise scenarios, pseudonyms and real names can be used with the permission of the individual and the individual's claim to copyright over their words is respected. In contrast, complete disguise involves no naming of groups, pseudonyms and other identifying features are changed (such as places, institutions, user names, domain names), verbatim quotes are not used if search mechanisms could link these quotes to the person in question, and some false details might be introduced deliberately so that a subject might not recognize themselves. In this way, someone seeking to uncover a subject's identity would be unable to do so. Clearly, the level of disguise depends on the aims and remit of the research project, the researcher's ethical philosophical position and recommendations from ethical committees. In some instances, following these procedures might ensure more thorough protection of research participants than is available through face-to-face means (cf. Johns *et al.*, 2004: 119), particularly owing to the added anonymity of the virtual realm. AoIR (Ess and the AoIR Ethics Working Committee, 2002: 7) have produced some general guidelines on the issue of informant confidentiality, stressing that this varies with the nature of the research arena. It is suggested that, generally, if internet participants' are understood as subjects (e.g. chatrooms, MUDs), then a greater sense of confidentiality is required. If the participants are understood as authors (weblogs, webpages, emails to large listservs) then there is less obligation to confidentiality. Indeed, authors of weblogs/webpages may not want subject confidentiality and not to refer to material by direct quotation and specific name would be considered an infringement of copyright. Thus, in order to respect individuals who share their ideas on public lists, the names of these participants should be properly attributed.

In terms of privacy, on the internet there is no clear agreement about what is public and what is private. The problematization of the simple binary division between public and private internet space has led to a vibrant debate surrounding privacy issues. For example, is a researcher ethically justified in using publicly available information as data for a research project, even if these data were provided by the internet user for private consumption? Should a researcher be able to 'data mine' from newsgroup postings and individual webpages? Hewson *et al.* (2003: 53) conclude that data that have been made deliberately and voluntarily available in the public internet domain (including on the World Wide Web and newsgroups) should be accessible to a researcher providing anonymity is ensured. Chen *et al.*'s (2004) research on using mailing lists and newsgroups for research purposes elicited responses from a variety of sensitive/controversial mailing lists. Many of the responses included animosity towards the 'research paparazzi' in cyberspace while other responses about online researchers were more welcome. Thus, it is important to remember that the specific arena of research is important when considering the privacy issue. This is linked to another key issue facing the online researcher – whether the individual or group considers their correspondence to be public or private. Ess and the AoIR Ethics Working Committee (2002: 7) suggest that if the participants of the research believe that their communications are made in private, or if they are understood as subjects participating in private exchanges via chatrooms/MUDs or MOOs, then there may be a greater obligation for the researcher to protect individual privacy. But, if the research focuses on publically accessible archives and inter/actions by authors/agents are public and performative (for example e-mail postings to large listservs or USENET groups, production of weblogs and home pages), then there may be less obligation to protect individual privacy. Barnes (2004: 219) notes that the situation for discussion lists is complicated – they may be considered both public and private – and here she cautions that the researcher must respect the specific privacy guidelines of the online group. Indeed, many discussion groups now state their privacy or citation policy when you join them, and the online researcher should check the welcome message of public discussion lists for guidelines on how to properly cite online messages.

Finally, in terms of informed consent, the principles of onsite research should also apply in the online environment. Participants must be made fully aware of the purpose of the research project. Generally, written information about the aims of the project, the roles of the participants and any potential risks should be provided, either as an email, on a dedicated website or bulletin board, or by conventional mail. However, while the issue of informed consent shows many similarities to onsite research, there are also some differences in the virtual realm. There is a particularly thorny issue regarding failure to gain informed consent for participant observation in the online environment. Deception involves researchers deliberately concealing the purpose of their study. In theory, any research should not involve deception but in practice there is a contested debate over the issue. Some researchers, for example Denzin (1999), argue that postings on bulletin boards are public so there is no need to disclose research activity, while Chen *et al*. (2004: 164) propose that 'lurking' is an important research act prior to gaining informed consent, in order to understand the topics and tone of exchanges in a mailing list or newsgroup before becoming involved. But although lurking as socialization into the online culture of a group may be considered an important prerequisite for research, Chen *et al*. (2004: 164) also found that moderators and group leaders generally disapproved of lurking as a data-collection method, so that observation without participation was generally considered unethical research practice. Eysenbach and Till (2001) support this view, contending that researchers 'lurking' in online communities might be perceived as intruders, that might damage some communities. They therefore suggest that the online researcher must tread very carefully in order to respect their participants' lives. Vigilance regarding informed consent is therefore essential throughout the research process, and while there is still much debate, there is an emerging consensus regarding informed consent. Generally speaking, for private or semi-private sources (mail, closed chat rooms), informed consent is considered essential whereas in open access fora (newsgroups, bulletin boards), it is suggested that informed consent may not always be essential. Ess and the AoIR Ethics Working Committee (2002: 5) recommend that the greater the acknowledged publicity of the venue, the less obligation there may be to protect individual privacy and the right to informed consent.

It is clear from the above discussion of the ethics of internet mediated research that some of the issues closely reflect the basic ethical principles of onsite research, but in other instances, specific issues arise from conducting research via the internet. Indeed, one of the problems of mapping existing research ethics onto the online world is the difficulty of deciding how to categorize the online spaces that are being observed: Are they public or private spaces? Are the virtual communications made in these online spaces public or informal? Are people in these spaces expecting to be anonymous or identified? Thus, in the process of putting together a research proposal, and at all stages during the process of using internet mediated research, the researcher will need to consider the ethics involved in doing online research (Box 13.3 summarizes some ethical questions for consideration).

There are some ethical guidelines for internet mediated research which can help frame this ethical enquiry. The forum on 'The Ethics of Fair Practice for the Collection of Social Science Data in Cyberspace' (Thomas, 1996) illustrates the variety of ethical positions held for online social-science research in the early days.

More recently, there have been moves towards a growing consensus as to what ethical research practice online might entail (Mann and Stewart, 2000; Buchanan, 2004; Ess, 2004) and greater recognition of the similarities between online and onsite research ethics (Ess, 2002; Thomas, 2004). These important moves have culminated in the Association of Internet Researchers (Ess and the AoIR Ethics Working Committee, 2002) making recommendations to inform and support researchers, organizations and academic societies responsible for making decisions about the ethics of internet mediated research.

Box 13.3   Some ethical questions to consider for internet mediated research

- How will you gain informed consent?
- How can the participant withdraw from the research? Is it easier or harder to withdraw compared to face-to-face research?
- Is deception a defensible research strategy? Can 'lurking' as socialization into the online culture of a group be an important prerequisite for research?
- How can confidentiality be assured and how might this vary with the nature of the research venue (e.g. chat rooms compared to weblogs, emails compared to large listservs)?
- How might subject anonymity be achieved in practice?
- How can you improve data security? Can you promise that your electronic information will not be accessed and used by others?
- Can you agree with the online community whether the data you collect is public or private in nature?
- How might participant expectations of privacy vary with the specific research method used and different virtual research venues?
- How will you respond to subjective readings of particular virtual cultures or specific expectations of online communities?
- What are the ethical implications of researching places that you have never been to and people you have never met?
- How does the digital divide limit who you can 'speak' to?
- Can you assume people can speak freely on the internet? What censorship issues should you consider?
- What language will you use in your internet mediated research? How might this limit whom you can research?
- What online inequalities might be significant in skewing your research findings?
- What ethical issues are raised by your research and how and why are these particular to the online environment?

Although there is an emerging consensus about ethical research practice online, equally there is a general understanding that such practices are contextual and so must be applied flexibly. This flexibility is essential because of the variety of internet mediated research available, the great range of research topic investigated and the many different disciplines that can be involved in online research. Moreover, the variety of virtual venues in which internet mediated research can occur and the varying expectations of the research subjects in those venues will further influence any pursuit of ethical research practice. Since research ethics are relational and contextual, different online methods and different online venues will produce different research relationships and so online research ethics will vary with methodology as well as virtual research context, the precedents of previous researchers, personal ideological ethical position, disciplinary background, institutional context, government and funding institutions, ethical committees and specific cultural interpretations and laws (Madge, 2007). Thus, online ethical guidelines should be thought of as a springboard for critical reflexivity rather than as a prescriptive set of rules (cf. Johns *et al.*, 2004: 108) and they need to be flexible given the ever-changing nature of cyberspace.

CONCLUSIONS

In this chapter, three main types of internet mediated research have been outlined and the debate surrounding online research ethics has been introduced. However, in an article entitled 'How virtual are virtual methods?', Phippen (2007: 1) proposes that 'while there is a significant body of social research that considers the online world, the methods used to research such things are what one might describe as traditional in nature'. He argues that the vast majority of research methods used to explore the virtual world to date have been 'conventional' in nature, and have involved a translation of onsite methods (including focus groups, surveys and interviews) to the online environment. He suggests there is now the need to explore more novel methods that might be used to examine the virtual world. This call, especially to examine Web 2.0 technologies, has been mirrored by Beer and Burrows (2007). The exciting potential offered by such Web 2.0 (and 3.0!) technologies for collecting geographical data are yet to be explored fully: for example the potential of using blogs as research diaries, social networking sites as locations of virtual interviews and digital story telling for performative social science. Maybe reading this chapter will stimulate you to be an online researcher of the future?

Summary

- Internet mediated research offers interesting potential for collecting geographical information and data.
- Its potential, however, should not be over-exaggerated: it is but one tool in the geographers' methodological toolkit.
- Online questionnaires, virtual interviews and virtual ethnographies are three main online research methods that can be used to collect data through the medium of the internet and each have advantages and limitations in comparison to face-to-face methods. The important point is that they should only be used when appropriate to the specific aims of your research project.
- Online research ethics raise some similarities and some differences in relation to conventional research ethics: such issues should be considered throughout the process of conducting your research and the AoIR Ethical guidelines should be considered as a starting point of ethical enquiry for internet mediated research.
- To date internet mediated research has involved a translation of onsite methods to the online environment. The methodological potential of Web 2.0 (and 3.0) technologies still remains to be explored.

Further reading

There are a growing number of books and articles written about conducting internet mediated research. I have highlighted some of the most important contributions below. The website 'Exploring online methods in a virtual training environment' (see http://www.geog.le.ac.uk/orm/) is also a good starting point.

- For design and use of online questionnaires, see: Best and Krueger (2004); Dillman (2007); Sue and Ritter (2007); and Russell and Purcell (2009).

- For online interviews, see: Hewson *et al.* (2003); Fielding *et al.* (2008); and James and Busher, H. (2009).

- For online ethnography, the basic source is: Hine (2000).

- For online ethics, clear sources of information are: Ess and AoIR Ethics Working Committee (2002); and Ess (2009).

Note: Full details of the above can be found in the references list below.

References

Ayling, R. and Mewse, A.J. (2009) 'Evaluating internet interviews with gay men', *Qualitative Health Research*, 19(1): 566–576.

Bampton, R. and Cowton, C.J. (2002) 'The e-interview', *Forum Qualitative Sozialforschung/Forum: Qualitative Social Research*, 3(2): http://www.qualitative-research.net/fqs-texte/ 2-02/2-02bamptoncowton-e.htm.

Barnes, S.B. (2004) 'Issues of attribution and identification in online social research', in M.D. Johns, S.S. Chen and G.J. Hall (eds) (2004) *Online Social Research: Methods, Issues, and Ethics*. New York: Peter Lang, pp. 203–222.

Baym, N.K. (1997) 'Interpreting soap operas and creating community: inside a computer-mediated fan club', in S. Kiesler (ed.) *Culture of the Internet*. Mahwah, NJ: Lawrence Erlbaum Associates, pp.103–120.

Beer, D. and Burrows, R. (2007) 'Sociology and, of and in Web2.0: some initial considerations', *Sociological Research Online*, 12(5): http://www.socresonline.org.uk/12/5/17.html.

Best, S.J. and Krueger, B.S. (2004) 'Internet data collection. Sage university paper 141'. London: Sage.

Beusch, D. (2007) 'Textual interaction as sexual interaction: sexuality and/in the online interview', *Sociological Research Online*, 12(5): http://www.socresonline.org.uk/12/5/14.html.

Bird, S.E. and Barber, J. (2002) 'Constructing a virtual ethnography', in M. Angrosino (ed.) *Doing Cultural Anthropology: Projects for Ethnographic Data Collection*. Prospect Heights, IL: Waveland, pp. 129–139.

Braunsberger, K. Wybenga, H. and Gates, R. (2007) 'A comparison of reliability between telephone and web-based surveys', *Journal of Business Research*, 60(7): 758–764.

Bruckman, A. (2002) 'Ethical guidelines for research online': http://www.cc.gatech.edu/~asb/ethics/

Buchanan, E. (2004) *Readings in Virtual Research Ethics: Issues and Controversies*. Madison, WI: University of Wisconsin-Milwaukee Press.

Chen, P. and Hinton, S.M. (1999) 'Realtime interviewing using the world wide web', *Sociological Research On-line*, 4(3): http://www.socresonline.org.uk/4/3/chen.html.

Chen, S.S., Hall, G.J. and Johns, M.D. (2004) 'Research paparazzi in cyberspace: the voices of the researched', in M.D. Johns, S.S. Chen and G.J. Hall (eds) *Online Social Research: Methods, Issues, and Ethics*. New York: Peter Lang, pp. 157–175.

Curasi, C.F. (2001) 'A critical exploration of face-to-face interviewing versus computer-mediated interviewing', *International Journal of Market Research*, 43(4): 361–375.

Davis, M., Golding, G., Hart, G., Sherr, L. and Elford, J. (2004) 'Reflecting on the experience of interviewing online: perspectives from the Internet and HIV study in London', *Aids Care*, 16(8): 944–952.

Denzin, N. (1999) 'Cybertalk and the method of instances', in S. Jones (ed.) *Doing Internet Research: Critical Issues and Methods for Examining the Net*. Thousand Oaks, CA: Sage, pp. 107–126.

Dillman, D.A. (2007) *Mail and Internet Surveys – the Tailored Design Method* (2nd edn). New York: Wiley.

Dillman, D.A., Phelps, G., Tortora, R., Swift, K. Kohrell, J., Berck, J. and Messer, B.L. (2009) 'Response rate and measurement differences in mixed-mode surveys using mail, telephone, interactive voice response (IVR) and the internet', *Social Science Research*, 38: 1–18.

Dodd, J. (1998) 'Market research on the Internet – threat or opportunity?', *Marketing and Research Today*, 26(1): 60–66.

Domínguez, D., Beaulieu, A., Estalella, A., Gómez, E., Schnettler, B. and Read, R. (2007) 'Virtual ethnography', *Forum Qualitative Sozialforschung/Forum: Qualitative Social Research*, 8(3): http://www.qualitative-research.net/fqs-texte/3-07/07-3-E1-e.htm.

Ess, C. (2002) 'Introduction', *Ethics and Information Technology*, 4: 177–188.

Ess, C. (2004) 'Epilogue: are we there yet? Emerging ethical guidelines for online research', in M.D. Johns, S.S. Chen and G.J. Hall (eds) *Online Social Research: Methods, Issues, and Ethics*. New York: Peter Lang, pp. 253–263.

Ess, C. (2009) *Digital Media Ethics*. Cambridge: Polity.

Ess, C. and AoIR Ethics Working Committee (2002) 'Ethical decision-making and internet research: recommendations from the AoIR ethics working committee': http://www.aoir.org/reports/ethics.pdf.

Estalella, A. and Ardèvol, E. (2007) 'Field ethics: towards situated ethics for ethnographic research on the internet', *Forum Qualitative Sozialforschung/Forum: Qualitative Social Research*, 8(3): http://www.qualitative-research.net/index.php/fqs/article/view/277.

Eysenbach, G. and Till, J. (2001) 'Ethical issues in qualitative research on internet communities', *British Medical Journal*, 323: 1103–1105.

Fay, M. (2007) 'Mobile subjects, mobile methods: doing virtual ethnography in a feminist online network', *Forum Qualitative Sozialforschung/Forum: Qualitative Social Research*, 8(3): http://www.qualitative-research.net/index.php/fqs/article/view/278/612.

Fielding, N., Lee, R.M. and Blank, G. (eds) (2008) *The Sage Handbook of Online Research Methods*. London: Sage.

Fox, F., Morris, M. and Rumsey, N. (2008) 'Doing synchronous online focus groups with young people: methodological reflections', *Qualitative Health Research*, 17(4): 539–547.

Gaiser, T. (1997) 'Conducting online focus groups: a methodological discussion', *Social Science Computer Review*, 15(2): 135–144.

Gaiser, T.J. (2008) 'Online focus groups', in N. Fielding, R.M. Lee and G. Blank (eds) *The SAGE Handbook of Online Research Methods*. London: Sage, pp. 290–306.

Gaiser, T. and Schreiner, T. (2009) *A Guide to Conducting Online Research*. London: Sage.

Green, N. (1999) 'Disrupting the field: virtual reality technologies and "multisided" ethnographic methods', *American Behavioural Scientist*, 43: 409–421.

Greschke, H.M. (2007) 'Logging into the field – methodological reflections on ethnographic research in a plurilocal and computer-mediated field', *Forum Qualitative Sozialforschung/Forum: Qualitative Social Research*, 8(3): http://www.qualitative-research.net/index.php/fqs/article/view/279.

Hamilton, R.J. and Bowers, B.J. (2006) 'Internet recruitment and email interviewing in qualitative studies', *Qualitative Health Research*, 16(6): 821–835.

Hewson, C., Yule, P., Laurent, D. and Vogel, C. (2003) *Internet Research Methods*. London: Sage.

Hine, C. (2000) *Virtual Ethnography*. London: Sage.

Hine, C. (ed.) (2005) *Virtual Methods: Issues in Social Research on the Internet*. Oxford: Berg.

Hughes, J. and Lang, K. (2004) 'Issues in online focus groups: lessons learned from an empirical study of peer-to-peer filesharing system users', *Electronic Journal of Business Research Methods*, 2(2): 95–110.

Illingworth, N. (2001) 'The internet matters: Exploring the use of the internet as a research tool', *Sociological Research Online*, 6(2): http://www.socresonline.org.uk/6/2/illingworth.html.

James, N. (2007) 'The use of email interviewing as a qualitative method of inquiry in educational research', *British Educational Research Journal*, 33: 963–976.

James, N. and Busher, H. (2006) 'Credibility, authenticity and voice: dilemmas in online interviewing', *Qualitative Research*, 6(3): 403–420.

James, N. and Busher, H. (2007) 'Ethical issues in online educational research: protecting privacy, establishing authenticity in email interviewing', *International Journal of Research and Method in Education*, 30: 101–113.

James, N. and Busher, H. (2009) *Online Interviewing*. London: Sage.

Johns, M.D., Chen, S.S. and Hall, G.J. (eds) (2004) *Online Social Research: Methods, Issues, and Ethics*. New York: Peter Lang.

Jones, S. (2004) 'Introduction: ethics and internet studies', in M.D. Johns, S.S. Chen and G.J. Hall (eds) *Online Social Research: Methods, Issues, and Ethics*. New York: Peter Lang, pp. 179–185.

Kagawa, F. (2007) 'Dissonance in students' perceptions of sustainable development and sustainability: implications for curriculum change', *International Journal of Sustainability in Higher Education*, 8(3): 317–338.

Kivits, J. (2005) 'Online interviewing and the research relationship', in C. Hine (ed.) *Virtual Methods: Issues in Social Research on the Internet*. Oxford: Berg, pp. 35–49.

Klein, E.E., Tellefsen, T. and Herskovitz, P.J. (2007) 'The use of group support systems in focus groups: information technology meets qualitative research', *Computers in Human Behavior*, 23(5): 2113–2132.

Leander, K. and McKim, K. (2003) 'Tracing the everyday "sitings" of adolescents on the internet: a strategic adaptation of ethnography across online and offline spaces', *Education, Communication and Information*, 3(2): 211–240.

Lefever, S., Dal, M. and Matthíasdóttir, Á. (2007) 'Online data collection in academic research: advantages and limitations', *British Journal of Educational Technology*, 38(4): 574–582.

Madge, C. (2007) 'Developing a geographers' agenda for online research ethics', *Progress in Human Geography*, 31: 654–674.

Madge, C. and O'Connor, H. (2002) 'Online with the e-mums: exploring the internet as a medium for research', *Area*, 34: 92–102.

Madge, C. and O'Connor, H. (2006) 'Parenting gone wired: empowerment of new mothers on the internet?', *Social and Cultural Geography*, 7: 199–220.

Madge, C., Meek, J., Wellens, J. and Hooley, T. (2009) 'Facebook, social integration and informal learning at university: "It is more for socialising and talking to friends about work than for actually doing work"', *Learning, Media and Technology*, 34(2): 141–151.

Mann, C. and Stewart, F. (2000) *Internet Communication and Qualitative Research*. London: Sage.

Meho, L.I. (2006) 'E-mail interviewing in qualitative research: a methodological discussion', *Journal of the American Society for Information Science and Technology*, 57(10): 1284–1295.

Murray, C. and Sixsmith, J. (1998) 'E-mail: a qualitative research medium for interviewing?', *International Journal of Social Research Methodology*, 1(2): 103–121.

O'Connor, H., Madge, C., Shaw, R. and Wellens, J. (2008) 'Internet-based interviewing', in N. Fielding, R.M. Lee and G. Blank (eds) *The Sage Handbook of Online Research Methods*. London: Sage, pp. 271–289.

O'Lear, R.M. (1996) 'Using electronic mail (e-mail) surveys for geographic research: lessons from a survey of Russian environmentalists', *Professional Geographer*, 48: 209–217.

Paccagnella, L. (1997) 'Getting the seat of your pants dirty: strategies for ethnographic research on virtual communities', *Journal of Computer Mediated Communication*, 3(1): http://www.ascusc.org/jcmc/vol3/issue1/paccagnella.html.

Parr, H. (2002) 'New body-geographies: the embodied spaces of health and medical information on the internet', *Environment and Planning D: Society and Space*, 20: 73–95.

Phippen, A. (2007) 'How virtual are virtual methods?', *Methodological Innovations Online*, 2(1): http://erdt.plymouth.ac.uk/mionline/public_html/viewarticle.php?id=43.

Rezabek, R. (2000) 'Online focus groups: electronic discussions for research', *Forum: Qualitative Social Research*, 1(1): http://www.qualitative-research.net/fqs-texte/1-00/1-00rezabek-e.htm.

Russell, B. and Purcell, J. (2009) *Online Research Essentials: Designing and Implementing Research Studies*. New York: Wiley.

Rybas, N. and Gajjala, R. (2007) 'Developing cyberethnographic research methods for understanding digitally mediated identities', *Forum Qualitative Sozialforschung/Forum: Qualitative Social Research*, 8(3): http://www.qualitative-research.net/index.php/fqs/article/view/282/620.

Schmidt, W.C. (2007) 'Technical considerations when implementing online research', in A. Joinson, K. McKenna, T. Postmes and U. Reips (eds) *The Oxford Handbook of Internet Psychology*. Oxford: Oxford University Press, pp. 461–472.

Schneider, S.J., Kerwin, J. and Frechtling, J. (2002) 'Characteristics of discussion in online and face-to-face focus groups', *Social Science Computer Review*, 20(1): 31–42.

Selwyn, N. and Robson, K. (1998) 'Using email as a research tool', *Social Research Update*, Issue 21. Guildford: University of Surrey.

Simona, I. (2007) 'Ethnography of online role-playing games: the role of virtual and real contest in the construction of the field', *Forum Qualitative Sozialforschung/Forum: Qualitative Social Research*, 8(3): http://www.qualitative-research.net/index.php/fqs/article/view/280/616.

Smith, C. (1997) 'Casting the net: surveying an internet population', *Journal of Computer Mediated Communication*, 3(1): http://jcmc.huji.ac.il/vol3/issue1/smith.html.

Soukup, C. (1999) 'The gendered interactional patterns of computer-mediated chatrooms: a critical ethnographic study', *The Information Society*, 15: 169–176.

Stieger, S. and Goritz. A.S. (2006) 'Using instant messaging for internet-based interviews', *Cyberpsychology and Behaviour*, 9: 552–559.

Sue, V.M. and Ritter, L.A. (2007) *Conducting Online Surveys*. London: Sage.

Sweet, C. (2001) 'Designing and conducting virtual focus groups', *Qualitative Market Research: An International Journal*, 4(3): 130–135.

Thomas, J. (1996) 'Introduction: a debate about the ethics of fair practice for collecting social science data in cyberspace', *The Information Society*, 12(2): 107–117.

Thomas, J. (2004) 'Re-examining the ethics of internet research: facing the challenge of overzealous oversight', in M.D. Johns, S.S. Chen and G.J. Hall (eds) *Online Social Research: Methods, Issues, and Ethics*. New York: Peter Lang, pp. 187–201.

Thurlow, C., Lengel, L. and Tomic, A. (2004) *Computer Mediated Communication: Social Interaction and the Internet*. London: Sage.

Vehovar, V. and Lozar Manfreda, K. (2008) 'Overview: online surveys', in N. Fielding, R.M. Lee and G. Blank (eds) *The Sage Handbook of Online Research Methods*. London: Sage, pp. 177–194.

Watson, M., Peacock, S. and Jones, D. (2006) 'The analysis of interaction in online focus groups', *International Journal of Therapy and Rehabilitation*, 13(12): 551–557.

Witte, J.C. (2009) 'Introduction to the Special Issue on Web Surveys', *Sociological Methods and Research*, 37(3): 283–290.

14   Diaries as a Research Method

Alan Latham

Synopsis

Diaries are pieces of autobiographic writing describing in a more or less systematic way a period of an individual's life. Diaries may be written for the express purposes of a research project either by a researcher recording their own observations as part of a participant observation study, or by a research 'subject' at the request of the researcher. This chapter is concerned with the latter or what are often referred to as respondent diaries. Respondent diaries are helpful because they allow the researcher to gain a sense of the routines, rhythms and texture of a person's life, or a part of their life, over a given period of time. The chapter is organized into the following sections:

- Introduction
- When are diaries useful?
- Different kinds of diaries
- The practicalities of organizing respondent diaries
- What kinds of material might you expect? And what to do with it?
- The limitations of respondent diaries

INTRODUCTION

As geographers we are often concerned with the everyday rhythms and textures of people's day-to-day lives. In particular, we often want to understand the spatial and temporal context within which particular social practices occur. Geographers, for instance, often want to know whether a particular practice tends to occur in a certain place, at a certain time of the day? Or a certain time of the week? Or month? Or year? Geographers are also often interested in the frequency and duration of particular social practices, and how these social practices are related to other events that are involved in structuring a person's daily, weekly, or monthly life path. To begin to understand British drinking patterns and their spatialities, for example, it would be crucial to have a sense of the ways the practice of drinking is related to the rhythms of the working week. Similarly, it would be difficult to make much sense of daily variations in traffic flow in a particular city without reference to the times most people begin and finish working, or when schools open and close. In many instances, research respondent diaries are a very productive way of generating such insights.

WHEN ARE DIARIES USEFUL?

There are many different research methods geographers can employ to explore the rhythms and textures of people's day-to-day lives. Interviews, questionnaires, participant observation and focus groups can all provide insightful material. However, in turn there are also limitations to each of these methods.

First, while the use of interviews, questionnaires and focus groups offers many obvious advantages (see Chapters 6 and 8, respectively) there are limitations to the extent to which such methods can provide the geographical researcher with rich insights into the everyday. It is often unreasonable, for example, to expect people to be able to reliably remember the frequency of activities that they carry out routinely on a day-to-day basis, and often without much thought, such as eating, drinking or smoking. Ask yourself, for example, how many times you have used a bus, bicycle, or car this week? Then try and list the time and destination of every individual journey you have made in the past week. Unless you have an exceptional memory (and most people do not) – or your routine is unusually predictable and unchanging – the chances are you will find it hard, if not impossible, to produce a reliable list without referring to your personal organizer or diary.

Second, as geographers we are often interested in the texture of people's mundane interactions in particular kinds of spaces – public spaces say, or within the workplace. Precisely because many of our everyday interactions are fleeting and mundane, or just plain routine and lacking in any obvious social consequence, potential research respondents will generally struggle to be able to say anything much about them. When asked in an interview or questioned as part of a survey people will rarely be able to cite specific instances of certain interactions, and without a concrete event to organize their account around, will frequently reply in vague generalities that are difficult to interpret. In other cases, people will simply not notice that they are involved in certain kinds of interactions at all. They will overlook the fact that they smile a hello at the newsagent cashier where they buy their morning newspaper, or that they always choose to sit at a particular place to eat their lunch, or that lunch always – or nearly always – consists of the same thing. Indeed, people are often quite shocked, and sometimes rather appalled, when they realize the degree that their lives are entrained in particular routines and rituals of behaviour! Participant observation might be one way of generating research material about these sorts of interactions. However, to do so would require you to be able to follow an individual over the course of several days, closely observing their daily routines. While some researchers have been fortunate enough to be able to follow their respondents through the routines of their day (see Laurier and Philo, 2003), to do so involves very significant commitments of time and energy on the part of both the researcher and the researched. However, most geography students and many researchers simply do not have the resources to be able to commit to research of this kind. Moreover, if the respondent spends relatively little time actually involved in the activity that is the focus of research, this might prove to be an unproductive (if very rigorous) research strategy.

Diaries produced by those involved in the flow of the social practice the researcher is seeking to understand can offer ways around the two difficulties outlined above. By asking people to note down when they are involved in a certain activity either during or immediately afterwards, research is not so beholden to the capriciousness of memory. Diaries can thus produce more detailed, more reliable and often more focused accounts than can other comparable qualitative methodologies. What is more, by asking research respondents to explicitly take note of and consider particular social practices or events such as drinking allows in effect the research respondent to stand in as a proxy for the researcher. The result is that the tracing out of someone's day, or week, or month, or whatever period of time, does not require the constant presence of the researcher. Instead the diary allows the researcher to 'virtually accompany' her or his research respondent as they go about their day-to-day routines without the intrusiveness and significant time demands that actually physically shadowing an individual would entail. In the same time period that it would take to conduct an in-depth participant observation study of the routines and habits of just one person, a diary-based study might be able to obtain data from four or five different diary respondents. However, it is also worth noting that the detail of the insights gained from participant observation may well be of a commensurately greater quality.

There are a number of further attractions to the use of respondent, or what are sometimes termed solicited diaries (see Meth, 2003). First, by asking people to attend to particular practices or events – whether that be journeys undertaken, people talked to, food eaten, or a whole range of other activities – people become more aware of the practices that they are involved in. So, for example, in a study on interactions in public, the respondent may become more aware of the degree to which they are in fact involved, and communicate implicitly with other people even though that involvement may not include explicit markers of engagement such as talking (e.g. smiling as you pass someone in the street, or holding open a door for a stranger). Similarly, a food diary might make a diarist aware in a way that they had not previously been of the degree to which the pleasure of certain kinds of leisure activities – television watching, say – is bound up with the consumption of certain kinds of foods. Second, in a manner that few other research techniques can match, diaries can provide respondents with a chance to reflect upon their lives in a systematic and sustained way. The completion of a diary offers a chance for respondents to quite literally 'write' a narrative of their day-to-day lives. Diaries also offer a chance for respondents to reflect on the wider meaning of the events and activities reported in their diaries, and it gives the enthusiastic diarist opportunities to place the ordinary events recounted in their diary within a broader biographical canvas. So, respondents might explain the pleasure of a certain kind of food within a narrative reflecting on what their mother used to cook when they were children. Or, the attraction of walking to work might be explained through the resonances such a walk has with childhood memories of walking to school. Third, if the remit of the diary is left reasonably open, the events the diaries narrate and the manner in which the diarist describes them may well suggest avenues of interest and concern that the researcher had simply not thought about, or had considered irrelevant or trivial. Fourth, as Felicity Thomas (2007) has highlighted in her work on HIV/AIDS in Southern Africa, diaries can provide an opportunity for respondents to explain and explore highly emotional and personally sensitive issues with a frankness and openness that face-to-face interactions might inhibit them from doing. Fifthly, and finally, in reflecting upon their day-to-day lives, and the specific elements they have been asked to attend to, respondents may begin to offer folk theories and vernacular explanations that may help the researcher better understand such activities and formulate their own theoretically literate accounts.

DIFFERENT KINDS OF DIARIES

Geographers have used research respondent diaries to study a wide range of different themes. Examples of diary use range from the experience of urban pedestrianism (Middleton, 2009), practices of food consumption (Valentine, 1999), experiences of violence (Meth, 2003), the sociality of cafés, bars and other hospitality spaces (Latham, 2003, 2006), New Age spirituality (Holloway, 2003), the mobility of blind people (Cook and Crang, 1995), children's journeys to school (Murray, 2009), consumer shopping decisions (Hoggard, 1978), the lives of street children (Young and Barrett, 2001), the relationship between fishermen and wildlife in Zimbabwe (McGregor, 2005), childhood in rural Bolivia (Punch, 2001), the lives of middle-class migrants (Conradson and Latham, 2007), the routines of homeless people (Cloke *et al.*, 2008) and the lives of women with HIV/AIDS (Thomas, 2007). This variety of topics indicates something of the versatility of respondent diaries as a research technique. The variety, however, also points to the diversity of ways in which respondent diaries can (and have) been used by human geographers.

In some of these examples the research relied solely on diaries. In other cases diaries were used together with a range of related methods such as participant observation and in-depth interviews. In yet other cases, the production of diaries was directly connected with follow-up diary-based interviews. There is also a great deal of variation in the kinds of diaries that were relied upon. In some cases, the researchers asked diarists only to attend to a very narrow range of parameters, while others simply asked diarists to describe what they felt was important, leaving the style and content of the diary entirely open to the respondent diarist's judgement. In some cases, people were not asked to keep a written diary but rather asked to provide a photographic diary of their lives, or provided with a video camera and asked to make a video diary. So, rather than being a single, easily definable, method or technique, respondent diaries in fact represent a quite broad set of research techniques. That said, it is possible to identify five basic types of respondent diary.

Diary-logs

A diary-log is simply a log-book where respondents are asked to note down as precisely as possible tightly defined details about certain key activities. Diary-logs are useful for generating quantifiable data. For example, data about travel patterns, or working hours. Diary-logs are highly prescriptive by design, providing little (or ideally) no scope for interpretation from the diarist (see Carlstein *et al.*, 1978; Schwanen *et al.*, 2008).

Written diaries

A written diary is the form of diary most commonly associated with respondent diaries. A written diary – usually simply referred to as a diary – is a description of a period of a research respondent's life, written by the person in question, and commissioned by the researcher. The remit of a diary can vary enormously. In some cases diarists are simply provided with a standard diary and asked to describe their day. In other cases, diarists are asked to focus only on certain kinds of activities, or activities that take place in particular places. A written diary may contain elements of a diary-log. For example, a researcher may be interested in the everyday movements of an individual and would like to use the material from the diary to produce a map of the diarist's daily movements. In this case the researcher may instruct the diarist to include very specific details about journeys undertaken, their timing, and purpose. Because the format of the diary is typically open-ended, the style, detail, focus and depth of the diaries produced within a single diary based project will often vary enormously from person to person.

Photographic diaries

Photographic diaries differ from written diaries in that rather than relying on written accounts, research respondents are asked to describe or illustrate elements of their lives through the medium of photography. Typically, respondents are given a disposable camera and asked to photograph that which the respondent feels is most relevant. Once the camera is full the diary is understood to be finished. This obviously limits the scope of the diary. However, digital cameras although much more expensive than disposable cameras offer a range of possibilities for widening the remit of the diarist – as they allow the diarist to take many more photographs, as well as allowing them to delete, edit and retake photographs that they are unhappy with. In some cases, diarists might be asked to note down when and where each individual photograph was taken, and why the photograph was taken. This is by no means universal. In most cases, upon the completion of the diary the diarist will be asked to talk through the photographs taken, with the researcher. An advantage of photographic diaries is that they do not require any degree of literacy. A further advantage is that, compared to a written diary, photographic diaries, in general, require much less of a time commitment. Rather than having to compose a written diary entry the diarist simply has to point a camera and take photographs. While photographic diaries represent a discrete form of diary production, in practice photographic diaries are often combined with written diaries.

Video diary

Video diaries are a relatively new, and at least within geography, a little explored form of diary production. Video diaries can be of two forms. They may involve the diarist simply talking to a static video recounting the events of a day. Or, as in the example of Murray (2009), the video might be used as a device to record key elements of a person's day; a child's journey to school, for example. The advantage of video is that through the ability to record significant blocks of time, and in catching movement, it provides an immediacy of context difficult to match in written and photographic diaries. That said, video diaries also have a number of disadvantages. They are more obtrusive than written or photographic diaries. The equipment necessary to make quality recordings is relatively expensive. Also, the usefulness of the diary is very dependent on the ability of the diarist to competently use the equipment provided for them to produce a diary.

Diary-interview

Strictly speaking this is not another form of diary production. Following the lead of American ethnographers Zimmerman and Wieder (1977) many geographers who have used diaries as a research technique have treated diary writing as an iterative process (see Holloway, 2003; Latham, 2006; Middleton, 2009). Research respondents are asked to write diaries. On the completion of the diary, the researcher then undertakes an in-depth interview with the diarist based on the diary. In the diary interview the diarist is asked to lead the researcher through the diary. This allows the diarists to explain ambiguities in their written account. It also provides the diarist with an opportunity to reflect upon and – if they feel necessary – to expand on the accounts presented in the diary. Additionally, the diary interview offers the researcher a chance to ask about the wider context of the events presented in the diary. Researchers have the opportunity to explore the diarist's relationship with key actors present within the respondent's diary, and they can explore the extent to which the research respondent feels the events recorded in the diary are representative of the respondent's life.

THE PRACTICALITIES OF ORGANIZING RESPONDENT DIARIES

Organizing the production of respondent diaries is an involved, drawn out and far from straightforward process. That said, following a few simple rules can significantly assist in the smooth running of a respondent-diary-based research project.

1   *Think carefully about what kind of information you want to generate from respondent diaries.* If your main interest is obtaining accurate details about when people engage in a certain activity, it may be superfluous to ask them to write about everything they do during a week. Similarly, if you are interested in the general texture of a person's day it may be inhibiting to demand that a diarist lists the exact time and date of everything recorded in the diary. In fact, it is a good idea to spend a few days or, better, a week filling out a diary in the manner that you intend to ask your respondents to. This will allow you to think about the details you would like your diarists to focus on, and the

appropriate strategies that need to be adopted to facilitate this focus. Writing you own diary also allows you to gain a sense of the time demands that diary writing is likely to make on potential diarists.

2   *Think carefully about who you want to recruit as diarists, and how you are going to recruit them.* The recruitment of diarists can be the most time-intensive part of the research process. If you are fortunate you may have an existing research contact who will be able to provide willing diarists. More commonly you will have to think creatively about how to recruit diarists. The most reliable technique is to ask acquaintances if they know people who fit the profile of the kinds of respondent diarists you are wishing to recruit. Advertisements in local or community newspapers, listing magazines such as London's *Time Out*, or notices on internet discussion boards, can also be effective. Similarly, do not be afraid to use wanted posters in places potential respondent diarists are likely to frequent. Once you have managed to recruit a number of initial diarists further recruitment through snowballing is often effective.

3   *Think about the competencies of the people you are recruiting as diarists.* One of the great strengths of respondent diaries is that they draw on the narrative skills of those producing them. This, of course, demands that if you are going to ask a certain population of people to produce a diary that they have those skills. If you are working with a social group with a low level of literacy written diaries may not be appropriate. Or, rather, you may only be able to gain diary accounts from relatively highly educated and privileged groups (although see Meth, 2003; Thomas, 2007). In this case, another form of diary keeping such as photographic diaries might be more appropriate.

4   *Provide diarists with a clear briefing of what you expect them to do.* Diarists need to have a good sense of what they are being asked to produce. They also need to have a reasonable sense of the purpose of the project to which they are contributing. Ideally the researcher should brief the diarist in person. This gives diarists the chance to clarify with the researcher just what they are being asked to do. Respondents should also be provided with a detailed instruction sheet. The instruction sheet should include information about who is undertaking the research, the institution the researcher is affiliated with, and contact details for both the researcher and others involved in supervising the research project. The instruction sheet should be firmly attached to the diary given to the diarist to complete.

5   *Provide your diarists with a notebook (or camera if it is a photographic diary) and a pen.* As the researcher is asking people to produce a diary the researcher must provide the diary. The diary should be robust, easy to carry around and have enough pages for the respondent to complete the task asked of them. Also it is good practice to provide a pen with which to write the diary.

6   *Devise a straightforward procedure for returning completed diaries.* Getting diaries back from diarists can be surprisingly time consuming. The most reliable way of getting diaries returned is to pick them up directly from the diarist. This has the advantage that you can ask the diarist about how they found the diary-writing process. If you are combining the diary with diary-interviews the diary pick-up also offers an opportunity to arrange a time for the interview. However, if you have a number of diarists writing at the same time, if your diarists are very

busy, or are dispersed over a large area, it may not be practicable to personally pick up each and every diary. In this case, you should provide diarists with a pre-paid self-addressed envelope and instruct the diarist to return by post.

WHAT KINDS OF MATERIAL MIGHT YOU EXPECT?
AND WHAT TO DO WITH IT?

The kind of research material generated through respondent diaries is dependent on the instructions given to the respondent diarists.

In the case of a diary-log the diarist will have produced a set of responses that are easily assimilated into a quantitative database (see Schwanen *et al.*, 2008). More open-ended diaries should be approached like any other set of qualitative data. As with a recorded interview, it is good practice to type out the text of the diary either into a word-processing document or a qualitative research program such as NVIVO or Ethnograph (see Chapter 28). It is important to recognize that the quality and detail of diaries may vary enormously (see Figure 14.1). Do not treat shorter, less detailed, diaries simply as failures that should be ignored. Indeed, it may be tempting to organize any research account primarily around diary material generated from the most loquacious and personable diarists. This temptation should be resisted. While the longer and more detailed diaries may offer more obvious sources for quotation and illustration, the shortness of other diaries might well point to equally important conclusions. Be prepared to recognize that there are multiple realities to any social situation, and work hard to construct research accounts that pay due respect to that.

In many ways the most challenging part of the analysis of diaries involves devising ways of (re)presenting diarists accounts that respect the texture of the lives recounted in the original diary material. Especially if the reason for using respondent diaries is a desire to understand the rhythms and routines of people's day-to-day lives, it is important to attempt to produce research accounts that express something of the vitality of those lives. This does not simply mean the research accounts produced from respondent diaries should quote liberally from the original diaries – although that may well be one appropriate strategy. It also suggests the need to experiment with different ways of narrating research material. This could involve a range of different strategies, from exploring new ways of diagramming time-space and 'mapping' everyday geographies (Figure 14.2) to simply letting diary material speak for itself.

THE LIMITATIONS OF RESPONDENT DIARIES

Respondent diaries can produce research material that is enormously productive. However, it is equally important to be aware of some of the limitations of diaries as a research technique.

DAY FOUR   Weds   DATE: 11-2-04

Time of contact	Your geographical location at time of contact (place, town/city)	First name of Person contacted* (add G: if part of a group)	Relation of person to you (e.g. friend, partner, family member, relative, colleague)	Geographical location of person at the time of contact (place, town/city)	Where did this person live at the time of contact? (area, town/city)	Mode of contact (in person, landline, mobile phone, email, text, MSN messenger, or other – please specify)	Nature of contact (brief description of communication or activity)	How long have you known this person? (approximate months/years)
10am	Worth Teddington	Ed	friend	Kingston London	Kingston London	text	brief chat	8mths
12.30	Work Teddington	Wayne	friend	Raynes Pk Teddington	Raynes Park London	Mobile phone	Advice about course	11yrs
1300	Cafe Teddington	Rebecca	Worker/friend	Cafe Teddington	?	In person	Introduction chat	2 weeks
8pm -10pm	Home R/Park London	Baloo G Bindu G	Friend	Raynes Park London	Wimbledon, London	In person	Cell group discussion	18mths 2 months
		Hannah G Andy G	" "	" "	" "	" "	" "	2 months 6 months
		Adam G Diane G Joy G	" " " "	" " " "	" " " "	" " " "	" " " "	4 months 1 year 18 months

Figure 14.1   Excerpts from two different types of diary
Source: The author

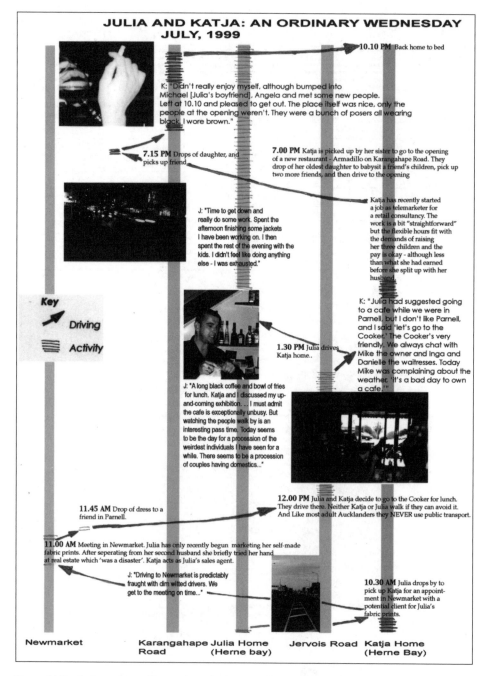

Figure 14.2 A time–space diagram based on respondent diaries

Source: The author

First, respondent diaries make significant demands on both the researcher and the research respondent. For the researcher the logistics of finding appropriate diarists, of distributing the diaries, ensuring that diaries are completed in a manner that is aligned with the project's research aims, ensuring the retrieval of completed diaries and arranging diary-interviews should not be underestimated. Of course, all good research is demanding of a researcher's time and intellectual resources. However, diaries make a much greater demand on a research respondent's time than more commonly employed techniques such as interviews or focus groups. That said, diaries are on the whole less demanding than a participant-observation-based study. What is more, respondent diaries require an on-going commitment from the diarist. This can make it difficult to recruit diarists and it may mean that many groups who would make ideal diary subjects will refuse to write diaries due to the time constraints they are under (it is likely to be very difficult to persuade the busy chief executive of a large company to keep a detailed diary of their working day, for example).

Second, as has already been mentioned, diary keeping assumes a certain set of personal competencies. Researchers such as McGregor (2006), Meth (2003) and Thomas (2007) who have used respondent diaries in non-western contexts have stressed the need to recognize the extent that the diary is in all sorts of ways a western technique of self-reflection deeply embedded in western traditions of self-hood. More prosaically, producing a diary requires certain basic skills: an ability to write, or an ability to use a camera, an ability to keep track of the passage of time, and so on. It might be assumed that these skills are generally distributed in most populations. However, even in highly educated societies like the United Kingdom, America, Canada or New Zealand, there are wide variations in people's levels of literacy, capacity for self-organization, and so on. Even highly competent individuals may find completing a diary intimidating, especially if in their normal day-to-day life they are rarely called upon to produce self-directed blocks of hand written text. Indeed, given the contemporary ubiquity of computer and keyboard use it may well be necessary to devise online forms of diary keeping as hand writing becomes for many an archaic – and thus alien – technique of communication.

Third, respondent diaries – whether written, photographic, or video based – can produce an enormous variety in the quality and depth of material generated. If one of the advantages of the diary technique is that it allows the respondent diarist to stand in for the researcher, the flip side of this is that many people are (sadly) rather poor observers and reporters. As with all qualitative techniques, it is tempting to focus on those respondents who produce the most detailed and compelling accounts. However, it is important to provide equal weight to the inarticulate and unobservant – not least because their accounts may well be more representative of the general state of affairs.

Summary

This chapter has provided an introduction to the use of respondent diaries.

- Respondent diaries are diaries commissioned by a researcher to provide information on some aspect of a person's life.
- Respondent diaries can be an excellent way of generating material about the rhythms and routines of people's day-to-day lives.
- There are a number of different approaches to producing respondent diaries. The content of respondent diaries may be highly prescriptive, or left entirely up to the prerogatives of the diarist.

- As well as written diaries, geographers and social scientists have also used respondent photographic and video diaries. Video and photographic diaries have in some cases been combined with written diaries.
- Organizing respondent diaries is an involved and sometimes complicated process. The researcher must provide clear guidelines about how they want diarists to approach their diaries. The researcher must also provide a concise sense of the purpose of the research the diary is being solicited for.
- Diary-interviews often provide an excellent supplement to a written or photographic diary.
- Research material generated through respondent diaries can be analysed in a range of ways using standard social-scientific analytic techniques.
- Research material generated through respondent diaries offers a range of distinctive opportunities for experimenting with different ways of diagramming and mapping the rhythms and textures of everyday life.

Further reading

The following articles and books provide good examples of respondent diaries in practice:

- Alaszewski (2006) offers a comprehensive overview of the many ways diaries have been used in the social sciences.

- Latham (2006) provides an account of everyday urban sociality based around photo-diaries, written diaries and diary interviews, and presents some interesting explorations of different ways to (re)present diary-based material. A detailed reflection on the development of the methods used to gather this material is also available in Latham (2004).

- Meth (2003) in a reflection on the advantages of using respondent diaries, provides a compelling example of the capacity of diaries to generate research material about emotionally sensitive issues.

- McGregor (2005) provides a wonderful example of the shift in perspective respondent diaries can provide. A study of the relationships various groups have to the Nile crocodile in Zimbabwe, the respondent diaries produced by Tonga fishermen provide a striking counterpoint to the accounts offered by scientists and those campaigning for the protection of crocodiles.

- Middleton (2009) is a thought-provoking study of the practice of walking to work. Demonstrates the usefulness of diaries for conveying a sense both of the texture of an experience, and how that experience is related to wider routines.

- Zimmerman and Wieder (1977) is an account of how two ethnographers developed the diary, diary-interview method as a way of obtaining observations about social spaces the ethnographer would normally have difficulty gaining access to. This paper has been influential in the work of many geographers who have employed diary-based methods.

Note: Full details of the above can be found in the references list below.

References

Alaszewski, A. (2006) *Using Diaries for Social Research*. London: Sage.
Carlstein, T., Parkes, D. and Thrift, N. (eds) (1978) *Timing Space and Spacing Time: Human Activity and Time Geography* (vol. 2). London: Edward Arnold.

Conradson, D. and Latham, A. (2007) 'The affectual possibilities of London: Antipodean transnationals and the overseas experience', *Mobilities*, 2(2): 231–254.

Johnsen, S., May, J. and Cloke, P. (2008) 'Imag(in)ing "homeless places": using auto-photography to (re)examine the geographies of homelessness', *Area,* 40: 194–207.

Hoggard, K. (1978) 'Consumer shopping strategies and purchasing activities', *Geoforum*, 9: 415–423.

Holloway, J. (2003) 'Make-believe: spiritual practice, embodiment and sacred space', *Environment and Planning A*, 35(11): 1961–1974.

Latham, A. (2004) 'Researching and writing everyday accounts of the city: an introduction to the diary-photo diary-interview method', in C. Knowles and P. Sweetman (eds) *Picturing the Social Landscape: Visual Methods and the Sociological Imagination*. London: Routledge, pp. 117–131.

Latham, A. (2006) 'Sociality and the cosmopolitan imagination: national, cosmopolitan and local imaginaries in Auckland, New Zealand', in J. Binny, J. Holloway, S. Millington and C. Young (eds) *Cosmopolitan Urbanism*. London: Routledge, pp. 89–111.

Laurier, E. and Philo, C. (2003) 'The region in the boot: mobilising lone subjects and multiple objects', *Environment and Planning D: Society & Space*, 21(1): 85–106.

Middleton, J. (2009) '"Stepping in time": walking, time, and space in the city', *Environment and Planning A*, 41(8): 1943–1961.

McGregor, J. (2005) 'Crocodile crimes: people versus wildlife and the politics of postcolonial conservation on Lake Kariba, Zimbabwe', *Geoforum*, 36(3): 353–369.

McGregor, J. (2006) 'Diaries and case studies', in V. Desai, and R. Potter (eds) *Doing Development Research*. London: Sage.

Meth, P. (2003) 'Entries and omissions: using solicited diaries in geographical research', *Area*, 35(2): 195–205.

Murray, L. (2009) 'Looking at and looking back: visualization in mobile research', *Qualitative Research*, 9(4): 469–488.

Punch, S. (2001) 'Multiple methods and research relations with young people in rural Bolivia', in M. Limb and C. Dwyer (eds) *Qualitative Methodologies for Geographers*. London: Arnold, pp. 165–180.

Schwanen, T., Kwan, M.-P. and Ren, F. (2008) 'How fixed is fixed? Gendered rigidity of time–space constraints and geographies of everyday activities', *Geoforum*, 39: 2109–2121.

Thomas, F. (2007) 'Eliciting emotions in HIV/AIDS research: a diary-based approach', *Area*, 39(1).

Valentine, G. (1999) 'A corporeal geography of consumption', *Environment and Planning D: Society and Space*, 17: 329–351.

Young, L. and Barrett, H. (2001) 'Adapting visual methods: action research with Kampala street children', *Area*, 33(2): 141–152.

Zimmerman, D. and Wieder, D. (1977) 'The diary: diary interview method', *Urban Life*, 5(4): 479–498.

Generating and Working with Data in Physical Geography

Getting Information about the Past: Palaeo and Historical Data Sources of Climate

15

Catherine Souch

Synopsis

The instrumental record of climate (where variables are measured directly) extends over only a tiny fraction of earth history and provides only very patchy spatial coverage. In order to understand past climatic conditions, and to provide a baseline against which to document and evaluate current and future climate change, reconstructions of past climatic and environmental conditions rely on proxy (or surrogate) data and numerical climate models. Proxy records are historical (documentary) archives or natural (physical, chemical or biological) systems that are dependent on climate and which incorporate in their structure some measure of this dependency so it is preserved through time. The challenge in interpreting the data is to separate the effects of climate (the signal) from all other non-climatic influences (the noise).

This chapter is organized into the following sections:

- Introduction
- Proxy data sources
- Analysis of proxy records
- Issues to consider with proxy data and reconstructions
- Conclusion

INTRODUCTION

Variability is an intrinsic property of the earth's natural environment. Atmospheric conditions, for example, vary over time periods which range from less than seconds to billions of years (the age of the earth). While some of these changes are periodic, linked to daily, annual and millennial astronomical cycles, others are more episodic, related to the processes and feedbacks internal to the earth-atmosphere system. Today, significant attention is focused on natural climate change and variability, not only to gain insight into past environments, but also to provide a baseline against which to document and evaluate the ever-increasing impacts of human activities (IPCC, 2007).

The instrumental record is generally accepted to be too short to provide a complete picture of climate variability and change (Figure 15.1). Direct observations of the environment, the atmosphere, hydrosphere, lithosphere and biosphere have been conducted for only a tiny fraction of the earth's ~ 4.5 billion-year history. Routine (instrument-based) observations of climate began in Western Europe during the late seventeenth and early eighteenth centuries, but many countries

Figure 15.1   Timescales of temperature change variability over the past 100,000, 10,000, 1,000 and 100 years

only established hydrometeorological networks in the late nineteenth century, after the Vienna Meteorological Congress in 1873. Today, surface data remain sparse for both polar regions, and regular measurements are available for only two-thirds of the surface of the southern hemisphere. Thus, instrumental data provide an inadequate perspective (both temporally and spatially) on natural and human-induced environmental change and variability and the development of conditions today.

If we are to understand how large natural climate change can be; how rapidly climate change can occur; the effects of climate change on other environmental processes at local, regional and global scales; and/or the extent of the effects of human activities, the record of climate information has to be extended back before the period of direct measurements. To do this we must rely on proxies – measurable properties of biological, chemical or physical systems – that provide quantitative information about past temperatures, rainfall, ice volumes, etc. Comprehensive reviews on this topic are provided by Bradley and Jones (1995), Bradley (1999), Jones *et al.* (2001), Ruddiman (2001), Alverson *et al.* (2003), Oldfield (2003), and the studies cited in NRC (2006) and IPCC (2007).

As in any field, the tools we use shape our understanding. Proxy data do not provide a direct record of climate; rather, they have acted as a filter transforming climatic conditions at a point/period in time, into a physical, chemical, or biological signature. Each line of evidence differs in many important regards – for example, the climate information it can yield, its spatial (geographic) coverage and representativeness, the period for which information can be reconstructed (temporal coverage), and its resolution (the ability to resolve events accurately in time). These are key issues to consider when evaluating proxy data and the reconstructions based upon them. Each is discussed further below.

Exactly which proxy data are used in a study depends on the research question being asked, the location of the study region (thus what proxies are available), and an array of logistical issues and constraints (access to equipment, laboratory facilities and technical expertise). Often, independent proxies are found in close proximity. Thus they can be, and should be, used in combination to provide complementary insight into past environmental conditions and to allow results to be cross-checked and uncertainties understood.

PROXY DATA SOURCES

The two general classes of climate proxies most commonly used to yield quantitative reconstructions of past climate are *biotic proxies*, which are based on the composition of plant and animal species or populations and/or measures of their growth rates, and *geological proxies*, which quantify changes (*physical* or *chemical*) in earth's materials that have accumulated through time, most commonly sediments in oceans or lakes, or ice in polar or alpine glaciers. These proxies, organized by the natural archives in which they are stored, are described in Table 15.1. Attention to where each proxy is found, and the length and detail of the records they can yield (their spatial and temporal coverage and resolution) is considered further below.

Where biologic material is preserved through time – for example through annual growth rings/layers in living systems such as trees and corals, or in lakes, bogs, estuaries, or ocean sediments – records of past biotic changes, which may

Table 15.1 Common sources of proxy data for paleoclimatic interpretations and their key characteristics. High-resolution proxy data sources provide information at daily, seasonal and annual timescales, often for specific locations and/or small regions. Low-resolution data, in contrast, provide integrated climate data for timescales of centuries to millennia

Archive	Proxy records	Climatic elements	Temporal resolution/coverage	Geographic coverage
High-resolution, local–regional scale patterns				
Historical	Written records of weather or related phenomena, or phenological information	Temperature, precipitation, snowfall, frost, floods	Detailed information (day, hour, season). Oldest records ~5000 y. Most records <10^2 y and discontinuous	Global. Longest records East Asia, Middle East and Europe. Local information
Tree rings	Ring widths, cell structure/density, isotopic composition	Temperature, rainfall, drought, run-off, cycles such as El Niño and Pacific Decadal Oscillation	Continuous records. Ring width resolution 1 year; may contain seasonal information (cell structure 1–5 weeks). Oldest records ~8000 y S.W. USA 10^{2-3} y common	Continental areas excluding desert and tundra regions. Mainly mid latitudes. Information on local/regional climates
Lake sediments	Varves (annual layers): thickness/composition; relative abundance of pollen & macrofossils; diatoms, ostracods & other aquatic biota and isotopic composition	Streamflow/snow melt, rainfall, temperature	Continuous records. Resolution $10–10^2$ y. Most records <10^4 y Oldest non-varved records ~ 10^5 y	Varves – need strong seasonality and absence of post-depositional mixing. Deep anoxic environments. Other lake records, most continents
Corals	Growth rates, isotopes, trace elements	Sea surface temperatures, adjacent continental rainfall & run-off, ocean circulation, tropical winds	Annual resolution. Coverage 10^4 y	Tropical/sub-tropical oceans
Ice cores	Ice Fabric, stable isotopes (H and O), gas content of air bubbles, trace element & micro-particle concentrations	Snowfall, temperatures, humidity, wind speed & atmospheric circulation, atmospheric composition	Continuous records. At best annual resolution for last 10^4 years. Longest record >750,000 y Antarctica	Glaciated regions. Polar and high mountain regions – primarily Greenland/Antarctica and ice-caps of Peru, Bolivia, China and Tibet

Table 15.1 (Continued)

Archive	Proxy records	Climatic elements	Temporal resolution/coverage	Geographic coverage
Terrestrial geomorphic evidence	Closed lake basin – lake levels; glacial & periglacial features; aeolian deposits & loess; speleothems; relict soils	Effective/net rainfall, run-off, temperature	Discontinuous records. Low resolution (10^2 y), often poor chronological control. Coverage varies, up to 10^6 y	Continental areas. Lake-levels – arid/semi-arid regions; glacial/periglacial features mid-high latitudes
Marine Cores	Isotopic composition, geochemistry (elemental ratios), mineralogy terrestrial (aeolian) dust & ice-rafted debris, floral & faunal abundance	Temperature (surface and deep water), ice volume, ocean circulation, aridity of continents, intensity and direction of winds	Continuous records. Resolution varies from about 100 to 2500 y depending on rate of deposition. Some coastal basins are varved (annual resolution). Temporal coverage up to 10^8 y	Data from virtually all oceans and latitudes (70% Earth's surface) – integrated global signal

Low-frequency, regional–global-scale patterns

be related to climate, can be interpreted. Given that fossils of plants tend to be more abundant than those of animals in geologic records of continents, vegetation plays a critical role in the reconstruction of ancient climates (millions to billions of years). Warmer climates tens of millions of years ago are inferred from the presence of palm-like trees (temperature sensitive species) in northern latitudes, for example (Ruddiman, 2001). For the younger continental record, the relative abundance of species, based on plant remains in the form of either macrofossils (cones, seeds, leaves) or pollen (which is very resistant to decay because of its protective coating), yields sensitive records of terrestrial climate (e.g. temperature and precipitation). In the oceans, the relative abundance and distribution of four main groups of shell-forming animal and plant plankton (formanifera, coccoliths, diatoms and radiolaria) are used for climate reconstructions.

Geological proxies rely largely on interpreting changes in the volume and/or physical/ chemical properties of earth materials (sediment or snow/ice) accumulating through time. Virtually all continental sedimentary deposits (aeolian, glacial, fluvial, lacustrine) convey some sort of paleoclimatic signal. However, it is often difficult to identify the unique combination of climatic events leading to the formation of such deposits. Much more detailed and precise information is derived where earth materials accumulate continuously through time, for example sediments in ocean and lake basins, and snow and ice in polar and alpine ice fields. Analyses of rates of accumulation, physical properties (size, shape) and geochemistry (mineralogy, trace element content, and isotopic composition) of the materials all yield information about past climates. The key sources of long-term climate history (thousands to millions of years) come from ocean cores (Bradley, 1999; Ruddiman, 2001). Most geographers, however, given their interest in natural climate variability on societal time scales and the impact of humans on the environment, are concerned with more recent, higher resolution geological records, where changes over years to centuries can be identified. Optimal records for such applications include ice cores (polar and alpine environments) and varved (annually layered) lake sediments (most commonly found in glaciated and arid environments).

Of emerging importance are proxies that rely on the chemistry of biological and physical entities which reflect well-understood thermodynamic processes that can be related to climate parameters such as temperature. Key examples include: oxygen (O) isotope ratios in coral and foraminiferal carbonate to infer past temperature and salinity; magnesium/calcium (Mg/Ca) and strontium/calcium (Sr/Ca) ratios in carbonate for temperature estimates; alkenone saturation indices from marine organic molecules to infer past sea surface temperature (SST); and oxygen and hydrogen isotopes and combined nitrogen and argon isotope studies in ice cores to infer temperature and atmospheric transport (for further information on all of these and their applications, see the comprehensive review volumes cited above).

Potential sources of historical paleoclimatic information also include ancient inscriptions, annals and chronicles, government records, private estate records, maritime and commercial records, personal papers (such as diaries and letters) and scientific and/or quasi scientific writing (such as non-instrumental weather journals). Contained within these, historical (documentary) evidence can be grouped into three main categories (Bradley and Jones, 1995): (1) direct observations of weather, for example the frequency and timing of frosts, or the occurrence of rainfall or snowfall; (2) records of weather-dependent phenomena, such as droughts, floods, river or lake freeze-up and break-up; and (3) phenological records, which describe weather-dependent biological phenomena, for example the flowering of shrubs or trees, the arrival of migrant birds, or crop yields. Stone inscriptions related to the Nile flood levels date back to ~5000 years BP (before present); Arabic chronicles of rainfall through the Middle East (Iraq, Syria) to ~1000 years BP. Documentary records do have a number of advantages: normally events are precisely dated, providing a high level of temporal resolution from daily to yearly time scales and often they describe events with important consequences for humans (floods, droughts, landslides, etc.). However, many records are discontinuous, very localized and may be strongly biased by individual observers. Moreover, such records often tend to focus on extreme events while long-term trends go undocumented. Thus, it is difficult to subject the data to sophisticated statistical analyses. However, historical records have been very important in documenting conditions throughout Europe, China and Japan for the last millennium, and have yielded detailed information about the cooling and warming phases through the Little Ice Age, a period of cooler temperatures from ~1550 to 1900 (Bradley and Jones, 1995).

ANALYSIS OF PROXY RECORDS

Regardless of the proxy used, commonly three steps are employed to extract the climate signal from extraneous noise (the influence of other factors). First, the physical, chemical and/or biological attributes of the system dependent on climate are identified. Second, the proxy data source is calibrated with contemporary climate data, often through statistical regression. This approach has the inherent assumptions that the modern relationships have operated unchanged through the period of interest (the principle of uniformitarianism) and that a modern analogue exists (conditions from the past can be found somewhere on Earth today). Third, the present-day relationships (often referred to as transfer functions) are then applied to the longer proxy record to infer past environmental/climatic conditions.

To illustrate this approach, the analysis of annual variations in tree-ring widths (dendroclimatology), one of the most extensively used proxy records in mid- and high-latitudes, is considered (Figure 15.2). The basic premise is that a tree's growth is a function of climatic conditions and this climate dependence is recorded in the width, density, or isotopic composition of the tree's annual growth rings. To collect raw data on ring widths, an increment borer is used to take tree cores (~0.5 cm in diameter) from multiple trees of the same species within a stand (Figure 15.2a). Analysis of proximal replicate trees aims to remove the effect of site-specific conditions that might influence growth rates of individual trees. All the individual tree cores are then cross-dated; i.e. all rings of the same age identified (Figure 15.2b). This is essential so the age of each ring is precisely known. The in-built chronology of tree rings, which allows annual or even seasonal resolution, is one of the greatest advantages of tree-ring analysis. If trees are collected from living and dead trees (in the ground, preserved in peat-bogs or incorporated into buildings, for example) continuous records may extend back centuries to thousands of years (see examples

Figure 15.2 Analysis of proxy records. The example of tree-ring analysis: (a) collecting the raw data – coring a tree; (b) cross-dating from multiple trees; (c) removing the growth trend from time series of ring widths (solid line) to generate a detrended index (dashed line). These data are averaged for the stand and calibrated with observed climate data to develop transfer functions to be applied to earlier tree-ring records to infer climate conditions beyond the period of instrumental records.

Source: All images and data provided by Dr Henri Grissino-Mayer

cited in Bradley, 1999). Next, growth effects are removed from the time series of ring widths. Smaller (younger) trees tend to have wider widths than larger (older) trees, both because they grow more rapidly and also because the circumference of a young tree is smaller, so a given amount of growth results in a thicker ring. This growth trend is usually removed by fitting some sort of mathematical function to the raw data (Figure 15.2c). Modified exponential functions, orthogonal polynomials, and cubic splines all are used, depending on whether the tree is growing in isolation or in a closed-canopy. The difference between the ring widths and fitted values averaged for the site is the proxy climate signal to be interpreted. Mathematical and/ or statistical procedures are used to derive an equation relating the tree ring width indices to a climate variable for a given period (referred to as calibration). Different methods, which range from simple to multiple to spatial regression, principal components analysis and canonical regression analysis have been used and are described by Bradley (1999) and Fritts (2001). Ideally, some climate data outside the calibration period are used to assess the performance of the fitted equation with independent data (verification), before the relations are applied to infer past climates. Clearly, careful attention needs to be directed to what climate data should be used and for what seasons. Best results tend to be obtained at sites where trees are growing close to ecological limits, thus growth rates are sensitive to climate. The two climatic stresses most commonly recognized are moisture (drought or flooding), the basis of studies of Bristlecone Pine in the arid southwestern United States, and temperature (see http://web.utk.edu/~grissino/ The Ultimate Tree-Ring web pages for specific examples).

ISSUES TO CONSIDER WITH PROXY DATA AND RECONSTRUCTIONS

Climatic interpretation

Often, there is a tendency to interpret proxy data in terms of single climatic variables, most commonly those used in numerical modelling (e.g. average annual temperature, July and/or January temperatures). However, proxies frequently represent the combined influence of climatic and non-climatic factors. For example, the isotopic composition of ice relates not only to local temperature conditions, but also to ice volume, the source regions of air masses, and an array of elevational/geographic effects. Tree-ring widths are more a function of growing season climates and conditions in preceding years than annual average temperatures.

Spatial coverage

The geographic coverage of proxy data depends both on where the methods are appropriate and where the data have actually been collected (Table 15.2). For example, tree rings, an extensively used source of high-resolution climate data in temperate and high latitudes, cannot be used widely in the Tropics where trees characteristically do not form annual growth rings. This is a significant issue given over half the surface area of the Earth lies in this latitudinal zone. As with the instrumental record, proxy data have not been analysed evenly across the globe. Rather, data which have been analysed are clustered in and around North America and Europe, with a notable paucity in the southern hemisphere. Jones *et al.* (2001) estimate that the number of records for the southern hemisphere is nearly two orders of magnitude lower than the northern hemisphere – in part because of the small landmass, but more so because less research has been undertaken. This is changing, however, with greater attention directed to the importance of global data sets (Figure 15.3).

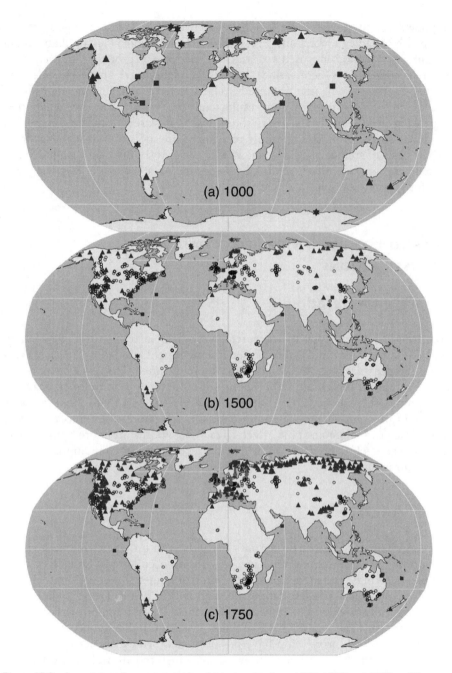

Figure 15.3   Locations of proxy records which date back to 1000, 1500 and 1750 – different symbols indicate instruments (red), tree rings (brown), borehole (black), ice cores (blue) and other low-resolution records (purple) – that have been used to reconstruct NH and SH temperatures or regional patterns

Source: IPCC (2007, Figure 6.11)

Spatial representativeness

The spatial scale of climate that the proxy data actually respond to varies: signals can reflect local, regional, or global conditions. For example, analyses of different attributes of ice cores provide records of climatic conditions at very different spatial scales: local information can be derived from stable isotope ratios (temperature) and snow accumulation; regional information on winds (intensity and direction) from wind blown salt and continental dust; and the gas content of trapped bubbles provides insight into global biogeochemical cycles. Increasingly sophisticated statistical techniques of upscaling allow different lines and scales of evidence to be integrated.

Temporal representativeness

The time period for which reconstructions can be made relates both to the nature of the proxy's response and the nature of the archive in which it is stored. Proxy data can respond to an event or can provide an integrated response to conditions/events over a number of years. Different proxy systems have different levels of inertia with respect to climate: some systems vary essentially in phase with climatic variations, whereas others lag by many centuries. For example, latitudinal/elevational boundaries of a forest may take hundreds of years to respond to marked changes in climate, whereas one winter with temperatures below a certain threshold will cause a species of beetle to die out (see further examples in Table 15.1). Clearly, the nature of the climate change, a rapid/step change versus gradual warming/cooling, and the extent to which the proxy is able to adapt to new conditions is relevant here also. For high-resolution proxy records, it is likely the parameter being used (tree ring growth, ice core isotopic value) actually reflects conditions in a particular growing or snowfall season of the year, not an integrative measure for the year. Historical records of variables such as grape flowering, harvest dates or snow cover duration are seasonally specific. Interpretations must take this into account. Stating that the seventeenth century was cooler than the twentieth century based on a greater number of days of snow lying on the ground may not be correct if the non-winter temperatures were warmer.

Issues related to the nature of the record are continuity, resolution and temporal coverage. Accumulation of ice, sediments, and so on, may be continuous through time, although rates may have varied, or be the result of a series of discontinuous events. This is often an issue with terrestrial geological records. For example, moraines represent maximum glacial extents and recessional positions, but give little information on conditions during periods of advance. Marine records, in contrast, tend to be continuous and for this reason commonly are the chronological and paleoclimatic frame of reference for long-term climatic fluctuations. Resolution relates to the smallest time period for which an event/change can be identified. In addition to the nature of the proxy's response (discussed above), this is dictated by the accumulation rate of a system, given that a minimum sample size (sediment, ice, biological material) is needed for analysis, and how fixed the

Table 15.2   Climate resolution for sediment archives in three environments. Duration of events that can be resolved relate to the rate of influx and accumulation of sediment, and the amount and activity of organisms burrowing in the sediments (bioturbation) which mix sediments that have accumulated, blurring the degree of detail that can be resolved in the permanent record. Numbers are approximate

	Influx rates	Depth mixing	Resolvable detail	Length record
Continental Lakes	1 mm per year	1–10 cm	10–100 years	$1-10^5$ years
Continental Shelves/	10 cm–1 m	10 cm–1 m	100–1000 years	$1-10^8$ years
Shallow Seas	per 1000 years			
Deep Ocean	1 cm per 1000 years	1–10 cm	1000–5000 years	$1-10^8$ years

Source: Adapted from Ruddiman (2001)

proxy signal remains with time. Many properties of sediments undergo changes (diagenesis) once buried. For example, isotopic/geochemical signatures diffuse, smearing the record, while sediments are often subjected to physical mixing or the effects of bioturbation, thereby reducing the resolution of the information that can be derived. A summary of the degree of resolution for sediments in three different environments (continental lakes, coastal oceans and shallow seas, and deep oceans) is provided in Table 15.2. In general, the greater the accumulation rate, the shorter the time period (temporal coverage) sample represents. Long records (10^6 plus years) tend to have poor resolution (for example, deep ocean cores); high-resolution records (providing information on seasonal/annual timescales) tend to cover only recent Earth history (for example, tree rings or historical data; Table 15.1).

Absolute chronology

Without reliable estimates of the ages of events in the past, it is impossible to know if changes occurred synchronously or if certain events led or lagged others. Nor is it possible to assess accurately rates of change. In general, dating accuracy gets weaker farther back in time and dating methods often have specific ranges where they can be applied. Few proxy data contain an absolute chronology inherently within their records. Notable and very important exceptions are tree rings, varves (annually laminated sediments) and ice cores. But is it important to caution that the age models associated with these are not always exact to a specific year. A wide array of absolute (radiometric and radiogenic), relative and correlative dating techniques exist. Special editions of the journal *Quaternary Science Reviews* are dedicated to innovations in methods and their application. Each radiometric system has ranges over which the system is useful, and palaeoclimatic studies almost always publish analytical uncertainties. Because there can be additional uncertainties, methods have been developed for checking assumptions and cross-verifying with independent methods.

Increasingly smaller samples can now be dated, resulting in more precise dates and thus higher-resolution records. Thus, for example, individual seeds can be dated rather than having to have large samples which resulted in bulk sediment dates integrated over some depth of core. However, the ability to analyse smaller samples means the question of what actually is being dated, its stratigraphic context and how it relates to the event(s) of interest becomes even more important to resolve.

A summary of characteristics of some of the proxy data organized by the key issues outlined above is presented in Table 15.1. Clear differences are evident: ocean sediments, for example, which theoretically can yield data for 70 per cent of the Earth's surface and may provide long continuous records, cannot be used under most circumstances to give high-resolution (years/decades) information about recent climates because of the low rate at which they accumulate (1 cm of core may integrate processes of a thousand years plus) and the problem of accurately dating the materials. At the other end of the spectrum, historical records can provide annual (or intra-annual) data for up to a thousand years in some areas. This proxy set tends, however, to provide localized information and, for most areas of the world, its potential has only been realized for the last few centuries.

CONCLUSION

Studies of past environments must begin with an understanding of the types of proxy data available and the methods used in their analysis. At the most qualitative level, proxy data provide information on wetter–drier and cooler–warmer periods. At the other extreme, precise quantitative predictions are possible. The most common climate variable reconstructed is temperature, although reconstructions of precipitation, evaporation, soil moisture (or some integrated hydro-meteorological measure), winds (direction and intensity) and features of the global circulation (e.g. the polar front, the inter-topical convergence zone) have all been developed.

Summary

- Proxy records are historical (documentary) archives or natural (physical, chemical or biological) systems that are dependent on climate and which incorporate in their structures some measure of this dependency so it is preserved through time.
- The challenge in interpreting these data is to separate the effects of climate (the signal) from other non-climatic influences (the noise).
- It is important to be aware of the assumptions and difficulties associated with each method. Only then is it possible to select the most appropriate method(s) for a given question and site, or to synthesize different lines of evidence to provide comprehensive insight into former climatic conditions.
- More, longer sequences of proxy data, combined with a better understanding of the climate in natural phenomena today, integrated with results of numerical modelling of paleoclimates and theoretical understanding, are still needed to provide critical insight into the patterns and causes of environmental change, and to provide a backdrop against which the effects of human actions can be assessed.

Further reading

- Bradley (1999) provides an excellent overview of proxy data sources with examples of how climatic variation in the past has been studied, and with Jones (Bradley and Jones, 1995) has edited a collection of papers on climate variations over the last 500 years, emphasizing the 'Little Ice Age' and climate of the twentieth century. The focus is on high-resolution data – documentary (historical), dendroclimatic, and ice core lines of evidence, as well as consideration of forcing factors (causes).

- Intergovernmental Panel on Climate Change (IPCC; Houghton *et al.*, 2001) was created by the World Meteorological Organization and the United Nations Environment Program to provide an assessment of all aspects of climate change, including how human activities can cause such change. Chapter 2 of that volume – 'Observed Climate Variability and Change' – provides an excellent, brief summary of high-resolution proxy data and results derived from their analysis, which provide context for more recent changes documented by instrumental records. The Panel updates its reports and principal themes – see 2007 report, below and the IPCC website.

- Jones *et al.* (2001) is one of seven papers in a special section of the journal *Science* on the Earth's past climate. Summarizes what is known about changes in temperature and two major ocean circulation systems (El-Nino – Southern Oscillation and the North Atlantic Oscillation) over the last 1,000 years.

- Lamb (1995) provides a comprehensive introduction to historical (documentary) sources of climate data; while Ruddiman (2001) is a useful introductory climate science textbook, which summarizes 550 million years of Earth's climate changes, including the impact by and on humans. Coverage extends to predictions about future climate changes. The text is accompanied by numerous, very useful, colour photographs and illustrations.

- Rutter and Catto (1995) is a thorough review of dating methods used for proxy data. For more recent developments, start with the journal *Quaternary Science Reviews*.

- Websites are now a primary source of up-to-date information and data. Key sites include:

IPCC: http://www.ipcc.ch/; world data center for paleoclimatology: http://www.ncdc.noaa.gov/paleo/ [a wealth of links are provided to key sites elsewhere]; PAGES, the international Geosphere-Biosphere Programme (IGBP) Core project: http://www.pages-igbp.org/; Ultimate Tree Ring web pages: http://web.utk.edu/~grissino/.

Note: Full details of the above can be found in the references list below.

References

Alverson, K.D., Bradley, R.S. and Pedersen, T.F. (eds) (2003) *Paleoclimate, Global Change and the Future*. International Geosphere Biosphere Programme Book Series. Berlin: Springer-Verlag.

Bradley, R.S. (1999) *Paleoclimatology: Reconstructing Climates of the Quaternary* (2nd edn). San Diego, CA: Harcourt Academic Press.

Bradley, R.S. and Jones, P.D. (eds) (1995) *Climate Since AD 1500* (2nd edn). London: Routledge.

Fritts, H.C. (2001) *Tree Rings and Climate*. Caldwell: Blackburn Press.

Houghton, J.T., Ding, Y., Griggs, D.G., Noguer, M., van der Linden, P.J. and Xiaosu, D. (eds) (2001) *Climate Change 2001: The Scientific Basis*. Cambridge: Cambridge University Press.

IPPC (2007) 'Climate Change 2007 – The Physical Science Basis', Working Group I Contribution to the Fourth Assessment Report of the IPCC (see particularly Chapter 6 of volume 1). Cambridge: Cambridge University Press.

Jones, P.D., Osborn, T.J. and Briffa, K.R. (2001) 'The evolution of climate over the last millennium', *Science*, 292: 662–667.

Lamb, H.F. (1995) *Climate, History and the Modern World* (2nd edn). London: Routledge.
NRC (National Research Council) (2006) *Surface Temperature Reconstructions for the Last 2,000 Years*. Washington, DC: National Academies Press.
Oldfield, F. (2003) *Environmental Change: Key Issues and Alternative Approaches*. Cambridge: Cambridge University Press.
Ruddiman, W.F. (2001) *Earth's Climate: Past and Future*. New York: W.H. Freeman and Company.

ACKNOWLEDGEMENT

As an undergraduate, my interest and understanding of this material was inspired by my advisor, Dr Jean M. Grove. This chapter is dedicated to her memory.

16 Making Observations and Measurements in the Field

Alice Turkington

Synopsis

Physical geography is traditionally a field-oriented science, and much of the research conducted involves the geographer working 'hands on' in the physical environment. There is a diverse assortment of questions addressed by physical geographers, encompassing the widest range of scales, from microscopic to planetary, and many of these lead geographers outdoors to observe, measure and record features of the environment. This chapter discusses some issues that are common to field-based studies: choosing the spatial and temporal context of the study; understanding the uniqueness of places and the importance of local factors in controlling environmental conditions or processes; and selecting appropriate methodologies for data collection.

The chapter is organized into the following sections:

- Introduction
- Fieldwork design
- Site selection and planning
- Making observations and measurements
- Conclusion

INTRODUCTION

For many physical geographers, doing fieldwork is the most enjoyable and rewarding part of their research, and for many students, field classes and primary research are the most memorable experiences in the undergraduate programme. Physical geographers are often driven by their curiosity about foreign places, their sense of adventure and the aesthetic pleasures of the outdoors, as well as the intellectual challenge of explaining the natural environment, how and why it has developed. While there are obvious attractions to exploring new places, geography is largely about a practical engagement with the world, and the only way to experience and understand the complexity of the real world is in the field (Roberts, 1992). Fieldwork is at the very core of physical geography, which focuses on understanding the Earth's physical environment; experiencing the environment first-hand through conducting fieldwork is a fundamental, perhaps essential, part of geographical research.

Fieldwork is often concerned with careful observation, measurements and recording of information. There are a wide range of methodologies available, and the choices of how, where and when to conduct fieldwork are based on the research questions being addressed. The diversity in field techniques used by geographers reflects the wide variety in geographical research objectives, which may also be manifold within sub-disciplines such as geomorphology, climatology and biogeography. Field methodology cannot be prescribed, but rather reflects a combination of specific research problems, opportunity, training and personal conviction (Bauer *et al.*, 2001). The value of fieldwork lies in the advancement of knowledge, which is in turn determined by the successful design and execution of field-based observations and measurements, and the meaningful interpretation of data collected. This chapter will discuss several of the issues inherent in fieldwork design and implementation. Importantly, no fieldwork should ever be planned or undertaken without an awareness of health and safety procedures, nor without a comprehensive analysis of risks and their management (see Chapter 4).

FIELDWORK DESIGN

Physical geography is essentially concerned with the morphological elements of the physical environment (forms), the processes operating in the environment and the materials upon which they operate. For example, fluvial processes operating on channel sediments may form point bars and meanders; ecological processes operating in certain environments may produce a plant community; atmospheric processes in a particular environment cumulatively produce a specific climate regime. The interactions between materials, processes and forms through time and across space are the focus of geographical research, which may address these at a range of spatio-temporal scales. The nature of the fieldwork to be conducted within individual projects is largely determined by the scale (spatial and temporal) of enquiry, and the research questions underlying the field project.

Environmental change is a function of a spatially and temporally continuous system, dependent on a wide range of factors, but most studies are concerned with the development of specific forms, the operation of specific processes, or the factors that control these changes in space and time. Geographers have often found advantage in reducing the time- and space-scales of inquiry, in observing and attempting to measure environmental change over short periods of time. Reduction of both time- and space-scales occurs in parallel as processes that operate over smaller space-scales often operate during shorter time-scales (Lane and Richards, 1997). Physical geographers have long recognized the problem of scale linkage, in that the results of research undertaken at one scale are not necessarily applicable to another scale. Rhoads and Thorn (1996) commented that reductionist approaches have generated more questions than solutions, and new conceptual issues emerge at every scale of interest.

The scale of enquiry of any research project is determined by the landform(s) and/or process(es) under investigation. For example, sediment transport within a fluvial channel may be examined within a particular reach or within a watershed, by researchers interested in the influence of bedforms on sediment transport mechanics or the impact of deforestation on sediment yield, respectively. The choices of how the field observations and measurements are to be made are concerned with issues of *space, time* and *place*. These are issues that are both logistical, referring to the choice of field sites and the length of time available for data collection, and are theoretical, as they impact on how the data may be interpreted and the conclusions that may be drawn.

A consideration of space is implicit in any decision to conduct field observations, through the choice of an appropriate field site. Consider an investigation into the controls on a beach profile. While seasonal changes to a beach profile may follow some predictable patterns, where the beach undergoes construction during summer months with low wave energy and then displays erosion and removal of sediment under stormy winter conditions, the profile form of an individual beach is also determined by the sediment characteristics, tidal range and local topographic conditions. This variation in environmental systems across space, or spatial contingency, is a reflection of geographical variability in environmental processes and controls (Phillips, 2007). The inferences that can be drawn from a series of profiles measured between seasons on a single beach are restricted to some extent, due to the importance of local factors in determining beach response to changing wave action. On the other hand, a comparison between beaches with differing sediment compositions or tidal ranges would allow conclusions to be drawn regarding the importance of these controls, if other variables can be assumed to be relatively constant.

Controlling for independent variables when taking measurements across space can facilitate an investigation of the influence of other factors impacting on landform response to physical processes. An illustration of the utility of this approach might be found in catenary relationships linked to hillslope processes. The relationship between soil-profile characteristics and distance downslope on a temperate hillslope might be examined, to test models of hillslope equilibrium (Selby, 1993). If the distance downslope were held constant, sampling on adjacent slopes with differing parent rocks would investigate the relationships between geology, soil characteristics and hillslope processes. If distance downslope and geology were held constant, observations on adjacent slopes with differing vegetation cover (e.g. forest and grass) would investigate the relationships between vegetation, soil characteristics and hillslope processes. This approach uses stratified sampling to compare environmental systems across space and identify patterns or distributions of forms, to determine the importance of controlling variables on process and response.

Environmental change over time may be observed directly through monitoring of the process of interest for a period of minutes, days, weeks or months. Collecting weather data is an example of environmental monitoring that may be performed at regular intervals; these data may be collected by the researcher in person by frequent visits to a weather station, or may be collected remotely by automatic loggers that record and store data. Measurement of stream discharge during rainfall events of differing magnitude and duration and measurement of rates of sand grain transportation across a windy beach are examples of monitoring of processes over time that may lead to understanding of their operation under changing environmental conditions. The time frame during which observations are made imposes some limitations on the interpretation of the data collected, however, as the changes observed are determined, in part, by historical contingency, or inheritance effects that are specific to each place and time. Inheritance involves features inherited from parent material or from previous environmental regimes, which differ from the contemporary environment (Phillips, 2007).

The influence of previous states of an environmental system might be usefully illustrated by returning to the question of distribution of soil profiles on a hillslope. Clearly, the expected relationship between soil depths and profile characteristics will not be identified if a transect down a hillslope crosses a recent landslide. In this case, the relationship between hillslope processes and soil development will be skewed by the recent occurrence of a large-scale erosion event. Eventually, the hillslope may 'recover' from this event, and an equilibrium-slope profile may be established; the period of time required for the impact of a high-magnitude event to be diminished is the relaxation time of the landform (Brunsden and Thornes, 1979).

Environmental monitoring is useful where processes operate at a fairly consistent rate, frequently and within a predictable range. Some physical processes, such as landsliding, are inactive for long periods, and when they do operate they can act very rapidly. Further, processes may vary markedly in their intensity and duration under contrasting climatic conditions. River discharges may vary in response to rainfall events in a humid environment within an order of magnitude, whereas fluvial systems in semi-arid regions may vary by several orders of magnitude (a factor of 100 or 1000) in response to large storm events (Summerfield, 1991). When investigating high-magnitude or low-frequency events, it is appropriate to examine the evidence for the process, and it may be possible to recognize a temporal sequence of system response.

Where geomorphic, ecological or historic evidence indicates a series of events of differing ages, it is possible to investigate changes through time using an ergodic approach, where sampling in space is substituted for sampling through time. For example, to investigate the dynamics of forest recovery from wildfire, stands of forest that have experienced wildfire events 10, 50 and 100 years ago can be examined, if it may be assumed that initial conditions (ecological composition) were similar. In an area prone to landslides, the nature and rate of hillslope recovery may be examined if historical evidence exists that allows scars to be accurately dated, and other structural or topographic factors can be held relatively constant. A sequence of rock weathering may be identified by examining gravestones in a cemetery of a range of ages that are cut from the same rock type and have the same aspect. The success of this approach will depend on the absence of controls other than time that may be responsible for landform change (e.g. lithology) and the existence of evidence to support the sequential development of adjacent landscape features.

In summary, there are three approaches to fieldwork design that are commonly used. First, samples or measurements may be taken across space, where the points of data collection are chosen to control for some independent variables. Second, change in a landform or system may be monitored over time. Third, observations may be made of landforms or systems of varying ages, an ergodic approach that uses space as a substitute for time. In many cases, a combination of the first three strategies may be used. A general survey may be undertaken initially at many sites to establish an overall pattern, and intensive measurements or monitoring may be performed at strategic sites within this broader framework. On a hillslope, soil depths may be measured at many locations on several downslope transects; a smaller number of soil pits may be examined to characterize soils at contrasting elevations within each transect. The usefulness of this combination of approaches lies in the integration of an extensive research design that identifies patterns, or responses, and a case-study approach that investigates specific processes.

SITE SELECTION AND PLANNING

Fieldwork is most often undertaken during a few short weeks or months and in a specific location, such as a hillslope, drainage basin, forest, beach or mountain. The researcher is faced with the prospect of choosing an appropriate field site and making pertinent observations during the time available to complete the project. The choice of field site is driven primarily by the research question to be addressed, but there are other considerations to be included in the decision, such as suitability, accessibility and feasibility. The field site must be representative of the environmental system under investigation, and should be adequately researched in advance. In many areas, previous studies of landscapes have been documented and observations or theories of landscape development are available to the researcher to provide background to the site. It is useful to gather other sources of information regarding the site, such as topographic or geologic maps, climate data, or aerial photographs. The suitability of the field site may be finally determined by a reconnaissance of the area, if it can be visited in advance of the fieldwork. Accessibility to an area may be restricted, so permission for access and for scientific study (if destructive methods are to be used or samples are to be collected) should be obtained. Accessibility is also dependent on the ability of the researcher to travel to the area, and in some inhospitable environments, such as high mountain, desert or Arctic environments, this may require considerable planning. The feasibility of research in these areas will depend on the researcher confirming that equipment can be transported to the area, that samples can be safely transported, and that the safety of the researchers is ensured. Research in remote locations is often possible through expeditions organized by groups of researchers; local fieldwork can be accomplished by one or two people, depending on the type of fieldwork being conducted.

Whether fieldwork is to be conducted during a discrete period, perhaps in a remote location, or by repeated visits to a local site, careful and thoughtful planning will significantly enhance the success of the endeavour. The researcher should prepare for fieldwork by identifying specific objectives and an estimated time-frame for completing them. A sampling strategy should be developed and prioritized, and a clear strategy for data collection and recording should be established. It may be useful to prepare worksheets for recording information and measurements, which standardizes measurements taken at different times and in different places. The appropriate equipment should be calibrated and tested, and be accompanied by replacement parts or batteries as required. Safety precautions should be taken as needed (Chapter 4) and all travel and logistical arrangements should be completed before fieldwork commences. Not all fieldwork is arduous or hazardous, but in all cases the researcher should try to prepare for unexpected changes in conditions, or risk associated with particular terrains.

Once in the field, it is likely that the researcher will re-evaluate the objectives of the fieldwork, the priorities of the work and the sampling strategies to be employed. It is not always the case that field observations will proceed entirely as planned, and there may be unforeseen circumstances or contingencies that require some adaptation of the plan of work. For example, my first experience as a field researcher was in Death Valley, USA, where I had naïvely hoped to survey alluvial fans at the foot of the Panamint Mountains. I soon discovered that surveying a ten-mile fan in such extreme heat was not feasible, and reassessed my research strategy! More importantly, as measurements and observations are collected the researcher may identify other interesting questions in the light of the findings as they are produced, and many geographers in the field follow their curiosity to new, productive avenues of enquiry that were not evident from literature-based research. As Burt (2003) suggests, one very important field tool is a curious eye. It is important to be flexible and revisit the research objectives as fieldwork proceeds in order to maximize the time and effort spent in data collection.

MAKING OBSERVATIONS AND MEASUREMENTS

A useful first step in fieldwork is description of the overall site, accompanied by photographic survey, sketches and notes. One element of this is to describe environmental features, which involves classification of types of landscape features and assumptions about the boundaries of the features. In some cases this is a simple enough task; recognizing a river channel, outwash plain or sand dune is fairly straightforward. In other cases it may not be easy, for example defining the boundary between two soil types or boundaries between plant communities. Field-site description is a prelude to choosing sites to make observations or measurements, and will improve with increasing familiarity with the field site. While it may be necessary to arbitrarily delimit the extent of a feature, if it is consistently defined, measurements will still be comparable.

Once the target features have been identified and delineated, the sampling strategy must be chosen (see Chapter 17). This is again driven by the research questions being addressed, but more pragmatic concerns, such as time and resources available in the field, will force some compromise between the ideal and the possible number of observations and/or measurements that can be made. If a large number of measurements are to be taken, a random or stratified sampling strategy will allow subsequent statistical analysis. If a case-study approach is to be adopted, following an intensive research design (Richards, 1996), it is important to select sample sites that allow testing of the hypotheses under investigation, and that allow detailed accounting of local conditions. Chapter 17 offers a comprehensive discussion of the type of sampling strategies that may be employed in geographic research.

When making observations and measurements, two considerations are important: the frequency of measurements in relation to process rate and the spatial distribution of measurements in relation to the distribution and size distributions of forms. Some examples of fieldwork methods will demonstrate the range of options that may be considered, and the type of choices to be made by the field researcher.

First, an investigation into geomorphic processes often derives information from analysis of the sedimentary body that has been transported by the process in question. Particle size is a key parameter to consider as it is a major contributor to particle resistance in sediment transport. As an example, the sediment size distribution on an alluvial fan may be investigated with a view to inferring the processes of sediment transport onto and down the fan. In this context, the sediment might be sampled along transects from the apex to the toe of the fan; particles all along the line might be measured, or quadrats established at regular intervals along the transect, within which a certain number of particles can be sampled. The number of transects, sample points and particles measured will be constrained by the time available, but should include sufficient numbers to facilitate statistical analyses. At each sample point, a random selection of particles can be chosen by forming a grid within the quadrat and sampling at each grid point; alternatively all particles within a small area might be sampled. The use of a well-defined strategy to choose clasts at random is important to avoid operator bias, the tendency of the observer to choose particles of a convenient size or an interesting appearance. Each particle can be measured along the a-, b- and c- axes with a ruler or calipers. As is often the case in fieldwork, more than one measurement method may be required, and the presence of fine sediments would necessitate the use of sieves (either in the field or in a laboratory) to ascertain sediment size distributions. Detailed methods for sediment analysis are available in geomorphology texts (e.g. Goudie, 1990; Kondolf and Piegay, 2003).

Second, the collection of meteorological data has many applications, such as agricultural meteorology or urban climatology. An example of a field-based study might be an investigation into the existence of an urban-heat island effect in a city, where the boundary between rural and urban areas exhibits a steep temperature gradient to the warmer city centre (Santamouris, 2001). To test this model, temperature measurements must be taken both within and around the city at similar times. The locations of the temperature measurements might follow a transect through the city to the surrounds, or be situated in a grid pattern across the urban area and adjacent countryside. The frequency of measurements will vary depending on the opportunity for remote measurement and automatic data logging as opposed to a researcher travelling across the city to measurement points. At each measurement point, care must be taken to avoid local factors that may influence temperature measurements, such as heat outflow from a building, and measurements must be taken and recorded in a consistent manner. The time at which temperature is recorded is also important; the urban-heat island is likely to be best developed under clear skies and light winds just after sunset, and data collected at other times during the day or night, or under adverse weather conditions, may not produce data suitable for the research question being addressed. General meteorology texts (e.g. Stull, 1988) are useful sources of methodological descriptions.

Third, a biogeographer may be interested in investigating the distribution of a plant species of community in an area. In biogeographical studies, there are many standardized methods available to the researcher (e.g. Tiner, 1999). One example of such a project might be to map the salinity gradient across an estuary, using distributions of salt-tolerant plants as an indicator of salinity. In this case, the researcher may traverse the area and select plots for analysis that are thought to be representative of stands of vegetation, and delineate plots of sufficient area to sample most of the species in the area. An alternative to preferential sampling is to select plots randomly, or within topographic units. Vegetation plots should be removed from obvious disturbances, roads and other infrastructure. The attributes to be measured in the plot might include areal cover, density, basal area, or frequency of plants; plots may be analysed by strata or point sampling may be performed. In ecological studies, the timing of the fieldwork is obviously important, due to seasonal variations in vegetation assemblages. Several visits may be required during the growing season to account for ecological changes, and it is also important to note that plants may respond to periods of dryness or wetness differently, in which case antecedent weather conditions may partially control the timing of fieldwork.

However measurements are made, the researcher should make every effort to minimize instrument error and observer error, and to maintain consistency in the measurement system. Observations may be recorded manually or digitally, and it is worthwhile to make copies of data, clearly labelled, as field observations are often irreplaceable. Making observations in the field is probably the most expensive (in terms of time, money and effort) part of geographical research; careful planning and execution will contribute to excellent research.

CONCLUSION

Physical geography has by tradition been a field-oriented science (Petch and Reid, 1988) and conducting fieldwork provides opportunities for learning and appreciating the natural environment that cannot be duplicated in the library, classroom or laboratory. For many geographers, fieldwork is the most rewarding part of their work, and great pleasure can be derived from working outdoors with colleagues and friends. Beginning a field-based research project is challenging, but provides an opportunity to be creative and satisfy intellectual curiosities, through engaging directly with the physical environment. When planning fieldwork, the researcher should always be cognizant of the theoretical framework underlying the project, and explicitly consider the issues of space, time and place in designing the study. A structured and rigorous approach to making observations and measurements will help to ensure that the data are of high quality, but the researcher should keep an open mind when in the field. In spite of our efforts to explain it, the natural environment can be full of surprises!

Summary

- Fieldwork design is determined by the research questions underlying the study and the spatial and temporal scale of enquiry.
- Field observations may be made across space, to eliminate variables from the investigation; observations may be made frequently to monitor change through time.
- Spatial and historical contingency, the unique geography and history of any place, must be considered when interpreting findings from field observations.

- The choice of field site, sampling strategy, measurement methods and sample size will reflect both theoretical and practical considerations.
- A structured, but flexible, approach to making field observations will contribute to collection of quality data and interesting outcomes from the study.
- No fieldwork should ever be planned or undertaken without an awareness of health and safety procedures, nor without a comprehensive analysis of risks and their management (see Chapter 4).

Further reading

- Gregory (2000) provides a comprehensive overview of the development of physical geography, the research that physical geographers conduct and the methods they use. The text discusses the nature of contemporary physical geography in detail, and suggests directions of future research in the twenty-first century. This is an excellent resource for physical geography students, and presents most of the literature central to the discipline.

- Goudie (1990) offers an extensive overview of the range of field techniques used by geomorphologists. The text is arranged thematically, including information on how to observe and measure form; collect evidence of processes; measure material properties; and reconstruct palaeoenvironments. This detailed examination of suites of field methodologies is extremely useful to geographers beginning a field-based research project.

- Trudgill and Roy (2003) have edited a volume that offers insights into why physical geographers do what they do. While this volume is short on practical help for the student of physical geography, it does offer some interesting debates on the nature of geographic science, and the sources of inspiration for physical geographers who have shaped the discipline over the course of their careers

Note: Full details of the above can be found in the references list below.

References

Bauer, B.O., Winkler, J.A. and Veblen, T.T. (2001) 'Afterword: a shoe for all occasions or shoes for every occasion: methodological diversity, normative fashions, and metaphysical unity in physical geography', *Annals of the Association of American Geographers*, 91: 771–778.

Burt, T. (2003) 'Realms of gold, wild surmise and wondering about physical geography', in S. Trudgill and A. Roy (eds) *Contemporary Meanings in Physical Geography*. London: Arnold.

Gardiner, V. and Dackcombe, R. (1983) *Geomorphological Field Manual*. London: Allen and Unwin.

Goudie, A. (1990) *Geomorphological Techniques* (2nd edn). London: Unwin Hyman.

Gregory, K.J. (2000) *The Changing Nature of Physical Geography*. London: Arnold.

Kondolf, G.M. and Piegay, H. (2003) *Tools in Fluvial Geomorphology*. London: Wiley and Sons.

Lane, S.N. and Richards, K.S. (1997) 'Linking river channel form and process: time, space and causality revisited', *Earth Surface Processes and Landforms,* 22: 249–260.

Petch, J. and Reid, I. (1988) 'The teaching of geomorphology and the geography/geology debate', *Journal of Geography in Higher Education*, 12(2): 195–204.

Phillips, J.D. (2007) 'The perfect landscape', *Geomorphology*, 84: 159–169.

Rhoads, B.L. and Thorn, C.E. (1996) 'Methodological issues', in B.L. Rhoads and C.E. Thorn (eds) *The Scientific Nature of Geomorphology: Proceedings of the 27th Binghamton Symposium in Geomorphology*. London: Wiley and Sons, pp. 145–146.

Richards, K. (1996) 'Samples and cases: generalization and explanation in geomorphology', in B.L. Rhoads and C.E. Thorn (eds) *The Scientific Nature of Geomorphology: Proceedings of the 27th Binghamton Symposium in Geomorphology*. London: Wiley and Sons, pp. 171–190.

Roberts, N. (2002) 'Not so remote sensing of our environment', in A. Rogers, H. Viles and A. Goudie (eds) *The Student's Companion to Geography*. Oxford: Blackwell Publishing, pp. 19–20.

Rogers, A., Viles, H. and Goudie, A. (2002) *The Student's Companion to Geography*. Oxford: Blackwell Publishing.

Santamouris, M. (2001) *Energy and Climate in the Urban Built Environment*. London: Cromwell Press.

Selby, M.J. (1993) *Hillslope Materials and Processes* (2nd edn). Oxford: Oxford University Press.

Stull, R.B. (1988) *An Introduction to Boundary Layer Meteorology*. London: Kluwer Academic Publishers.

Tiner, R.W. (1999) *Wetland Indicators: A Guide to Wetland Identification, Delineation, Classification and Mapping*. London: CRC Press.

Trudgill, S. and Roy, A. (2003) *Contemporary Meanings in Physical Geography*. London: Arnold.

17 Sampling in Geography

Stephen Rice

Synopsis

Geographers recognize the value of both extensive statistical sampling and intensive 'case-study' sampling for exploring an uncertain world. A benefit of extensive sampling is that a set of techniques known as inferential statistics can be applied to make probabilistic statements about the population from which the sample is drawn. Sampling is therefore a powerful tool, but geographical research frequently engages with heterogeneous phenomena that require careful sampling in order to maximize the accuracy of conclusions. Thoughtful design of the sampling programme is therefore crucial and is driven by both the research aims and available resources. The chapter is organized into the following sections:

- Introduction
- Samples and case studies
- What makes a good sample?
- Designing a sampling programme
- Statistical inference
- Sample size
- Conclusion

INTRODUCTION

Sampling is the acquisition of information about a relatively small part of a larger group or population, usually with the aim of making inferential generalizations about the larger group. Sampling is necessary because it is often not possible, practicable or desirable to obtain information from an entire population. For example, it is essentially impossible to measure all of the sand grains on a beach to ascertain their average size; impracticable, in the course of a normal day's work, to question every person on the beach to determine the variety of their views about personal use of public spaces; and undesirable, never mind unethical, to stress the fish community of a seashore rock-pool by catching and examining all of its members. Moreover, in quantitative research, a set of procedures known as inferential statistics can be applied to sample data in order to make generalizations, validated by probability statements, about the entire population from which the sample was drawn. Thus, it is not *necessary* to interrogate a whole population to make useful generalizations about it.

In one form or another, sampling is the basis of almost all empirical research in both physical and human geography and is widely relied upon. However, such a powerful methodological tool comes with a set of health warnings: samples are only as valuable as they are representative of the larger population; at best, bad sampling leads to imprecision; at worst, bad sampling yields incorrect or prejudiced results. In the extreme, lack of sampling rigour makes inferences meaningless: 'The tendency of the casual mind is to pick out or stumble upon a sample which supports or defies its prejudices, and then to make it the representative of a whole class' (Walter Lipman, journalist, 1929). This remark may be directed at the dangers of using isolated examples to make unfounded inferences in everyday life rather than in academic research, but it nicely highlights the important association of weak, non-rigorous sampling with bias and inaccuracy.

This association, the difficulties that are frequently encountered in order to collect representative samples and the impenetrable nature of 'stats' for some students breed a scepticism about sampling and statistical inference that can run deep. It is fairly common to hear the dismissive statement 'Well, you can prove anything with statistics.' Benjamin Disraeli's remark that 'There are three types of lies: lies, damned lies and statistics' has been a convenient aphorism of opposition politicians concerned with the scruples of their counterparts for over a century. This scepticism is, however, unfortunate because geographers often seek to understand spatially diverse, highly complex phenomena that can be efficiently accessed by sampling and usefully examined using statistical inference.

While questions about motivation and affiliation (political, epistemological, etc.) have a place in the critical assessment of all research, for students of method the value of statistical inferences must be judged simply against the quality of the sample data and the quality of the analysis applied to them. These issues are the focus of this chapter. It reviews the use of sampling by geographers; considers the criteria by which samples should be judged; discusses the design and implementation of sampling schemes; introduces some of the basic elements of statistical inference; and suggest some methods for defining sample size. The overall message is that sampling is a powerful tool that most geographers need, but if research methodology aims to be as impartial and free of error as possible, then sampling must be done thoughtfully and rigorously.

SAMPLES AND CASE STUDIES

Not all geographers seek to make quantitatively robust inferences about large populations or develop theories and models of universal validity. Nevertheless, a principal aim of most, if not all, geographical research is to make useful generalizations – that is, to seek out and explain patterns, relations and fluxes that might help model, predict, retrodict, or otherwise understand better, the human and physical worlds around us. Thus, some geographers may restrict their attention to small areas, short periods of time, small groups or even to individual places or people, but the underlying approach remains nomothetic – it focuses on identifying the general rather than the unique, and in turn most geographical research involves some form of sampling.

Geographical systems are complex and affected by historical and geographical contingencies, indeterminism or singularity (Schumm, 1991). That is, while there are general similarities between objects, there is also inexplicable (or at least, as yet unexplained) variation in any population, so that each item is a little different from the others. This means that while we may be able to develop generalizations that are valid for a whole population, estimating the behaviour or character of any single individual is difficult and prone to error. This is as relevant to predicting river discharge as it is to commenting on the global reach of large corporations or people's views on street architecture. Thus, the geographical world is an uncertain world.

Geographers have adopted a variety of strategies to search for general understanding while recognizing this uncertainty. A useful distinction can be made between approaches that intensively examine a small number of examples and those that sample extensively (Harvey, 1969; Richards, 1996). In the extreme, geographers have made substantial use of single case studies (samples of one) to learn about both physical and human phenomena. While case studies provide detailed information, a fundamental criticism of the approach is that the generality of the case is unknown. That is, there is no formal basis for substantiating inferences made about a population on the basis of a single sample, such that extrapolating the findings of the case to the population remains merely a matter of intuitive judgement on the part of the investigator. This is problematic, because most geographers would (and should) be sceptical of such subjectivity. In contrast, the collection of large, extensive samples provides the opportunity to utilize statistical methods which, at least when sampling is appropriately conducted, offer a means of assigning objectively derived conditional statements to population inferences.

While recognizing the limitations of case studies and the benefits of statistical inference it is important not to denigrate the value of intensive investigations, not least because there is no single, simple model that defines how geographical knowledge can or should be obtained and ratified (see Chapters 16, 18 and 19). Case studies may not provide a basis for making wide-ranging inferences about a population but inferences based on case studies are not necessarily false or unreasonable. Rather, they stand to be substantiated and the detailed information gathered in a case study may reveal general structures or relations that can be used to generate or modify models or hypotheses (Harvey, 1969). Similarly, case studies may present unique opportunities for understanding the mechanisms that underlie empirical observations. In geomorphology, Richards (1996) argues that as long as the location of field sites is carefully planned, case studies offer important advantages over extensive approaches. This is because case studies provide an opportunity to ask fundamentally different questions in a fundamentally different way. Case studies often aim to explain the mechanisms that generate patterns observed in extensive studies (Blaut, 1959), and as such, case studies should not be judged by their representativeness (or lack thereof) but by the quality of the theoretical reasoning that they generate (Richards, 1996). Similar arguments, based on alternative views of generalization and theory validation, are used by qualitative researchers to explain the use of case studies (e.g. Miles and Huberman, 1994).

While statistical theory guides sample collection in extensive approaches, the selection of case studies for intensive study is based on less well-established criteria. Richards (1996) highlights the importance of carefully selecting cases that have those properties which facilitate rigorous tests of the hypotheses under consideration, and meticulous definition of local conditions. Curtis *et al.* (2000) review case-selection criteria formulated for qualitative research which include ethical considerations and relevance to a theoretical framework. Using examples from medical geography, they find that it can be difficult to reconcile these two criteria such that ethical considerations are at odds with the selection of the most useful or theoretically relevant cases.

The collection of information within a case study, especially in quantitative research, will typically require extensive sampling *internally*, albeit on relatively small temporal or spatial scales. If useful internal inferences are to be made then this sampling must be rigorous and statistically accountable. In turn, as Richards (1996) points out, the practical distinction between 'case studies' and 'extensive studies' is somewhat fuzzy in geomorphology, and largely semantic. In practice, many geographers are typically engaged in studies that shift between these two styles, using their complementary opportunities to explain geographical phenomena. The literature reviewed above provides a starting point for further consideration of case studies as a form of sampling. The remainder of this chapter focuses on extensive sampling and classical statistical inference.

WHAT MAKES A GOOD SAMPLE?

When sampling is used to make generalizations about a larger population, the aim of sampling is to obtain a 'representative characterization' of whichever aspects of the population one is interested in. Characteristics of a population are referred to as parameters, while those of a sample are called statistics. A good sample may then be defined as one that satisfies two criteria: it must provide an unbiased estimate of the parameter of interest, and it must provide a precise estimate of the parameter of interest.

Accuracy, precision and bias

The very nature of sampling means that, in repeated sampling of the same population, different items (individuals, businesses, households, etc.) will be drawn. Thus, statistics vary between samples and, in each case, differ to a greater or lesser degree from the population parameters that they estimate. The difference between sample estimates and the true population value is referred to as the accuracy of the sample. Accuracy is gauged in terms of its two components: bias and precision.

These terms can be illustrated by representing repeated sampling of a population by a darts competition in which each dart represents a single sample estimate and the bull's eye represents the true population value (Figure 17.1). Bias refers to

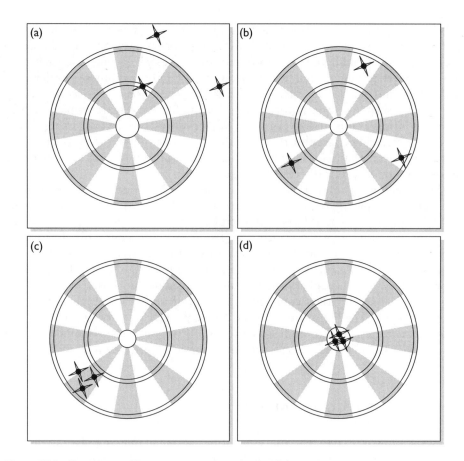

Figure 17.1   Precision and bias represented as a game of darts

the systematic deviation of sample statistics from the true value. A set of sample statistics that vary around the true value without any discernible pattern is said to be unbiased (Figure 17.1b and d) whereas a tendency to be consistently different (for example too high or too low) reveals bias (Figure 17.1a and c). Lack of bias ensures the representativeness of a sample and is a fundamental requirement of statistical inference. In principle, lack of bias is achieved by sampling randomly from a population and in practice this becomes the main challenge in sample collection. Precision refers to the size of the deviations between repeated estimates of a given statistic. It describes the degree to which repeated estimates are clustered together: are they tightly bunched (Figure 17.1c and d) or widely dispersed (Figure 17.1a and b)? In real sampling problems, bias and precision cannot be observed directly because the value of the population parameters are unknown (otherwise there would be no need to sample) and, typically, only a single sample is collected. When sampling we should then aim to maximize precision and minimize bias so that we can have faith that the single sample we collect is accurate (Figure 17.1d).

Minimizing bias

Bias is introduced into a sample in one of three main ways. First, the actions of the person collecting the data may introduce operator bias. Two different individuals asked to sample the same population may produce samples that are not just different (this is expected), but *systematically* different from one another. For example, in approaching passers-by to interview on a city street, an operator may knowingly or inadvertently select individuals of a certain age, gender or ethnicity, thereby over-representing those groups in the sample. In collecting pebbles from a beach to determine their average size, an operator may tend to pick up pebbles that are more distinctive in colour and more easily seen. Since colour depends on lithology and lithology affects size, the sample could be unrepresentative of the range of sizes present. Operator bias is reduced by careful training and adherence to a standardized set of procedural rules consistent with the sampling method (see below). Thus, the street interviewer may be asked to minimize selection-based bias by approaching every tenth person who passes by and the pebble picker may be instructed to collect the pebble that lies beneath the metre marks on a measuring tape laid down on the beach surface. However, the propensity for humans to ignore rules or simply to make mistakes means that operator bias is difficult to remove entirely. Thus, looking down at the correct spot on the tape, the pebble picker can choose between that pebble that lies to the right or left of the tape and it is common for operators to preferentially select pebbles that fit comfortably in their hand, potentially ignoring large or small clasts and thereby biasing the sample (see Marcus *et al.*, 1995 for a fuller discussion of this case). Additional procedural rules can be made to try to minimize such errors, but ultimately it may be difficult to eradicate them entirely. The potential for operator bias increases with the number of operators involved so that minimizing the number of operators helps to reduce this source of inaccuracy.

Second, bias can be introduced by faulty or misused measurement devices that systematically misrepresent the characteristic of interest – note that this is different from non-systematic measurement error discussed below. This is fairly clear in the case of instruments that measure some chemical, physical or biological property and have the potential to be mis-calibrated or broken. Mis-calibration is a problem because many measurement devices do not measure the property of interest directly but via some more easily assessed quality that is quantitatively related to the target property. For example, stream velocity is routinely measured using devices that generate a small magnetic field in the flowing water and detect changes in the electrical current that is produced as water flows through it. The current is proportional to the water's velocity, so that velocity can be determined by measuring the output current. However, care must be taken to correctly calibrate and set-up such instruments to ensure that no systematic error produces consistently deviant velocity estimates. In the widest sense, a question on a questionnaire, for example, is also a type of measurement device that can return a biased response by asking leading or biased questions (see Chapter 6).

Third, and perhaps most commonly, bias can be introduced during the design of the sampling programme, particularly poor definition of the population, or choice of an inappropriate sampling method (the method by which individual sample items are drawn from the population). In geographical research the content of the population is apt to vary in space, in time and with the scale of interest, often in a systematic way (see Chapter 24). Unfortunately, this means that it is relatively easy to collect a sample that is not representative of the intended population (these issues are discussed at length below).

Maximizing precision and minimizing non-systematic measurement errors

Precision is largely a function of three things: the number of observations that make up a sample; the heterogeneity (variability) of the characteristic of interest within the population; and non-systematic errors that arise from the technical limitations of the measuring procedure. We shall return to the first two below, suffice to say here that the larger the sample size and lower the population heterogeneity, the more precise sample estimates will be.

Measurement errors, sometimes called pure errors, exist in all measurement operations because of technical limitations within the measuring system (the people and instruments involved). These are the irreducible errors that one has to be willing to accept in making any set of observations. Wherever possible, every effort should be made to minimize them and, of course, one must endeavour to ensure that the errors are free of bias. A brief example from geomorphology illustrates several relevant issues.

In the field, we can use an instrument called a total station to survey topography – for example, exposed gravel bars in an alpine river channel. The instrument is set up at a base station and measures distances, declinations from horizontal and directional angles to survey points across the bar surfaces. A target prism attached to a pole of known length is used to mark each survey position and its reflective properties allow distance and angle measurements to be made to it. These measurements are used with simple trigonometry to derive the elevation of each point and its position within a Cartesian coordinate system. The instrument measures declinations from horizontal with an accuracy of ±5 seconds (where 1 second is 1/3600th of a degree), and distances of over 1 km with an accuracy of ±3 mm. This kind of instrument is expensive and the measurements it makes are incredibly refined. Nevertheless a number of measurement errors can be identified:

- As indicated by the manufacturer's specifications, repeated measurements of exactly the same target position return declination values that vary by as much as 5 seconds either side of the true value. In practice this error is very small, introducing a deviation of no more than 5 mm into the calculation of the elevation of a position 1 km away.
- Holding the target pole still, especially in cold, windy, weather is difficult. Despite every effort it is common for the target to move a few millimetres back and forth over a period of seconds. In turn, repeated measurements with the pole in the same position will yield very slightly different distance estimates.

- The survey aims to characterize bar topography but our measurements are affected by the smaller scale, gravelly surface texture of the topography being surveyed. Thus, choosing to place the target pole 1 cm to the left or 1 cm to the right of a particular spot can mean measuring the elevation of a hole between two pebbles or the elevation of the top of a pebble, in which case our estimate of bar elevation at that position can vary by tens of millimetres.

In practice, the first two sources of error are of little concern because they introduce only a very small amount of uncertainty into the results. They could be reduced further, for example, by moving the base station closer to the survey positions and using a tripod rather than a cold assistant to hold the target pole. The third error is more worrying because it is slightly larger. However, there is no bias involved because the assistant selecting each target position is instructed to place the pole randomly rather than, for example, consistently selecting pebble tops. This ensures that while individual points may be a few centimetres higher or lower than the average bed elevation in their vicinity, the overall surveyed surface is neither consistently above, nor consistently below the average. Most importantly, all of the errors combined are very small relative to the variations in bar topography that we aim to characterize (millimetres compared with metres). This means that we can be confident that what little uncertainty the errors do introduce does not affect our ability to describe the overall shape of the bar surfaces. Clearly, this kind of decision-making is dependent on the job at hand, its aims and whether or not the overall precision achieved is sufficient to meet the objectives. Making such a decision always depends on appreciating the measurement errors involved and, therefore, that every effort should be made to characterize them.

An alternative example from ecology illustrates how small amounts of imprecision can be important and can lead to scientific misunderstanding. It was recently reported that, in addition to growing longer as they grow older, Galapagos marine iguanas shrink during periods of reduced food availability, for example in response to major wet episodes like El Niño climatic events (Wikelski and Thom, 2000). These arguments were based on catch and release studies in which many individually identifiable animals were periodically recaptured and measured over a number of years. Shrinkage involved more than simply losing weight, with animals becoming shorter by up to 20 per cent, possibly by bone absorption. This 'bidirectional' growth phenomena is highly unusual, so several researchers set out to see whether it was present in other types of reptile, including snakes. Examination of a large catch and release dataset for 16 species of West African snakes revealed shrinkage in up to 6 per cent of cases for some species (Luiselli, 2005). However, careful examination of these data and the measurement errors within them revealed that the proportion of shrinkage cases was strongly correlated with the 'measurability' of different species. Large, vigorous, aggressive and highly venomous snakes are understandably more difficult to measure than smaller, docile, non-biting, non-poisonous snakes and there was a clear correlation in the dataset between greater handling difficulty and a higher incidence of apparent shrinkage. The implication of this is that shrinkage is not real, but an artefact of measurement error, and Luiselli (2005) concludes that snakes do not shrink but that snakes which are difficult to handle have a higher probability of being measured incorrectly so that there is a greater probability that they will appear to shrink (either because their length was overestimated to begin with or underestimated during a subsequent measurement). Although the individual measurement errors are unbiased (measurements are equally likely to be too long or too short) they are imprecise and irreducible, in the sense that it would take substantially more effort than it is feasible to expend to obtain more accurate measurements. This lack of precision is important in this case because it could have led to unreasonable conclusions (snakes shrink) if it had not been for the additional, careful analysis. This example highlights the importance of always making every effort to understand and characterize measurement errors during a sampling campaign.

DESIGNING A SAMPLING PROGRAMME

In any project there are two main controls on the design of the sampling programme: the research aims and the resources available (time, money, person power). While the research objectives should drive the sampling design, more often than not it is resource issues that limit the sampling programme, and compromises have to be made. Two key issues are the definition of the population of interest and the choice of sampling method.

Defining the target population

Defining the target population is a critical step, and begins with a clear definition of the unit of study (the items about which generalizations are to be made and that will be sampled). In social geography this might be an individual, a household, or an organization. In fluvial geomorphology it might be a channel cross-section, a bar, a hydrological link or a river basin. One of the things that makes geography such a fascinating discipline and makes sampling necessary is that the character of these units and therefore the content of the population is apt to vary, often systematically, in space, time and with the scale of interest. The population must be defined with this heterogeneity in mind, while at once satisfying the needs of the research aims and working within resource limitations. Defining the population is then an iterative process in which several questions are asked: How do the population characteristics of interest vary spatially and temporally? Which variations are important for the study and which are not? How can important sources of variation be included and unimportant sources of variation excluded? Can the research aims be modified to accommodate practical difficulties? Answering these questions depends on careful investigation of published research, consideration of what one might reasonably expect and a clear understanding of the research aims. In turn, the spatial and temporal character of the intended population should be stated and used to guide the design of the sampling programme.

Failing to accommodate temporal and spatial variability, by targeting too narrow a slice of the possible population, will produce a sample that is unrepresentative. For example, questioning households in only one enumeration district about their leisure activities is unlikely to yield results that are applicable to the city as a whole because one district is unlikely to encapsulate the range of economic, ethnic and age-related factors that influence use of leisure time across the city. Similarly, sampling suspended sediment concentration in a stream only during the rising limb of a flood is likely to yield an average value that is too high for the flood as a whole because of temporal variations in sediment availability over the course of the event. Equally, it is possible to target too much of the possible population in terms of its spatial extent, temporal boundaries or internal structures. This not only spreads precious resources thinly with implications for sample precision (see below), but may also add sources of variability that are not of direct interest and that obfuscate or dilute critical information. Thus, it is often necessary to exclude sources of variability from a sampling programme and focus attention on particular objects, places, times or patterns.

Choice of sampling method

Having clarified the spatial, temporal and structural dimensions of the target population, the next problem is to determine the best way of sampling from this target population. A variety of sampling methods are used by geographers and fall into two basic groups: non-probability methods and probability-based methods. Non-probability methods cannot be used to make statistical inferences about the population from which they are drawn. In choosing to adopt non-probability methods one must therefore accept that statistically rigorous representativeness is not a primary issue in the research design (which may be the case, for example, in some research utilizing case studies). If the intention is to make generalizations about a larger population then non-probability methods should only be used with extreme caution and it is in this context that such methods are briefly reviewed here.

In *accessibility sampling*, units are selected on the basis of convenience, such that one selects the most accessible units from the population. Such samples are likely to yield a biased sample. An example from biogeography illustrates the potentially serious consequences if this sampling flaw remains unrecognized. Reddy and Dávalos (2003) examined the spatial distribution of 3,504 sites in sub-Saharan Africa where passerine (perching) bird species were observed between the 1800s and 1970. Datasets of animal distributions like this, compiled in a large number of studies over many years, are important because they provide information over large areas that are used to define biodiversity hotspots and priority locations for conservation. What Reddy and Dávalos (2003) found, however, was that the location of sampling was strongly influenced by accessibility, with sampling sites concentrated in a non-random pattern close to cities, roads and rivers (Figure 17.2). The use of these data to identify conservation priorities is, therefore, problematic because the information is biased. It may be that some of the apparent hotspots targeted for conservation are less rich in species than other locations that are less accessible (a long way from roads, rivers and cities) where little or no information exists. The implications for conservation biogeography are that greater effort is required to collect information in less accessible locations and to develop methods for correcting accessibility bias where such fieldwork is impracticable.

Figure 17.2   Map of sub-Saharan Africa showing approximately 3,504 locations where sampling has been conducted for passerine birds between the 1800s and 1970 (light grey dots). Major rivers (dark grey lines) and cities (large black dots) are also shown. In many regions, sampling locations tend to be located relatively close to cities and rivers; a pattern that is confirmed using formal testing. This illustrates accessibility bias in the selection of sampling sites. That is, sites close to rivers and cities are over-represented because they are relatively easy to access. See Reddy and Davalos (2003) for a full discussion and details of the data sources

In *judgemental* (also referred to as *purposive*) *sampling*, units are selected subjectively by the researcher on the basis of prior experience. This is problematic because the researcher's previous experience may be limited and his or her own prejudices, derived from his or her expectations and viewpoint, become an integral part of the selection process. Even if, by chance or skill, a judgemental approach yields an unbiased sample, it is difficult to prove that this is the case and therefore difficult to convince critics of the value of any generalizations that are made. *Quota sampling* aims to be more representative by attempting to produce a sample that replicates the general structure of the population. Predefined quotas based on factors like age, gender and class are filled, thereby imposing some useful control on the selection of units, but the choice of individual items within each quota group is still subjective. Kitchin and Tate (2000) suggest that this method can yield representative samples but should only be used in situations where prior work has shown this to be the case.

In contrast, *probability-based sampling* methods aim to preclude bias and produce representative samples. Their common characteristic is that the sampling units are selected by chance and the probability of any unit being selected can be determined. Probability-based methods must be used if one intends to use inferential statistics to generalize from the sample to the population. These methods require that a sampling frame exists or can be developed. A sampling frame is a list or other representation of the target population from which units can be drawn (for example an electoral roll, a catalogue of discharge gauging stations, an aerial photograph, a map, or a street directory).

Table 17.1 illustrates several probability-based methods. The two basic methods are *simple random* and *systematic sampling*. Their common feature is that there is an equal probability of selecting each and every unit within the sampling frame. Two issues are worth considering when adopting these methods. First, if systematic sampling is applied within a sampling frame that includes a repetitive structure and the sampling interval that is chosen coincides with that structure, then bias will be introduced. For example, many alluvial rivers exhibit repetitive pool-riffle-bar morphology in which the spacing between units is typically five to seven times the channel width. If water depth or grain size or stream velocity are systematically sampled using a similar interval, it is possible that measurements will be biased toward the characteristics of pools or riffles. Second, with a target population where the characteristic of interest is heterogeneous but also exhibits some internal pattern, it is important to obtain uniform coverage of the sampling frame without any gaps. Simple random sampling may not do this as well as systematic sampling because it is possible for sampled units to be unevenly distributed, as illustrated for the case of river sediment characterization by Wolcott and Church (1991).

In a *stratified sample* a number of homogeneous sub-groups or strata, differentiated by some relevant characteristic, are recognized within the population. In contrast to the simple and systematic methods, the probability of selecting an individual unit from the sampling frame varies, depending upon the stratum that the unit belongs to. Three common reasons for utilizing stratified sampling illustrate its value. First, it can be used to ensure that the number of units drawn from distinctive strata is in proportion to their true size in the population. This is known as *proportionate stratified sampling*. Simple random and systematic sampling will achieve this by default if the sampling frame is appropriate, comprehensive and accurate, which should be the case if the sampling frame is developed for the research project. However, it is not uncommon for the sampling frame to be obtained from a source that compiled the frame for purposes other than those for which it is now intended. Such frames may be biased in favour of one or other strata. Similarly, instrument malfunction at a particular time or place, or non-responses to questionnaire surveys may yield a sample that is known to be biased. In either case, if the true population proportions are known then each strata can be randomly sub-sampled in those proportions to obtain an unbiased sample. Second, it may be uneconomical or unfeasible to sample strata of very different

Table 17.1 Basic sampling methods

	Description	Physical illustration	Human illustration
a) Simple Random	Within the sampling frame each unit is assigned a unique number or position. Numbers and thence units are selected at random from the sampling frame.	**Sediment size** *Aim:* ascertain average size of the sediment particles on a river bar. *Population:* all particles on the river bar. *Unit:* a sediment particle. *Frame:* a map of the bar surface located in an arbitrary cartesian space. *Measurement:* using a size template. A random number generator is used to pick x and y coordinates. These coordinates locate particles for measurement.	**Street safety** *Aim:* ascertain views of university students on campus safety. *Population:* all students at the university. *Unit:* an individual student. *Frame:* a list of students and their addresses. *Measurement:* by questionnaire. Each student on the list is assigned a unique number. A random number generator is used to pick numbers and the corresponding people are sent questionnaires.
b) Systematic	A sampling interval is defined (e.g. every 10 m, every fourth individual, every 60th second). The first unit is randomly selected as in (a) and subsequent units are selected systematically according to the sampling interval.	The bar is approximately 40 m² and a sample of 100 is required. A sampling interval of 2 m is defined. From an arbitrary origin, a grid of 2 m squares is projected onto the sampling frame map. Grid intersections locate particles for measurement.	The list contains 500 names and a sample of 100 students is required. An interval of 4 units is defined. One name is randomly selected as above. Subsequently, every fourth student is selected. If the end of the list is reached, counting continues at the beginning.
c) Stratified	Mutually exclusive sub-groups (strata) are identified and sampled randomly or systematically in one of two ways: *Proportionate:* each stratum is sampled in proportion to its true population proportion. This is necessary if the sampling frame is inadequate. *Disproportionate:* an equal number of units are sampled from each stratum irrespective of their true population proportion. This is necessary when comparisons between strata are required.	Four strata corresponding to distinct facies (areas of homogeneous sedimentary character) are evident. In this case the frame is adequate and there have been no measurement problems. Simple random and systematic sampling are adequate. Suppose one wishes to compare size in facies 1 (a small area) and facies 4 (a larger area). An equal number of particles should be selected from each. Thus, disproportionate sampling is necessary (note this will yield a biased sample of the population so weighting is required).	It is suspected that gender is an important factor in determining views on campus safety. Suppose the supplied list is for students in only one faculty. Different faculties typically exhibit distinct gender distributions. In this case the list is not representative of gender distribution across the university. Proportionate sampling is required: stratify (male, female) and randomly sample in each group to obtain numbers that yield the female:male ratio for the university as a whole.

sizes in proportion to their size (total area, number of units, etc.). A more efficient method is often to collect a random sample of common size within each stratum, then weight the statistics obtained for each strata according to the stratum's size within the population, and combine them appropriately in order to generate population estimates. Sampling the same number of units from strata of different size is referred to as *disproportionate stratified sampling*. Third, individual research projects may ask questions about the strata, often requiring that comparisons are made between them. In this case it is necessary to obtain equally precise samples for each stratum, which means selecting a similar number of units from each. Simple random or systematic sampling does not do this, but rather selects a number of units from each stratum that is in proportion to the stratum's size. With disproportionate stratified sampling this problem is overcome by randomly selecting the same number of units from each stratum, irrespective of their true relative sizes. In using this method it is important to remember that as far as the population as a whole is concerned, one has created a biased sample so that if estimates of population parameters are required, strata estimates must be combined using appropriate weighting techniques.

A final example of a probability-based method is the *multi-stage* or *hierarchical sample* in which the sample is selected in several stages that usually relate to spatial or temporal scale. For example, if the campus safety study (Table 17.1) was extended to a global scale the aim might be to sample 100 universities from around the world. First, ten countries might be randomly selected, then within each country five cities, and ultimately within each city, two universities. Multi-stage surveys are an efficient method when faced with a very large population in space or time.

Choosing between these various probability-based methods (and the many others that have been suggested) requires some prior knowledge or reasoned judgement concerning any spatial or temporal structures within the population, a thorough understanding of the sampling frame and a clear set of aims. Without a good appreciation of these it is possible to inadvertently choose a sampling method that systematically favours some parts of the population over others, in which case the characteristics of interest are not properly represented. This basic point has been stressed by several authors who have considered the specific details of applying standard sampling methods to spatial data (e.g. Berry and Baker, 1968; Harvey, 1969). Haining (1990) suggests that systematic sampling is superior where the underlying spatial variation is random. Wolcott and Church (1991) find that a particular combination of grid and random sampling known as stratified systematic unaligned sampling (cf. Smartt and Grainger, 1974; Taylor, 1977) performs well for areally structured data. They point out that it avoids the primary problems with each of random and systematic sampling: the possibility that random sampling is unevenly distributed thereby missing small spatial structures, or that the data contain spatial structures that have the same spacing as the grid spacing, thereby introducing bias.

In summary, non-probability methods are less desirable than their probability-based counterparts and certain probability methods are more appropriate than others in certain circumstances. Nevertheless, it is important to recognize that the vagaries of empirical research often make meeting the ideal difficult (if not impossible) with the result that the target population and the sampled population differ

(Krumbein and Graybill, 1965). This might be because the resources necessary are not forthcoming. It may be that accurate information about the population is not available to guide programme design or that there are unknown and hidden sources of variation within the population. It may be that an appropriate sampling frame does not exist or that we are forced to accept an accessibility sample because only certain people will talk with us or only certain places can be reached. In cases like these it is incumbent on the researcher to make it very clear exactly how sampling was conducted and for him or her to interpret his or her results in light of suspected sampling weaknesses.

Analytical requirements

Finally, in designing a sampling programme it is important to think ahead to the analytical stage of the research and identify any restrictions or requirements that the intended analysis imposes on the sampling strategy. For example, it may be that the inferential statistics used require a minimum number of samples or that a laboratory machine requires individual samples to be of a particular mass. It is certainly the case that any hypothesis being tested will require the data to be collected in a particular manner. In experimental and some observational projects the experimental design will be an integral part of designing the correct sampling programme. It is therefore crucial to identify the analytical procedures that will be used in the laboratory or at a desk *before* setting out with clipboard or shovel.

STATISTICAL INFERENCE

We have already noted that geographical enquiry must deal with uncertainties. Hicks (1982: 15) defines inferential statistics as 'a tool for decision making in the light of uncertainty', and geographers have certainly found inferential statistics to be a valuable tool. Inferential statistics use sample data to make probabilistic statements about the population from which they are drawn. Statements can be made about the characteristics of the population, which is referred to as parameter estimation, and also whether a particular supposition about the population is true or false, which is referred to as hypothesis or significance testing.

Numerous text books are available that explain the principles and practical application of the great array of inferential statistical techniques used by geographers. These include specifically spatial techniques that extend statistical analysis to the examination of patterns in space (e.g. Norcliffe, 1977; Williams, 1984; Haining, 1990; Shaw and Wheeler, 1994; Fotheringham *et al.*, 2000; Rogerson, 2006). Particular attention should always be paid to the data assumptions that these procedures have and whether so-called parametric or non-parametric techniques are most appropriate. There are also some specifically geographical issues to be aware of too, particularly spatial autocorrelation. This refers to the propensity for the value of a variable at one location to be related to the value of that same variable in a nearby location. It is problematic, because inferential statistical techniques often require that each sample measurement is independent of all others. In spatial data, autocorrelation is common (otherwise location would not matter) such that the performance of standard methods may be degraded and there is the potential for misinterpretation. It is possible to measure the significance of spatial autocorrelation in a data set (see, for example, Kitchen and Tate, 2000) and standard inferential procedures can be adapted to minimize its impact (see, for example, Cliff and Ord, 1975; Fotheringham *et al.*, 2000; Rogerson, 2006). Chapter 26 provides alternative means of describing and exploring spatial associations.

There are many introductory texts that can provide a detailed step-by-step introduction to inferential statistical methods. The aims of this section are limited to explaining the apparent incongruity of statistical inference – how can one make statements about a population based on a single sample drawn from it, even though one knows that no two samples would ever be exactly the same? – an apparent leap of faith that brings to mind Jean Baudrillard's comment that, 'Like dreams, statistics are a form of wish fulfilment' (Baudrillard, 1990: 147). The simple answer is that, although we know our sample to be unique, statistical theory allows us to assess the reliability of sample estimates (called statistics) such as the sample mean. It is, therefore, possible to ascertain the likely difference between a sample statistic and the equivalent population parameter *without knowing* the value of the population parameter. In turn, the differences between sample statistics, for example mean values from different groups, can be compared with one another to test the hypothesis that they come from different populations. The following exposition of these ideas is necessarily very brief and non-technical and focuses on ascertaining reliability rather than hypothesis testing. The reader is directed to one of the above texts (Shaw and Wheeler, 1994; Rogerson, 2006) for a fuller account.

Probability and the 'normal' distribution

A basic understanding of probability distributions is necessary before continuing. A probability distribution describes the changing frequency with which particular values of a variable of interest are measured. It is commonly visualized as a histogram in which the ordinate shows the number of occurrences (the frequency) with which groups of values occur. For example, one can describe the frequency distribution of beach pebble sizes in a 100-pebble sample by indicating the number of particles in each of several consecutive 10 mm grain-size classes. Frequencies can be represented as absolute numbers or as relative proportions, in which case they represent the empirical probabilities of measuring a value in each class. Thus, if 35 of the 100 pebbles were found to be between 40 and 50mm in diameter, it follows that there was a probability of 0.35 (a 35 per cent chance) of finding a pebble in that size range on the beach. Probability distributions for measured phenomena take a wide variety of forms, but a typical situation is that values close to the mean are common and those further away are proportionately less common. Specifically, many phenomena exhibit an approximately 'normal' distribution (sometimes referred to as a Gaussian distribution after the mathematician who first defined it) with its characteristic bell-shaped curve, centred on the mean.

The properties of the normal distribution, and in turn empirical probability distributions that approximate it, are at the heart of basic statistical inference. Any normal distribution can be described in a standardized form in which raw empirical data are transformed into so-called z-values. These numbers express changes in the measured values as multiples of the data's standard deviation. In standardized form the mean of the distribution is zero and the standard deviation of the distribution is 1.0. Because the mathematical form of the standardized distribution is known, the probabilities of observations falling within any given range of z values can be calculated and most statistical text books contain tables that give the probability associated with specific z ranges. Thus, there is a 0.68 probability (0.34 either side of the mean) of a standardized observed value falling in the range z = −1.0 to z = 1.0; i.e. within one standard deviation of the mean. Similarly, 95.45 per cent of observations will be within two standard deviations and 99.73 per cent within three standard deviations of the mean. This is true of any normally distributed variable which means that we can apply such reasoning to a wide variety of phenomena in physical and human geography. By using such tables in reverse, the values of z that are associated with selected probabilities can be ascertained. For example, 95 per cent of the values (47.5 per cent either side of the mean) in a normally distributed phenomenon will have z values that are in the range ±1.96. Equally, sampled values with z values outside this range have a probability of being observed 5 per cent of the time or less.

Confidence statements about sample statistics

If repeated samples are drawn from the same population and in each case the mean is calculated, the mean values will vary from sample to sample but will tend to cluster around the true mean of the population. Such a collection of sample means (or indeed any other sample statistic) is called a sampling distribution. A piece of mathematical theory called the Central Limit Theorem (CLT) proves that sampling distributions are normal with a mean value equal to the value of the true population parameter (e.g. the true population mean) and that this holds irrespective of the population distribution. Thus, even for a phenomenon that does not exhibit a normal distribution, the sampling distribution of the mean is normal. The standard deviation of a sampling distribution is known as the standard error and it has the same general properties as the standard deviation of any normal distribution so that, for example, 95 per cent of the sampling distribution lies within 1.96 standard errors of the true population mean.

Standard errors can be determined empirically by repeated sampling of a given population, but this is rarely plausible. It is of significant consequence, then, that standard errors can be calculated on the basis of collecting only a single sample. For example, the standard error of sample means (σ_x) can be calculated as

$$\sigma_x = s / \sqrt{n}$$

where s is the standard deviation determined from a single sample and n is sample size. Armed with this value and our knowledge of the normal distribution it is possible to make statements about the reliability of the sample mean; that is, to say how confident we are that the true population mean is within a given interval about the sample mean. Remembering that the probabilities in a z table indicate that 95 per cent of a normal distribution lies within 1.96 standard deviations of the mean, we can say that there is a 95 per cent chance that the sample mean lies within 1.96 standard errors of the true population mean. This is equivalent to saying that there is a 95 per cent chance that the population mean lies within 1.96 standard errors of the sample mean.

For a given case, the interval can be specified in the original data units and is known as a confidence interval. So, for example, for a sample of pebble diameters with a standard deviation of 20 mm and n = 100, there is a 95 per cent chance that the sample mean lies within $1.96 \times \sigma_x = 1.96 \times (20/\sqrt{100}) = \pm 3.9$ mm of the population mean. This is commonly interpreted as meaning that in 95 samples out of 100 the sample mean would lie within ± 3.9 mm of the population mean, although more precisely it says that if 100 samples were used to construct 100 confidence intervals the true population mean would be included within 95 of them. Confidence intervals for any probability can be constructed using the appropriate z value, so that at 0.99 probability the confidence intervals in the above example are $2.58 \times (20/\sqrt{100}) = \pm 5.2$ mm. An important caveat for the reader to investigate further is that while large samples always have normal sampling distributions, irrespective of the population distribution, small samples (n < 30) have distorted distributions with a form that is a little different from 'normal'. Small samples tend to yield statistics that are distributed according to the t-distribution, sometimes known as 'Student's t-distribution'. This has similar properties to the normal distribution and it is used in the same way to determine the reliability of sample estimates, except that probability values from published t-tables, rather than z-tables, are used.

The CLT and standard errors are so important because they allow us to make rigorous statements about the reliability of the statistics we derive from sample data – that is, to accurately quantify the uncertainty that is inherent in a sample. In turn, they provide a basis for making rigorous comparisons between samples and thence for testing hypotheses. Just as confidence intervals are used in assessing reliability, so-called significance levels, denoted by α, are used to attach probability statements to the decisions made in hypothesis testing. There is always the chance that a given decision is incorrect and levels of significance define the probability that one incorrectly rejects a true hypothesis. Significance levels are set by the researcher as part of the testing procedure. Usually, we are only willing to accept low-levels of error, so significance levels are set to 5 or 1 per cent, though smaller, more stringent values can be used. The important point to make at the end of this section is that statistical inference, beyond the mathematical formulation of the various procedures and tests, involves commonplace ideas of confidence and significance *not* certainty. It allows us to attach probability statements to estimates and decisions but crucially, statistical techniques do not provide binary, 'black and white', yes and no answers. It is always the responsibility of the researcher to choose levels of confidence and significance, and to interpret results thoughtfully in light of these choices.

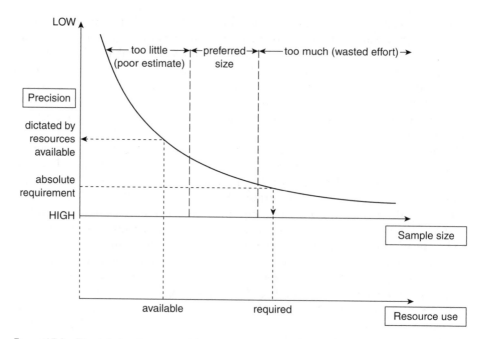

Figure 17.3   The relation between estimate precision, sample size and sampling resources

SAMPLE SIZE

A frequently asked question is, 'How big should my sample be?' The answer reflects a compromise between the desired precision of the sample estimates and the resources available, because maximizing precision and thence the significance that can be attached to statistical inferences (see above) requires the collection of large samples, but also demands greater resource expenditure.

In general, precision improves with sample size in a curvilinear fashion. As sample size increases precision improves rapidly to begin with but then more slowly (Figure 17.3). In this case, we can imagine that in any sampling procedure there is an optimal sample size corresponding to that region where the curve in Figure 17.3 begins to flatten out. Beyond this point additional gains in precision are small and do not warrant the additional sampling effort (resources) required. Before this point, the sample size is too small to yield reasonable estimates of the population characteristics and it is worth expending small amounts of effort to obtain significantly better estimates. Ideally, we want to obtain samples of a size somewhere in this optimal zone. Several ways of doing this are discussed below, but it is first worth noting two cases where the option of seeking the optimal solution does not arise.

First, a common situation is that the resources needed to collect a sample of the ideal size are simply not available so that low sample precision is inevitable. Similarly, where secondary rather than primary data is being used, the second-ary data may be less voluminous than that desired. When designing a sampling programme it is always necessary to carefully consider the allocation of resources in light of this problem. It may be that sacrifices can be made in one part of the programme in order to improve precision elsewhere. For example, rather than obtaining low-precision estimates of water quality in twenty lakes, it may be advisable to seek quality estimates for five lakes, especially if the retained lakes are carefully selected to test one or more hypotheses.

Second, it may be that a specified level of precision is required by the research aims in which case there may be no option but to collect inefficient, large samples. Work with legal or health implications often has an absolute level of precision that is required by the project objectives. A related issue that constrains sample size is the possibility that a particular physical or statistical technique used to analyse the sample data has sample size requirements. It is therefore important to know how your data will be analysed before the sampling programme is finalized.

In any given study, the relation between sample size and precision is driven by the variability (heterogeneity) of the characteristic of interest within the population. For a given sample size, precision is worse in populations that exhibit greater spread or variability and the more heterogeneous a population, the greater the sample size required to obtain a given level of precision.

Formulae exist for calculating the sample sizes needed to obtain specified levels of precision for a given statistic. For example, in the case of estimating the population mean μ, precision can be thought of as the error, δ, that we are willing to accept – that is, the acceptable difference between a sample mean, x and μ ($\pm \delta$ units). δ is equivalent to half of a confidence interval and a confidence interval has length $2.(z.\sigma_x)$, where σ_x is the standard error of the mean and z is the tabled value associated with the chosen significance level α. Thus,

$$\delta = z.\sigma_x$$

therefore:

$$\delta = z.(s / \sqrt{n})$$

where s is the sample standard deviation, and

$$n = (z^2.s^2) / \delta^2$$

This gives the sample size n, needed to obtain an estimate of the mean that is within δ units of the population mean with a $100.(1-\alpha)$ per cent level of confidence.

Similar formulae can be developed for estimating other statistics or for use in hypothesis testing. A device known as an Operating Characteristic Curve can also be used to determine optimal sample sizes in hypothesis testing. The operational problem with these methods is quite simply that usually we do not know the sample standard deviation beforehand. This can be overcome by a two-phase sampling procedure or by estimating the standard deviation from previously published work. An additional problem is that researchers seldom find it easy to define an acceptable error, δ.

Empirical approaches to sample size determination may then be useful. As sample size increases from one, the value of any statistic will vary significantly as successive population values are added, but will gradually achieve a degree of stability. This indicates that the sample has incorporated most of the variance evident within the population (see again Figure 17.2). If it is possible to monitor the value of the statistic of interest as the sample is collected, sampling can be curtailed when values for successive n become relatively stable. This can be an especially effective method if the same type of sample is to be obtained from a number of strata or discrete sampling frames where it is anticipated that there is little change in the population variance between those strata or frames. A pilot exercise conducted in one case can then be used to inform sample size for the whole programme. For example, in the case of an insect survey consisting of many discrete quadrat samples where the aim is to examine variations in number of taxa present, it may be worthwhile to conduct a pilot exercise in which one monitors the changing number of taxa as the size of the quadrat is gradually increased. A graph can be plotted of area against number of taxa and the stabilization point will reveal the optimal quadrat size, to be utilized throughout the survey, for obtaining a reasonable estimate of taxa number (e.g. Chutter and Noble, 1966; Elliot, 1977: 128). More elaborate empirical methods can also be used to examine sample precision and identify optimal sample sizes, for example a technique called bootstrapping (e.g. Rice and Church, 1996), though these require very large data sets and the ability to invest resources in a significant pilot study.

CONCLUSION

A principal aim of most geographical research is to make useful generalizations that might help to model or otherwise understand better the uncertain human and physical words that are the geographer's realm. Because it is usually impossible or impractical to observe all instances of variation, a smaller number of instances (the sample) are used, from which the 'population' characteristics can be estimated. Achieving this in a reliable and reproducible fashion is the basis of sampling theory and sampling design.

Geographers recognize the value of both extensive statistical sampling and intensive 'case-study' sampling. A benefit of extensive sampling is that a set of techniques known as inferential statistics can be applied to make probabilistic statements about the population from which the sample is drawn. Sampling is therefore a powerful tool, but geographical research frequently engages with very heterogeneous phenomena that require careful sampling in order to maximize the accuracy of inferential conclusions. Careful design of the sampling programme is crucial and is driven by both the research aims and available resources. The overall message is that sampling is a tool that most geographers need, but if research methodology aims to be as impartial and free of error as possible, sampling must be done thoughtfully and rigorously.

Summary

The most important aspects of sampling are as follows:

- Ensuring the quality of the sample by maximizing the precision and minimizing the bias in any measurements or observations.
- Relating individual observations and sample sets to the observed or expected geographical patterns forming the population by the correct sampling design (choosing the sampling method and the sample size).
- Assessing the significance of sample estimates using graphical and statistical methods, including the use of inferential hypothesis testing.

Further reading

- Most basic textbooks on statistical analysis include an introductory section on sampling and two that are not overly technical are Shaw and Wheeler (1994) and Rogerson (2006). The latter is useful in terms of addressing autocorrelation issues.

- Kitchin and Tate (2000) is a general book on research methods in human geography that contains some useful examples of sampling schemes and a lot more besides, and can be recommended to physical geographers for its coverage of basic statistical techniques.

- Haines-Young and Petch (1986) contains a brief but lucid overview of measurement errors and statistical inference in Chapter 11. Harvey (1969) and Richards (1996) provide useful, and in the latter case advanced, discussions of the role of case studies in geography.

- For a detailed consideration of the benefits of alternate spatial sampling techniques look at Wolcott and Church (1991). Several hypothetical examples used here have been drawn from my own experience of sampling sediments. For anyone embarking on a project involving sediment sampling, Bunte and Abt (2001) provides a plethora of valuable information.

Note: Full details of the above can be found in the references list below.

References

Baudrillard, J. (1990) *Cool Memories* (English translation by Chris Turner). London: Verso.

Berry, B.J.L. and Baker, A.M. (1968) 'Geographic sampling' in B.J.L. Berry and D. Marble (eds) *Spatial Analysis: A Reader in Statistical Geography*. London: Prentice Hall.

Blaut, J.M. (1959) 'Microgeographic sampling', *Economic Geography*, 35: 79–88.

Bunte, K. and Abt, S.R. (2001) 'Sampling surface and subsurface particle-size distributions in wadable gravel- and cobble-bed streams for analysis in sediment transport, hydraulics and streambed monitoring', *General Technical Report RMRS-GTR-74*, USDA Forest Service, Rocky Mountain Research Station, Fort Collins, Colorado.

Chutter, F.M. and Noble, R.G. (1965) 'The reliability of a method of sampling stream invertebrates', *Archive Hydrobiologia*, 62(1): 95–103.

Cliff, A. and Ord, J.K. (1975) 'The comparison of means when samples consist of spatial autocorrelated observations', *Environment and Planning A*, 7: 725–734.

Curtis, S., Gesler, W., Smith, G. and Washburn, S. (2000), 'Approaches to sampling and case selection in qualitative research: examples in the geography of health', *Social Science and Medicine*, 50: 1001–1014.

Elliot, J.M. (1977) 'Some methods for the statistical analysis of samples of benthic invertebrates', *Freshwater Biological Association, Scientific Publication* 25. Windermere: Freshwater Biological Association.

Fotheringham, A.S., Brunsdon, C. and Charlton, M. (2000) *Quantitative Geography: Perspectives on Spatial Data Analysis*. London: Sage Publications.

Haines-Young, R. and Petch, J. (1986) *Physical Geography: Its Nature and Methods*. London: Harper and Row.

Haining, R. (1990) *Spatial Data Analysis in the Social and Environmental Sciences*. Cambridge: Cambridge University Press.

Harvey, D. (1969) *Explanation in Geography*. London: Edward Arnold.

Hicks, C.R. (1982) *Fundamental Concepts in the Design of Experiments* (3rd edn). New York: CBS College Publishing.

Kitchin, R. and Tate, N.J. (2000) *Conducting Research Into Human Geography: Theory, Methodology and Practice*. Harlow: Prentice Hall.

Krumbein, W.C. and Graybill, F.A. (1965) *An Introduction to Statistical Models in Geology*. New York: McGraw-Hill.

Luiselli, L. (2005) 'Snakes don't shrink, but "shrinkage" is an almost inevitable outcome of measurement error by the experimenters', *Oikos*, 110: 199–202.

Marcus, W.A., Ladd, S. and Stoughton, J. (1995) 'Pebble counts and the role of user-dependent bias in documenting sediment distributions', *Water Resources Research*, 31(10): 2625–2631.

Miles, M. and Huberman, A. (1994) *Qualitative Data Analysis*. London: Sage.

Norcliffe, G.B. (1977) *Inferential Statistics for Geographers*. London: Hutchinson and Company.

Reddy, S. and Dávalos, L.M. (2003) 'Geographical sampling bias and its implications for conservation priorities in Africa', *Journal of Biogeography*, 30: 1719–1727.

Rice, S.P. and Church, M. (1996) 'Sampling surficial fluvial gravels: the precision of size distribution percentile estimates', *Journal of Sedimentary Research*, 66(3): 654–665.

Richards, K. (1996) 'Samples and cases: generalisation and explanation in geomorphology', in B.L. Rhoads and C.E. Thorn (eds) *The Scientific Nature of Geomorphology*. Chichester: John Wiley and Sons.

Rogerson, P.A. (2006) *Statistical Methods for Geography: A Student's Guide* (2nd edn). London: Sage Publications.

Schumm, S.A. (1991) *To Interpret the Earth: Ten Ways to be Wrong*. Cambridge: Cambridge University Press.

Shaw, G. and Wheeler, D. (1994) *Statistical Techniques in Geographical Analysis* (2nd edn). London: David Fulton Publishers.

Smartt, P.F.M. and Grainger, J.E.A. (1974) 'Sampling for vegetation survey: some aspects of the behaviour of unrestricted, restricted, and stratified techniques', *Journal of Biogeography*, 1: 193–206.

Taylor, P.J. (1977) *Qualitative Methods in Geography*. Boston: Houghton-Miffin.

Wolcott, J. and Church, M. (1991) 'Strategies for sampling spatially heterogeneous phenomena: the example of river gravels', *Journal of Sedimentary Petrology*, 61(4): 534–543.

Wikelski, M. and Thom, C. (2000) 'Marine iguanas shrink to survive El Niño', *Nature*, 403: 37–38.

Williams, R.B.G. (1984) *Introduction to Statistics for Geographers and Earth Scientists*. London: Macmillan.

18 Analysing a Natural System

Ellen Wohl

Synopsis

Geographers use conceptual models to organize disparate observations about complex natural systems. This chapter summarizes the most widely used conceptual models in physical geography and provides examples of their application to natural systems including hillslopes, rivers, soils, glaciers, coastlines and deserts. When a community of scientists shares a conceptual model, the model guides the thinking of the community and helps to define the type of questions asked by individual investigators. Physical geographers tend to conceptualize natural systems as open systems that continually exchange matter and energy with the surrounding environment. Individual components of each system interact in a manner that can promote the stability described by equilibrium, or the directional tendencies described by extremal hypotheses. These interactions are seldom linear, but instead involve delays known as lag times, and abrupt changes known as thresholds.

The chapter is organized as follows:

- Introduction
- Conceptualizing a natural system
- Natural Systems as ecosystems
- Natural systems with a past
- Challenges to conceptualizing from direct observation
- Conclusions

INTRODUCTION

The formulation of research hypotheses and the application of research methods are driven by conceptual models of how a natural system works. A natural system, in the context of this chapter, is a portion of a landscape, such as a river, a lake, a hill slope, or a glacier, or the broader landscape that integrates numerous individual components. A conceptual model results from a series of assumptions about how the natural system works. These assumptions may have been tested; all investigations of natural systems on Earth, for example, assume that gravity is present and will influence the movement of materials across the Earth's surface. Or, the assumptions underlying a conceptual model may not yet have been rigorously tested; sometimes the intent of research is to collect data to further test the assumptions underlying a conceptual model prevalent in a given discipline. Regardless of the certainty of the assumptions on which a conceptual model rests, a scientist who can clearly articulate these assumptions is more likely to be able to effectively pose a hypothesis and collect data to test that hypothesis. At a minimum, being able to clearly articulate assumptions helps the scientist to think about what assumptions are included in the conceptual model, and which of these assumptions are more likely to require further testing.

A conceptual model can be qualitative, as when a scientist investigating landslides collects a great deal of data on soil thickness because he/she assumes that soil thickness exerts a primary control on landslide occurrence. A quantitative conceptual model is one in which the relations among different components are expressed mathematically. Bull (1964) developed a quantitative conceptual model of the relation between the surface area of alluvial fans, A_f, and the area of the contributing drainage basin, A_d, which he expressed as

$$A_f = cA_d^b,$$

where the values of b and c are supplied by a regression equation for the best-fit line of field values of A_f and A_d. The conceptual model underlying the equation assumes that the primary control on the surface area of an alluvial fan is the area of the contributing drainage basin, and it expresses this assumption mathematically. A quantitative conceptual model can be more precise than a qualitative model, but it is not necessarily more accurate.

Church and Mark (1980) use the equation above as an example of inductive reasoning applied to natural systems. Deductive reasoning leads to necessary statements embodied in theories. Inductive reasoning expresses formal relations, or relations that are not analytically rigorous, among phenomena that have been observed empirically. Such empirical correlations are scientifically useful only if they can provide insight into the underlying mechanisms that determine why the correlation occurs. The correlations are not deductively rigorous because there is insufficient theoretical basis or because theory leads to intractable questions. Such relations can be a starting point, however, for theoretical proposals.

A continual feedback usually exists between posing and testing hypotheses and developing conceptual models of a natural system. Sometimes a hypothesis will start from an apparent relationship between two variables, such as soil thickness and landslide occurrence. If subsequent tests support the hypothesis, the relationship between the two variables then becomes part of a conceptual model that is often broader in scope. Thin soils in a particular landscape may reflect relatively recent and frequent landsliding, for example, and occur most commonly on straight or convex slopes. The higher frequency of landslides on these slopes exposes fresh bedrock to renewed weathering and thus results in higher rates of weathering than on adjacent, concave slopes that are covered by thicker soils. A conceptual model of slope development might therefore be based on the assumption that over sufficiently long spans of time, the straight and convex portions of bedrock slopes weather and erode more rapidly than the concave portions of slopes, so that adjacent slopes become more uniform. The assumption underlying this conceptual model can of course be tested as a hypothesis, although this hypothesis is broader in scope and more complex than the original hypothesis of the correspondence between soil thickness and landslide occurrence.

Posing and testing hypotheses and conceptual models can in practice be extraordinarily difficult, partly because of the temptation to reduce complex natural systems to a limited number of relevant variables. Bull's equation, for example, relates the size of an alluvial fan to a single dominant control variable yet, as Bull himself acknowledged, the lithologies present in the source drainage basin the climate of the area, land use in the source basin and the tectonic setting of the basin in which the alluvial fan is formed can also exert important influences on the size of an alluvial fan.

T.C. Chamberlin (1897) proposed that investigators rely on the method of multiple working hypotheses as a means to avoid over-simplifying complex natural systems. Applying this method to the example above, a scientist studying controls on the area of an alluvial fan would develop several potential explanations for fan size, such as 'The dominant control on fan area is upstream drainage area', 'The dominant control on fan area is upstream rock type', 'The dominant control on fan area is land use', 'The size of an alluvial fan does not reflect a single dominant control, but rather a combination of several factors', and so forth.

Many contemporary investigators use a modified form of Chamberlin's method when they pose null and alternative hypotheses. The null hypothesis proposes something initially presumed to be true. Often this assumption takes the form that there will be no differences between two groups for the variable being compared. Alternative hypotheses describe potential differences between the groups. When trying to evaluate the effects of timber harvest on stream flow, for example, the null hypothesis would be that there will be no statistically significant differences in discharge between otherwise similar drainage basins with timber harvest and those without timber harvest. An alternative hypothesis would be that peak flow will increase significantly in basins with timber harvest because of declines in infiltration and increases in runoff.

Individual hypotheses and the conceptual models that arise from them can be tested using a variety of research methods that include: measurements of natural systems using ground-based or remote-sensing technologies; laboratory experiments that allow investigators to simplify a system such as a river channel and then selectively manipulate aspects of that system in order to observe how other aspects respond to these manipulations; numerical simulations that allow an investigator to test his understanding of the system by comparing simulated behaviour to behaviour observed in real natural systems; and statistical analyses that facilitate detection of correlations among numerous variables in complex natural systems.

This chapter focuses on existing conceptual models for natural systems, rather than on how those models were developed or tested. The chapter deals with physical geography and uses landscape examples, with an emphasis on watersheds, because a majority of the conceptual models presently used in physical geography were developed to explain properties of watersheds and river channels. The most widely used conceptual models in physical geography assume that natural systems are best characterized as open systems that exhibit tendencies toward a form of stability which represents some balance between the driving and resisting forces acting on the system. Attaining and maintaining stability is complicated by the existence of lag times, thresholds, feedbacks among system components, historical legacies, and variations in system operation across different scales of time and space. Any natural system such as a coastline or glacier is also most realistically considered as an ecosystem that operates within a global context.

CONCEPTUALIZING A NATURAL SYSTEM

Open and closed systems

At the most fundamental level, natural systems such as rivers, coastlines, or desert sand dunes are open systems. Chorley (1962) borrowed the concept of a system from physics, and applied the distinction between closed and open systems to features of the landscape. A system is a set of objects together with the relationships between the objects and between their attributes. A closed system possesses clearly defined closed boundaries across which no import or export of materials or energy occurs. Under these conditions, the system exhibits a progressive increase in entropy. An open system, in contrast, has a constant supply and removal of matter and energy. Think of a valley glacier, for example. Energy streams onto the surface of the glacier in the form of solar radiation and, if the glacier is in Iceland or somewhere else with a high geothermal gradient, in the form of subterranean heat. Mass is added to the glacier from aeolian dust, rockfalls or landslides, erosion of the underlying substrate and precipitation. Mass is lost from the glacier during melting, sublimation and evaporation of water or ice, and during deposition of sediment.

The operation of a closed system has an irreversible character. Initial system conditions are very important, and there are no intermediate states of equilibrium as entropy increases once the system is closed. In contrast, open systems exhibit a tendency toward steady-state conditions resulting from self-regulation among interacting components; a concept referred to as dynamic equilibrium or, for rivers, as being at grade. Recognizing that natural systems are open systems tends to focus attention on possible relationships between form and process, and allows investigators to characterize inputs, outputs, storage and fluxes of matter and energy within a system. As an example of the latter, Dietrich *et al.* (2003) use mathematical statements known as geomorphic transport laws to express mass flux or erosion caused by one or more processes in a manner that can be parameterized from field measurements and tested in physical models. Existing transport laws quantify processes including mass conservation for soil- or sediment-mantled landscapes underlain by bedrock, sediment transport in rivers and bedrock incision by rivers.

The concept of open systems also recognizes the fact that most phenomena of interest to physical geographers reflect the influences of multiple variables. In essence, the conceptual model of open systems reminds physical geographers that the landscape component of interest is intimately connected to a much greater landscape assemblage.

The application of general systems theory to natural systems started during the 1950s with work by investigators such as A.N. Strahler and J.T. Hack, but R.J. Chorley formally codified this application in his 1962 paper. Many investigators have subsequently noted that the ideas of two of the great nineteenth-century investigators of physical geography exemplified implicitly thinking about natural systems as being closed or open. W.M. Davis did not use the phrase 'closed system', but his view of landscapes progressing from youthful terrains of high relief and steep rivers through mature regions of rolling hills and into old landscapes of low relief implies a linear progression through time that exemplifies a closed system. This Cycle of Erosion was the prevailing conceptual model among physical geographers during Davis' lifetime (1850–1934), but has subsequently been largely replaced by the views of his contemporary G.K. Gilbert (1843–1918). Gilbert emphasized that natural systems could exhibit periods of little change with continual feedbacks and adjustments among multiple interacting variables; characteristics which exemplify an open system.

The importance of time

Davis and Gilbert developed distinctly different conceptual models of natural systems in part because of the relative time-scales that they chose to emphasize. In the absence of continuing inputs of tectonic energy, for example, a drainage basin may progressively lose mass and elevation as a result of weathering and erosion which exports weathered materials beyond the drainage basin. This would be an example of Davis' progressive decline in relief, even though the system remains open. Such trends may only be apparent over tens or hundreds of millions of years, however, because isostatic rebound of the Earth's crust as mass is removed from the drainage basin effectively counteracts the elevation lost through mass removal. At shorter time-scales, the drainage basin may be most effectively characterized as being in equilibrium, with very little net change in variables such as relief or elevation.

Schumm and Lichty (1965) systematically characterized the effects of considering natural systems over different time-scales. They noted that the distinction between cause and effect in the development of landforms is a function of time and space (area) because the factors that determine the character of landforms can be either dependent or independent variables as the limits of time and space change. During moderately long periods of time, for example, river channel morphology is a dependent variable; morphology depends on geology and climate as these factors influence the supply of water and sediment reaching the river, and the resistance of the substrate over which the channel forms. During shorter spans of time, channel morphology becomes an independent variable with respect to geology and climate because these latter factors are effectively unchanging. Over these shorter time-spans, channel morphology influences hydraulics, however, so hydraulics remain a dependent variable even at relatively short time-spans. The concept of considering variables of natural systems to be independent or dependent as a function of time and, to a lesser extent, space is now well established in physical geography and influences the manner in which hypotheses are posed and tested.

The importance of space

The spatial scale at which a natural system is examined can also influence whether a specific variable is effectively dependent or independent. River channel morphology is dependent at the scale of an entire drainage basin. Morphology varies across the basin in relation to topographic relief, sediment supply, resistance of the channel substrate and discharge. Considering only a 100 m-long segment of a river, however, the morphology can effectively be an independent variable that influences patterns of water and sediment movement and aquatic habitat.

Spatial considerations are also important in that many natural systems have systematic spatial variations. A glacier, for example, can be divided into zones of accumulation and ablation. Mass is progressively added to the glacier in the zone of accumulation, which is generally the central portion of an ice cap or continental ice sheet, or the upstream portion of a valley glacier. Mass is progressively lost from the ablation zone along the margins of the ice cap or ice sheet, or at the downstream portion of the valley glacier. The boundary between these two zones can shift with time as the entire mass budget of the glacier changes in response to climate, but the relative positions of the zones remain the same. Conceptualizing a glacier as having a central or upper segment that is gaining mass and a marginal or lower segment that is losing mass allows an investigator to focus on the mechanics of ice movement or the quantities and sources of mass gained and lost.

Similarly, a drainage basin can be conceptualized as having an uppermost portion that is the primary source of sediment and water. Schumm (1977) referred to this as the zone of production (Figure 18.1). Processes and morphology in this zone are dominated by tectonic regime, climate, geology and land use. Sediment and water produced in the upper portion of the basin move progressively downstream through the transfer zone, where the input of sediment can equal the output if the drainage basin is in equilibrium. Eventually the sediment will reach the lower depositional zone, where processes and morphology reflect base-level and tectonic regime. Using this conceptual model to spatially partition dominant processes in a drainage basin can help a scientist to examine feedbacks that operate between different portions of the basin. If sediment yield from the production zone increases, for example, how do process and morphology in the transfer zone change in response?

The spatial differences in dominant processes within a drainage basin can also be applied to differences in channel process and morphology at spatial scales of tens to hundreds of metres. Montgomery and Buffington (1997) differentiated headwater streams into source, transport and response reaches based on general differences in the ratio of sediment transport capacity to sediment supply (Figure 18.2). Source reaches are the uppermost portions of a drainage network, where channels can be dominated by mass movements such as debris flows that episodically deliver large quantities of sediment to the channel network. Transport reaches are the steeper segments of the channel network, which commonly have step-pool or plane-bed morphology, where more flow energy is available to transport sediment than is commonly supplied. When a debris flow or other large input of sediment occurs, the sediment is usually moved efficiently down through the transport reaches. Response reaches have lower stream gradients and pool-riffle or regime morphology.

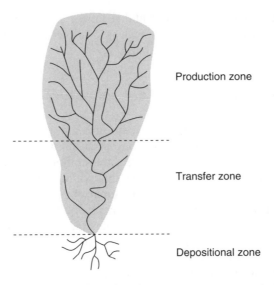

Production zone

Transfer zone

Depositional zone

Figure 18.1   Schematic illustration of the three zones common to most drainage basins
Source: After Schumm (1977)

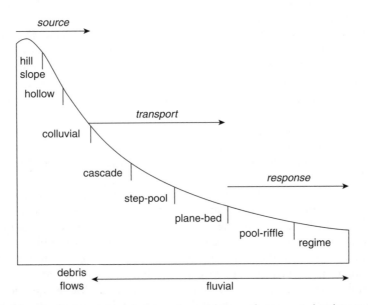

Figure 18.2   Longitudinal zonation of stream channel form and process in headwater channels
Source: After Montgomery and Buffington (1977, Figure 4)

These reaches of the stream have less energy to transport sediment, and large increases in sediment will more likely result in sediment accumulation within the channel – hence the designation 'response' for these stream reaches. This conceptual model, in which the spatial distribution of channel segments at the watershed-scale governs network response to changes in water and sediment supply, provides a powerful framework for predicting characteristics such as locations of greatest channel change in response to land use or climatic variability.

Interactions between components of natural systems

Another implication of viewing natural systems as open systems is that each system includes multiple components. Conceptual models have also been developed to describe the interactions among these components. One of the most widely used is the idea that natural systems can exhibit various forms of stability, or equilibrium. Equilibrium is defined in dictionaries as a state of balance between opposing physical forces. In physical geography, equilibrium generally refers to a condition with no net change. An active valley glacier continually flows down slope, for example, yet it can be in equilibrium if there is no net gain or loss of mass, or net change in the position of the glacier terminus, over the time-span of interest. Time-span is critical in defining equilibrium. A river that undergoes substantial erosion during two days when a large flood is occurring can still be in equilibrium if considered over ten years because subsequent smaller flows replace the sediment eroded during the flood. In recognition of the importance of time-span when defining equilibrium, physical geographers sometimes refer to different types of equilibrium as a function of time (Figure 18.3).

Equilibrium implies that multiple interacting components within a natural system can reach a state of stability. It also implies that if the energy or material supplied to the system changes, the system will change in response (Howard, 1982). The simplest form of change would be an immediate, linear response such that an external change produces a proportional change in the natural system. An increase in the size and energy of waves reaching a coast line, for example, causes an immediate change in sediment transport and coastal configuration if the response is linear. Many natural systems, however, are better characterized by delayed and non-linear responses, as embodied in the concepts of lag times and thresholds.

Lag time is the interval between an external change and the response of the system. This interval occurs because of the size and mass of natural systems (Howard, 1982). Returning to the example above, although the size and energy of waves gradually increase, the beach may be composed of cobble to boulder-size sediment that does not begin to move for a period of days until wave energy becomes sufficient to start moving the coarse sediment on the beach. The larger and more complex the system, the greater may be the lag time. Movement along a fault that produces vertical displacement of the ground surface can initiate a knickpoint along a bedrock river, for example, yet it may take hundreds or even thousands of years as that knickpoint migrates upstream along the river before the longitudinal profile of the river is completely adjusted to the instantaneous deformation caused by movement along the fault. As implied in this latter example, lag time is generally considered to include the time between the external perturbation and any initial response in the system, as well as the time during which the system is changing in

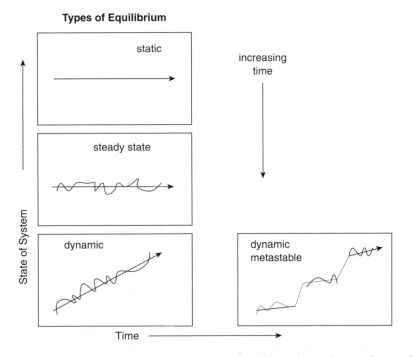

Figure 18.3   Schematic illustration of different types of equilibrium. In each case, the straight line represents the net tendency of the system through time, and the squiggly line represents shorter-term fluctuations in mean condition. A particular feature, such as elevation of the streambed within a reach, is unchanging over very short time intervals of perhaps hours to days or weeks (static). At slightly longer intervals of days to months or years, streambed elevation fluctuates as floods alternate with lower discharges, but the mean elevation of the streambed does not change (steady state). At longer intervals, the mean elevation of the streambed increases as sediment accumulates progressively (dynamic) or episodically (dynamic metastable)

response to the external perturbation. Bull (1991) distinguished between reaction time (before any response starts), relaxation time (system is changing), response time (reaction + relaxation times) and persistence time (the time during which the system is stable and unchanging) in conceptualizing changes in natural systems through time (Figure 18.4).

The lag time for a large or complex system to respond to an external change can be hundreds of thousands of years. Consequently, it is often difficult to demonstrate that a system is in equilibrium or moving in a certain direction because of internal adjustments among system components. A large body of work describes theoretical directional tendencies for the adjustment of river channels. Descriptions of these tendencies are sometimes called extremal hypotheses because they predict that a river channel will adjust to maximize or minimize a given parameter. Langbein and Leopold (1964) initiated this type of approach to understanding the adjustment of natural systems when they proposed that channels adjust so as to balance the two opposing tendencies of minimum total rate of work and uniform

Figure 18.4   Schematic illustration of changes in streambed elevation through time as examples of threshold (T) and equilibrium (E) conditions, as well as reaction, relaxation, response and persistence times. Vertical arrows at top of diagram indicate external disturbances to the system

Source: After Bull (1991, frontispiece)

distribution of energy expenditure throughout the channel network. Subsequent work has focused on tendencies including minimum energy dissipation rate (Yang, 1971), maximum friction factor (Davies and Sutherland, 1980) and maintaining critical flow (Grant, 1997).

The point at which operation of a system changes is referred to as a threshold. Schumm (1979) distinguished extrinsic thresholds, which delineate abrupt land-form change as a result of progressive change of external controls, from intrinsic thresholds, which delineate a condition at which there is significant landform change without a change of external controls such as base level or climate. How can a landform change without any change occurring in external controls? One example comes from the semi-arid river channels studied extensively by Schumm. Semi-arid regions commonly have sufficient precipitation to move sediment from hill slopes into river channels, but not enough precipitation to generate the stream flow necessary to move the sediment all the way through the channel network. Consequently, sediment can accumulate through time in limited reaches of the network, causing specific channel segments to grow progressively steeper. These gradual changes occur over a period of tens to hundreds of years, until some chan-nel segments become over-steepened and unstable. A headcut initiated at these unstable reaches can then migrate upstream along the channel until the gradient of the channel segment is lowered and once again stable. Semi-arid channel networks around the world appear to go through repeated episodes of alternating aggradation, incision and stability over hundreds to thousands of years, whether or not external factors such as climate or land use change (Schumm and Hadley, 1957; Schumm and Parker, 1973; Betts *et al.*, 2003; Eriksson *et al.*, 2006).

Schumm (1980) noted that many landforms do not necessarily evolve progressively through time. Instead, erosion and deposition is episodic in nature, with periods of stability separated by relatively unstable periods of active landform adjustment. An example of episodic activity in a natural system comes from a study that demonstrated that desert soils form relatively rapidly during periods of enhanced input of aeolian dust, with very little change in soil properties during intervening periods of reduced aeolian input (Chadwick and Davis, 1990).

The examples thus far describe thresholds in how a system operates through time. The concept of thresholds can also be used to delineate spatial domains for particular processes (Kirkby, 1980; Church, 2002). Sediment moving along a braided channel can remain in transport within a sub-channel or be deposited on an island or bar, and a threshold of sediment movement separates these domains. Valleys and interfluves on hill slopes, soil-covered and bedrock hillslopes, sand dunes and inter-dune playas, and beaches and vegetated backshore dunes provide additional examples of spatial domains separated by a threshold in sediment transport processes.

Physical geographers can quantify thresholds in terms of a ratio between driving forces and resisting forces. Driving forces promote movement within a natural system. Gravity drives mass down slope; precipitation that infiltrates hill slope materials adds mass that facilitates down slope movement; and seismic vibrations reduce the friction between particles on the slope – all of these factors are examples of driving forces. Surface friction between particles, as well as tension created by small amounts of interstitial water, resists down-slope movement and is an example of resisting force. These and other effects are incorporated in the factor of safety for a hillslope, which is the ratio between resisting forces and driving forces. When this ratio is greater than 1, the slope should be stable. As the ratio approaches 1, the slope becomes increasingly unstable and prone to adjustments in the form of mass redistribution through processes including soil creep, debris flows, or landslides.

Another example of quantifying a threshold comes from Bull (1979), who described a threshold of critical power in streams. Stream power (Ω) can be quantified as

$$\Omega = \gamma \, Q \, S$$

where γ is the specific weight of water, Q is discharge, and S is stream gradient. Critical power (Ω_c) is the stream power necessary to transport the sediment supplied to the stream. The threshold of critical power occurs where

$$\Omega/\Omega_c = 1$$

When the ratio exceeds 1, the stream has greater power than is needed to transport the available sediment, and this extra power is used to erode the stream bed. When the ratio is less than 1, the stream does not have sufficient power to transport the sediment supplied, and deposition occurs. Graf (1983) quantified such a threshold for the Salt River of central Arizona, USA, where floods with a recurrence interval of less than five years produce no channel migration or downcutting, whereas less frequent floods destabilize the channel.

Conceptualizing a natural system as reflecting the balance between driving forces and resisting forces is particularly powerful because it provides a context to understand and predict the nature of change. An increase in driving forces without a corresponding increase in resistance should produce a redistribution of material within the system, for example, and the greater the increase in driving forces, the greater should be the response of the system. The proportionality between perturbation and response of a natural system cannot be taken for granted, however, because of the complex interactions among the multiple components of natural systems. These interactions can involve feedbacks that either enhance or dampen system response. A self-enhancing feedback occurs when an initial perturbation triggers a system response that grows greater with time. For example, a landslide deposits sediment on what had been bare bedrock along a hillslope. This increases infiltration capacity at the site, providing water and soil to support vegetation that traps additional sediment moving down slope, which further increases the tendency for sediment to accumulate at the site (Bull, 1991). In contrast, an initial perturbation can produce very little change in a natural system if self-arresting feedback mechanisms occur. An increase in discharge during a flood can promote entrainment of sediment from a streambed with a mixed grain-size distribution. As the finer grains are preferentially removed, the resulting coarser sediments form a resistant armour layer that limits further streambed erosion.

The idea that system response can be exacerbated or dampened by interactions among variables is closely related to the idea that a single external perturbation can initiate a series of responses within a system. Schumm and Parker (1973) developed the conceptual model of complex response to explain river adjustment across time and space following the initial perturbation of a base-level fall (Figure 18.5). When base level falls, a headcut forms at the downstream end of the river and migrates upstream with time. Upstream migration of the headcut produces increased sediment load to downstream points that the headcut has already moved past, and these lower segments begin to aggrade in response. Eventually the headcut reaches its farthest upstream location and the supply of sediment to downstream reaches declines, causing these reaches to once again erode, and potentially creating a second headcut that migrates upstream. As a result of these dynamics, spatial differences in channel response to the initial base level fall can be dramatic; the upstream portion of the river can be eroding while the downstream portion is simultaneously aggrading, or vice versa. Also, any given point in the channel can go through contrasting responses to the initial base-level fall over a period of time; first eroding, then aggrading, then finally eroding and aggrading again before the streambed elevation once more becomes stable.

The manner in which a natural system undergoes complex response or is influenced by lag times, thresholds and feedbacks reflects characteristics such as size

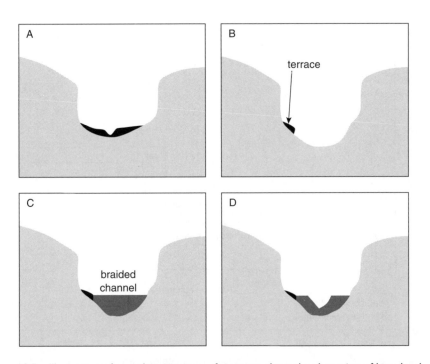

Figure 18.5   Illustration of complex response of a stream channel to lowering of base level. (A) View looking upstream of valley (lighter shading) and stream alluvium (black) which was deposited before lowering of base level. (B) After base-level lowering, channel incises into alluvium and bedrock of valley to form a terrace. Following incision, bank erosion widens channel and partially destroys terrace. (C) An inset alluvial-fill is deposited as the sediment discharge from upstream increases. The broad, shallow channel is braided and unstable. (D) A second terrace forms as the channel incises slightly and becomes deep and narrow in response to reduced sediment supply

Source: After Schumm and Parker (1973, Figure 1)

and complexity of the system. Knighton (1998) illustrated this schematically for river systems (Figure 18.6). The concepts of landscape or landform sensitivity and resilience embody differences between landscapes. Sensitivity involves the system's response to external influences in terms of either spatial variation in the ability of landforms to change, or susceptibility of a system to disturbance (Allison and Thomas, 1993). Sensitivity focuses on the potential and likely magnitude of change within a natural system and the ability of the system to resist change. A sensitive landscape could be sensitive to small changes, but have only minor responses, or it could have thresholds such that only a small change in key input variables produces a major change in the landscape.

Extremely cold and dry environments are often described as being sensitive to climatic changes. A 20 per cent increase in mean annual precipitation is less likely to produce substantial and persistent changes in weathering, soil development, sediment supply, and stream flow in humid temperate or tropical regions, for example, than in arid or semi-arid regions. Similarly, a 10 per cent increase in mean annual temperature

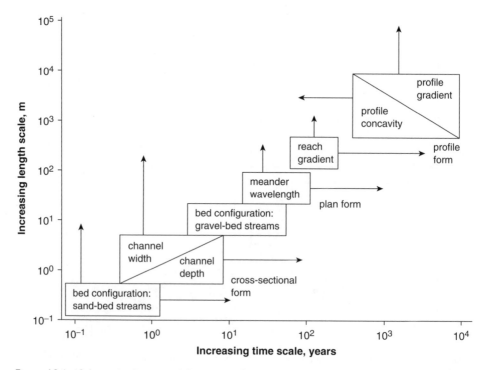

Figure 18.6   Schematic diagram of the time and space scales across which various components of river channels adjust

Source: After Knighton (1984, Figure 4.1)

is likely to have less effect in middle and lower latitudes than in higher latitudes or high-altitude environments. Much of our knowledge regarding such examples of landscape sensitivity comes from sedimentary and geochemical records of landscape response to climatic variability during the past two million years (Figure 18.7).

Resilience refers to how quickly a system recovers following disturbance. It is thus analogous to response time (Bull, 1991). Natural systems that recover relatively rapidly are considered to be resilient. Related to the idea of resilience is the concept of characteristic form-time for landforms. A persistent landform is one that lasts longer than the recurrence interval of the event that created or modified it. A transient landform is one that has a shorter duration than the event which created it. Extreme storm winds might mobilize widely dispersed sand along a coast line and create climbing dunes along the coastal cliff. If these climbing dunes persist until the next extreme storm of similar magnitude, they are persistent landforms. If winds of lower intensity subsequently rework the sand and remove the climbing dunes before the next extreme storm, then the dunes are transient landforms. Wolman and Miller (1960) initiated the idea of evaluating the relative importance of large, infrequent events versus smaller, more frequent events in shaping landforms when they quantified the geomorphic work (defined in terms of sediment transport) performed by floods of differing magnitude. Wolman and Gerson (1978) extended this analysis for rivers to geomorphic effectiveness, defined in terms of the persistence of the landforms; floods that create persistent features are geomorphically effective.

A

Climatic variables

++ Temperature
+ Precipitation
+ Rain/snow

Hillslope subsystem

−− Periglacial processes

++ Vegetation density

+ $\dfrac{\text{Chemical weathering}}{\text{physical weathering}}$

++ Soil thickness

−− Sediment yield

Stream subsystem

+ Peak flood discharge

Threshold of critical power greatly exceeded

A + indicates an increase and A − indicates a decrease in the adjacent variable, with the relative magnitude of the change indicated by the number of symbols.

B

Climatic variables

++ Temperature
+ Precipitation
+ Rain/snow

Hillslope subsystem

++ Periglacial processes

++ Vegetation density

+ $\dfrac{\text{Chemical weathering}}{\text{physical weathering}}$

+ Fluvial erosion

− Soil thickness

+ Sediment yield

Stream subsystem

+ Peak flood discharge

Threshold of critical power probably not exceeded

Figure 18.7 Process-response models for hillslopes of the Charwell River basin, New Zealand for a change from semi-arid to wetter conditions. (A) Watershed from basin mouth to present treeline (450–1200 m elevation). (B) Watershed from present treeline to summit (1200–1600 m elevation)

Source: After Bull (1991, Figure 5.15)

NATURAL SYSTEMS AS ECOSYSTEMS

Thus far, this chapter has emphasized conceptual models of physical processes operating in natural systems. Physical processes, however, do not operate in isolation from biological communities, and it is more appropriate to conceptualize natural systems as resulting from interacting physical, chemical and biological components.

A simple yet profoundly important case study provides an example. In the Negev desert of Israel, snails less than 1 cm in length gouge the rocks at the surface, leaving 0.5–2 mm deep grooves of fresh rock (Jones and Shachack, 1994). The snails do this to get at layers of green algae and fungi in the upper portion of the rock, but they digest only 5 per cent of what they eat; the remainder is excreted in small coils of powdered rock and undigested lichen. Approximately 20 snails per square metre are present in the Negev, and their activities expose fresh rock, and minerals that the lichens need. The snails annually graze 4–7 per cent of the total rock surface area in the Negev, turning about 900 kg of rock into soil per hectare of desert. By comparison, aeolian input accounts for 250–470 kg of dust per hectare. The snails also fertilize the soil; approximately 5 per cent of the dry weight of snail faeces is nitrogen, which adds 0.9 kg of nitrogen per hectare annually. Soil development in the Negev desert is thus profoundly influenced by the activities of these small creatures that most human visitors to the desert would probably not notice.

One of the fundamental properties of ecosystems is that, although investigators define spatial boundaries for the ecosystem such as the limits of a particular rainfall regime or the margins of a mountain range, the ecosystem is intimately connected to the entire planet and its atmosphere. Coral reefs are marine systems composed of living organisms that create a skeletal framework of calcium carbonate in waters less than 30 m deep where water temperature is 22–29 °C and sufficient sunlight penetrates to allow photosynthesis at the surface of the reef. The presence of the reef both reflects and influences local patterns of water chemistry, sediment transport and wave energy, and it might seem that the reef can thus be conceptualized solely in terms of local influences. Recent studies of coral reefs in the Caribbean Sea, however, indicate that aeolian dust falling on the region suppresses the ability of reef-building organisms to photosynthesize, and may allow attached fungal spores and bacterial cysts to invade reef communities. Scientists have identified an African soil fungus as the cause of widespread death among sea fans across the Caribbean. African dust also carries pesticides such as chlordane and DDT to the Americas. In 2001 a cloud of Saharan dust enriched iron concentrations in the surface waters off Florida by 300 per cent. The iron spike fuelled a 22,500 km² bloom of red algae (*Karenia brevis*) that poisoned local shellfish and made them inedible by humans, as well as killing millions of fish and hundreds of manatees. This is but one example of the global-scale processes that also influence coral reefs.

Another example of the broader influences on natural systems comes from rivers. Any river has at least six degrees of connection to the adjacent environment. Longitudinal connections occur as water, sediment, contaminants, nutrients and organisms move upstream and downstream. It would be futile to attempt to understand complex response to lowered base level, for example, without understanding that headcuts can erode upstream and sediment from migrating headcuts can be transported downstream. Lateral connections occur as movements of energy, matter and organisms between the river channel and the adjacent floodplains. Water requires only about 60 days on average to move down the Amazon River from its headwaters in the Andes Mountains to the Atlantic Ocean, whereas sediment requires an average of many hundreds to a few thousand years to make the same journey (Mertes, 1994). Transport of sediment requires much longer because during the large floods that transport most of the suspended sediment, more than 80 per cent of the sediment that flows onto floodplains from the main channel or tributaries settles from the water and remains on the floodplain at least temporarily. Sediment eventually returns to the main channel through bank erosion. More of the approximately 1.5 billion tons of sediment eroded annually along a 2,000 km reach of the Amazon comes from bank erosion than from tributaries or from transport down the channel. Most sediment thus repeatedly cycles between the floodplain and the main channel (Mertes *et al.*, 1996).

Lateral connections between the river and hill slopes also occur when a landslide delivers sediment and organic matter to the river. Conceptual models of the spatial configuration of river networks in the forested mountains of the northwestern United States emphasize the role of debris flows in creating morphological heterogeneity of rivers through localized erosion and deposition (Benda *et al.*, 2003).

Vertical connections between the river and the atmosphere occur as water evaporates from the river, precipitation and aeolian dust falls on the river, and aquatic insects emerge from their larval stage in the river as winged adults. Conceptual models of stream flow, energy, sediment transport and river morphology in the Congo River basin, for example, will be inadequate unless they include the distinctive precipitation patterns of the region. Precipitation in turn reflects the seasonal movement of the Intertropical Convergence Zone – a band of intense atmospheric convection and precipitation – from latitude 5° N during December–February, to 20° N during June–August.

Vertical exchanges in the other direction connect water and organisms in the river with the hyporheic zone immediately beneath the river channel, and with ground water aquifers at greater depth. A conceptual model that guided river restoration along the Carmel River of California was based on the assumption that stream banks would become more resistant to erosion if riparian vegetation could be established. Willows planted along the stream could not survive without supplemental irrigation, however, because the water table had dropped below historical levels as a result of intensive regional pumping of ground water (Kondolf, 1986).

These examples demonstrate that any conceptual model that fails to place a natural system within a broader, regional or even global context, may not be adequate to explain observed phenomena.

NATURAL SYSTEMS WITH A PAST

A final important consideration when understanding a natural system, and attempting to conceptualize that system, is that historical influences may persist for very long periods of time. Three case studies from rivers provide examples. The first example involves a historical land use. During the period 1868–1940, logs cut in the mountains of Wyoming were floated downstream to be used as railroad ties. The logs were usually cut during the winter and stacked beside the river channel until snowmelt created high flows. Splash dams created by driving logs vertically into the stream bed, ponded water and logs upstream, and a blast of dynamite then sent a surge of logs and water downstream to collection points. The streams were also modified to facilitate transport of the logs by blocking off overbank areas and removing natural obstructions within the channel. The tie drives only lasted a few years along most streams, yet a study conducted a century later found that there were still statistically significant differences in stream characteristics includ-ing pool volume, fish habitat, and age and diversity of riparian vegetation between otherwise analogous streams that had experienced tie drives and streams that had not (Young *et al.*, 1994). For these streams, a conceptual model that attempted to explain variability among individual streams solely in terms of natural controls such as drainage area, lithology, climate, or topography would miss a key historical event that continues to influence inter-stream variability.

The second example involves systems with long-period variability. Examin-ing the records of flash flooding in the United States, Hirschboeck (1987) found decadal-long alternations between periods of enhanced flash floods across the country, and periods of relatively few floods. She tied these variations to the char-acteristics of the jet stream. When the jet stream has a relatively straight path from west to east across North America, flash floods decrease in occurrence and magnitude. During periods of meridional circulation, when the jet stream takes a much more sinuous path across the continent, cyclonic eddies generate persistent storm tracks in anomalous locations and facilitate flooding.

Another example of long-period variations comes from incision of the Indus River within the Himalaya over periods of thousands of years. Burbank *et al.* (1996) found that if rapid river incision oversteepens adjacent hillslopes, the resulting increase in landslides supplies so much sediment to the river that the rate of incision decreases until the sediment has been evacuated and the river can once again incise into bedrock. Examining either the occurrence of flash floods or the rate of bedrock incision by the Indus River over time periods of a few years, without realizing that these rates fluctuate dramatically over long periods of time in response to controls outside of the river channel, would result in inadequate conceptual models of each system.

The third example involves prolonged and continuing response to a natural perturbation. Analysing a region with large variations in rate of tectonic uplift over small distances, Merritts and Vincent (1989) found that larger streams could incise with sufficient rapidity to keep pace with uplift, and maintain a smoothly concave longitudinal profile. Smaller streams did not have enough stream power to incise at a rate comparable to uplift, and became progressively steeper with time. Incision of these bedrock rivers in response to uplift, however, had lag times in the order of 150,000 to 400,000 years. This implies that this landscape is still responding to external changes that occurred during the Pleistocene.

CHALLENGES TO CONCEPTUALIZING FROM DIRECT OBSERVATION

The existence of lag times and thresholds in system response, feedbacks among sys-tem components, different behaviour by system components as a function of time and space, and historical legacies all help to explain why it is challenging to analyse any natural system. Conceptual models are useful so long as they serve to organize and explain numerous disparate observations, but these models can become impediments to understanding if they become so ingrained in thinking that scientists are largely unaware of the assumptions held about system operations and no longer critically challenge these assumptions by considering alternative explanations for observed patterns and processes. Because the time period of direct human observation of any natural system is brief relative to many of the processes occurring in that system, it is critical that scientists bear in mind what Schumm (1991) described as ten sources of potential error in using modern conditions as a basis for extrapolation. These ten potential problems are: time (absolute duration and relative time spans); space (scale and size); location (site of concern within a natural system); convergence (the production of similar results from different processes and causes, which is known as equifinality); divergence (the production of different results from similar processes and causes); efficiency (variable efficiency and work accomplished by a process); multiplicity (multiple explanations that combine to influence and cause natural phenomena); singularity (natural variability among like things); sensitivity (susceptibility of a system to change); and complexity (complex behaviour of a system that has been subject to altered conditions). Each of these issues presents challenges, but each is inherent in the operation, and thus the conceptualization, of natural systems.

CONCLUSIONS

Conceptual models both reflect and guide how physical geographers think about landscape forms and processes. Conceptual models reflect contemporary thinking in that they provide a framework within which to organize disparate observations made by many individual investigators across a range of environments, and insofar as they express the consensus of the scientific community. Conceptual models guide contemporary thinking in that the structure which they create implies that some potential questions are of more relevance and higher priority for investigation than are other potential questions. Conceptual models are likely to be most effective when individual investigators explicitly acknowledge the existence and characteristics of the simplifying concepts being used to understand infinitely complex natural systems.

Summary

The key points raised in this chapter are as follows:

- Physical geographers attempting to understand complex natural systems use conceptual models to organize disparate observations and to provide a framework for developing hypotheses.
- Natural systems are open to the surrounding environment, with which they continually exchange matter and energy.
- The manner in which geographers conceptualize the behaviour of a natural system depends strongly on the scales of time and space over which the system is being examined.

- Equilibrium describes the relative stability of natural systems, and the adjustments among multiple components. These components can be differentiated as driving and resisting forces, and a change in the ratio of these two types of forces will produce a corresponding change in the system if a threshold is passed. This change may not occur until some lag time has passed, and the change may be enhanced or arrested by interactions among the system components.
- Biological and chemical, as well as physical, processes are a vital part of the operations of natural systems.

Further reading

- Conceptual models commonly applied in physical geography are described in greater detail in the following books. Schumm (1977) uses the idea of production, transport and depositional zones within drainage basins as a means to explain and organize a wide variety of data from river systems around the world. The volume edited by Allison and Thomas (1993) provides numerous site-specific examples of how to evaluate and quantify landscape sensitivity. Bull (1991) describes examples from around the world of how natural systems respond to climatic variability, and evaluates these examples within the conceptual model of equilibrium and the associated ideas of lag times, thresholds, and driving and resisting forces.

- Among the classic papers that laid out the foundation for most contemporary conceptual models used in physical geography are those by: Wolman and Miller (1960) on evaluating the relative importance of large, infrequent events and smaller, more frequent events; Schumm and Lichty (1965) on the importance of temporal and spatial scales when evaluating interactions among the components of natural systems; Schumm and Parker (1973) on complex response; and Langbein and Leopold (1964) on directional tendencies of river adjustment. Church and Mark (1980) provide a nice discussion of how deductive and inductive reasoning are used to understand natural systems.

- Schumm (1991) is a short text that thoroughly explores the challenges resulting from using modern conditions to understand operations of natural systems over much longer periods of time.

Note: Full details of the above can be found in the references list below.

References

Allison, R.J. and Thomas, D.S.G. (1993) 'The sensitivity of landscapes', in D.S.G. Thomas and R.J. Allison (eds) *Landscape Sensitivity*. Chichester, John Wiley and Sons.

Benda, L., Veldhuisen, C. and Black, J. (2003) 'Debris flows as agents of morphological heterogeneity at low-order confluences, Olympic Mountains, Washington', *Geological Society of America Bulletin*, 115: 1110–1121.

Betts, H.D., Trustrum, N.A. and de Rose, R.C. (2003) 'Geomorphic changes in a complex gully system measured from sequential digital elevation models, and implications for management', *Earth Surface Processes and Landforms*, 28: 1043–1058.

Bull, W.B. (1979) 'Threshold of critical power in streams', *Geological Society of America Bulletin*, 90: 453–464.

Bull, W.B. (1991) *Geomorphic Responses to Climatic Change*. Oxford: Oxford University Press.

Burbank, D.W., Leland, J., Fielding, E., Anderson, R.S., Brozovic, N., Reid, M.R. and Duncan, C. (1996) 'Bedrock incision, rock uplift and threshold hillslopes in the northwestern Himalayas', *Nature*, 379: 505–510.

Chadwick, O.A. and Davis, J.O. (1990) 'Soil-forming intervals caused by eolian sediment pulses in the Lahontan Basin, northwestern Nevada', *Geology*, 18: 243–246.

Chamberlin, T.C. (1897) 'The method of multiple working hypotheses', *The Journal of Geology*, 5: 837–848.

Chorley, R.J. (1962) 'Geomorphology and general systems theory', *U.S. Geological Survey Professional Paper* 500-B, 10 pp.

Church, M. (2002) 'Geomorphic thresholds in riverine landscapes', *Freshwater Biology*, 47: 541–557.

Church, M. and Mark, D.M. (1980) 'On size and scale in geomorphology', *Progress in Physical Geography*, 4: 342–390.

Davies, T.R.H. and Sutherland, A.J. (1980) 'Resistance to flow past deformable boundaries', *Earth Surface Processes and Landforms*, 5: 175–179.

Dietrich, W.E., Bellugi, D.G., Sklar, L.S., Stock, J.D., Heimsath, A.M. and Roering, J.J. (2003) 'Geomorphic transport laws for predicting landscape form and dynamics' in P.R. Wilcock and R.M. Iverson (eds) *Prediction in Geomorphology*. Washington, DC: American Geophysical Union Press.

Eriksson, M.G., Olley, J.M., Kilham, D.R., Pietsch, T. and Wasson, R.J. (2006) 'Aggradation and incision since the very late Pleistocene in the Naas River, south-eastern Australia', *Geomorphology*, 81: 66–88.

Graf, W.L. (1983) 'Flood-related channel change in arid-region river', *Earth Surface Processes and Landforms*, 8: 125–139.

Grant, G.E. (1997) 'Critical flow constrains flow hydraulics in mobile-bed streams: a new hypothesis', *Water Resources Research*, 33: 349–358.

Hirschboeck, K.K. (1987) 'Catastrophic flooding and atmospheric circulation anomalies' in L. Mayer and D. Nash (eds) *Catastrophic Flooding*. Boston, Allen and Unwin.

Howard, A.D. (1982) 'Equilibrium and time scales in geomorphology: application to sand-bed alluvial streams', *Earth Surface Processes and Landforms*, 7: 303–325.

Jones, C.G. and Shachack, M. (1994) 'Desert snails daily grind', *Natural History*, 103(8): 56–61.

Kirkby, M.J. (1980) 'The stream head as a significant geomorphic threshold', in D.R. Coates and J.D. Vitek (eds) *Thresholds in Geomorphology*. London: George Allen and Unwin.

Knighton, D. (1984) *Fluvial Forms and Processes*. London: Edward Arnold.

Kondolf, G.M. (1996) 'A cross section of stream channel restoration', *Journal of Soil and Water Conservation*, 51: 119–125.

Langbein, W.B. and Leopold, L.B. (1964) 'Quasi-equilibrium states in channel morphology', *American Journal of Science*, 262: 782–794.

Merritts, D. and Vincent, K.R. (1989) 'Geomorphic response of coastal streams to low, intermediate, and high rates of uplift, Mendocino triple junction region, northern California', *Geological Society of America Bulletin*, 101: 1373–1388.

Mertes, L.A.K. (1994) 'Rates of flood-plain sedimentation on the central Amazon River', *Geology*, 22: 171–174.

Mertes, L.A.K., Dunne, T. and Martinelli, L.A. (1996) 'Channel-floodplain geomorphology along the Solimões-Amazon River, Brazil', *Geological Society of America Bulletin*, 108: 1089–1107.

Montgomery, D.R. and Buffington, J.M. (1997) 'Channel-reach morphology in mountain drainage basins', *Geological Society of America Bulletin*, 109: 596–611.

Schumm, S.A. (1977) *The Fluvial System*. New York: John Wiley and Sons.

Schumm, S.A. (1979) 'Geomorphic thresholds: the concept and its applications', *Transactions of the Institute of British Geographers*, 4: 485–515.

Schumm, S.A. (1980) 'Some applications of the concept of geomorphic thresholds', in D.R. Coates and J.D. Vitek (eds) *Thresholds in Geomorphology*. London: George Allen and Unwin.

Schumm, S.A. (1991) *To Interpret the Earth: Ten Ways to Be Wrong*. Cambridge: Cambridge University Press.

Schumm, S.A. and Hadley, R.F. (1957) 'Arroyos and the semiarid cycle of erosion', *American Journal of Science*, 255: 161–174.

Schumm, S.A. and Lichty, R.W. (1965) 'Time, space, and causality in geomorphology', *American Journal of Science*, 263: 100–119.

Schumm, S.A. and Parker, R.S. (1973) 'Implications of complex response of drainage systems for quaternary alluvial stratigraphy', *Nature*, 243: 99–100.

Wolman, M.G. and Gerson, R. (1978) 'Relative scales of time and effectiveness of climate in watershed geomorphology', *Earth Surface Processes*, 3: 189–208.

Wolman, M.G. and Miller, J.P. (1960) 'Magnitude and frequency of forces in geomorphic processes', *Journal of Geology*, 68: 54–74.

Yang, C.T. (1971) 'On river meanders', *Journal of Hydrology*, 13: 231–253.

Young, M.K., Haire, D. and Bozek, M.A. (1994) 'The effect and extent of railroad tie drives in streams of south-eastern Wyoming', *Western Journal of Applied Forestry*, 9: 125–130.

Numerical Modelling: Understanding Explanation and Prediction in Physical Geography

19

Stuart N. Lane

Synopsis

This chapter introduces the use of numerical modelling for understanding environmental systems. Numerical models are used extensively throughout society, and hence by physical geographers, for dealing with situations where the researcher is 'remote' from what they are studying: where the events that are of interest have occurred in the past (e.g. reconstruction of past climates); may occur in the future (e.g. patterns of inundation associated with future flood events); or where they are occurring now, but cannot be measured or studied using other methods. The potential of numerical models aside, the problems of modelling environmental systems mean that models are commonly scientifically wrapped crystal balls, whose predictions must be treated with caution at best and scepticism at worst. This chapter comprises:

- Introduction: why model?
- Fundamental aspects of environmental modelling
- What a model can and cannot do
- Conclusion

INTRODUCTION: WHY MODEL?

In *Winnie-the-Pooh*, A.A. Milne demonstrates clearly the basic reason why we need to consider using numerical models. Pooh Bear finds a jar labelled 'hunny' with a yellow substance in it. Being a 'good' scientist, he cannot be sure that it is 'hunny' until he has done a proper scientific experiment to test that his hypothesis is valid. This has to be grounded in direct observation of what is in the jar and involves him tasting it. However, he cannot be sure that it is 'hunny' until he has tasted all of it, right to the bottom of the jar as, according to Pooh bear, his uncle had once said that someone once put cheese in a 'hunny' jar for a joke. The view that *all* jars labelled 'hunny' contain 'hunny' has already been proven false, and so each individual jar that Pooh bear finds has to be subject to Pooh-type assessment, in which the entire contents of the jar are consumed. If we apply this to environmental systems, we find ourselves visiting a series of great environmental disasters that have been created because as a society we have been unwilling to accept certain evidence (e.g. that of a jar labelled 'hunny' with a yellow substance in it) until there is definitive observational evidence that confirms that a partial observation or a theory is correct. The best example of this is provided by depletion of stratospheric ozone concentrations. The potential that chlorofluorocarbons might lead to long-term ozone depletion was demonstrated in the early 1970s (Molina and Rowland, 1974). It was not until the direct observation of a Spring ozone hole over Antarctica in 1985 (Farman *et al.*, 1985) that this theoretical notion was accepted and subsequent environmental policy was developed. The same scenario has emerged in relation to global climate change: we have a theoretical basis to expect that atmospheric greenhouse gas accumulation has the potential to change climate; much of the critique of this theory is based around the fact that there is no demonstrable evidence to confirm this hypothesis (e.g. Michaels, 1992). We will not do anything about it until we have actually observed it.

The problem with this view is also illustrated by Milne. When Piglet goes to visit his heffalump trap, he finds what looks like a heffalump. Being only a small animal, and with the trap being very deep, he needs to get closer to see if the heffalump (Pooh bear with an empty 'hunny' jar stuck on his head) is indeed a heffalump. This mirrors classic observation-based science, in which we search for a better understanding of an apparent phenomenon, through more in-depth observation. However, if Piglet discovers that what he thinks is a heffalump is actually a heffalump then, being a small animal, he is likely to be in a lot of trouble as the heffalump may attack him. We do not want, through more intensive investigation based upon observation, to discover things (e.g. heffalumps, ozone holes, global warming, severe species loss, serious organic pollution). However, sometimes, we can only confirm that what we think is a matter of concern is actually so through more in-depth assessment.

The numerical model is one of the tools that the geographer might use to break out of the circularity that Pooh bear and Piglet find themselves in. It provides a tool for investigating things that are inaccessible. Inaccessibility arises because we are commonly interested in: (i) environments in the past, before records began, or where environmental reconstruction is unreliable or impossible; (ii) environments in the present, from where we cannot obtain measurements, perhaps because those environments are inaccessible, whether because they are remote, or too large to measure, or too small to measure; and (iii) environments in the future, where we are concerned about the possible impacts of decisions made now for future generations. This chapter seeks to introduce the basics of environmental modelling as part of geographical enquiry, but also to reflect upon the challenges and problems that result. Thus, the first section of the chapter introduces the fundamentals of environmental modelling, in terms of conceptual, empirical and physically based approaches. The second section seeks to evaluate models critically, by considering what a model can do and what it cannot do.

FUNDAMENTAL ASPECTS OF ENVIRONMENTAL MODELLING

There are two distinct approaches to mathematical modelling: (i) empirical; and (ii) physically based. However, as this section will note, these two distinct approaches are not actually that distinct and both of them draw fundamentally upon the idea that there is a resilient and defensible conceptual model of the system that is being considered.

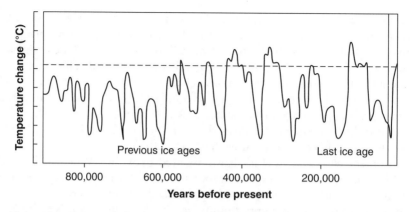

Figure 19.1   Globally averaged temperature change over the last million years. The dashed line indicates (approximately) the 1880 temperature

Source: Houghton *et al.* (1990)

The conceptual model

Without a conceptual model of the system that is being modelled, it is impossible to develop either an empirical or a physically based mathematical model. Conceptual models involve a statement of the basic interactions between the components of a system (see also Chapter 18). If we think about the understanding of past climates, we can start to think about a simple conceptual model for climate change. Empirical evidence has shown that the earth's climate has fluctuated between periods that were much colder than present (glacial periods) and periods that were slightly warmer than present (Figure 19.1). Why has this occurred? The initial conceptual model might assume that it is due to an external forcing, and we know that a good candidate in this respect is the nature of the earth's rotation around the sun (see Imbrie and Imbrie, 1979), which varies over a number of different time-scales. There is good support for some of this cyclical behaviour being explained by orbital forcing (e.g. Imbrie and Imbrie, 1979), but there is also much evidence to question the conclusion that this is the only explanation (e.g. Broecker and Denton, 1990). Thus, researchers have begun to recognize that processes internal to the earth-atmosphere system may exert an important conditioning role upon the way in which these external forcing factors affect climate. This is where we can develop a simple conceptual model to illustrate what form this system might take.

A system is made up of components that are connected together by links. Flows between components are driven by processes in a way that depends upon both the components and the nature of the links. Thus, in the case of glacial cycles, we may start to build a system by considering three components: (i) albedo which relates to the way in which a surface reflects incoming solar radiation; (ii) temperature; and (iii) ice-sheet growth. As a first approximation, these may be linked together (Figure 19.2). The links are specified very simply as positive and negative: (i) a negative link between albedo and temperature, which reflects the fact that as albedo (earth surface reflectivity) goes up, more short-wave radiation will be reflected and

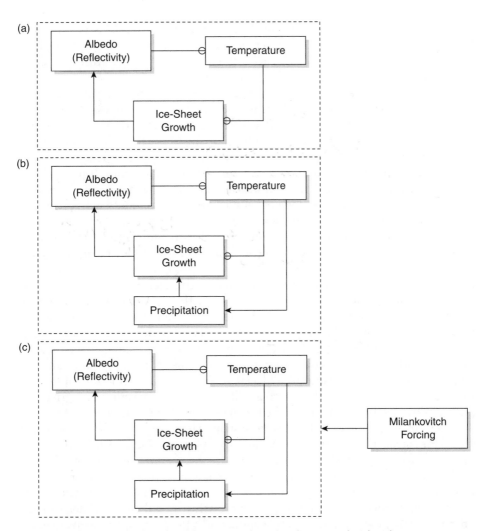

Figure 19.2   Three examples of simple systems that are relevant to glacial cycles

hence temperature will go down; (ii) a negative link between temperature and ice-sheet growth, which reflects the fact that as temperature goes down, ice-sheet growth will go up; and (iii) a positive link between ice-sheet growth and albedo, which reflects the fact that as ice-sheet growth goes up, so albedo goes up as ice cover tends to reflect more short-wave radiation than other types of land cover. Put more generally, a positive link involves the effect variable responding in the same direction as the cause variable; a negative link involves the effect variable responding in the opposite direction to the cause variable.

Figure 19.2a illustrates a very important property of feedback, which occurs in all environmental systems and which arises from the combined interaction of these links between components: as temperature goes down, ice-sheet growth goes up; as ice-sheet growth goes up, albedo goes up; and as albedo goes up, temperature will go down. Thus, the net effect of the system in Figure 19.2a is positive feed-back, where an initial reduction in temperature would be amplified through the ice-sheet growth and albedo links to result in further temperature reduction. Positive feedback causes a system to change or evolve.

In Figure 19.2b, precipitation is introduced as an additional and important component of glacier growth and decay: precipitation is required to add mass to an ice sheet. Precipitation levels will be governed by temperature, as this controls evaporation. For ice-sheet growth, the global oceans act as the major water source. As temperature goes down, precipitation will go down, due to less evaporation (i.e. the link is positive as both changes are in the same direction). If precipitation goes down, ice-sheet growth will go down. Thus, if we examine the links from ice-sheet growth to albedo to temperature to precipitation and back to ice-sheet growth we have a negative feedback: as ice-sheet growth goes up, albedo goes up, temperature goes down, precipitation goes down, and hence ice-sheet growth goes down. Negative feedback is a self-limiting feedback, which causes a system to resist change.

With these basic ideas about positive and negative feedback, we have a means of understanding the dynamics of glacier growth and decay in the system in Figure 19.2. If negative feedbacks dominate in a system, then the system will be maintained in the state that it is in, either as a glacial period or as an inter-glacial period. If positive feedbacks dominate, then we may have change, which may be rapid, and which may cause a system to evolve, either from a glacial to an inter-glacial or from an inter-glacial to a glacial. Thus, if there is some sort of external forcing, then the effects of that forcing will depend upon whether the system dis-plays positive feedback and hence the forcing causes change, or negative feedback, in which the forcing is absorbed by the system. Thus, we have a basic conceptual model for considering the effects of Milankovitch forcing (Figure 19.2c): the orbital driving factors are either enhanced or reduced by the system to which they are applied. In practice, the system is much more complex than this. Feedbacks may be delayed, and simple components like 'temperature' and 'precipitation' need to be disaggregated in relation to glacier growth (e.g. enhanced winter snowfall (increased precipitation) and reduced summer melt (lower temperatures) are gener-ally considered to be the most conducive to glacier growth).

Thus far, we have only considered the environment as changing through time. However, the environment has a critical spatial and vertical dimension. Thus, feedbacks do not simply operate through time, but also through space: if you change one part of the environment, feedbacks may not be limited to just that part of the environment, but may also affect processes operating in other places in the environment. For instance, in the case of glacial cycles, it is well established that ice-sheet growth in the high latitudes can have major environmental impacts at both the mid- and the low latitudes. As the ice sheets extended to lower lati-tudes than present, so cold polar air extended to lower latitudes, causing the mid-latitudinal zones where warm and cold air mix and the westerly jet stream to move equatorwards. In the American south-west, this caused substantial increases in precipitation during the last glacial maximum (e.g. Spaulding, 1991). Similarly, as more water was locked up in the ice sheets during the last glaciation, and with changes in atmospheric and oceanic circulation, the tropical latitudes were some-what cooler and drier than today. This has resulted in suggestions that tropical plant and animal communities retreated into refugia, although this is contested, with suggestions that continuous forest cover was maintained throughout the Pleistocene (e.g. Colinvaux *et al.*, 2000). This provides an important reminder of the connected nature of the environment, and the way in which changes in one part of the environment can affect other parts of the environment.

It follows that the way that the environment is modelled needs careful consideration so that these feedbacks are properly incorporated. This allows us to note what an effective model of the environment must allow for: (i) the proper representation of the spatiality of processes, especially because many processes are driven by gradients that are spatial (e.g. the slope of a hill controls the velocity with which water moves over it); (ii) vertical gradients of process (e.g. flow at the bed of a river is generally faster than at the water surface); (iii) the system to evolve through time in response to the operation of these two- and three-dimensional processes; and (iv) feedbacks, manifest as change in the process drivers in response to the operations of processes themselves (e.g. as water moves over a hillslope, it may cause erosion (or deposition) and so change the slope so that processes at future times operate in different ways. The conceptual model allows us to specify (i) through (iv) and hence: (a) what is excluded explicitly in the model; (b) what is included, but represented in a simple way; and (c) what is excluded. As we will see later, this means that models are rarely generic representations of whole systems, but truncated representations of reality, where model predictions are partly defined by how the modeller builds the model (Lane, 2001). Regardless of the conceptual model chosen, it is possible to conceive of a continuum from models that are largely built around a set of observations (empirical models) through to models that are largely built around a set of laws (physically based models). This is an artificial continuum as most empirical models have an implicit theoretical justification and most physically based models have an explicit need for empirical relationships. However, it provides a useful categorization for exploring the nature of environmental models.

Empirical approaches

Mathematical models based upon empirical approaches involve making a set of observations of a number of phenomena and then using these observations to construct relationships amongst them. Thus, this approach is heavily dependent upon either field or laboratory measurements to provide the data necessary to construct the models. Central to empirical approaches are statistical methods. The key assumption is that one or more forcing (independent) variables cause changes in a response (dependent) variable. A good example of this is the eutrophication of shallow lakes. Eutrophication is a natural process associated with a progressive increase in lake primary productivity as lakes are ineffective at removing accumulated nutrients. Research in the 1970s (e.g. Schindler, 1977) established that the key limiting nutrients for eutrophication are phosphates and not nitrates, as is commonly assumed, because certain primary producers are capable of fixing

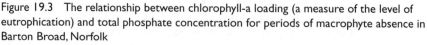

Figure 19.3   The relationship between chlorophyll-a loading (a measure of the level of eutrophication) and total phosphate concentration for periods of macrophyte absence in Barton Broad, Norfolk

Source: Lau (2000)

atmospheric nitrogen. Figure 19.3 illustrates this for Barton Broad, using a dataset when no submerged plants (macrophytes) are present: as the total phosphate concentration of the water column rises, so the level of primary productivity increases. Thus, we have a conceptual model in which we assume that levels of eutrophication are determined by levels of phosphate in a given lake ecosystem. By synthesizing a set of data from different lake ecosystems, Cullen and Forgsberg (1988) were able to build simple statistical rules that allow levels of chlorophyll-a, and indicator of the level of primary productivity in the system, to be predicted from phosphate concentrations. In practice, eutrophication is also affected by predator–prey interactions and seasonality effects, and more complex models can be developed for predicting the level of eutrophication in a lake by using more than one forcing variable (e.g. Lau and Lane, 2002). This is not a straightforward task if the forcing variables are themselves related to one another or, as is commonly the case with ecological data obtained for a single ecosystem (e.g. a lake), individual observations are correlated through time or across space. Additional information on empirical modelling of quite complex environmental systems is provided in Pentecost (1999).

The use of an empirical approach to modelling requires a number of assumptions. First, the empirical relationship must have a justifiable theoretical basis in the form of an appropriate conceptual model. Whilst the use of statistical methods allows assessment of how good the model is (e.g. by assessing the goodness of fit of the empirical relationship – see Chapter 21) this is not normally a sufficient test of a model: empirical relationships may be spurious, and hence have a poor predictive ability. They may also be strongly affected by individual observations, especially at the extremes of a variable, which can affect the model's predictive ability in relation to extreme situations. Unfortunately, many of the environmental issues of most concern are those that are connected with extremes (e.g. floods and droughts).

Second, many empirical models perform poorly when used to predict beyond the range of observations upon which they are based. For instance, Cameron *et al.* (2002) compare traditional (using probability distributions and kinematic routing) and neural network approaches to flood forecasting. The latter is based upon using artificial intelligence methods to construct an empirical relationship between forcing parameters (e.g. upstream rainfall) and the key response variable (e.g. water level). Cameron *et al.* (2002) found that these models are only good at predicting water levels associated with patterns of forcing parameters that have happened before. Questions over temporal validity also appear in relation to spatial validity. In the case of lake eutrophication, there is a good reason to assume that lakes are generally phosphate limited, but research has shown (Kilinc and Moss, 2002) that in certain situations lakes can be nitrate limited. Thus, a generalized model (e.g. Figure 19.3) will not hold everywhere. There are two important implications of this problem: empirical models do not always hold through time; and they often transfer poorly to different places. At the root of these problems is the poor generalizability of empirical models. It is often argued that this is because they do not necessarily have a good physical basis, as they are grounded upon statistical interactions rather than fundamental physical processes.

Third, however the relationship is constructed, it is important to remember that empirical relationships involve statements of uncertainty. The scatter around the line in Figure 19.3 means that, for a given phosphate loading, there is a range of possible levels of eutrophication. The greater the scatter, the greater the uncertainty, and it is necessary to provide predictions with an uncertainty attached to them, normally in the form of a standard deviation (i.e. prediction ± standard deviation) or confidence limits about a regression fit (dotted lines in Figure 19.3). When predictions from more than one empirical model are combined, it becomes especially important to propagate the uncertainty associated with each prediction. For instance, it is common to estimate the sediment load produced by a catchment from the product of an estimated discharge (predicted on the basis of an empirical relationship between discharge and the more readily measured parameter, water level) and an estimated suspended sediment concentration (predicted on the basis of an empirical relationship between suspended sediment concentration and discharge). Both of these have uncertainties that will tend to magnify error when those uncertainties are combined. In some situations, this uncertainty can be propagated mathematically (e.g. Taylor, 1997). However, as the assumptions required for mathematical propagation of error are commonly violated, it is more common to propagate error statistically using techniques such as Monte Carlo analysis.

Physically based numerical models

Empirical approaches have a physical basis insofar as they have some sort of conceptual model that justifies the form of the relationship developed. Physically based numerical models take this one stage further by using the conceptual model to define links between fundamental physical, chemical and occasionally biological principles, which are then represented mathematically in computer code. At the scale of the natural environment, we are fortunate in having a number of key principles, largely deriving from Newtonian mechanics: (i) rules of storage; (ii) rules of transport; and (iii) rules of transfer. Rules of storage are based upon the law of mass conservation: matter cannot be created or destroyed, but only transformed from one state into another. Thus, in relation to the prediction of flood inundation extent, and assuming that evaporation and infiltration losses are negligible, an increase in river discharge must result in one or more of: (i) an increase in flow velocity; (ii) an increase in water level; or (iii) transfer of water onto the floodplain. Mass conservation underpins almost all numerical models in some shape or form. However, it is rarely sufficient to predict how a system behaves. Thus, in the case of flood inundation, how an increase in discharge is divided between changes in flow velocity, water level and floodplain transfer depends upon what is conventionally labelled a force balance, and which is normally based upon further Newtonian mechanics: for example, every body continues in its same state of rest or uniform motion unless acted upon by a force. In the case of a river, this partitioning involves consideration of the pressure gradients and potential energy sources which drive the flow and the loss of momentum, due to friction at the bed and due to turbulence, which slows the flow: a rule of transport. For instance, in the simplest of terms, with a rougher bed, and subject to the shape of the river channel, an increase in discharge is less likely to lead to an increase in flow velocity than an increase in water level. Finally, rules of transfer allow for the possibility that chemical reactions cause a change in the state of an entity. For instance, phosphate bound to aluminium or iron may become soluble in an eutrophic lake, and hence available for fuelling eutrophication, if there are oxygen deficits sufficient for reducing conditions to develop.

Following from the discussion of conceptual models above, the key to the operation of these rules is to allow feedback between them and the parameters that describe them. For instance, in the case of the river example above, an increase in discharge was noted to lead to more of an increase in water level for a rougher bed than a smoother bed. However, if water level rises, the effect of bed friction will be reduced, making it more easy to translate increases in discharge into increases in velocity. Here we see one of the fundamental advantages of numerical models over empirical models: they are more likely to capture the dynamics of the system through incorporating feedbacks between system parameters, something that our consideration of conceptual models identified as being crucial.

Stages in the development of a physically based numerical model

Stages in the development of a mathematical model are shown in Figure 19.4 for the case of a model of eutrophication processes in a shallow lake (from Lau, 2000). This introduces a number of important components in model building. First, it demonstrates the dependence upon a proper conceptual model. As noted above, the conceptual model is 'closed' as the boundaries of the system that the model will address must be defined. Ideally, all relevant processes will be included, and the closure will not exclude processes that might matter. In practice, processes are

(a)

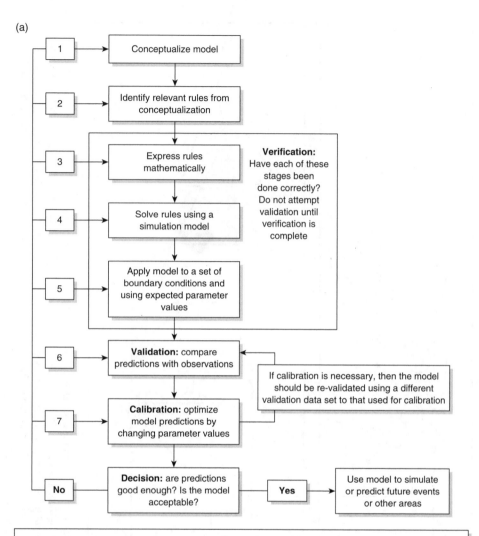

1 Is the model properly conceptualized? Are there enough processes represented? Are there too many processes represented? Are the relevant processes represented?

2 Have the correct rules been identified? Are the rules properly specified? What assumptions have been introduced here? Are they accurate?

3 Is the mathematical expression of the rules correct? Have any unsupportable assumptions been introduced during expression?

4 Are the rules being solved correctly? Is the computer program free of programming error? Is the numerical solver sufficiently accurate? Is the model discretization robust?

5 Are the boundary conditions properly specified? Is there error in the boundary conditions that is affecting the model? Is the poor model performance due to a lack of necessary boundary conditions?

6 Are the observations being used to validate the model correct? Are the observations equivalent to model predictions? Are they representative of model predictions? Where (in space and time) is the model going wrong when compared with observations and predictions?

7 Has the calibration process produced realistic parameter values? Does the model produce more than one set of parameter values that are equally good at predicting when re-validating the model? Is the model overly sensitive to parameterization?

(b)

Figure 19.4 A general approach to model development (19.4a) and the conceptual model (19.4b) applied to Barton Broad, Norfolk, England. This shows that the model had three major components: the algal component (phytoplankton), a nutrient component (phosphorous) and an algal grazing component (zooplankton). There were then interactions both within and between these three major components. For instance, phytoplankton are re-generated according to nutrient availability and die naturally. They are also grazed by zooplankton. As phytoplankton are generated, die and are grazed, so phosphorous is moved around the various components in which it can be stored

Source: Lau (2000)

excluded for two reasons: (i) if they do not matter for the particular system being studied; or (ii) if there are limitations on the possibility of including a particular process. Situation (i) can arise for a number of reasons. First, the geography or the history of the problem may allow processes to be excluded (i.e. they may not be occurring in a particular place or at a particular time, and so can be ignored). Second, there may be time-scales of space-scales that are not relevant to the system that is being considered. For instance, if we wish to model the spatial patterns of flood inundation over large lengths of river, it is not necessary to include a sophisticated treatment of turbulence, a short time-scale aspect of the process, as uncertainties in other aspects of the model will tend to dominate. This reflects the common situation that time and space are coupled (e.g. Schumm and Lichty, 1965), where processes of interest over large spatial scales are commonly associated with longer time-scales. Unfortunately, numerical modelling faces a serious challenge when processes couple across time and space scales, and a fine resolution of process representation (in space and time) is required in order to get an adequate system representation. Situation (ii) follows from (i), when there are limits placed upon process representation because of limits to computation, but also more generally where either the process knowledge is missing, or process representation is difficult. For instance, in eutrophication modelling, there is a major difficulty in coupling the aggregate behaviour of phytoplankton to the species specific behaviour of an individual fish, in relation to its life cycle. This situation is commonly dealt with by using a more simple version of a process's effects. In the case of lake eutrophication, the system may be driven by both nutrient limitations and food chain interactions. The latter are controlled by the presence of bottom growing vegetation which act as refugia for zooplankton that graze upon algae. It is not necessary to model plant life cycles in most shallow lakes as these are relatively straightforward functions of seasonality. Thus, they may be dealt with using simple parameterizations (see below).

The second component of model building involves taking the conceptual model, identifying appropriate process rules and transforming these into a simulation model that can solve the equations. This can be the most difficult stage of model development as many equations do not lend themselves to easy solution. This is well illustrated for the case of predicting the routing of flood waters through a drainage network. The discharge entering the network varies as a function of time. In simple terms, the rate at which it moves through a part of the network depends upon the water surface slope in that part of the network (steeper slopes mean faster flows). Thus, there is a spatial dependence. The water surface itself will evolve as the water moves through the network. This leads to two basic problems. First, the combined space–time dependence of flood routing means that the dominant equations are partial differentials (as they contain derivatives in time and space). This is common to almost all environmental models, as we are interested in how things move through space. Moving through space takes time, and hence all models should contain both space and time. Partial differentials are very difficult to solve. Second, all models require some form of initial conditions. In this case, we need starting values for water level and discharge throughout the drainage network. We will also need to know some combination of discharge, velocity and depth at the inlet and/ or the outlet. Hence, models have a crucial dependence upon the availability of data to initialize them. Solution of the governing equations commonly requires us to introduce parameters. This may be because during the conceptualization process, we chose either to exclude certain processes, or to represent them in a simplified way. In algal modelling, lake vegetation effects were treated in a simple way (a presence or absence conditioned by seasonality). Flood routing commonly ignores the lateral and vertical movements of water, and flow turbulence. These affect the routing of discharge, but their inclusion in a model may make solution impossibly time consuming if we are interested in drainage-network-scale flood routing. However, parameterization also results from the fact that whilst equations can have a good physical basis, as they are simplified, new terms can appear whose physical basis is less certain and, most commonly, whose field or laboratory measurement is especially difficult. Parameterization can also be required in situations where a process has been excluded from a model, with its effects being represented through one or more parameters. A good example of this in flood routing studies is the use of a roughness parameter which not only represents the effects of friction at the bed upon flow hydraulics but also turbulence. The process of parameterization is rarely conducted independently from check data, in which the model is optimized by changing parameter values such that the difference between check data and model predictions is minimized. This can result in parameters taking on values that are very different from what they might appear to be if measured in the field. For this reason, they may be called *effective* parameters, ones required to allow a model to predict known behaviour, rather than derivable from fundamental process analysis. Importantly, this causes us to question the supposed generality of models as the results of parameterization may not be guaranteed to hold beyond the range of conditions for, and the location at, which parameterization has been undertaken.

Model assessment involves two important stages: verification and validation. Verification is the process by which the model is checked to make sure that it is solving the equations correctly. This may involve debugging the computer code, doing checks upon the numerical-solution process and undertaking sensitivity analysis. The latter may be used to make sure that the model behaves sensibly in response to changes in boundary conditions or parameter values. Validation is the process by which a model is compared with reality. This normally involves the definition of a set of 'objective functions' that describe the extent to which model predictions match reality (see Lane and Richards, 2001). This is where there can be confusion with regard to validation and parameterization. Once an objective function has been determined, a model may be developed to reduce the magnitude of the error defined by the objective function. This may involve changing parameters or checking boundary conditions through the process of optimization described above. It may involve a more radical redevelopment of the model through the incorporation of new processes or alternative treatment of existing processes. Figure 19.5 shows default and optimized predictions of chlorophyll-a concentrations obtained for the eutrophication model described above, as applied to Barton Broad. Figure 19.5a shows the default predictions and Figure 19.5b shows the optimized predictions, in which parameterization was used to maximize the goodness of fit between the model and independent data. Unfortunately, Figure 19.5b is not an example of validation, as we have no evidence that the fit between the model and the measurements is due to anything other than the fact that we have forced the model to fit the measurements. To claim, from Figure 19.5b, that

Figure 19.5   The default predicted (19.5a), optimized using 1983–1986 data (19.5b) and predicted using 1987–1993 data (19.5c) and chlorophyll-a concentrations for Barton Broad
Source: Lau (2000)

the model is validated is tantamount to cheating. Modellers get around this problem through one of a number of means. First, it may be possible to do a split test, in which some data are not used for model parameterization, but held back for independent use for validation. An example of this is shown in Figure 19.5c, in which the optimized model (Figure 19.5b for 1983 to 1986 data) is used to predict for the period 1987 to 1993. This demonstrates why the optimization undertaken for 1983 to 1986 does not constitute validation: when the model is applied in a predictive mode, it is clear that there is a progressive divergence between observations and model predictions and the model is not capturing some part of the system. An optimized model is not necessarily a validated model. Second, validation can be undertaken by taking a model that has been parameterized for one location and, under the assumption that the parameters can be transferred, applying it to a second location in order to provide independent validation data.

The main purpose of validation is to place confidence in the extent to which a model can be used to simulate or to predict beyond the range of conditions for which it is formulated. As we noted in the introduction, this is the core purpose of a model: to extend the bounds of space and time. Here we face some serious difficulties, and these cause us to look very critically at deterministic mathematical models and what they can and cannot do.

WHAT A MODEL CAN AND CANNOT DO

The above examples illustrate four basic reasons why we might wish to develop numerical models and, connected to each of these, four reasons why models might not deliver what we hope. The first relates to the need to understand whole systems (see Chapter 18). This relates to the idea that there is a potentially large number of processes that might operate in a system, that these connect with one another through both positive feedbacks (which amplify change) and negative feedbacks (which slow or even stop change), and that these processes operate over space and through time. Designing either field or laboratory experiments to understand such systems is a serious challenge. The numerical model provides the opportunity to integrate processes together in order to understand whole-system behaviour. The problem here is that numerical models are no different to field or laboratory experiments in that they have been subject to closure. Models require assumptions to be made about what does and does not matter. For instance, climate models vary in the extent to which they include feedbacks associated with core components of the earth system (e.g. vegetation feedbacks). As noted above, they also involve simplifications over how processes are included. Also in climate models, the effects of clouds upon the energy and water balance of the atmosphere have to be dealt with using simplifications as the spatial scale of cloud physics is typically smaller than that at which models are operating. Thus, models are closed representations of reality and a model is partly dependent upon how the modeller sets up the model (i.e. defines the closure). This is similar to the way in which the experimental design associated with fieldwork (e.g. choice of field site, measurement methods) and laboratory work (e.g. definition of controlled experimental conditions) partly determines the results that are obtained (Lane, 2001).

The second important role of a model relates to system understanding through sensitivity analysis. This is where the model is used in the same way as a laboratory experiment: model predictions derived from models with different structures, process representations or parameter values are compared in order to understand which processes and parameters have most effect upon the system. This tells us which aspects of a system are most important for further research, for field work, or for laboratory analysis. In a three-dimensional numerical model of flow over a rough gravel bed, model predictions were highly insensitive to changes in bed roughness, but highly sensitive to bed surface structure (Lane *et al.*, 2002, Figure 19.6). Bed roughness affects the momentum balance at the bed. Bed-surface structure affects both the momentum balance and mass conservation (i.e. flow blockage). The poor sensitivity to bed roughness demonstrates that the effect of bed-surface structure will not be properly represented through a bed-roughness term, and requires us to reconsider how bed-surface structure is represented in models of flow over gravelly surfaces. The problem with this type of sensitivity analysis is similar to the whole system problem. Do the results hold in a more general sense, or are they a product of the nature of the model as it has been developed and as it is being applied? Unfortunately, this is not a simple question to answer. In the gravel-bed river-flow model, the treatment of turbulence is known to be inappropriate for this kind of flow. Turbulence has a major effect upon momentum transfer at the bed. Thus, whilst the conclusion that blockage affects associated with bed-surface structure matter does hold, the extent to which this is the case must remain uncertain as turbulence effects are not being included.

The third role of a model is the one that is of most interest to environmental managers: the prediction of system properties either at future time periods or for locations where reliable measurements cannot be made. This rationale allows us to break out of the 'hunny' jar and heffalump problems described in the introduction by using technology (a properly validated numerical model) to try to make statements about what might happen before they actually happen. We live with this type of modelling on a day-to-day basis. For instance, short-term weather forecasting is based upon regional predictive models of weather systems. Flood warning systems are commonly driven by a predictive model that indicates where water is expected to reach for a range of river discharges, and which can be used to issue warnings to potentially affected properties when a particular discharge occurs. The main issue here is the extent to which trust can be placed in a model's predictions as a result of uncertainties in model predictions. These uncertainties can be classified into seven broad headings (Table 19.1). Table 19.1 identifies three core issues that we must ev aluate when considering models as a strategy within physical geography. First, uncertainty will be an endemic characteristic of all modelling efforts. Attempts to reduce uncertainty are tempting, through the adoption of more sophisticated modelling approaches, or the improved specification of boundary conditions. Unfortunately, research has shown that attempts to deal with uncertainty tend to introduce yet more uncertainty, in the way that Wynne (1992) conceptualizes the dynamics of science as being the generator of uncertainty rather than the eliminator of it. For instance, Lane and Richards (1998) argue that three-dimensional models of river flow are required in tributary junctions in order to represent properly the effects of secondary circulation upon flow processes. Lane and Richards (2001), having used a three-dimensional model for this purpose, demonstrate the significant new uncertainties that have emerged from: (i) the difficulty of specifying inlet conditions in each of the tributaries in three dimensions; (ii) problems of designing a numerical mesh that provides a stable numerical solution that minimizes numerical

Figure 19.6 Model predictions of flow over a rough gravel-bed surface. (a) shows the digital elevation model and (b) shows velocity vectors in a cross-section taken along the dashed line in (a) (i.e. vectors show the cross-stream and vertical components of velocity). w is width, h is depth and l is channel length. y is position along the section, x is position downstream and z is position in the vertical

Source: Based on Lane et al. (2002)

diffusion; (iii) uncertainties over the performance of a roughness treatment in three dimensions; and (iv) problems with finding an appropriate turbulence model. The continual creation of uncertainty keeps the science of this sort of modelling alive. However, when models need to be put to practical use for forecasting, or process understanding, obvious questions emerge as to whether or not numerical models are little more than computationally intensive crystal balls.

Second, Table 19.1 emphasizes that when models are to be used in an applied sense, it is vital that the associated uncertainty is communicated along with those model predictions. This is necessary to avoid false faith being placed in model predictions. This requires methods for: (i) determining what the uncertainty is; (ii) representing the uncertainty in a manner that has meaning to the user of a model's predictions (as well as the modeller themselves); (iii) communicating uncertainty; and (iv) persuading those that use model predictions to accept both uncertainties in model predictions and uncertainties in determining the uncertainty of model predictions. Before commenting on these, it is worth

Table 19.1   Model uncertainties

Type of uncertainty	Explanation
Closure uncertainties	This relates to uncertainties that arise because certain processes have been included or excluded during model development. A common strategy to deal with this is to include processes, test their effects, and to ignore them or to simplify them if they seem to have only a small effect upon model predictions. The problem with this is that the effect of a process can often change as the effects of other processes (i.e. system state) changes. Thus, it is impossible to be certain that a given process will always be unimportant. Note that problems of closure are not just associated with models, but are an inherent characteristic of all science (see Lane, 2001). In general, closure uncertainties are easy to spot in any model application. We often forget that the same criticism of closure applies equally to almost all other aspects of scientific method.
Structural uncertainties	These arise because of uncertainties in the way in which the model is conceptualized in terms of links between components. A good example of this is whether or not a component is an active or a passive component of a model. For instance, in a global climate model, the ocean may be assumed to be a passive contributor to atmospheric processes: it acts as a heat and moisture source, but does not itself respond to atmospheric processes. This type of structural uncertainty will define limits to model applicability. For instance, the ocean has a high specific heat capacity, and so will respond slowly to atmospheric processes. Thus, specification of the ocean as a passive source of heat and moisture is acceptable if the model is being applied over a timescale (e.g. days) when the ocean can assume to be in steady state. As with closure uncertainties, structural uncertainties limit a model's applicability and remind us of the importance of evaluating a model in relation to the use to which it will be put. Any model can be criticized on the grounds of structural uncertainties without a purpose-specific evaluation.
Solution uncertainties	These uncertainties arise because most numerical models are approximate rather than exact solutions of the governing equations. Commonly, numerical solution involves an initial guess, operation of the model upon that initial guess, and its subsequent correction. This continues until operation of the model on the previous corrected guess does not alter by the time of the next guess. This process can result in severe numerical instability in some situations, which is normally easy to detect. However, more subtle consequences, such as numerical diffusion associated with the actual operation of the solver, can be more difficult to detect. Following guidelines in relation to good practice can help in this respect.
Process uncertainties	These arise where there is poor knowledge over the exact form of the process representation used within a model. A good example of this is the treatment of turbulence in models of river flow. Turbulence can have an important effect upon flow processes as it extracts momentum from larger scales of flow and dissipates it at smaller scales. Most river models average turbulence out of the solution, but then have to model the effects of turbulence upon time-averaged flow properties. Turbulence models vary from the simple to the highly complex and it can be shown that different turbulence treatments are more or less suitable to different model applications. Thus, the process representation in a model, as with structural aspects of a model, needs careful evaluation with reference to the specific application for which the model is being used.
Parameter uncertainties	These arise when the right form of the process representation is used but there are uncertainties over the value of parameters that define

(Continued)

Table 19.1   *(Continued)*

Type of uncertainty	Explanation
	relationships within the model. Particular problems can arise when parameters have a poor meaning in relation to measurement. This can arise in two ways. First, some parameters are difficult to measure in the field as they have no simple field equivalent. Second, during model optimization, parameters can acquire values that minimize a particular objective function, but which are different to the value that they actually take on the basis of field measurements. A good example of this is the bed roughness parameter used in one-dimensional flood routing models. It is common to have to increase this quite significantly at tributary junctions, to values much greater than might be suggested by the bed shape of the river or the bed grain-size. In this case, there is a good justification for it, as one-dimensional models represent not only bed roughness effects but also two- and three-dimensional flow processes and turbulence through the friction equation. Roughness is therefore representing the effects of these other processes in order to achieve what Beven (1989) labels 'the right results but for the wrong reasons'.
Initialization uncertainties	These are associated with the initial conditions required for the model to operate. They might include the geometry of the problem (e.g. the morphology of the river and floodplain system that is being used to drive the model) or boundary conditions (e.g. the flux of nutrients to a lake in a eutrophication model).
Validation uncertainties	Given the above six uncertainties, a model is unlikely to reproduce reality exactly, and validation is required to assess the extent to which there is a reasonable level of agreement. However, validation data themselves have an uncertainty attached to them. This is not simply due to possible measurement error, but also when the nature of model predictions (their spatial and temporal scale, the parameter being predicted) differs from the nature of a measurement. A commonly cited example of this is validation of predictions of soil moisture status in hillslope hydrological models, when point measurements of soil moisture status (in space and time) are used to validate areally integrated predictions. This creates problems for modelling in two senses. First, apparent model error may actually be validation data error. Second, if validation data are then used for model optimization (and note that data used for model optimization should not then be used for validation), uncertainty will be introduced into model predictions as the data that the model are optimized to may be incorrect. Following Beven (1989) this means that we may get the wrong results for the wrong reasons.

remembering the four-fold division introduced by Wynne (1992) of different types of uncertainty. Wynne argues that all uncertainties can be given one of four labels: (a) risks, or quantifiable uncertainties; (b) unquantifiable uncertainties; (c) uncertainties that we are ignorant about, but which we may find out about through further experience or investigation; and (d) indeterminacies, or uncertainties that cannot be determined through any form of investigation, prior to them happening. Thus, the determination of uncertainty is largely about the determination of risk, or quantifiable uncertainty, and the representation of that uncertainty accordingly. However, as uncertainty is communicated, it is vital to include unquantifiable uncertainty, ignorance and indeterminacy. As an example, consider estimation of the uncertainty in patterns of floodplain inundation. Sensitivity analysis can be extended to uncertainty analysis through assessing the implications of model uncertainties (e.g. Table 19.1) for model predictions. As Binley *et al.* (1991) have demonstrated, we need formal methods for doing this (e.g. General Least-squares Uncertainty Estimation, or GLUE). This is not a straightforward task as a result of the large number of parameters that we are typically uncertain about. Nonetheless, uncertainty bands can be determined in the form of probabilities of floodplain inundation for events with different return periods (e.g. Romanowicz *et al.*, 1996). These provide a partial account of uncertainty. Actual predictions of flood inundation will be affected by unquantifiable uncertainties (e.g. the difficulties of determining likely run-off generation given a particular combination of antecedent moisture conditions and rainfall patterns), ignorance (e.g. incorrect assumptions built into the model, such as that extreme flood events only occur when a catchment is fully saturated) and indeterminacy (e.g. aspects of floodplain management, such as culvert maintenance and repair, that cannot be determined in any realistic way before an event occurs). Unfortunately, we are not good at communicating these uncertainties *and* accepting these uncertainties as a natural aspect of the science of environmental management. The latter is compounded by the problem that uncertainty is measured on a continuous probability scale whereas decisions over floodplain management have to be made on a discrete scale involving action (e.g. improve floodplain defences, refuse flood insurance protection) or no action (do not improve floodplain defences, allow flood insurance protection).

Third, we need to be sensitive to those who are the ultimate judge of model predictions and for whom the inevitable errors in what a model predicts translate into real day-to-day impacts. One of the best examples of this are the online flood maps of the Environment Agency of England and Wales (www.environment-agency. gov.uk) which are accessible to anyone and which identify 'indicative floodplains', which are overlain on basic Ordnance Survey maps, allowing anyone to find out whether or not they live in a floodplain. However, these maps are bound up with the process of modelling and the way in which flooding is represented in those models in complex ways. First, the models follow standard methodologies for predicting flood inundation using one-dimensional and two-dimensional models. Such models are commonly evaluated using contingency tables (e.g. Table 19.2) that evaluated predicted and modelled inundation for a particular flood event. In Table 19.2, we would hope that the diagonal cells marked with the * would add up to 100 per cent. They

Table 19.2 An example contingency table for flood-risk assessment

	Predicted flooded	Predicted as not-flooded
Observed flooded	43%*	12%
Observed not-flooded	13%	32%*

never do, and values of 70 to 80 per cent are commonly considered acceptable. Thus, a model which is an acceptable model overall may actually have specific locations within the model that are wrong. However, it is commonly the specific locations of the models that are used in decision-making: solicitors doing searches for potential buyers commonly report on whether or not a property is in the indicative floodplain; insurance companies may change premiums to reflect possible flood risk. Thus, model predictions at the scale of the individual property, a scale at which the models are known to be wrong, may have a material impact upon the value of that property and those who live in it. Second, the model predictions are actually of the indicative flood inundation expected in the absence of flood defences. Hence, many properties defended by levées, for instance, and at flood risk only if the levées fail or are breached, are actually protected. The presence of defences has a critical impact upon predicted inundation areas. Both of these issues have resulted in the flood maps, and so the models that underpin them, becoming contested. This dispute is aided by both: (i) the growing ease with which model predictions, especially those with a spatial content, can be disseminated using online methodologies and in an era of growing freedom of information; and (ii) the growth within society more generally of distrust of both experts and expertise. Model uncertainties are starting to take on a material form, giving those with expertise that is traditionally excluded from the modelling process, the means of questioning and interrogating the model predictions and models to which they are increasingly subject. The externalization of model predictions will lend them up to scrutiny and will change the way in which we use models as part of the process of making difficult decisions.

The final role for modelling is for simulation, in which we ask 'what if?' questions about environmental behaviour. In many senses, this is the combination of the above three reasons for using numerical models: use a representation of the whole system (reason 1), in which we vary system parameters (sensitivity analysis, reason 2), to make predictions (reason 3) that allow us to improve environmental management. The need to do this was emphasized in the introduction: awaiting confirmed observational evidence from a noisy system means that significant damage may have occurred before action can clearly be justified.

CONCLUSION

Harré (1981) identified three different roles for scientific investigation: (i) as formal aspects of methodological investigation; (ii) to develop the content of a theory; and (iii) in the development of technique. Table 19.3 applies these to numerical models in an attempt to demonstrate what models can and cannot do. In practice, this sort of classification should be treated as a fuzzy one, and this table is included as a basis for discussion, perhaps in a tutorial. For instance, following A1 in Table 19.3, models may be used to explore the characteristics of a naturally occurring process, but the extent to which this can be done will depend upon the confidence that can be placed in a model and, as noted above, there will always be uncertainty in the 'natural' characteristics that the results from a model may suggest. Similarly, strictly, models may not be used to provide negative results (A6), as we never know whether

Table 19.3   What models can and cannot do

Models as formal aspects of method		Can a numerical model help?
A1	To explore the characteristics of a naturally occurring process	The classic role of numerical simulation – asking 'what if?' questions
A2	To decide between rival hypotheses	Used as part of sensitivity analysis in which rival hypotheses are tested using a model
A3	To find the form of a law inductively	X Generally not possible as laws are required to get a model to work, so generating laws from model predictions runs the risk of circular argument (but what is a law?)
A4	As models to simulate an otherwise unresearchable process	A critical function of models and perhaps where they are most powerful but also most problematic (are a models' results a reflection of reality or a reflection of the way the model has been set up?)
A5	To exploit an accidental occurrence	But more to understand possible accidental occurrences that might occur (e.g. what if there was a dam break?)
A6	To provide negative or null results	X A major problem for models – is the negative or null result a true property of the system, or simply because a model has been used?
Models in the development of the content of a theory		
B1	Through finding the hidden mechanism of a known effect	X Can't find hidden things if they are not included in the model
B2	By providing existence proofs	Models can provide corroboration of things observed using other methods
B3	Through the decomposition of an apparently simple phenomena	X Not possible
B4	Through demonstration of underlying unity within apparent variety	Identification of general patterns
In the development of technique		
C1	By developing accuracy and care in manipulation	X Not possible
C2	Demonstrating the power and versatility of apparatus	X Not possible

Source: After Harré (1981)

or not the null result is due to the model as constructed or a real characteristics of the system that the model is being used to represent. However, a model may provide an indication of a null result that causes us to look elsewhere in order to find additional supporting evidence. This discussion of both A1 and A6 emphasizes a key theme in the use of models: a model's effectiveness largely depends upon the ability of a modeller to engage with a broad spectrum of methods, including those that are field- and laboratory-based. This is where *Winnie-the-Pooh* again provides us with key guidance. During the flood, Pooh bear finds a 'missage' in a bottle.

Being unable to read, he has to get to Owl. Surrounded with water he uses a classic piece of reasoning (i.e. a model): if a 'missage' in a bottle can float, then a bear in a 'hunny jar' can float. Having got the skeleton for his model, he develops his model by trying various positions on the jar, until he finds one that is stable. Here we see the crucial iteration between model development and empirical observation. He now has an optimized model, which he proceeds to validate by successfully floating off to Owl. Finally, he demonstrates the transferability of his model by going to rescue Piglet, with Christopher Robin, in an upside-down umbrella. This requires much less development and represents the common transition of a model, through time, from being a developmental piece of science to a practical part of technology. Unfortunately, Pooh bear gives us little guidance as to when a model as a piece of science is sufficient for us to make it a practical piece of technology. However much we, as modellers, might debate the philosophical and methodological aspects of what we do, we cannot escape the fact that the believability of a model is no longer solely the domain of the academic or the policy-maker.

There is a bigger picture here, in relation to the role that we expect of models more generally. Our collective understanding of the past is that it is full of surprises, things we wish to avoid. Models become a means of making the future predictable. Futures that are predictable can be avoided and so controlled. It is not surprising that models, then, increasingly permeate our day-to-day existence in ways that we rarely appreciate. However, there are occasions when our reliance on models emerges, often during times when they unsettle, undermine or simply fail to deliver the secure and predictable futures that we wish. During such occasions, not only do models come under scrutiny, but so do the very knowledge principles and practices of which they form a part, as we struggle to come to terms with the uncertainty that is endemic to any understanding of the future. This makes it very exciting to be a modeller in the twenty-first century, as it is one of those areas of scientific endeavour that increasingly needs the kind of interdisciplinary working so characteristic of geographical research.

Summary

- Numerical models are used by physical geographers for dealing with situations where the researcher is 'remote' from what they are studying: where the events that are of interest have occurred in the past (e.g. reconstruction of past climates); may occur in the future (e.g. patterns of inundation associated with future flood events); or where they are occurring now, but cannot be measured or studied using other methods.
- Without a conceptual model of the system that is being modelled, it is impossible to develop either an empirical or a physically based mathematical model. Conceptual models involve a statement of the basic interactions between the components of a system.
- The way that the environment is modelled needs careful consideration, so that feedbacks are properly incorporated.
- Empirical approaches have a physical basis in so far as they have some sort of conceptual model that justifies the form of the relationship developed. Physically based numerical models take this one stage further by using the conceptual model to define links between fundamental physical, chemical and occasionally biological principles, which are then represented mathematically in computer code.

- Model assessment involves two important stages: verification; and validation. Verification is the process by which the model is checked to make sure that it is solving the equations correctly. Validation is the process by which a model is compared with reality.
- Modelling has an important role in understanding the operation of environmental systems. This is achieved by using models to integrate multiple processes over many time-scales; to undertake experiments of sensitivity; to predict outcomes; and through simulation to ask 'what if?'-type questions.
- Models and the modelling process demonstrate many of the fundamental aspects of scientific research. The transition from scientific model to a useable technology for environmental management requires that it is transferrable between case studies, and reflects the need for qualitative, as well as quantitative judgements.
- Models are increasingly subject to scrutiny, which is changing the way that they are used in decision-making.

Further reading

- Beven (1989) is a very useful introduction to critical thinking in relation to numerical models. Beven introduces a set of ideas that challenged an emerging paradigm regarding the power of numerical models and provides a critical framework for evaluating the role of modelling in this case for the hydrological sciences, but with implications for the modelling of environmental systems more generally.

- Rather like the Beven paper, Anderson *et al.* (2001) is useful because it brings together a very wide range of theoretical and methodological perspectives in relation to numerical modelling, albeit around the hydrological sciences.

- Kirkby *et al.* (1992) is a good general introduction to numerical modelling in physical geography. It is especially strong on the way modelling is done and has some easy but useful examples of models that can be coded to illustrate principles of model building.

- Beven (2000) is definitely the best book around on modelling in hydrology in general, with excellent coverage of material specific to hydrology but illustrating environmental modelling in general. Likewise, Huggett (1993) is very effective on conceptual modelling across the environmental spectrum and how to apply conceptual models using a range of modelling techniques.

- Jakeman *et al.* (1993) is useful for understanding modelling over a range of spatial scales and especially at the global scale.

Note: Full details of the above can be found in the references list below.

References

Anderson, M.G. and Bates, P.D. (2001) (eds) *Model Validation: Perspectives in Hydrological Science*. Chichester: John Wiley & Sons.

Beven, K.J. (1989) 'Changing ideas in hydrology: the case of physically-based models', *Journal of Hydrology*, 105: 157–172.

Beven, K.J. (2000) *Rainfall-runoff Modelling: The Primer*. Chichester: John Wiley & Sons.

Binley, A.M., Beven, K.J., Calver, A. and Watts, L.G. (1991) 'Changing responses in hydrology: assessing the uncertainty in physically-based model predictions', *Water Resources Research*, 27: 1253–1261.

Broecker, W.S. and Denton, G.H. (1990) 'What drives glacial cycles?', *Scientific American*, 262: 42–50.

Cameron, D., Kneale, P. and See, L. (2002) 'An evaluation of a traditional and a neural net modelling approach to flood forecasting for an upland catchment', *Hydrological Processes*, 16: 1033–1046.

Colinvaux, P.A., De Oliveira, P.E. and Bush, M.B. (2000) 'Amazonian and neotropical plant communities on glacial time-scales: the failure of the aridity and refuge hypotheses', *Quaternary Science Reviews*, 19: 141–169.

Cullen, P. and Forgsberg, C. (1988) 'Experiences with reducing point sources of phosphorous to lakes', *Hydrobiologia*, 170: 321–336.

Farman, J.C., Gardiner, B.G. and Shanklin, J.D. (1985) 'Large losses of total ozone in Antarctica reveal seasonal CLOx/Nox interaction', *Nature*, 315: 207–210.

Harré, R. (1981) *Great Scientific Experiments:* Twenty Experiments that Changed Our View of the World. London: Phaidon Press Limited.

Houghton, J.T., Jenkins, G.J. and Ephramus, J.J. (1990) (eds) *Climate Change: the IPCC Scientific Assessment.* Cambridge: Cambridge University Press.

Huggett, R.J. (1993) *Modelling the Human Impact on Nature.* Oxford: Oxford University Press.

Imbrie, J. and Imbrie, K.P. (1979) *Ice Ages: Solving the Mystery.* London: Macmillan.

Jakeman, A.J., Beck, M.B. and McAleer, M.J. (1993) *Modelling Change in Environmental Systems.* Chichester: John Wiley & Sons.

Kilinc, S. and Moss, B. (2002) 'Whitemere, a lake that defies some conventions about nutrients', *Freshwater Biology*, 47: 207–218.

Kirkby, M.J., Naden, P.S., Burt, T.P. and Butcher, D.P. (1992) *Computer Simulation in Physical Geography.* Chichester: John Wiley & Sons.

Lane, S.N. and Richards, K.S. (2001) 'The "validation" of hydrodynamic models: some critical perspectives', in P.D. Bates and M.G. Anderson (eds) *Model Validation: Perspectives in Hydrological Science.* Chichester: John Wiley & Sons, pp. 413–438.

Lane, S.N. and Richards, K.S. (1998) 'Two-dimensional modelling of flow processes in a multi-thread channel', *Hydrological Processes*, 12: 1279–1298.

Lane, S.N., Hardy, R.J., Elliott, L. and Ingham, D.B. (2002) 'High resolution numerical modelling of three-dimensional flows over complex river bed topography', *Hydrological Processes*, 16: 2261–2272.

Lane, S.N. (2001) 'Constructive comments on D. Massey Space-time, 'science' and the relationship between physical geography and human geography,' *Transactions of the Institute of British Geographers*, NS26: 243–256.

Lane, S.N. (2003) 'Environmental modelling', forthcoming in A. Rogers and H. Viles (eds) *The Student's Companion to Geography.* Oxford: Blackwell.

Lau, S.S.S. (2000) 'Statistical and dynamical systems investigation of eutrophication processes in shallow lake ecosystems', thesis submitted in partial fulfilment of the requirements for the PhD degree, University of Cambridge.

Lau, S.S.S. and Lane, S.N. (2002) 'Biological and chemical factors influencing shallow lake eutrophication: a long-term study', *Science of the Total Environment*, 288: 167–181.

Michaels, P.J. (1992) *Sound and Fury: The Science and Politics of Global Warming.* Washington, DC: Cato Institute.

Molina, M.J. and Rowland, F.S. (1974) 'Stratospheric sink for chlorofluoromethanes: chlorine atom-catalysed destruction of ozone', *Nature*, 249: 810–812.

Pentecost, A. (1999) *Analysing Environmental Data.* Harlow: Longman.

Romanowicz, R., Bevan, K.J. and Tawn, J. (1996) 'Bayesian calibration of flood inundation models', in M.G. Anderson, D.E. Walling and P.D. Bates (eds) *Floodplain Processes.* Chichester: John Wiley & Sons, pp. 333–360.

Schindler, D.W. (1977) 'Evolution of phosphorous limitation in lakes', *Science*, 195: 260–262.

Schumm, S.A. and Lichty, R.W. (1965) 'Time, space and causality in geomorphology', *American Journal of Science*, 263: 110–119.

Spaulding, W.G. (1991) 'Pluvial climatic episodes in North America and North Africa – types and correlations with global climate', *Palaeogeography, Palaeoclimatology and Palaeoecology*, 84: 217–227.

Taylor, J.R. (1997) *An Introduction to Error Analysis: The Study of Uncertainties in Physical Measurements* (2nd edn). Sausalito, CA: University Science Books.

Wynne, B. (1992) 'Uncertainty and environmental learning: reconceiving science and policy in the preventive paradigm', *Global Environmental Change*, 2: 111–127.

20 Using Remotely Sensed Imagery

Paul Aplin

Synopsis

The aim of this chapter is to provide a broad introduction to the field of remote sensing. To understand remotely sensed imagery and its uses fully, it is necessary to appreciate the underlying scientific principles facilitating remote sensing, to know something of the historical background to remote sensing and to consider the uses to which remote sensing is applied. This chapter first defines the term 'remote sensing' and then summarizes the field's historical development. It then goes on to explain the key role of electromagnetic radiation in providing a measurable parameter for remote sensing and reviews different remote sensing platforms and instruments. Image characteristics are described next, followed by a summary of image analysis. Finally, a brief discussion on the current status of remote sensing is provided, identifying the main application areas and outlining the key developments that continue to affect the field.

The chapter is organized into the following sections:

- Introduction
- Principles of remote sensing
- Remote-sensing platforms and instruments
- Image characteristics
- Image analysis
- Conclusion

INTRODUCTION

Standard definitions of remote sensing can be quite unnerving for the non-technical novice: e.g. 'the analysis and interpretation of measurements of electromagnetic radiation that is reflected from or emitted by a target and observed or recorded from a vantage point that is not in contact with the target' (Mather, 2004: 1). (As such, some would-be practitioners have adopted 'alternative' definitions: e.g. 'staying as far away from the problem as possible' (Wright, pers. comm., 1994).) In simple terms, and broadly speaking, remote sensing just involves the collection of information about distant objects. The first point to raise, then, is that remote sensing is not as abstract a concept as may be imagined. For instance, human vision is a form of remote sensing, since objects are seen at a distance. Similarly, to take a photograph is to acquire a remotely sensed image. In the context of geographical study, though, remote sensing extends further than basic sight or photography. In its academic context, remote sensing, also known widely as Earth Observation,

Figure 20.1   Remotely sensed image of Kruger National Park, South Africa, acquired by the 'QuickBird' satellite sensor

Source: Original data copyright: 2002 DigitalGlobe

refers to the acquisition of images of parts of the Earth's surface (usually) from an overhead perspective, using specialized sensing instruments (e.g. Figure 20.1).

The history of remote sensing stretches back nearly 150 years, when pioneer photographers ascended in hot-air balloons to take snapshots of the landscape. For around a century, remote sensing was almost exclusively the domain of aerial photography (Lillesand *et al.*, 2007). Then, in the 1960s, satellite reconnaissance emerged, providing a powerful new platform for remote-sensing activities. Around this same period, a second revolutionary development in remote sensing occurred as electronic sensors were developed to generate digital images, in contrast with traditional analogue photographs (Campbell, 2007). With these innovations remote sensing entered the computer age, and subsequent development, as with almost any ICT computer technology-related field, has been rapid. However, it is not only technological progress that has shaped recent development in remote sensing; political change has also played a part. While the foundations of remote sensing were principally in the military and meteorology (television weather forecasters still prize their 'satellite pictures'), the last three decades have seen widespread adoption of remote sensing by civilian users (Birk *et al.*, 2003). Nowadays, remote sensing is used as commonly by house buyers to view likely locations (e.g. using http://www.about myplace.co.uk) as it is by military strategists to deploy resources. The implications for geographers are immense. Never has there been such a ready supply of spatial data, the central component of so much geographical analysis (see Chapter 25).

PRINCIPLES OF REMOTE SENSING

It has been stated above that remote sensors generate images of the Earth's surface. What has not been explained is how they do this, so a simple explanation follows. For a sensor to generate an image, it must measure some parameter related to the feature being represented. The parameter used is electromagnetic radiation, defined as a 'system of oscillating electric and magnetic fields that carry energy from one point to another' (Rees, 1999: 55). Although this terminology may be a little perplexing to non-physicists, the most common example of electromagnetic radiation is blindingly straightforward: solar radiation, or sunlight. Put simply, a remotely sensed image is a representation of reflected sunlight (Figure 20.2). In fact, this explanation is perhaps a little too simple, since it infers that 'all' sunlight is reflected by the Earth's surface and received by the remote sensor, and that reflected sunlight is the 'only' source of electromagnetic radiation used by remote sensors. Neither of these inferences is true, as will be shown in the following paragraphs.

The most important point to raise in objection to the preceding explanation of remote sensing is that sunlight occurs over a range of wavelengths (the electromagnetic spectrum), but not all of these are useful for remote sensing purposes.

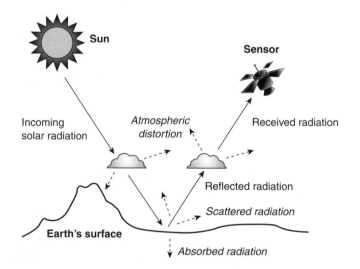

Figure 20.2   Reflected sunlight is the most common source of electromagnetic radiation received by remote sensors to generate images of the Earth's surface, although some radiation is absorbed or scattered at the surface and in the atmosphere

(It should be borne in mind that sunlight does not refer simply to light that is sensitive to human vision (visible light). On the contrary, visible light accounts for only a tiny portion of the electromagnetic spectrum.) That is, while the electromagnetic spectrum extends from short wavelength Gamma rays to long Radio waves and includes X-rays and ultraviolet waves, only parts of the visible, infrared and microwave wavelengths are commonly used in remote sensing (Elachi and van Zyl, 2006).

Of the sunlight that is used by remote sensing, a considerable portion never reaches the sensing instrument. Sunlight is first affected by the Earth's atmosphere (light waves are scattered or absorbed at various wavelengths depending on atmospheric conditions), distorting the remotely sensed image of the Earth's surface (Figure 20.2). To help understand this concept, consider the colour of the sky. During the day, when the sun is directly overhead and sunlight only has to travel a short vertical path through the atmosphere, short wavelength (blue) light is scattered resulting in a blue sky. In the evening, when the sun is near the horizon and sunlight has to travel a longer oblique route through the atmosphere, longer (red) wavelengths are scattered causing a red sunset.

Where sunlight does pass through the atmosphere and reach the Earth's surface, a certain proportion of the incoming radiation is absorbed or scattered, rather than reflected, depending on the surface feature (Figure 20.2). In fact, this is the basis for discriminating between features using remotely sensed images and is, therefore, a very useful property. For instance, water features often appear dark on remotely sensed imagery because water absorbs almost all of the sunlight it receives, while sand tends to appear light because most sunlight is reflected by this surface type (see Figure 20.1).

Finally, sunlight is not the only source of electromagnetic radiation used in remote sensing. Certain 'active' remote sensors emit electromagnetic radiation and record the amount of radiation scattered back from the Earth's surface. The most common of these are 'radio detection and ranging' (radar) devices which emit and receive microwave radiation (Woodhouse, 2005; radars for tracking ships, aircraft and so on operate in the same way), and 'light detection and ranging' (LiDAR) devices which use laser technology (Wynne, 2006). In contrast, 'passive' remote sensors, by far the majority of remote sensing instruments, simply receive electromagnetic radiation from an external source (primarily reflected sunlight, as shown in Figure 20.2).

REMOTE-SENSING PLATFORMS AND INSTRUMENTS

Any remote-sensing system comprises two main components: the 'platform' on which the sensing instrument is installed, and the 'sensing instrument' itself. There are three main categories of remote sensing platforms: satellites (which also include manned space shuttles), aircraft (including aeroplanes, helicopters and even balloons) and ground-based devices (e.g. handheld structures and hydraulic platforms). Different platforms offer different advantages. Satellites orbit at a relatively high altitude, enabling wide, synoptic views, and they can also offer a continuous data supply. However, satellite-sensor images can be affected by, for instance, cloud cover (see Figure 20.1). Aircraft operate at a relatively low altitude, providing fine spatial detail, but image supply is irregular, depending on when the aircraft flies. Ground-based remote sensing is performed for various reasons, but commonly it is used to measure the spectral properties of surface features or to calibrate airborne or satellite-based imagery to correct for atmospheric distortion.

A wide variety of remote-sensing instruments exist. As mentioned above, early remote sensing involved traditional analogue cameras, often acquiring large-format photographs. Although now largely superseded by digital technologies, analogue photographs do offer certain benefits, including fine spatial detail and a faithful representation of the Earth's surface, unlike that provided by pixelated digital images. Also, since large archives of aerial photographs exist, re-used widely in historical and monitoring projects, traditional aerial photography will continue to be used in remote sensing for some time to come.

Several types of digital sensor are used in remote sensing. Panchromatic sensors capture reflected light at one part of the electromagnetic spectrum (known as a spectral waveband) and use this to create a single image, often shown in black and white (or more correctly, 'grey scale'). A little more advanced, and quite common in remote sensing, multispectral sensors capture light at several spectral wavebands and present these as separate image layers or bands (Figure 20.3). These multiple bands, combined to create colour-image displays, tend to offer a greater amount of discriminating information than single-band panchromatic sensors. While multispectral sensors tend to have a few spectral wavebands, hyperspectral sensors have tens or hundreds of narrow spectral wavebands. Hyperspectral image bands can offer very specific discriminating information, useful in certain circumstances, but the large volume of data is impractical and often unnecessary for routine remote-sensing analysis. Other types of digital sensor include relatively simple digital cameras and videos, and active radar and LiDAR remote-sensing instruments.

IMAGE CHARACTERISTICS

Remotely sensed imagery is produced in many forms, depending on the specifications of the sensing instrument. Certain characteristics of remotely sensed data are particularly significant since these can influence how the images are used. Three main image characteristics will be discussed here: spectral properties, spatial resolution and temporal aspects.

Spectral properties

The spectral properties of remotely sensed imagery are perhaps the most significant characteristic in terms of recognizing features from images (spectral properties refer to the part or parts of the electromagnetic spectrum used to generate an

Figure 20.3   Multispectral image of Skukuza, Kruger National Park, South Africa
Source: Original data copyright: 2002 DigitalGlobe

image). The important concept to grasp is that features reflect electromagnetic radiation in different ways at different parts of the spectrum. Therefore the representation of features in remotely sensed imagery varies according to which part of the spectrum (spectral waveband) is used. For instance, vegetation may appear dark in an image generated using a 'visible light' waveband, but bright using a 'near infrared' waveband (see Figure 20.3).

As mentioned above, most remote-sensing devices use either a single spectral waveband (panchromatic images), several wavebands (multispectral images) or many wavebands (hyperspectral images). The benefit of increasing the number of wavebands is that, since features appear different at different parts of the spectrum, an increasing amount of information is provided to help distinguish between features. However, each waveband produces a layer of data, so increasing the number of wavebands also increases data volume and computer processing time.

The defining feature of panchromatic images is their grey-scale representation. Much aerial photography is grey scale and some digital sensors generate panchromatic images. Although useful for basic remote-sensing analysis, panchromatic images have a low-information content compared with other types of remotely sensed image. Multispectral images, on the other hand, have a relatively high information content and can be represented in numerous colour combinations (e.g. Figure 20.3). Essentially, multispectral images are multi-layer data sets, where each layer corresponds to a specific spectral waveband and all layers share a common grid of pixels. Typically, multispectral remote sensing devices use between three and seven spectral wavebands, from the visible and infrared parts of the electromagnetic spectrum. For instance, the Landsat-5 satellite's Thematic Mapper (TM) sensor (the longest-serving instrument in the widely used series of the National Aeronautics and Space Administration's (NASA) Landsat satellite sensors) uses three visible wavebands, one 'near' infrared waveband, two 'mid' infrared wavebands and one 'thermal' infrared waveband. (It is a common misconception that the term 'infrared' is associated singularly with heat. In fact, there are three main categories of infrared radiation, only one of which provides an indication of temperature. The other two represent different forms of electromagnetic radiation.)

Automated remote-sensing analysis can be performed simultaneously on 'all' layers of a multispectral image. However, only 'three' layers or fewer can be used for visual display, either on screen or when printed out. This is because computer-based display involves only three basic colours: red, green and blue, which are combined to create a multicolour presentation. Since any layer of a multispectral image can be displayed using any of these three colours, different representations (colour combinations) are possible for images with more than three layers. For instance, using Figure 20.3 as an example, at first glance, the right-hand image may seem to provide an 'unnatural' representation of the landscape since vegetated features appear red and impervious surfaces appear blue, rather than the more familiar green or white. In fact, this is simply the result of displaying an infrared layer with two visible layers in a combination commonly known as a 'false colour composite'. The inclusion of an infrared layer enables certain characteristics of the landscape to be highlighted. A more conventional view of the landscape is provided by the 'true colour composite' (Figure 20.3, left-hand image), but in this image it is quite hard to distinguish dark vegetated features from the rather dull background.

Despite fundamental differences between panchromatic and multispectral remote sensing, they are often used in combination. In fact, several remote-sensing devices generate coincident panchromatic and multispectral images, such as the French Satellite Pour l'Observation de la Terre (SPOT) satellite's High Resolution Geometry (HRG) sensor and the US commercial IKONOS satellite sensor.

In contrast to multispectral remote sensors that use a few broad spectral wavebands, hyperspectral instruments use tens or hundreds of very fine wavebands. The benefit of using hyperspectral devices, such as NASA's EO-1 satellite's Hyperion sensor, is that certain features can be identified very precisely according to their specific spectral properties. For instance, hyperspectral images are commonly used for identifying minerals (Chen *et al.*, 2007) or determining water quality (Giardino *et al.*, 2007).

Finally, while most passive remote sensing instruments use spectral wave*bands* as described above, active remote sensors use individual wave*lengths*. For instance, radars emit and receive electromagnetic radiation at an exact wavelength, not over a range of wavelengths (i.e. a waveband). The European Space Agency's ENVISAT satellite's Advanced Synthetic Aperture Radar (SAR) instrument, for example, uses a wavelength of 5.6 cm. The product of radar imaging is a single layer of data, similar to a grey-scale panchromatic image. As such, radar images are fairly limited in terms of spectral information.

Spatial resolution

Spatial resolution refers to the size of features discernible from remotely sensed data. Various different technical definitions exist, but spatial resolution is often used to refer simply to the 'pixel size' of remotely sensed images (Fisher, 1997). The importance of spatial resolution for image interpretation is obvious. High (or fine) spatial resolution images enable relatively small features to be identified, whereas low (or coarse) spatial resolution images can only be used to identify fairly large features (Aplin, 2006). For instance, Figure 20.4 presents a section of the River Sabie, South Africa, viewed over a range of spatial resolutions. The relatively fine (2.8 m) spatial resolution provided by QuickBird enables the identification of small features such as individual trees, roads and buildings. These features become progressively blurred as spatial resolution is coarsened to 32 m, equivalent to imagery from the international Disaster Monitoring Constellation of remote sensing satellite sensors. At a spatial resolution of 250 m, equivalent to NASA's Moderate Resolution Imaging Spectroradiometer, no discernible features can be depicted.

The spatial resolution of remotely sensed imagery varies widely, depending on the data source. At the finest scale, images with a spatial resolution of a few centimetres can be generated by low-altitude airborne instruments. In contrast, certain satellite sensors orbiting hundreds of kilometres above the Earth's surface generate images with a spatial resolution of several kilometres. For instance, the 18th National Oceanic and Atmospheric Association satellite's (NOAA-18 or NOAA-N) Advanced Very High Resolution Radiometer (AVHRR) has a spatial resolution of 1.1 km. However, other satellite instruments are able to generate relatively fine spatial resolution imagery. For instance, the panchromatic sensor on board the commercial WorldView-1 satellite acquires imagery with a spatial resolution of 0.5 m.

Associated with the spatial resolution of an image is its 'area of coverage' (the area on the Earth's surface represented by the image). Generally, the finer the spatial resolution of an image, the smaller the area covered. For instance, a fine (1 m) spatial resolution IKONOS scene covers an area of 11 km × 11 km, while a medium (30 m) spatial resolution Landsat TM scene measures 185 km × 185 km and a NOAA-18 AVHRR scene measures 2,600 km × 2,600 km.

Figure 20.4 The effect of varying spatial resolution on an image of the River Sabie, Kruger National Park, South Africa. (ASTER = Advanced Spaceborne Thermal Emission and Reflection Radiometer, DMC = Disaster Monitoring Constellation, MODIS = Moderate Resolution Imaging Spectroradiometer)

Source: Original data copyright: 2002 DigitalGlobe

Temporal aspects

Time is an important factor in much remote-sensing analysis. Many studies use images acquired at different times (multitemporal images) to investigate dynamic features on the Earth's surface (Coppin *et al.*, 2004). For instance, meteorologists require images every few hours to update weather predictions, or at a longer time-scale, agricultural bodies need images every few weeks to monitor crop health (Blaes *et al.*, 2007). The availability and nature of multitemporal images depends on the 'temporal resolution' (the frequency with which an image is acquired at a constant location) of a remote-sensing device.

Satellites have a significant advantage over aircraft when it comes to acquiring multitemporal images. Satellite sensors are in a constant orbit and therefore provide a 'continuous' data supply. In contrast, airborne remote sensing occurs sporadically (as and when a decision is made to fly an aircraft) such that data are available only on an occasional basis. Thus, while the temporal resolution of airborne-sensing instruments is highly irregular, it is routine for satellite sensors. However, that is not to suggest that all satellite sensors have the same temporal resolution. On the contrary, the temporal resolution of such instruments varies considerably, depending on each instrument's area of coverage and tilting capabilities. Generally, sensors with a wide area of coverage acquire images of a constant location more frequently than those with a narrow area of coverage. However, some sensors can tilt obliquely enabling them to target locations more frequently than sensors pointing vertically downwards. For instance, the temporal resolution of Landsat TM, which covers an area 185 km wide but has no tilting capabilities, is 16 days. In contrast, the temporal resolution of IKONOS, which only covers an area 11 km wide but can tilt to ±45°, is less than four days.

IMAGE ANALYSIS

There are many different types of remotely sensed images, and these are used for many purposes (Narayan, 1999). Consequently there are numerous methods of image analysis. While manual interpretation remains in use for analogue aerial photographs (Philipson, 1997), most image analysis is now digital, involving specialized image-processing software.

Remote sensing studies have varying image processing requirements, depending on their objectives. Common in many studies, however, is the need to pre-process the imagery prior to the main part of analysis. Pre-processing is performed to correct data distortions or standardize images. For instance, atmospheric correction is often carried out to remove atmospheric distortion such as haze from images. Geometric correction is performed to correct for geometric distortion such as that caused by curvature of the Earth, or to register an image to a known map coordinate system. The effects of geometric registration are evident on Figure 20.1. The thin black wedges along the sides of the image are characteristic of the rotation often involved in geometrically registering satellite-sensor images. This rotation is necessary to compensate for the rotation of the Earth that occurs during the (line by line) image-acquisition process.

Following pre-processing, it is often useful to enhance the presentation of an image to aid interpretation. Commonly, contrast stretching is performed to increase contrast throughout an image, making features more distinct and identifiable (Figure 20.5, top-left image). In fact, many modern image processing systems automatically contrast stretch images during display. While contrast stretching generally involves altering image display without affecting pixel values, more advanced enhancement techniques can involve some operation altering pixel values to generate useful information. Such techniques are too numerous to cover in detail here, so three example procedures will be described: image filtering, image arithmetic and classification.

Figure 20.5   Image-processing operations performed on an image of Skukuza, Kruger National Park, South Africa. Top left, contrast stretched image; top right, low pass filtered (smoothed) image; bottom left, NDVI (Normalized Difference Vegetation Index) image; bottom right, classified image (green = vegetation, brown = bare soil, blue = water, white = bright surfaces)

Source: Original data copyright: 2002 DigitalGlobe

For further information on image processing, see Richards and Jia (2007) in the first instance, and Mather (2004) for more advanced instruction.

Image filtering is used to selectively stress or suppress information within an image. For instance, by passing an edge detection filter over an image, a new image is created that emphasizes linear features (e.g. roads or field boundaries). Alternatively, where general patterns are sought rather than fine detail, a smoothing filter (also known as a low pass filter) may be used to blur sharp features. For instance, Figure 20.5 compares the original image of Skukuza, South Africa with a low pass filtered image of the same area. Broad patterns, such as the differentiation between residential areas (characterized by irrigated vegetation, displayed red in this false colour composite image) and semi-natural savanna (areas of non-irrigated vegetation and bare soil) appear dark grey.

Image arithmetic is used to generate new, useful images from one or more multispectral and/or multitemporal image layers. Essentially, simple arithmetic operations such as addition, subtraction and so on are performed by associating corresponding pixels in different data layers. Various types of image arithmetic are possible, but one common example is image differencing (subtraction) for change detection. By subtracting an image acquired at a certain time from another image acquired at a different time, areas of change are highlighted. Other more complex arithmetic procedures can also be performed. One fairly common example is the normalized difference vegetation index (NDVI), which highlights areas of healthy vegetation using an arithmetic combination of the near infrared and red wavebands of a multispectral image:

$$NDVI = \frac{NIR - R}{NIR + R}$$

where NIR = near infrared waveband and R = red waveband (Oindo *et al.*, 2002). The NDVI is particularly useful for this purpose because healthy vegetation has a predictable spectral response at the infrared and red parts of the electromagnetic spectrum. To illustrate this, the near infrared and red wavebands of a QuickBird image of Skukuza, South Africa were used to generate an NDVI image (Figure 20.5). NDVI images are typically displayed in grey scale, where vegetation increases with brightness. The white areas in the image represent irrigated gardens, while the circular cricket pitch and golf fairway stripes towards the left side of the image are also obvious.

Finally, land-cover classification involves the segmentation of images into meaningful categories. To achieve this, pixels with similar spectral properties are grouped, forming general classes (Tso and Mather, 2001). Classification has many uses, including generating environmental inventories and identifying areas suitable for development. To illustrate the process, the original multispectral QuickBird image of Skukuza, South Africa was classified into four general classes: vegetation (green), bare soil (brown), water (blue) and bright features such as sand and concrete (white; Figure 20.5).

CONCLUSION

The field of remote sensing has undergone significant development since the advent of aerial photography 150 years ago, and even since the emergence of satellite reconnaissance in the 1960s and 1970s. Thirty years or so ago, remotely sensed data were scarce and lacked detail (both spectral and spatial), and image analysis was limited by technological constraints. Today, in contrast, an extensive range of detailed data and advanced analytical techniques are available for geographical study.

Remotely sensed imagery is used in a very wide range of application areas, preventing full discussion here. These applications include, but are by no means limited to, agriculture, cartography, environmental studies, forestry, geology, hydrology, meteorology, military, oceanography, planning and urban studies. Various remote-sensing textbooks provide strong coverage of applications, including Drury (1998) and Narayan (1999), and more recently, Cracknell and Hayes (2007) and Lillesand *et al.* (2007).

The field of remote sensing continues to benefit from technological developments. Recently, for instance, the launch of WorldView-1 has provided half-metre spatial resolution imagery from space for the first time. Meanwhile, image suppliers are becoming more sensitive to customer needs (perhaps due to the strong commercial market) and provide targeted 'information' products (e.g. orthorectified images, NDVI maps), instead of uncalibrated imagery that often proved difficult for users to process. Image supply and storage has been revolutionized by new media and, in particular, the internet. In addition to simplifying the process of viewing and ordering imagery, the internet has brought remote sensing to a much wider audience through ventures such as Google Earth (http://earth.google.com/).

One negative development in remote sensing is uncertainty over the continuation of medium spatial resolution image acquisition. Notably, NASA's Landsat series, consistent providers of 30 m spatial resolution multispectral imagery for the past two to three decades, has experienced problems over the last few years, and future data acquisition is uncertain. Nonetheless, in the current technological climate, there remains an insatiable demand for data. It has been recognized widely that the 'spatial' domain provides an excellent means of integrating data, due at least in part to the growing popularity of GIS. Consequently, by providing an extensive, continuous and accurate supply of spatial data, the field of remote sensing should continue to prosper.

Summary

- In the context of geographical study, remote sensing refers to the collection of images of parts of the Earth's surface using specialized instruments, commonly aerial cameras and satellite sensors.
- Remote-sensing technology may be separated into two components: the 'platform' on which the sensing instrument is installed, and the 'sensing instrument' itself.
- There are three main categories of remote-sensing platforms satellites, aircraft and ground-based devices.
- Remote-sensing instruments include cameras and digital sensors, video and recently developed 'light detection and ranging' (lidar) devices.
- Generally, remote-sensing devices use either a single spectral waveband (panchromatic images), several wavebands (multispectral images) or many wavebands (hyperspectral images).
- The spatial resolution of remotely sensed data varies widely, depending on the data source. At the finest scale, images with a spatial resolution of a few centimetres can be generated by low-altitude airborne instruments. In contrast, certain satellite sensors orbiting hundreds of kilometres above the Earth's surface generate images with a spatial resolution of several kilometres.
- Following pre-processing, it is often useful to 'enhance' the presentation of an image to aid interpretation. Commonly, 'contrast stretching' is performed to increase contrast throughout an image, making features more distinct and identifiable. Other procedures include 'image filtering', 'image arithmetic' and 'classification'.
- Computers provide the opportunity for advanced, accurate and very fast image analysis, and a wide range of 'digital image processing' (DIP) techniques have been developed. Specialized DIP systems are available for use, but many geographical information systems (GIS) also provide capabilities for DIP. Image supply and

storage has been revolutionized by new media and, in particular, the internet. In addition to simplifying the process of viewing and ordering imagery, the internet has brought remote sensing to a much wider audience through ventures such as Google Earth (http://earth.google.com/).

Further reading

- Cracknell and Hayes (2007) is a strong, if at times rather technical, introduction to remote sensing. Early chapters on theory are complemented well with a lengthy concluding chapter on applications.

- Lillesand *et al.* (2007) is the classic textbook on remote sensing, but because of its origins in the 1970s (1st edition) it is a little dated in parts. Comprehensive discussions are provided on aerial photography and early satellite sensing.

- Mather (2004) provides an excellent, thorough overview of digital image processing, although it may be rather technical for complete beginners. A CD of image-processing software and exercises is provided.

- Richards and Jia (2007) is a good first introduction to image processing, and also covers much of the basics of remote sensing. Designed principally for newcomers to the field, each chapter includes problems for the reader to consider.

Note: Full details of the above can be found in the references list below.

References

Aplin, P. (2006) 'On scales and dynamics in observing the environment', *International Journal of Remote Sensing*, 27: 2123–2140.

Birk, R.J., Stanley, T., Snyder, G.I., Hennig, T.A., Fladeland, M.M. and Policelli, F. (2003) 'Government programs for research and operational uses of commercial remote sensing data', *Remote Sensing of Environment*, 88: 3–16.

Blaes, X., Holecz, F., Van Leeuwen, H.J.C. and Defourny, P. (2007) 'Regional crop monitoring and discrimination based on simulated ENVISAT ASAR wide swath mode images', *International Journal of Remote Sensing*, 28: 371–393.

Campbell, J.B. (2007) *Introduction to Remote Sensing* (4th edn). New York: Guilford Publications.

Chen, X.F., Warner, T.A. and Campagna, D.J. (2007) 'Integrating visible, near-infrared and short-wave infrared hyperspectral and multispectral thermal imagery for geological mapping at Cuprite, Nevada', *Remote Sensing of Environment*, 110: 344–356.

Cracknell, A.P. and Hayes, L. (2007) *Introduction to Remote Sensing* (2nd edn). Boca Raton: CRC Press.

Coppin, P., Jonckheere, I., Nackaerts, K. and Muys, B. (2004) 'Digital change detection methods in ecosystem monitoring: a review', *International Journal of Remote Sensing*, 25: 1565–1596.

Drury, S.A. (1998) *Images of the Earth: a Guide to Remote Sensing* (2nd edn). New York: Oxford University Press.

Elachi, C. and van Zyl, J. (2006) *Introduction to the Physics and Techniques of Remote Sensing* (2nd edn). Hoboken: John Wiley & Sons.

Fisher, P. (1997) 'The pixel: a snare and a delusion', *International Journal of Remote Sensing*, 18: 679–685.

Giardino, C., Brando, V.E., Dekker, A.G., Strombeck, N. and Candiani, G. (2007) 'Assessment of water quality in Lake Garda (Italy) using Hyperion', *Remote Sensing of Environment*, 109: 183–195.

Lillesand, T.M., Kiefer, R.W. and Chipman, J.W. (2007) *Remote Sensing and Image Interpretation* (6th edn). New York: John Wiley & Sons.

Mather, P.M. (2004) *Computer Processing of Remotely-Sensed Images* (3rd edn). Chichester: John Wiley & Sons.

Narayan, L.R.A. (1999) *Remote Sensing and its Applications*. Hyderabad: Universities Press.

Oindo, B.O. and Skidmore, A.K. (2002) 'Interannual variability of NDVI and species richness in Kenya', *International Journal of Remote Sensing*, 23: 285–298.

Philipson, W.R. (ed.) (1997) *Manual of Photographic Interpretation*. Bethesda: American Society for Photogrammetry and Remote Sensing.

Rees, G. (1999) *The Remote Sensing Data Book*. Cambridge: Cambridge University Press.

Richards, J.A. and Jia, X. (2007) *Remote Sensing Digital Image Analysis: An Introduction* (4th edn). Berlin: Springer-Verlag.

Tso, B.C.K. and Mather, P.M. (2001) *Classification Methods for Remotely Sensed Data*. New York: Taylor & Francis.

Woodhouse, I. (2005) *Introduction to Microwave Remote Sensing*. London: Taylor & Francis.

Wynne, R.H. (2006) 'Lidar remote sensing of forest resources at the scale of management', *Photogrammetric Engineering and Remote Sensing*, 72: 1310–1314.

Representing and Interpreting Geographical Data

21 Data Handling and Representation

Richard Field

"What is to be sought in designs for the display of information is the clear portrayal of complexity. Not the complication of the simple; rather the task of the designer is to give visual access to the subtle and the difficult − that is, the revelation of the complex."

Tufte (1983: 191)

Synopsis

Both to understand the complex 'real world' and to communicate research findings, the ability to work with and present numerical data is essential. The enormous range of techniques available to geographers gives great scope for flexibility and power, but also for misunderstanding and deception. Unbiased handling of data and honest graphical and numerical presentation of findings should be an integral part of quantitative study. This chapter focuses on the manipulation and presentation of data so as to aid interpretation. It first considers the process of abstraction in data collection, interpretation and presentation and then considers various means of presenting data (tables, graphs and maps). The organization of data is then reviewed, followed by a discussion of data types and distributions. Finally, the chapter considers the explanation and interpretation of bivariate and multivariate graphical plots.

The chapter is organized into the following sections:

- Introduction
- Abstraction
- Types of data
- Presentation of data
- Examining the distribution of data
- Examining how one variable affects another
- Explanation and interpretation
- Conclusion

INTRODUCTION

Ever since people started thinking about the world, they have needed ways of coping with its inherent complexity. Early scientists relied primarily on logic: simplifying the world into a series of principles. Leonardo Da Vinci (1452–1519), among others, was instrumental in shifting emphasis towards data collection and experiment. A period followed in which many leading figures, including some of the great 'natural philosophers' such as Lyell and Darwin, meticulously collected large amounts of information from observation and experiment. However, during much of this period, there was no generally accepted, objective way of drawing inferences from data. Instead, the subjective judgement of experts was relied upon, and this could vary considerably from expert to expert.

Tables were used quite routinely for much of this time, but were typically very lengthy: good for presenting detail but of little use as summaries. It seems that graphs did not occur to the Greeks or the Romans, nor to the likes of Newton and Leibnitz. Graphs were not invented until the great work of Descartes, who set up the Cartesian system in his book *La Géométrie*, published in 1637 (Spence and Lewandowsky, 1990). After that, the use of graphical means to present data did not really advance until the seminal work by Playfair (1786, 1801), who invented many of the popular statistical graphs still in use today, including the histogram, the line graph and the pie chart (Spence and Lewandowsky, 1990). These methods became more and more popular during the nineteenth century, and were linked to the rise of statistical thinking at the same time (Porter, 1986). Much important work was done in medicine. For instance, Dr John Snow, a London anaesthetist and Queen Victoria's obstetrician, used inductive methods to work out the cause of cholera. His classic piece of work, 'On the mode of communication of cholera', was enlarged from a pamphlet published in 1849 to a book published in 1854. Florence Nightingale, too, systematically gathered data on illness and fatalities and presented them graphically in order to make some sense of them. Nightingale was meticulous in her attention to detail and careful to ensure that her premises were backed up statistically (Kopf, 1916) – that is, to grasp the essential, emergent trends from a mass of individual cases.

Probably the greatest period of development of statistical methods was towards the end of the nineteenth century and early in the twentieth century. Probability theory (developed to a considerable extent because gambling-loving French nobles had employed great mathematicians like Pascal to give them an edge) was taken and applied to data. The work of Jevons, Pearson and Fisher was instrumental during this period. During the twentieth century, the field of statistics was becoming increasingly theoretical and inaccessible to non-statisticians, until the considerable input of the great American statistician, John Tukey. Tukey argued strongly that data should be explored and visualized as much as possible, as well as being subjected to formal statistical analysis. He pointed out the fuzzy nature of real-world data, which do not usually comply with theoretical ideals. He also passionately believed that statistical methods should be more accessible to ordinary people, and many of the techniques he developed go a long way towards achieving this goal. More recently, other major developments have taken place in statistical theory, such as the rise of Bayesian statistics, which takes a fundamentally different approach to data analysis.

ABSTRACTION

In trying to make sense of 'real-world' complexity, we go through several processes of abstraction (Figure 21.1). The first is the dataset itself. In most cases in geography,

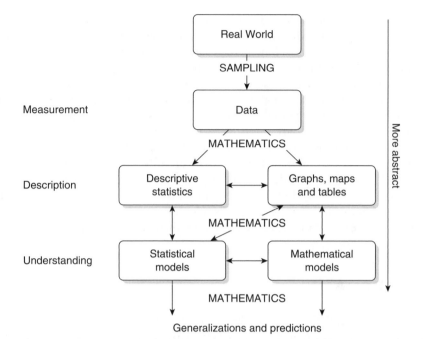

Figure 21.1   Abstraction of 'real-world' complexity in a quantitative study. The diagram is idealized and simplified, but illustrates the point that as you move down through the levels of abstraction, complexity is reduced. Detail (information) is lost, but understanding is enhanced – or at least, that is the aim. Note that mathematics is the language that translates between the different levels of abstraction

data are collected via a sampling process (see Chapter 17) and are therefore not the entire population of relevant units. These data consist of measurements that are subject to error and that cannot retain all possible information about the units sampled, no matter how careful and comprehensive the data collection. Thus, the immense complexity of the 'real world' is reduced to a carefully chosen sample for which certain attributes are measured.

In order to describe the information obtained in a more meaningful way than a large set of numbers, the raw data can be plotted on graphs or maps and summarized via descriptive statistics (e.g. averages and measures of variability), often arranged into tables. These methods represent further levels of abstraction because they summarize, but also lose, information. Graphs and maps (see Chapter 22) are powerful tools in that they can simultaneously allow both an instant impression of the data and the display of detail. However, for all except the simplest datasets, each graph or map depicts only a subset of the variables measured. In the case of maps, Geographical Information Systems (Chapter 25) provide powerful tools for displaying and analysing several variables at the same time. Descriptive statistics allow an instant impression of the data but, again, do not convey all the detail present in the dataset.

The results of statistical analyses represent a higher level of abstraction still: the estimation of parameters within the 'real-world' population, based on the data collected, and the evaluation of hypotheses that aim to make generalizations about the 'real world'. Mathematical simulation-based models occupy a similar level of abstraction because they incorporate estimated parameters in attempting to simulate real-world processes (see Chapter 19).

Graphical (including cartographic) and numerical means of presenting data can therefore sit in between the 'real world' (which is typically too complex to understand immediately) and the highly abstracted results of mathematical models and statistical analyses (which are typically of little use unless interpreted in the light of the data). Such techniques are of fundamental importance to most geographical enquiry, and it is essential that they are used carefully and competently. As seen in Figure 21.1, mathematics is the 'vocabulary' and technique for translating between the levels of abstraction; thus numeracy and a basic grasp of mathematical principles are essential skills for geographers. However, one can perform high-quality quantitative research without being a good mathematician (see Chapter 23).

TYPES OF DATA

Most standard statistical textbooks have chapters explaining the differences between types of data (e.g. Ebdon, 1985). Various labels are attached to these types but the primary distinction is between categorical and continuous data. Continuous data are those that are measured on a continuous scale and so may have any value, including fractions. The most important thing about continuous data is the relational element: a meaningful progression as the numbers increase. For example, a length of 5 mm is twice the length of 2.5 mm and 6 mm less than 11 mm. This differs fundamentally from categorical data in which each value signifies membership of a particular group. An example should help to illustrate this point. Table 21.1 is a dataframe. Dataframes are essentially spreadsheets consisting of rows and columns and are the best way of organizing data in most cases – in fact, almost all statistical packages on computers require that the data are in this format. Each row represents a 'case' – i.e. a unit of sampling. In this example, each case is an individual person but it might also be a quadrat from a vegetation survey or a pebble on a shingle beach. Each entry in the row represents an attribute of that unit: an individual person's age, sex, income or occupation. The complete set of values for an attribute (one value for each case) makes up one column in the dataframe. This is called a 'variable', because the values vary between cases. Two of the variables in this dataframe are continuous (age and income) and the other two categorical (sex and occupation).

Notice how there are no unique values for the categorical variables: members of the same group have the same value (e.g. 'lawyer'), by definition. In contrast, most or all of the values for a continuous variable are unique (depending on the level of precision used for recording the data). It is important to realize that categorical variables such as sex or occupation *are* variables – there is a value assigned to each unit – they just differ in data type from continuous variables. The value assigned can be text (e.g. 'female') or numeric (e.g. a value of 1 used as code for 'female'). Statistical packages on computers tend to use the numeric

Table 21.1　An example of a dataframe

This dataframe consists of 18 units (often called 'cases'; in this example they are individual people) and four variables (sex, age, occupation and income). 'Name' is not a true variable, but is a list of case identifiers. Notice how the cases are rows and the variables are columns. This is the standard way of organizing data, and is the format used by almost all statistical packages. Two of the variables are continuous and two categorical. The data in the table are fictional

Name	Sex	Age (yrs)	Occupation	Income (£)
David	M	36.0	Doctor	55741
Justin	M	22.7	Social worker	19569
Lindsay	F	46.0	Doctor	42183
Vicki	F	60.3	Farmer	28293
Madeleine	F	59.6	Doctor	49658
Mark	M	63.0	Social worker	22485
Shelley	F	18.7	Lawyer	48627
Lizzie	F	37.1	Social worker	24630
Jessica	F	58.6	Lawyer	45268
Philip	M	24.5	Farmer	39228
Charles	M	29.5	Farmer	44165
Steve	M	20.1	Doctor	55182
Katherine	F	19.5	Lawyer	40677
Nicola	F	25.7	Farmer	40607
Charlotte	F	28.3	Lawyer	61191
Nicole	F	18.8	Doctor	50598
Nicholas	M	31.4	Farmer	44048
Daniel	M	34.1	Social worker	15878

form (internally at least), but some (such as PASW/SPSS, Chapter 26) allow both numbers and text labels to be used simultaneously. Note that, if a categorical variable is recorded as numbers, these numbers are not relational like they are in a continuous variable: 'female' could be recorded as 1 and 'male' as 2, but it makes no sense to say that 'male' is one more than (or indeed twice) 'female'!

Continuous data can be reduced to groups, e.g. fertilizer application recorded as 'low', 'medium' or 'high'. This loses information and so is not generally recommended. Further, it is important to realize that, while exact measurements can easily be converted to this form, the reverse is not true: data collected in this form cannot be converted into exact measurements. What exact amount, in grams, corresponds to a fertilizer application recorded as 'high'? Many of the most powerful types of statistical analysis, such as regression, require data in continuous form and so, as a general rule, it is better to record exact measurements when collecting data. There are three main exceptions to this rule. The first is trivial: some variables can only sensibly be measured as categories. In Table 21.1, sex and occupation are examples of this. Second, there are times when the apparent precision of an exact measurement does not reflect the accuracy of that measurement. A common example is the recording of people's ages in questionnaire surveys: in many situations, it has been shown that people have a greater propensity to lie about their age when asked to give a figure than when asked which age category they belong to. In such cases, accuracy can actually be improved by the loss of apparent precision (the gain is in the reduced bias). The third major exception to the rule is when a better overall impression of the sampled population is likely to be gained by collecting a lot of categorical, rather than a few continuous data. For example, recording exact percentage vegetation cover in quadrats is immensely time consuming and difficult. One alternative is to record a rough estimate, done quickly by eye. While this gives an 'exact' value its accuracy is very dubious. Another alternative is to use a categorical scale, such as recording the vegetation as '<1%', '1–4%', '4–10%', '10–25%', etc. (Many such methods exist, two of the best known being the Domin and the Braun-Blanquet scales.) This kind of method allows the rapid collection of data without producing values that are misleading in their apparent precision.

Other data types include discrete data (or integers), which are like continuous data but cannot include fractional numbers, and categories that can be ordered (as in vegetation cover measured using the Domin scale) – so that the difference between a '1' and a '2' is more meaningful than for purely categorical data. Some data are bounded: values cannot lie above a certain number and/or below another. A good example is percentage data (for a fuller discussion of types of data, see standard statistical textbooks, e.g. Ebdon, 1985).

It is a good idea, whenever you are thinking about data, to think first about what type of data they are. This helps understanding of the study or dataset concerned and of the context of any manipulation of the data that is performed. Knowing what type of data you are dealing with is an essential prerequisite to deciding what statistical analysis is suitable. Always *think* about the numbers in a study, both when reading reports of other people's work and when conducting your own quantitative work. How were the numbers measured or arrived at? What do they mean? Are they what you expected? Do the results of an abstraction technique (e.g. a statistical analysis) make sense, or could there have been an error somewhere in the data manipulation process? Such thought processes also help develop a feeling of being comfortable with numbers, which itself is extremely useful both for spotting errors and for spotting points of interest in the results.

PRESENTATION OF DATA

Most quantitative geographical studies involve a lot of data – too much to make much sense of simply by looking at the raw data. In almost all cases, it is a good idea to start by plotting the data graphically and cartographically and using simple descriptive statistics (see Chapters 23 and 26), often arranged in tables. These techniques help to identify underlying structure or pattern, and allow researchers to develop an understanding of their data. Once the data have been analysed and/ or modelled, and we have an idea what the data are actually telling us, we need ways of presenting the results in an effective and efficient way. Because graphs, maps and tables also help in interpreting the results from statistical analysis or mathematical modelling, they are usually used for this purpose. Thus they are used both to explore data and to present results. Before examining these tools in more detail, it is worth illustrating how important our choice of presentation

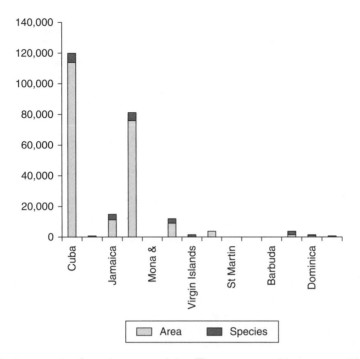

Figure 21.2   An example of poorly presented data. There are many things wrong with this chart. Some of the more important ones are: 1. Axis labelling – neither axis is labelled; not all of the islands are named on the x-axis and no units are given for the y-axis. 2. It is not clear, but the point of interest here is the way in which the number of plant species increases as island area increases (a classic relationship in biogeography) for islands of the Caribbean. This type of graph is not appropriate at all for showing such a relationship. 3. The scale of the y-axis is such that most of the data are lost in bars too small to discern. The overall result is that we struggle to understand what message we are being given

Source: Data taken from Frodin (2001)

technique can be. At its most basic, the way that data are presented makes the difference between results being meaningful or not.

Look at Figure 21.2. What does it show? It is very unclear, for several reasons – not least that the poor labelling makes it hard to work out what the point of the graph is in the first place. The most serious problem is that the chart type is completely inappropriate for the type of relationship being shown. The graph plots data on numbers of plant species and areas of islands. The relationship in question is a correlation – what we need to see is what happens to the number of plant species as island area increases. Because the numeric scale is the same for both the variables plotted and the bars are stacked, it would be very hard to discern this relationship from the chart, even if all the bars were big enough to see. What we need is a bivariate scatterplot, otherwise known as an x–y scatter graph. Figure 21.3 is such a plot, and displays the same relationship far more clearly. The axes are properly labelled, and the log–log scale allows us to see the variation in the data much more clearly. What emerges is

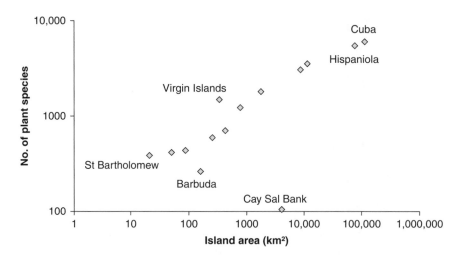

Figure 21.3   Better presentation of the data in Figure 21.2. Here, the same data are shown as in Figure 21.2 but the message is much clearer. The number of species tends to increase with island area in quite a predictable manner, though there are exceptions. These exceptions, which are labelled (along with the largest and smallest islands), can be examined in more detail to see if an explanation is apparent. The type of graph is appropriate for the relationship being shown, and the log scale allows the variation within the data to be seen much more clearly than in Figure 21.2

a straight-line relationship (on these log–log axes), with a few notable exceptions – islands that deviate from the general trend. These islands are labelled (along with those at the extremes of the plot) as they are of particular interest. The immediate question emerging is: why there are such obvious exceptions to an otherwise strong relationship? I return to this in the next section, but the important point here is that both the relationship and the exceptions are immediately apparent in Figure 21.3 but are completely obscured in Figure 21.2. Some basic, guiding principles for presenting data are explored below.

Maps

Although a type of graphical display, maps deserve separate consideration in a geographical text! Most geographical data are spatial in some way or other. Typically, each datum is associated with two spatial co-ordinates, such as latitude and longitude, often obtained from global positioning systems. This spatial information can be very important for understanding the data, the relationships between the variables and many apparent outliers. Figure 21.4a shows an example: the species–area relationship for reptiles and amphibians (herptiles) in some Caribbean islands. There is a remarkably strong relationship between species richness and island area for all islands except Trinidad. The obvious question is: why does Trinidad have so many more herptile species than the other Caribbean islands shown? Various answers are suggested by the map (Figure 21.4b), including the most likely one. Trinidad is much

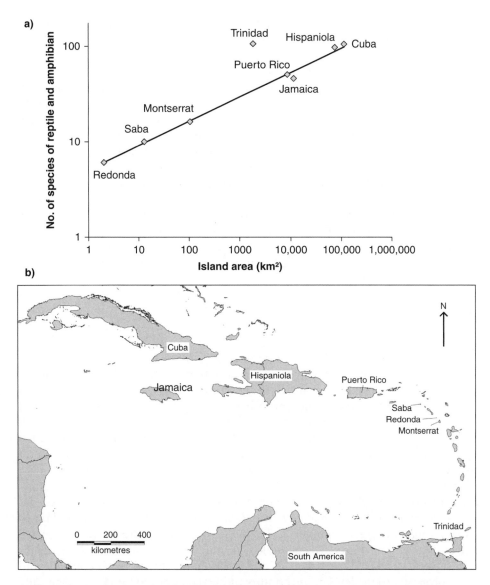

Figure 21.4   Species–area data for West Indian herptiles: (a) Scatterplot of the species–area relationship of herptiles (reptiles and amphibians) for some Caribbean islands. The fit-line shown relates to all islands except Trinidad, which appears to have far more herptile species than expected for its size. (b) Map showing the locations of the islands in the dataset. The map immediately suggests two possible explanations for the relatively high herptile species richness of Trinidad. The first is that it is further south than the other islands, and therefore nearer to the equator: it is a well-known trend for species richness to increase towards the equator. The second is that Trinidad is very close to the South American mainland, while the other islands are not

Sources: Data taken from MacArthur and Wilson (1967); base map taken from MapInfo Professional version 7.0 in-built map nearer to the South American mainland than the other islands are (and was probably joined to the mainland when sea levels were lower). South America, being very large, has a very large number of herptile species, and acts as a source of immigrants to the Caribbean islands. However, it is difficult for terrestrial organisms to disperse long distances over sea. Many amphibians in particular are poor at dispersing over salt water because they have semi-permeable skin and inundation in salt water leads to rapid dehydration. So it is very likely that Trinidad has more species than expected for its area because it has received and continues to receive more immigrants from South America than the other islands in the dataset. The spatial information is vital to this explanation, which is very obvious from looking at the map.

Maps are also extremely useful for displaying results from statistical analysis or mathematical modelling of geographical data. For example, model predictions can be mapped. Residuals from statistical models (the errors: the difference between modelled values and actual values) should routinely be mapped as part of checking their assumptions and adequacy, and doing so often leads to improvements to the models. The relationship between mapping and graphicacy is explored in more detail in Chapter 22.

Tables

Tables are typically instruments for summarizing large amounts of information, and when used as such should present information in a way which is as concise as possible, while being accurate, honest and clear. There are also circumstances (e.g. tables of raw data in the appendices of dissertations or research project reports) where very long, detailed tables are required. These are briefly considered, before discussing summary tables in more detail. The guiding principle advocated is to consider the needs of the reader as paramount. Think what the purpose of the table is and try to ensure that all attributes of the table help that purpose.

Long, detailed tables such as tables of raw data are usually intended to allow the reader to access the detail, if required. This could be for the purpose of checking work, looking for anomalies in the data, or perhaps to use the raw data to work on something else completely. The table should therefore be well organized and as clear and uncluttered as possible. The number of decimal places (or significant figures) displayed for values in any one column should be (i) constant and (ii) appropriate to the level of uncertainty of the data (see Taylor (1982, especially Chapter 2) for a good explanation of uncertainty and how to report it). Rows and columns should be properly labelled and explained. When you have worked on something for some time, you become very familiar with it and it is extremely easy to take things for granted. Another reader is likely to be much less familiar with the data than you are, and so needs help in working out what everything means. Where graphs are used, all axes should be labelled, including the units of measurement used. Maps should include a north arrow and have the scale clearly marked. Further explanation of graphs, maps and tables should be given in the legends/ captions of these illustrations. It is better to err on the side of giving too much information than to leave the reader guessing. Note that, when you are short of words, these legends/captions can convey useful information without adding to your word count, as long as they directly describe the illustrations they support. Generally, the supporting information (legends, captions, labels, etc.) for tables, graphs, maps and other illustrations should be comprehensive enough that each illustration is understandable without reference to the main text.

Summary tables are central to the presentation of data in most quantitative studies. The same principles of organization, explanation, clarity and precision apply as discussed above for detailed tables. However, in summary tables, more thought is required to select what should be included and what excluded. Also, summary tables cannot be completed until some form of data processing has been done. This might involve descriptive statistics. For example, Table 21.2 is an example of a simple form of summary table which gives the mean and standard deviation of pollutant concentrations measured for three streams on ten different days. Why give the standard deviations? Typically, pollutants are not a problem at very low concentrations but, above some threshold level, they may become toxic. Legislation often defines threshold levels, below which pollution levels are acceptable but above which they are not. Let us assume that the important threshold pollution concentration in our example is 10 units. The mean (or average) level of pollutant is clearly important, therefore. But it is not enough simply to know the mean level. In Table 21.2, if we were only to look at the means, we would conclude that the average state of each stream is within the acceptable limit for pollution. We might be concerned, though, with Stream B whose mean pollution level is only just acceptable. Streams A and C would seem to be well within acceptable limits. These would be incorrect conclusions! Look at the raw data (given in the legend of Table 21.2). While the average pollution level of Stream B is quite high, the 10 measurements are remarkably consistent, and based on these data a pollution concentration of 10 on any given day seems really quite unlikely. In contrast, while the average pollution levels of Streams A and C are much lower, the measured values are much more variable. The highest recorded value for Stream A is 9.5, but based on the data it seems quite likely that values of 10 or more might occur reasonably often. Stream C, which has the lowest mean concentration, actually has two measured values above the critical concentration of 10, suggesting that unacceptable concentrations of the pollutant occur very often in this stream. If we work out the actual probabilities (assuming a normal distribution of the data), it turns out that the chance of finding a concentration exceeding 10 on any given day is about 7 per cent for Stream A, about 16 per cent for Stream C, but only about 0.0003 per cent for Stream B. Our conclusions obtained simply from looking at the means are therefore completely wrong.

The example just given concerns a general principle: an associated measure of either variability or uncertainty should always be given when sample statistics (such as averages) are presented. Whether you present a measure of variability or one of uncertainty depends on the context. In the example in Table 21.2, standard deviation measures variability, which is appropriate here because the important thing is whether or not the stream is likely to suffer toxic pollution levels. When statistical analyses are performed, the focus is usually on the reliability of the parameter estimate, and so the appropriate measure is one of uncertainty, such as

Table 21.2   An example of a simple summary table

Here, a set of sample means is given (pollutant concentration in arbitrary units), along with their associated standard deviations (std dev.) and sample sizes (n), for a set of three streams, each sampled on ten different days. A measure of variability or uncertainty should always accompany a sample statistic, as is demonstrated here. The fictional raw data behind these statistics are as follows. Stream A: 5.4, 3.4, 6.5, 8.6, 8.4, 9.5, 1.6, 5.5, 8.2, 3.8; Stream B: 9.1, 8.7, 9.2, 9.2, 9.2, 9.4, 8.8, 9.1, 9.1, 9.0; Stream C: 8.7, 3.1, 5.4, 3.3, 10.2, 6.0, 14.5, 4.4, 0.5, 0.0

	Stream A	**Stream B**	**Stream C**
Mean	6.1	9.1	5.6
Std dev.	2.6	0.2	4.5
n	10	10	10

a confidence interval or the standard error of the mean. Further discussion of how to measure uncertainty is beyond the scope of this chapter, but it is important to note that all parameter estimates have associated measures of uncertainty. A good place to start learning about measuring and using uncertainty is Taylor (1982) and Chapter 17 of this volume.

It is very common to use summary tables to describe the results of statistical analyses, the outputs of mathematical modelling, or other quite complex procedures. *A key point is that the same understanding of statistical procedures necessary for the analyses themselves should be used in deciding what to include in any summary table reporting their results.* When designing summary tables, always ask yourself what information is necessary and what is not. Important information that is commonly left out includes sample sizes and/or degrees of freedom, significance values ('p-values') and measures of uncertainty. When you have decided what should go in the summary table, ask, what is the most efficient way of presenting it? Table 21.3 is an example of a table summarizing statistical analysis, in this case for a dataset used in the remainder of this chapter.

Graphical display

Graphical display is an excellent way of communicating results. When done properly, it is concise, memorable, persuasive and honest. For these reasons, graphics are almost always preferable to tables in oral presentations (Ellison, 2001). However, tables are generally more useful when exact values are important, and for certain types of information summary (e.g. reporting numerous statistical models simultaneously, as in Table 21.3). Graphical displays serve three main purposes: (i) preliminary exploration of patterns within data; (ii) checking the quality and assumptions of statistical and mathematical models; and (iii) communicating results to the reader or audience. The first two of these require accurate, revealing depiction of the data, but do not require great quality of presentation. They should be quick and easy to produce, and easy to interpret. Communicating results to the target audience, on the other hand, requires high-quality graphics. Again, clarity and ease of interpretation are important, though a relatively high

Table 21.3   An example of a summary table reporting the results of statistical analyses

This table reports the results of a set of simple, bivariate statistical analyses. In each case, the analysis aims to account for variation in plant growth. The dataset concerned simulates an experiment investigating the effect of the mean temperature, rainfall, soil pH (all measured as continuous variables), fertilizer addition and light level (both measured as categorical variables: fertilizer added or not; low, medium and high light levels). This dataset is used for many of the graphs in the following sections, and it is given in full in the Appendix. In the table, each statistical analysis (model) is summarized with the following information: % of variation in the growth data accounted for by the variable (r^2); significance value for the model (p), number of data points in the analysis (n), number of degrees of freedom used by the model (df) and the parameter estimates of the model ± 95% confidence interval (CI). Continuous variables were analysed using regression, which estimates the slope and the intercept of the best-fit line. Categorical variables were analysed using an analysis of variance (ANOVA), which estimates differences between means; the intercept for the fertilizer ANOVA is the mean growth of plants with fertilizer added and the difference is this value minus the mean growth of plants without fertilizer added

					Parameter estimates	
Model	r^2	p	**n**	**df**	**Slope ± 95% CI**	**Intercept ± 95% CI**
Temperature	0.185	0.000	60	1	5.0 ± 2.7	3.5 ± 48.0
Rainfall	0.000	0.414	60	1	–	–
pH	0.000	0.550	60	1	–	–
					Difference ± 95% CI	*Intercept ± 95% CI*
Fertilizer	0.859	0.000	60	1	87.4 ± 9.2 g/yr	135.7 ± 6.5
Light	0.025	0.182	60	2	–	–

Dashes indicated parameter estimates that are not significant

level of complexity is often acceptable and necessary. For all these purposes, it is important to understand the principles that underlie graphical depiction of data.

Tukey (1977), Cleveland (1985) and Tufte (1983, 1990) provide comprehensive discussions of the principles of graphing. A good summary can be found in Ellison (2001: 38):

> The question or hypothesis guiding the experimental design also should guide the decision as to which graphics are appropriate for exploring or illustrating the dataset. Sketching a mock graph, without data points, *before* beginning the experiment usually will clarify experimental design and alternative outcomes... Often, the simplest graph, without frills, is the best. However, graphs do not have to be simple-minded... and they need not be assimilated in a single glance... Besides the aesthetic and cognitive interest they provoke, complex graphs that are information-rich can save publication costs and time in presentations.

Four principal guidelines for graphics are recognized:

- Patterns of interest should be shown without adversely affecting the integrity of the data.
- Graphics should be 'honest' – i.e. they should not distort, censor or exaggerate the data.

- It should be as easy as possible for the reader to read the data (as numbers) off the graphic.
- Figures should be efficient. That is, ink should be used only to show relevant information, and not unnecessary special effects. Try to attain a high 'data-to-ink ratio'.

Some of these principles are mutually reinforcing. In particular, efficient figures allow the patterns of interest to be shown most clearly. They also allow more space for annotations. Depending on the target audience, it may be desirable to annotate figures to point out particular features of interest; if this is done, it should again be efficient, and avoid the use of gimmicks. All these principles apply to the mapping of data just as much as to other types of graphic. Tufte (1983: 51) summarizes the main principles of graphical data display: 'Graphical excellence is that which gives the viewer the greatest number of ideas in the shortest time with the least ink in the smallest space ... And graphical excellence requires telling the truth about the data.'

Different types of graphic are, then, suitable for different types of data and should be chosen according to the type of relationship we are trying to show. Two common examples involve examining the distribution of data and exploring correlation and causation.

EXAMINING THE DISTRIBUTION OF DATA

Graphs like those shown in Figure 21.5 depict the distribution of values of a *single variable* (usually a continuous one) within a sample. They are typically used for data exploration and model checking, but there are times when such graphs need to be presented formally. (The simulated dataset used for Figures 21.5–21.12 can be found in the Appendix to this chapter.)

The histogram

The most common graph type used to examine the distribution of data is the histogram (Figure 21.5a, b). Histograms are often confused with bar charts (see below) because, to the beginner at least, they look similar. However, they are used to convey fundamentally different information. A histogram consists of categories of data along the x-axis (i.e. the horizontal axis) and the number of data points in each category on the y-axis. The categories are often referred to as 'bins', using the analogy of physically sorting the data into bins, each of which is for a different range of values. The number of data points in each category is known as the 'frequency'. In Figure 21.5a, the data for plant growth rate are distributed normally. More formally, we say that the difference between the distribution of the data and the theoretical normal distribution is not significant. To allow ease of comparison, the theoretical normal distribution curve is superimposed onto the histogram. Theoretical distributions can be represented as curves like this

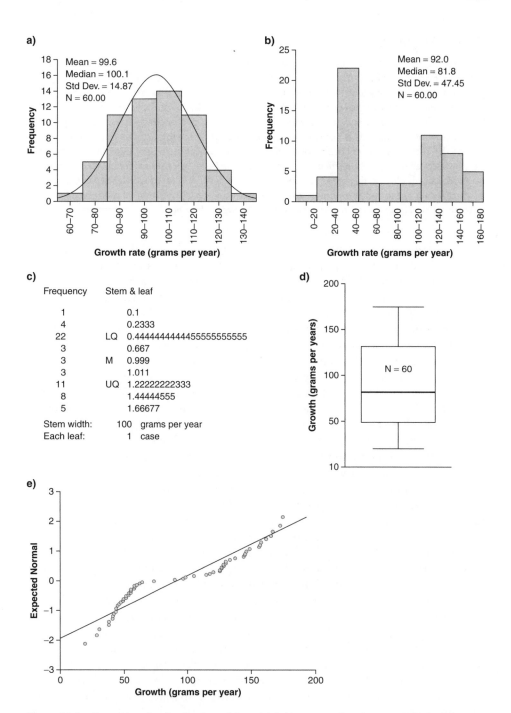

Figure 21.5 Examining the distribution of data: (a) A histogram showing an approximately normal distribution of plant growth rate data – a normal curve has been added to ease comparison between the distribution of the data and the theoretical normal distribution. Histograms suffer several disadvantages. In particular, they hide the raw data and the division into categories ('bins') is arbitrary but can affect interpretation (Ellison, 2001). Hiding the

(Figure 21.5 Continued)
raw data means that data processing cannot be done from a histogram – such as calculation of summary statistics (though the histogram can be annotated to include these, as here).
(b) A histogram showing a distinctly non-normal distribution – in this case very bimodal.
(c) A stem-and-leaf plot (shown here for the same data as (b)), which is variant on the histogram. The main difference is that the raw data are included, not hidden. However, again the division into categories is arbitrary. (d) A box-and-whisker plot, used to display the same data as (b). This is an efficient and informative way of displaying such data, but can tend to hide certain features of the data, particularly bimodality – as here. See text for further explanation. (e) A probability plot (in this case the plot type is called a Q-Q plot). Probability plots display the actual data against what would be expected from a given distribution (the diagonal line), so as to show clearly any pattern in the differences between the two.

because an exact probability can be calculated for each possible value of the variable. Such curves are called probability density functions and can be thought of as histograms with infinitesimally small bin categories. In fact, the term 'density plot' is a more general one, which includes both histograms and curves representing theoretical distributions.

A key assumption of most parametric statistical techniques is that the distribution of sample statistics can be assumed to conform to one of the known theoretical distributions. Because the exact probability function of each of these theoretical distributions is known, all sorts of probability statements can be made – *as long as the assumption holds*. Many of the more familiar statistical techniques (such as regression, ANOVA and t-tests) assume that the data conform to the normal distribution. Hence, the value of tools like the graphs shown in Figure 21.5, and also numerical statistical tests for normality (Sokal and Rohlf, 1995), which we can use to establish the validity of this key assumption. But herein lies a common misunderstanding.

Exactly what should we test for normality? The assumption of the relevant statistical techniques is that the *errors (or residuals)* are normally distributed. Before expanding on this point, some technical terms need explaining. The response variable in a statistical analysis is the one under investigation – i.e. the one whose variation we are trying to explain. It is also known as the dependent variable, y-variable or ordinate. Each variable we use to try to account for variation in it is called an 'explanatory variable', also known as an independent variable, x-variable or abscissa. The 'errors', otherwise known as the 'residuals', are the differences between the actual, observed values of the response variable and the modelled values; there is one such difference for each data point. Normal distribution of the raw data for a response variable will often mean normal distribution of residuals from a statistical model and, for this reason, many people examine the distribution of variables prior to statistical analysis. However, results from such an exercise should be treated with caution, as the errors are often normally distributed when the response variable is not.

In Figure 21.5b, the distribution of the data is clearly not normal. It is bimodal (i.e. two modes, or concentrations of values along the x-axis scale). When performing statistical modelling, many departures from normality of errors can be cured by transforming variables, but this may not be possible with bimodal errors. Fortunately, however, bimodality of distribution of the raw data of a response variable is usually due to underlying structure in the data, and if that can be accounted for the bimodality disappears. For example, the distribution of people's heights is typically bimodal because men tend to be taller than women; but when men and women are analysed separately, the heights tend to be normally distributed. This simple example neatly illustrates the folly of only examining the distribution of the raw data. As we saw on p. 331, the data in Figure 21.5b–e represent a similar scenario.

Stem-and-leaf plots

Figure 21.5c–e are different ways of depicting the distribution of the same data as Figure 21.5b. Figure 21.5c is a 'stem-and-leaf' plot. This is similar to a histogram (on its side), but has the advantage that the raw data are shown. Stem-and-leaf plots can be very useful, once you have got used to reading them. As with a histogram, the length of the 'bars' represents the number of data in the category or bin. Each of these data points is not only counted (as in a histogram) but also evaluated. The 'stems' are the first column of numbers, and represent the first figure(s) of the data values in the row. The 'leaves' are to the right of the stems, and comprise the next figure of the data value. In Figure 21.5c, the stems are in hundreds, and the leaves are tens. Thus, the lowest value in the dataset is somewhere between 10 and 20 (0 hundreds and 1 ten). Its actual value is 19.2. The highest value reads off as somewhere between 170 and 180; its actual value is 174.2. As with a histogram, labels denoting the categories containing the mean, median and/or quartiles can be added to the plot, to provide further information, as in Figure 21.5c. However, most statistical packages do not offer the opportunity to superimpose normal distribution curves over stem-and-leaf plots, as they do for histograms.

Box-and-whisker plots

Figure 21.5d is a 'box-and-whisker' plot (or just 'boxplot'). The box in the middle spans the range between the 25th and 75th percentiles (respectively, the lower and upper quartiles or 'hinges') – in other words, the box shows the inter-quartile range. The line in the middle of that box shows the value of the median (the 50th percentile). The whiskers extend a further 1.5 times the inter-quartile range, away from the box. Any points lying outside the whiskers (beyond the 'inner fence') are 'outliers': points that are considerably different from the rest of the data. These outliers are sub-divided into two types. The less-extreme outliers lie no further from the end of the whisker than 1.5 times the inter-quartile range (the 'outer fence'). The more-extreme outliers are any points that are further still from the box. The two types of outlier are usually distinguished by different plotting symbols. Boxplots are extremely useful, information-rich graphics and are excellent ways of summarizing entire variables or categories within variables. However, as ways of showing distribution of data, they suffer from the disadvantages of hiding data and tending to obscure bimodality (as demonstrated by Figure 21.5d).

Probability plots

Figure 21.5e is a probability plot. This is a very common way of examining how closely the distribution of a set of data conforms to a theoretical distribution. It is extremely useful for examining how well a statistical model conforms to the assumption of normality of errors. Probability plots are constructed so that the theoretical distribution plots as a straight, diagonal line, and the data are plotted as points. If the data were to conform perfectly to the theoretical distribution, all the points would lie on the line. In practice, of course, this does not happen. The key thing indicating difference from the theoretical distribution is systematic departure from the line, especially within the range in which most of the data lie (i.e. not so much at the two extremes). Some random scatter either side of the line is acceptable, but when the data clearly form a non-linear pattern (such as an S-shape, as in Figure 21.5e), it is a sure sign that the distribution is not normal – and that there is something wrong with the model, if the plot is being used to check errors for normality. Often, the pattern is caused by the influence of something not accounted for by the model, such as an influential but unmeasured variable (as here, because the 'model' being tested is the null model).

EXAMINING HOW ONE VARIABLE AFFECTS ANOTHER

Bivariate analysis

Most quantitative studies in geography are concerned with trying to infer causation: we want to know what causes the patterns we see. The simplest case is the bivariate one: analysing the effect of one variable on one other. The pattern being examined is within the response variable, and the possible cause of that pattern is measured via the explanatory variable. If both of these variables are continuous, by far the most useful type of graphic is the scatterplot (Figure 21.6). The explanatory variable goes on the x-axis and the response variable on the y-axis. Modelled relationships can be shown on scatterplots as lines, one of the simplest examples being a linear best-fit line. Scatterplots can also be used where no causation is being inferred – the focus instead being simply on whether there is any correlation between the two variables.

When one variable is continuous and the other categorical, scatterplots are not particularly helpful as they obscure many of the data (Figure 21.7a). If both variables are categorical, then there is usually little point in attempting to plot them against each other – devices such as tables of frequencies of each category combination tend to be more useful. Figure 21.7 shows perhaps the most common type of scenario: where the response variable is continuous and the explanatory variable categorical. Categorical explanatory variables are often called 'factors'.

There are many different ways of plotting factors. Scatterplots can be adapted to cope with factors by moving data points left and right slightly by adding random x-axis variation (known as 'jitter'). This has been done in Figure 21.7b, in which the data are still clearly categorized, but the points are much less obscured

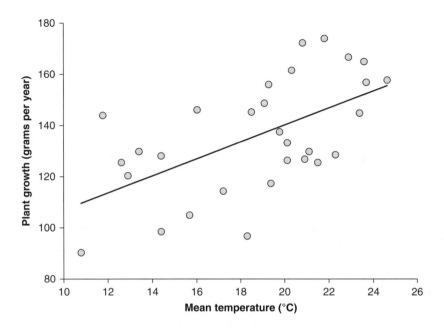

Figure 21.6   Scatterplot. This is one of the simplest and most useful of bivariate plots and one of the most commonly used. Values of one variable are spread along the x-axis, and those of another are spread along the y-axis. Both variables are usually continuous. The convention is that the explanatory variable is plotted as the x-axis, and the response variable as the y-axis (in cases where any sort of causality is inferred). A best-fit line from a simple linear regression is shown. This plot only shows data for plants to which fertilizer was added

than in Figure 21.7a. Another way of displaying factors, which tends to be more satisfactory, is the boxplot. In Figure 21.7c, a separate boxplot is shown for each category within the factor. Bar charts can be used to display the means, with error bars added to show either the variation in the data (usually the standard deviation) or the reliability of the estimates of the means; in Figure 21.7d, the 95 per cent confidence intervals of the means are shown. Strictly speaking, the bars on the bar chart are unnecessary and could be replaced with points, but it is common practice to include them. Pie charts (Figure 21.7e) and stacked bar charts (Figure 21.7f) are more often used for percentage data, but are not recommended even for that: it is not easy to read the data from them, and it is hard to display any measures of uncertainty or variability on them. Indeed, Ellison (2001: 57) goes as far as to say: 'I can think of no cases in which a pie chart should be used.'

Multivariate analyses: accounting for multiple causes

In most cases, more than just a single influence will cause a pattern. Analyses can be multivariate in the sense that they try to account for multiple influences on a given pattern: several explanatory variables are used to try to explain patterning

Figure 21.7 Ways of plotting factors. A factor is a categorical explanatory variable. All the graphs show the same relationship, which is that between fertilizer addition and plant growth. See text for further explanation: (a) scatterplot; (b) scatterplot with jitter; (c) boxplot – see above for explanation; (d) bar chart with error bars (the error bars represent the 95 per cent confidence interval); (e) pie chart; (f) stacked bar chart

in a single response variable. An example could be the run-off rates of different streams in an area. The amount of rainfall, timing of rainfall, land-use types in the catchment and bedrock characteristics are just some of the things that are likely to affect run-off rates, and it would be unrealistic to expect any one of these variables to explain all patterning in them. The example of the growth of plants is also a case in point. Sometimes, the pattern itself can be multivariate: too complex to measure as one single response variable. Instead, several attributes can be measured, to try to gain an overall picture. For example, we could measure a whole suite of attributes that could be construed as being related to the concept of quality of life, such as average income, equality of income, amount of free time, prevalence of disease and access to services.

Numerous statistical techniques have been developed to allow analysis of multivariate patterns, including dimensionality-reducing procedures such as principal components analysis. These are beyond the scope of this chapter, but the reader is referred to statistical texts such as Pallant (2007), Scheiner and Gurevitch (2001), or the manuals of most standard statistical packages – the S-Plus Guide to Statistics (Insightful, 2005) is particularly useful (and see Chapter 26). Multiple influences on a single response variable can be analysed using extensions of the simple, bivariate techniques of ANOVA and regression. Again, it is not the purpose of this chapter to discuss these statistical methods, but some of the issues related to such analyses are relevant to this section. Two issues are especially important: (i) accounting for other influences when illustrating any given effect; and (ii) how to deal with interactions between explanatory variables.

The scatterplot matrix

When a pattern results from more than one cause, a simple bivariate plot of the response versus one of the explanatory variables will not give the true picture of the effect we are trying to show. We must try to account for variation explained by other influences before we can show the 'true' modelled relationship. An extremely useful tool for exploring higher-dimensional data is the scatterplot matrix (Figure 21.8). This comprises a bivariate scatterplot of each possible combination of two variables from the list of those under consideration. Usually, each of these plots is displayed twice – one way above the diagonal and the corresponding plot with the axes transposed below the diagonal.

Partial plots

Scatterplot matrices are very useful for exploring data and informing statistical analysis and mathematical modelling, but they do not achieve the goal of accounting for other causes when displaying the modelled effects of one influence on the response variable. To achieve this, other graphical methods are commonly used. One is the partial plot (Figure 21.9), in which the response variable is re-calculated, 'correcting for' the modelled effects of other significant explanatory variables. This is very useful for interpreting the influences of particular variables in relatively complex statistical analyses, and as such one of its main uses is in model development. It can also be an effective way of presenting the meaning of a model to a viewer, but the scale of the response variable needs careful explanation.

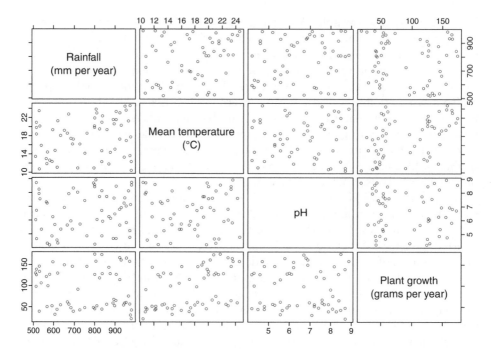

Figure 21.8   Scatterplot matrix. As its name suggests, this type of plot is a matrix of bivariate scatterplots. The variable names are listed in the diagonal, and apply to the y-axes of the plots in the row in which the name is located, and the x-axes of the plots in the column containing the variable name

3-D plots and the separation of categories

3-D plots can be used to good effect, but only allow the plotting of one extra variable in a multiple-cause situation, so their use is limited. Also, it is typically hard to read off the data from such plots, because of the difficulty of projecting a 3-D plot onto the 2-D medium of a page or a screen. Another common method is the separation of categories in graphs. This can be achieved by plotting separate graphs for the different categories of a factor (e.g. separate graphs for men and women), or different symbols on the same graph with separate best-fit lines if appropriate (Figure 21.10). Using different symbols on the same graph is often better for comparing the categories.

The use of different plotting symbols on the same graph for different categories of a factor is also a good way of illustrating an interaction between a continuous and a categorical explanatory variable – a form of statistical interaction. Statistical interactions are significant when the effect of an explanatory variable on the response variable depends on the level of another explanatory variable. Figure 21.10 shows an example of this, in which the relationship between rainfall and plant growth depends on whether or not fertilizer has been added. Where fertilizer has not been added, there is no relationship between rainfall and plant growth, but there is a positive relationship for plants given fertilizer. This interaction is statistically significant ($p = 0.002$). These results would be consistent with a situation in which nutrients, not water, represent the primary limitation on plant growth.

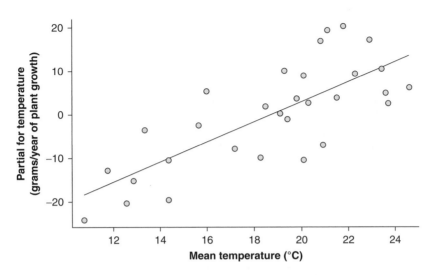

Figure 21.9   Partial plot. A partial plot shows the relationship between the two variables, after accounting for the modelled effects of other explanatory variables on the response variable. The response variable here, as before, is plant growth. This is a partial plot from a model in which the explanatory variables are rainfall, temperature and light, and in which only the plants given fertilizer are considered. The modelled effects of rainfall and light have been corrected for, and the y-axis therefore displays the change in plant growth directly related to temperature. This plot is directly comparable to Figure 21.6 – note how the best-fit line now describes the data better than in Figure 21.6

Clustered graphs and conditioning plots

Where a statistical interaction involves two or more categorical explanatory variables, clustered graphs are often the best way of displaying the interaction. Figure 21.11 is an example of a clustered boxplot, which shows an interaction between the effects of light and fertilizer on plant growth.

Finally, some statistical interactions involve two or more continuous explanatory variables. By extension of the idea of the best-fit line on a 2-D scatterplot, a two-way interaction of this sort can be displayed as a best-fit surface within a 3-D scatterplot. Again, this can suffer problems of difficulty in reading-off data and only allowing for one extra dimension. In addition, such a plot tends to imply, usually unrealistically, that all parts of the surface represent the valid range of the model; in fact, some parts of graphs like this are often data deficient. Instead, conditioning plots (or 'coplots') are often better ways of showing these interactions. Figure 21.12 is an example of a coplot. The range of values for one of the explanatory variables is split into segments (rather like the bins of a histogram), and the data for each segment are plotted as a separate 2-D scatterplot. This shows the way that the relationship between the x and y variables changes across the range of the other explanatory variable. In this case, there is no significant interaction; if there were, the fit lines would have different slopes from each other, indicating that the relationship between temperature and plant growth depends on the amount of rainfall. Advantages include the ability to add more explanatory variables, thereby being able to show higher-order interactions (e.g. three-way). Disadvantages include the fact that such plots focus

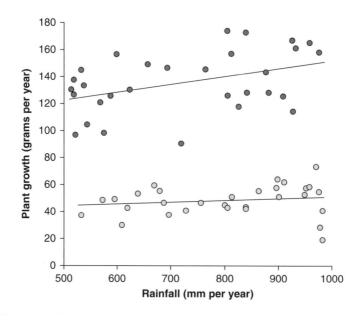

Figure 21.10   Using different symbols to account for the influence of a factor. Here, the relationship between rainfall and plant growth is shown separately for plants to which fertilizer has (closed circles) and has not (open circles) been given. If all the points were plotted with the same symbols and separate best-fit lines were not shown, it would appear that there is no relationship between rainfall and plant growth

on one of the explanatory variables, rather than being an equal depiction of all simultaneously.

Many other types of plot exist – triangular and time-series plots to name but two. The interested reader should refer to more specialist texts such as Tufte (1983) and Waltham (1994) to learn more. Note also that some of the graphical techniques discussed above are not mutually exclusive. For instance, in trying to illustrate a very complex statistical model, we could draw separate graphs (to deal with the categories of one explanatory variable), each of which is a partial plot with different plotting symbols. All of these devices are aimed at illustrating most effectively the significant effects found in the data analysis. These effects are our best guess about the cause(s) of the pattern in which we are interested. In other words, when presenting results in graphs and tables, we try to adhere to the Shaker maxim: form follows function (Ellison, 2001).

EXPLANATION AND INTERPRETATION

Figure 21.1 illustrates how our attempts to understand real-world complexity by quantitative study involve various levels of abstraction, all linked together by the 'language' of mathematics. Figure 21.1 also suggests that graphs, tables and maps lie right at the heart of quantitative treatment of data. When trying to explain

Figure 21.11   Clustered boxplot. This shows an interaction between the light and fertilizer treatments in their effect on plant growth. In other words, the amount of increase in plant growth with increasing light levels depends on whether or not fertilizer has been added. The corollary is that the degree of difference in plant growth between the two fertilizer treatments depends on the light level. The interaction is statistically significant (p = 0.0004)

patterns, present and interpret results, it makes sense to use these illustrative tools to maximum effect. Graphs, in particular, can allow us to combine raw data, descriptive statistics and the output of statistical or mathematical models within a single illustration. (Note that the output of statistical modelling has been included in some of the graphs above in the legends – e.g. Figure 21.11 where the interaction was noted as statistically significant, with a p-value of 0.0004.)

Each quantitative study is unique, and so a considerable amount of thought is required to interpret the results and their implications and to consider the generality of the findings. So far, our emphasis has been on the use of data in analysis and presentation of information. The flip side of this is our reaction to illustrations presented by others. An understanding of data-handling processes (including the principles of good illustrative practice) is critical if we are to be able to interpret the meaning and the context of work presented to us. We need to be able quickly to judge how well quantitative work has been done, what the value and implications of the results are and, indeed, whether we have been given enough information to make such judgements. When we are presented with a graph, we should seek to interpret it in as full and critical a way as possible. In doing this, we should consider the extent to which form follows function. Recall that illustrations are typically used to imply that the pattern we see (the form) results from the causes depicted (the function).

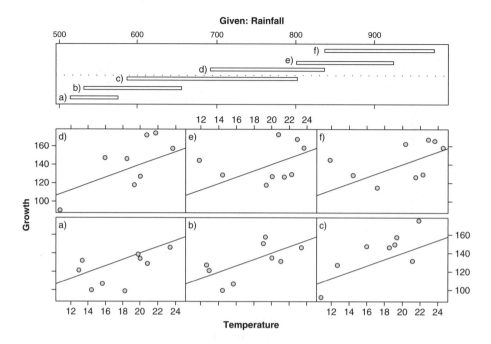

Figure 21.12   Conditioning plot (or coplot). This examines the relationship between
temperature and plant growth, for the plants with fertilizer added, and how this relationship
varies with rainfall. The plot comprises separate scatterplots of plant growth versus
temperature, each of which corresponds to a different range of rainfall. The rainfall ranges
are indicated by the top panel, which shows that the bottom-left graph in the bottom panel
(labelled 'a)') only includes sites that have annual rainfall between about 515 and 570 mm/year.
The top-left scatterplot (d) only includes sites with rainfall in the range 690–835 mm/yr, and so
on. The dotted line in the top panel indicates the move from the graphs in the bottom row of
the lower panel (a–c) to those at the top (d–f). There is no significant interaction between the
effects of temperature and rainfall on plant growth for these data, so the fit lines drawn are for
the overall regression – i.e. they are the same in all the graphs

   In Figure 21.13a, the implication is that greater area of islands tends to cause
greater plant species richness. The best-fit line goes much further. It suggests that
the increase of plant species richness with area is highly predictable: it follows a
linear relationship on a log–log scale. What does this actually mean? Looking at
the graph suggests that a ten-fold change in island area is associated with much
less than a ten-fold change in species richness. The equation of the line quanti-
fies this change in plant species number as 2.35-fold, which is consistent with the
rule-of-thumb used by many conservationists that a 90 per cent loss of habitat
leads to a 50 per cent loss of species. Figure 21.13b shows the same data on linear
axes. The solid line is the same best-fit line as in Figure 21.13a and represents a
power relationship. The dashed line shows an alternative model for describing the
data: a logarithmic function. There has been debate for decades in the ecological
literature about which of these two models better represents the species–area
relationship.

Further exploration of the two curves shown in Figure 21.13b gives greater insight into the relative merits of the two models on which they are based – for this dataset, at least. In statistical terms, the logarithmic model provides a slightly better fit with the data, accounting for 91 per cent of the variation in plant species numbers, compared with 90 per cent for the power model (though we must bear in mind that the sample size is small). But how do the two models compare on theoretical grounds? It is clear from the graph that the two models predict vastly different numbers of species (in absolute terms) for large islands. It is also clear

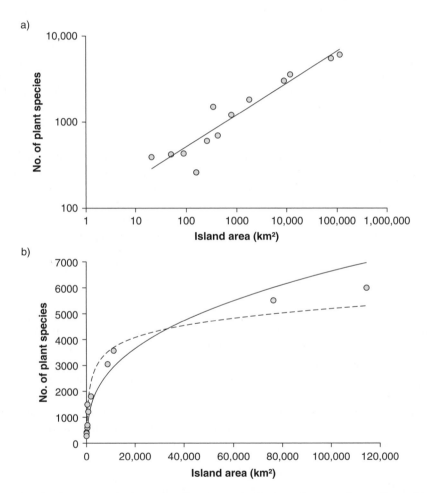

Figure 21.13 Species–area relationship. These two graphs show the same data as Figure 21.3 (plant species numbers plotted against island area for Caribbean islands), except that Cay Sal Bank has been removed as an outlier. Plot (a) is on log–log axes, while (b) is on untransformed axes. The solid line in both graphs shows the best-fit power model. The equation for this is $y = 92.434x^{0.371}$, and the model accounts for 90 per cent of the variation in the data ($r^2 = 0.896$). The dashed line in (b) is the best-fit logarithmic model: $y = 693.3 \ln(x) - 2810.5$. This model accounts for 91 per cent of the variation in the data ($r^2 = 0.913$)

that, if we extrapolate beyond the range of the data to even larger islands, this difference will increase rapidly. Such differences could be very important if we wish to use these data to predict for other islands or if we wish to predict the effect of changes in habitat area on species numbers in nature reserves. What is less clear from Figure 21.13b is what happens if we extrapolate beyond the range of the data in the other direction: to smaller islands. The equation for the logarithmic model yields negative species numbers for islands smaller than 57 km²; for a 1-ha island, the predicted number of plant species is –6003. That is clearly nonsense, and suggests that extrapolation in this direction is highly dubious in the case of the logarithmic model. The power model, in contrast, predicts 17 plant species on an island of 1 ha, and this model reaches 0 species at an area of 0 km². Such predictions are at least plausible! On the basis solely of these results, we might conclude that the power model is more theoretically sound and more likely to be useful as the basis for a general model of the species–area relationship than the logarithmic model.

This example raises a number of issues. For a start, it brings to the fore the differences between interpolation, extrapolation and prediction. Relationships that are qualitatively very different can look very similar in certain parts of their ranges. Figure 21.14 shows fundamentally different relationships, all of which look very similar in the highlighted range.

The species–area example also raises the issue of what to do with outliers. Notice the difference in data between Figure 21.3 and Figure 21.13: Cay Sal Bank has been excluded from the data in Figure 21.13. It was clearly an outlier in Figure 21.3, but this alone is not sufficient reason to exclude it when we are interested in trying to explain the real world. In fact, there are good theoretical reasons to omit the Cay Sal Bank data. This set of islands in the Bahamas is little more than coral reef and so is both qualitatively different from the other data points, and has a high degree of uncertainty attached to the measurement of its area. The important point here is that the exclusion of this data point makes a big difference to the models' fit to the data.

Perhaps most importantly, the species–area example also demonstrates the important difference between a statistical fit (e.g. a best-fit line) and an underlying causal relationship. The fit of both models to the data is strong, but this does not prove that there is a causal relationship. Is the geographic pattern symptomatic of one or more geographic processes? In the above analysis, no mechanisms linking changes in area to changes in species numbers have been investigated, so the answer to this question is not clear. What is clear, though, is that we have to be very careful in deciding what we can, and cannot, conclude from any given study.

CONCLUSION

The way that data are presented can seriously affect our inferences and conclusions. It is common knowledge that 'statistics' can be very misleading (the word 'statistics' here is usually used to mean facts and figures, rather than parameter estimates or probability statements). The same is true of models, graphs, maps and tables.

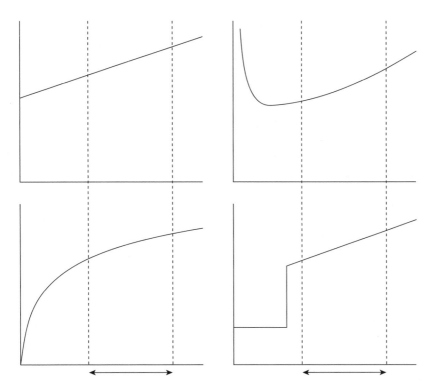

Figure 21.14   Qualitatively different relationships that look similar over a given data range. The four relationships shown are fundamentally different: one is linear (top left), two are curvilinear but of very different form (top right and bottom left) and the other involves a threshold (bottom right). However, all four relationships look very similar indeed over the range between the dotted lines. If the data were only collected over this range, it would be almost impossible to distinguish between these relationships (and many others not shown). In practice in this situation, we would use Occam's Razor and accept the linear relationship because this is the simplest explanation, but if the true underlying relationship were not linear, we would be wrong. This highlights how dangerous it can be to extrapolate beyond the range of the data

Whenever we consider a model, we should ask searching questions about what its purpose is – is it for prediction or merely description, for instance – and about exactly what it tells us. We should also think about what alternatives there might be. Would a different model work better – perhaps a non-linear rather than a linear relationship, or vice-versa? How robust is the model? (Good practice in both mathematical and statistical modelling involves sensitivity analysis, in which the effects of changes in assumptions or input variables are evaluated to inform us about this, see Chapter 19.) There are many different ways of modelling and presenting data, and each one is both an abstraction of the 'real world' (a construct of the person doing the modelling) and a simplification of reality. Standard, repeatable techniques are used where possible in quantitative study, but there are always subjective decisions involved. Models and illustrations can be used for deliberate deception or distortion of results, perhaps for a political purpose (Tufte, 1983, has many examples of this kind of deception), but in academic study it is vital that we avoid such purposeful distortion. However, misleading representation of data often occurs unintentionally through incomplete or incorrect analysis of data. The risk of this can be minimized by keeping in mind the following guidelines:

- Plotting the data in a variety of ways, in conjunction with calculation of summary statistics, forms the basis of exploratory data analysis, a very powerful way of examining trends in data quickly and easily.
- Statistical analysis should typically include the use of graphs and maps to help examine the nature of the model fit and to check the validity of the underlying assumptions.
- Graphical display of the results is often by far the best way of communicating the information in an elegant, concise and accurate way. This graphical display should typically incorporate elements of the statistical analysis, such as error bars (or other measures of uncertainty or variability) and indications of statistical significance on the graphs.

Whatever we do, we need to be aware of the fact that all our data-presentation tools are constructs, and all our models are wrong to at least some extent. The important question is: how far from the truth are they?

Summary

- The ability to work with data is essential to achieving and communicating understanding of 'real-world' complexity. Data-handling skills are both a prerequisite for statistical analysis and allow presentation and interpretation of numerical results.
- When presented with a large mass of data, a fundamental starting point is the realization that there are different levels of abstraction from the 'real world': data collection, graphical and numerical representation of data, and generalization from statistical or mathematical modelling.
- Different types of data are best presented using different numerical and graphical methods. Data (re)presentation can, therefore, affect understanding, judgement and inference and is a vital part of the examination of geographical pattern and process, and of the application of techniques such as interpolation, extrapolation and prediction.

Further reading

- John Tukey was instrumental in developing and promoting useful, accessible ways of handling data. His book (Tukey, 1977), although a bit idiosyncratic, is still well worth a read. The book *Modern Methods of Data Analysis,* edited by Fox and Long (1990), covers the topics I have discussed in much more detail. Tufte (1983) is a clearly written book containing much wisdom about the use and presentation of data. It encourages careful, honest and imaginative display of information and is a very interesting and informative read. A more up-to-date treatment of graphics is provided by Ellison (2001). This short chapter is much more focused on research (with particular emphasis on experimental work in ecology) than Tufte (1983). Although I do not agree with everything written in Ellison (2001), I strongly recommend it, along with much of the rest of the book containing it (Scheiner and Gurevitch, 2001) – even for geographers with little interest in ecology. Wainer (2005) is written for a more popular (American) audience; it covers the history of graphical display, as well as colourful examples of its use in the modern world, and is well worth reading.

- Taylor (1982) provides a good introduction to error (uncertainty): how to estimate it, cope with it and report it. Even though it is written for physicists, it is a valuable read for undergraduate geographers. Taylor also has a useful section on dealing with outliers (Chapter 6, 'Rejection of data').

- Although an old book, Thornes and Brunsden (1977) contains much that is still important, including a good, short section on interpreting graphs with respect to the way that physical systems operate (in Chapter 7). Sayer (1992) provides a useful discussion on abstraction – from a social-science viewpoint.

- Finally, it is very important to get at least some grasp of statistical methods. There are numerous books dealing with this, though not many specifically focus on geography. Crawley (2005) provides an excellent introduction to statistical techniques, starting from first principles and using a freeware statistical analysis package, R. This is a general book, not just for geographers, but in my opinion it is hard to find a better statistics text. Many students just starting statistics and using SPSS find Pallant (2007) and the previous editions very helpful. Rogerson (2006) has some useful material, especially on spatial statistics.

Note: Full details of the above can be found in the references list below.

References

Cleveland, W.S. (1985) *The Elements of Graphing Data*. Monterey, CA: Wadswork Advanced Books & Software.

Crawley, M.J. (2005) *Statistics: An Introduction Using R*. Chichester: Wiley.

Ebdon, D.S. (1985) *Statistics in Geography* (2nd edn). Oxford: Blackwell.

Ellison, A.M. (2001) 'Exploratory data analysis and graphic display' in S.M. Scheiner and J. Gurevitch (eds) *Design and Analysis of Ecological Experiments* (2nd edn). Oxford: Oxford University Press.

Fox, J. and Long, J.S. (eds) (1990) *Modern Methods of Data Analysis*. London: Sage.

Frodin, D.G. (2001) *Guide to Standard Floras of the World: an Annotated, Geographically Arranged Systematic Bibliography of the Principal Floras, Enumerations, Checklists, and Chorological Atlases of Different Areas* (2nd edn). Cambridge: Cambridge University Press.

Insightful Corporation (2005) *S-Plus 7 Guide to Statistics, Volumes 1 and 2*. Seattle, WA: Insightful Corporation. These are freely available in pdf form at http://www.insightful.com/support/splus70win/statman1.pdf and/ tatman2.pdf.

Kopf, E.W. (1916) 'Florence Nightingale as a statistician', *Journal of the American Statistical Association*, 15: 388–404.

MacArthur, R.H. and Wilson, E.O. (1967) *The Theory of Island Biogeography*. Princeton, NJ: Princeton University Press.

Pallant, J. (2007) *SPSS Survival Manual* (3rd edn). Milton Keynes: Open University Press.

Playfair, W. (1786) *The Commercial and Political Atlas*. London: Corry.

Playfair, W. (1801) *Statistical Breviary*. London: Wallis.

Porter, T.M. (1986) *The Rise of Statistical Thinking, 1820–1900*. Princeton, NJ: Princeton University Press.

Rogerson, P.A. (2006) *Statistical Methods for Geography: a Student's Guide* (2nd edn). London: Sage.

Sayer, A. (1992) *Method in Social Science: a Realist Approach* (2nd edn). London: Routledge.

Scheiner, S.M. and Gurevitch, J. (eds) (2001) *Design and Analysis of Ecological Experiments* (2nd edn). Oxford: Oxford University Press.

Sokal, R.R. and Rohlf, F.J. (1995) *Biometry*. New York: W.H. Freeman.

Spence, I. and Lewandowsky, S. (1990) 'Graphical perception', in J. Fox and J.S. Long (eds) *Modern Methods of Data Analysis*. London: Sage.

Taylor, J.R. (1982) *An Introduction to Error Analysis*. Mill Valley, CA: University Science Books.

Thornes, J.B. and Brunsden, D. (1977) *Geomorphology and Time*. London: Methuen.

Tufte, E.R. (1983) *The Visual Display of Quantitative Information*. Cheshire, CT: Graphics Press.

Tufte, E.R. (1990) *Envisioning Information*. Cheshire, CT: Graphics Press.

Tukey, J.W. (1977) *Exploratory Data Analysis*. Reading, MA: Addison-Wesley.

Wainer, H. (2005) *Graphic Discovery: a Trout in the Milk and Other Visual Adventures*. Princeton, NJ: Princeton University Press.

Waltham, D. (1994) *Mathematics: a Simple Tool for Geologists*. London: Chapman & Hall.

APPENDIX

This table gives the simulated dataset that is used for much of this chapter. Each row represents a site, for which the attributes in the columns were measured. Units are mm per year for rainfall, °C for mean temperature and grams per year for plant growth. Fertilizer is recorded as applied to the plants or not, and light levels are recorded as low (shade), medium (semi-shade) or high (no shade).

Rainfall	Temperature	pH	Fertilizer	Light	Plant growth
696	17.2	8.3	no	low	37.9
686	17.4	5.9	no	low	46.2
803	19.7	4.2	no	low	43.5
609	14.6	5.6	no	low	30.3
728	16.0	5.7	no	low	41.1
980	10.4	8.7	no	low	19.2
976	17.7	7.1	no	low	28.3
755	14.0	7.4	no	low	46.7
811	22.9	6.0	no	low	51.4
638	18.0	5.3	no	low	53.5
898	20.6	7.6	no	medium	64.0
955	16.9	7.0	no	medium	58.2
837	14.6	8.1	no	medium	43.5
595	12.3	4.8	no	medium	49.3
678	22.6	7.9	no	medium	55.3
899	15.0	6.4	no	medium	51.3
950	24.3	7.6	no	medium	57.8
533	10.7	7.9	no	medium	37.6
668	18.5	5.6	no	medium	59.6
573	11.8	7.2	no	medium	48.8
798	10.8	8.7	no	high	44.9
947	13.2	7.2	no	high	53.1
838	12.3	4.8	no	high	42.1
968	21.1	8.1	no	high	73.3
974	13.5	5.2	no	high	55.0
619	11.0	4.5	no	high	43.0
908	23.3	8.4	no	high	62.3
981	12.2	8.5	no	high	40.7
862	19.2	7.7	no	high	55.1
894	21.5	4.6	no	high	58.0
575	14.4	4.3	yes	low	98.3
543	15.7	5.3	yes	low	105.0
839	22.3	5.9	yes	low	128.3
906	21.5	4.6	yes	low	125.5
924	17.2	5.6	yes	low	114.4
718	10.8	8.0	yes	low	90.2
521	18.3	6.7	yes	low	96.8
624	21.1	4.3	yes	low	129.9
826	19.4	7.7	yes	low	117.4
803	20.1	8.4	yes	low	126.5
533	23.4	8.2	yes	medium	144.7
880	14.4	6.1	yes	medium	128.2
923	22.9	6.8	yes	medium	166.5

(Continued)

Rainfall	Temperature	pH	Fertilizer	Light	Plant growth
692	16.0	5.3	yes	medium	145.9
975	24.6	4.7	yes	medium	157.5
764	18.5	4.8	yes	medium	145.1
520	19.8	8.7	yes	medium	137.5
518	20.9	4.6	yes	medium	126.7
568	12.9	7.3	yes	medium	120.4
514	13.4	6.0	yes	medium	130.0
957	23.6	5.8	yes	high	165.0
837	20.8	6.7	yes	high	172.1
656	19.1	6.9	yes	high	148.5
810	23.7	8.9	yes	high	156.8
802	21.8	8.5	yes	high	174.2
873	11.8	6.2	yes	high	144.0
537	20.1	7.5	yes	high	133.2
600	19.3	6.2	yes	high	156.1
930	20.3	6.9	yes	high	161.6
587	12.6	4.2	yes	high	125.4

22 Mapping and Graphicacy

Chris Perkins

Synopsis

Maps are very powerful tools for representing our ideas and knowledge about places. As such, the skills of producing and reading maps are important within the discipline of Geography. Cartography deals with the development, production, dissemination and study of maps in a wide variety of forms, whereas graphicacy is the skills of reading and constructing graphic modes of communication, such as maps, diagrams and pictures. This chapter introduces the differing roles played by mapping and the changing social significance of the map, discusses the availability of mapped information, explains how maps work and offers advice about how to design maps.

The chapter is organized into the following sections:

- Introduction
- The contested terrain of mapping
- Finding the map
- How maps work
- Practical suggestions for design
- Conclusion

INTRODUCTION

The map is a powerful medium for the representation of ideas and the communication of knowledge about places. It has been used by geographers to store spatial information, to analyse and generate ideas and to present results in a visual form. Maps are not just artifacts; mapping is a *process* reflecting a way of thinking. The quality of a printed map or map display on a screen is a reflection of the 'graphicacy' of its authors and readers. Over forty years or so ago, Balchin and Coleman (1966) argued that graphicacy should be placed alongside numeracy, literacy and articulacy as educational prerequisites – this chapter echoes that call and argues that technological, social and intellectual changes have made graphicacy increasingly important (see also Chapter 21). This chapter explains the changing social significance of the map, exploring roles maps might play and discussing how these may be relevant for the geographer. It introduces sources of mapping by discussing the availability of mapped information and introducing practical ways of finding out what maps exist.

With desktop mapping packages and web-served mapping, we can increasingly create our own maps, but if we are to realize the creative power of the medium it is important to understand how maps work. Half a century of cartographic research has led to some consensus, but there is still debate about issues of graphical quality. Most researchers accept the continuing need for a holistic and artistic approach to design.

The chapter concludes with some practical suggestions of how you might design better maps that use graphicacy in a creative and useful way. It reasserts the importance of mapping for anyone studying or researching in Geography.

THE CONTESTED TERRAIN OF MAPPING

The ability to construct and read maps is one of the most important means of human communication, as old as the invention of language and as significant as the discovery of mathematics (Borchert, 1987). The mapping impulse seems to be universal across cultures, time and environments (Blaut, 1991). Histories of cartography have charted changing production technologies, from the earliest surviving map artifacts dated to 3500 BC. They have described changing world views, reflected on accuracy and design and increasingly examined the social context of these images (Harley and Woodward, 1987).

Contemporary official definitions encompass a wide diversity of maps but also extend the scope of cartography well beyond earlier narrow concerns with the technology of map making. In 1995, for example, the International Cartographic Association defined a map as 'a symbolized image of geographic reality resulting from the creative efforts of cartographers and designed for use when spatial relationships are of special relevance'. The ICA also defined cartography as 'the discipline dealing with the conception, production, dissemination and study of maps in all forms' (International Cartographic Association, 1995: 1). This chapter also adopts a very catholic view of mapping (see Figure 22.1).

Following the influential work of Arthur Robinson, academics have sought to classify maps according to content and scale (Robinson *et al.*, 1995). Thematic maps focus on one particular kind of information, such as solid geology or voting patterns. In contrast, general-purpose maps include diverse information ranging from large-scale planimetric coverage of a building, through official medium-scale topographic mapping, to smaller-scale maps of the world in reference atlases.

Others have sought to classify by format of publication. Maps were formerly only issued as printed paper publications. Mapping is now also available as digital data delivered on a wide variety of media and held locally on CD-ROM, hard drive or network server, but increasingly also distributed over the internet (Kraak and Brown, 2000). Producers no longer control content or design; users can play a much more active role, and there is an increasing appreciation that maps might also be constructed to communicate by sound or be read by touch. For example, tactile maps are being created for visually impaired people (Perkins, 2002).

Distinguishing maps by content becomes less significant when users can create their own maps. Emphasizing content also completely misses an essential aspect: it can be argued that all maps have a theme and every one represents an 'interest' (Wood, 1992). Understanding this interest may be much more useful than listing

Figure 22.1   The world of maps

what the map appears to show. This view implies that maps might best be understood as propositions, instead of being seen as representations (Wood and Fels, 2008). The complexity of the medium is clarified if you appreciate the different roles played by maps (see Box 22.1).

Box 22.1   Some of the possible roles played by maps and mapping

Practical tools
Models
Language and representation
Inventories and databases
Visualizations
Cultural artifacts and practices
Imaginings
Political devices
Metaphors for scientific knowledge
Persuasive icons
Contested texts
Geography

Mapping is above all else a practical form of knowledge creation and representation, and is usually carried out with an end in mind. Cartography is a useful art, and the map is a tool to be used for informing, navigating, describing places, analysing spatial relationship, or many other purposes. Maps work as tools by simplifying and by serving as guides to the much greater complexity they represent. This recognition was used by geographers in the 1970s to develop scientific approaches that focused upon how maps work as a form of communication (e.g. Board, 1984) and in the search for optimal map designs.

Maps communicate spatial information in a series of codes and these codes not only work within the map but also allow us to use maps at a social level (Wood, 1992; Pickles, 2004). By the 1990s, there was an increasing concern that a narrow scientific view of the map as part of a cartographic communication system failed to reflect the diversity of roles it played. Instead, more recent research has treated the map as a representation and sign system functioning at different levels (MacEachren, 1995).

The map is also an efficient way of storing large amounts of spatial information (Tufte, 1983: 166). This 'visual inventorying' role has come to be supplanted by digital databases that no longer need to have a visual expression; we can follow the instructions given by the sat nav, and navigate from a to b without seeing the bigger picture of the map. But even the simplest visual display still stores and communicates information very effectively: a picture is still 'worth a thousand words'.

In the last two decades of the twentieth century, the development of the digital computer and increasingly complex software allowed graphical images to be manipulated, and more interactive map use became possible (see Figure 22.2). Mapping also increasingly moved towards the private end of the use axis and became increasingly collaborative in design and use. In the 1990s, scientific visualization began to reintegrate map design with science (MacEachren *et al.*, 1992).

The development and uses of mapping reflect changing social and economic contexts (Thrower, 1996). All maps are cultural, crafted works of the human mind. They are artifacts imbued with the cultural values of the society producing them. An Australian aboriginal bark painting maps out aboriginal environmental relations and might be incomprehensible to a western European (Turnbull, 1989). The tube map serves as a navigational aid for a tourist visiting London but might not be understood by a bushman. Recent research has increasingly shown how mapping practices are fundamental in understanding the significance of the medium (Dodge *et al.*, 2009). Web-served applications now make collaborative mapping much easier than was the case, and crowd-sourced, shared resources such as OpenStreetmap.org disseminate community-sourced and free mapping. Meanwhile, everyday use of maps is increasing and modern artists frequently deploy mapping in their work (Cartwright *et al.*, 2009).

The way maps work in different cultures reflects individual imagined geographies. The power of the imagination is obvious in properly designed maps that reflect intuitive and individual artistic judgement (Keates, 1984): maps can decorate as well as inform. Edward Tufte (1983) has argued that this artistic aspect of design implies a quest for quality.

It has also been argued that maps are all imbued with power and that the work they carry out is always political (Harley, 1989a). The map often stands for control

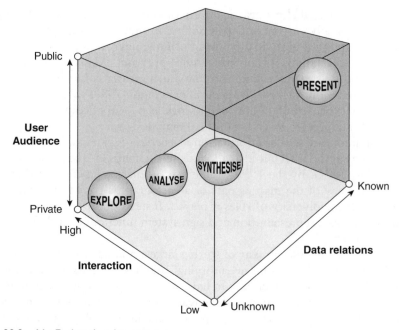

Figure 22.2   MacEachren's cubic map space

and acts as a synonym for order. As a form of power knowledge, the map has often represented the interests of elite groups in society and served to reinforce social norms (Black, 1997). The power of many western maps resides in their apparent objectivity (Turnbull, 1989). The map appears to show everything, to offer a neutral 'view from above' (Cosgrove, 2001) that indexes the world and allows the unknown to be known. But while they may stand for science and factual knowledge, maps may also have an inherent ability to persuade – all the evidence suggests they are more likely to be believed than words. Mapping is widely used in political propaganda, in advertising, in cartoons and in the mass media where maps are associated with news stories and reinforce the narrative of the story line (Monmonier, 1996). Maps can also add authority to people associated with them: military leaders and politicians often give their press briefings in front of maps. Others have argued that all maps are inherently persuasive, that maps are best read as rhetoric and that the map is a text that needs to be deconstructed and interpreted (Harley, 1989a). The problem is how to interpret such persuasive icons (Pickles, 2004). Maps interact with other discourses, may be read in conflicting ways by different social groups and say different things to different people.

Perhaps above all else, mapping is still seen as something that distinguishes Geography from other disciplines. The map in the journal article or the geography dissertation sends a signal to the reader that the work is geographical (Harley, 1989b) and there is strong evidence that the discipline is returning to mapping (Dodge and Perkins, 2008), so deploying maps with your dissertation sends powerful messages.

FINDING THE MAP

In the UK in 1996, there were 250 publishers releasing printed mapping (Perkins and Parry, 1996) and over 2,500 were listed worldwide in Parry and Perkins (2000). By 2000, Mapquest.com had already delivered more maps from its web-based server than any other publisher in the whole history of cartography (Peterson, 2001). In September 2009 over 25,000 mapping sites were indexed on the definitive list of cartography websites (Oddens, 2009). Since 2005, there has been a widespread adoption of virtual globes such as Google Maps, or Google Earth, together with a profusion of mashup sites combining these backdrops with additional datasets. Quite simply, there are more maps available now than at any time in human history. These quantities make it difficult to find the right map – a problem exacerbated by the increasing diversity of types and by the infinite design possibilities offered by digital mapping. How do you decide which to use and how do you find that map?

It is difficult to find out what printed fixed-format maps have been published. Publishers provide descriptive information about their mapping, and increasingly these data are available over the web. So one option is to search the home pages of mapping organizations. A few detailed guides to mapping have been published for some nations, notably for the UK (Perkins and Parry, 1996). A basic but comprehensive introduction is provided in World Mapping Today, including lists of URLs and publisher details, pen portraits of the state of the art in different countries, simple graphic indexes and bibliographic information (Parry and Perkins, 2000).

When searching through these sources, you might be interested in such factors as the spatial and temporal coverage, resolution, currency, consistency and reliability. But a more complex appreciation also helps. For example, who produced the map for which market? What does it include or omit? Are there legal restrictions on use? How is copyright enforced? Is the map affordable? Perhaps above all else, is it available? Despite globalization the nation-state and its legal framework continue to influence public civilian availability (Barr, 2001). Figure 22.3 shows that, although most terrestrial areas have been well mapped, it is very difficult to get access to larger scales of official mapping. Maps may still be reserved for military use (e.g. in India, much of the Islamic world and even in Greece) or priced at a market cost recovery rate that maximizes revenue and makes individual access expensive (e.g. in the UK).

Having identified the best printed source, tracking down a copy involves consultation in a map library or purchase from the publisher or map seller (Parry, 1999). The quality of map libraries varies enormously according to funding, the nature of their parent organization and how they are staffed (Parry and Perkins, 2001). In the UK, many schools of Geography or Earth Science departments hold collections of printed mapping. The currency and scope of many university collections are declining but a more comprehensive range of cartographic resources is to be found in major and national library collections, such as the six copyright libraries in the UK (British Cartographic Society, 2009). In North America, most significant map collections are to be found in central libraries on university campuses, where they

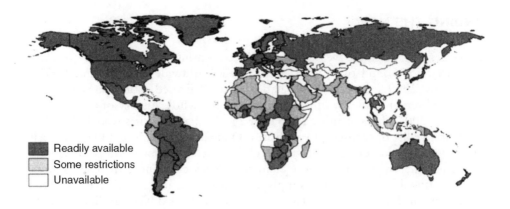

Figure 22.3   Map availability

serve as Federal depositories and where digital datasets are increasingly also being archived. National libraries such as the Library of Congress in Washington, DC, the National Library of Australia in Canberra, or the British Library in London offer the most significant map holdings and may be the best sources for the most difficult items. Information about the scope of major map collections is to be found in Loiseaux (2000) and via the map collections section of the Oddens list on the web (Oddens, 2009). Descriptive information may increasingly be found in web-based online public access catalogues and may be used to discover whether the library holds a copy of the map you need. It may be necessary to visit the collection to seek advice from professional curatorial staff.

Parry (1999) suggests that buying a printed map may well be the last resort for students if resources are not available in library systems. In Great Britain and North America, mapping is relatively easy to acquire from many retailers. Wider ranges are available from specialist map shops (such as Stanfords), most of whom maintain mail-order operations and e-businesses with online ordering facilities (see Box 22.2 for a list of the most significant of these dealers). For more information about the relative merits of these competing sources, see Parry and Perkins (2000).

Box 22.2   Map dealers

GeoPubs
4 Highland Grove, Billericay, Essex, CM11 1AF, UK
Tel: +44 1277 632454
Fax: +44 1277 632056
Email: sales@geopubs.co.uk
URL: http://www.geopubs.co.uk/

The Map Shop
15 High Street, Upton-on-Severn, Worcs WR8 0HJ, UK

Tel: +44 1684 593146
Fax: +44 1684 594559
Email: themapshop@btinternet.com
URL http://www.themapshop.co.uk/

Elstead Maps
The Threshing Barn, Woodhayes, Luppitt, Honiton, Devon EX14 4T, UK
Tel: +44 1404 45 400
URL: http://www.elstead.co.uk/

East View Cartographic Inc.
10601 Wayzata Blvd., Minneapolis, MN 55305 USA
Tel: +1.952.252.4551
Fax: +1.952.252.1202
Email: eastview@eastview.com
URL: http://www.cartographic.com

ILH Internationales Landkartenhaus
Stuttgart, Schockenriedstrasse 44, D-70565, Stuttgart, Germany
Tel: +49 711.490.722.10
Fax: +49 711.490.722.11
Email: ilhINFO@ilh-stuttgart.de
URL: http://www.ilh-stuttgart.de/

Map Link Inc.
30S la Pactera Lane, Unit 5, Santa Barbara, CA 93117, USA
Tel: +1 805 692 6777
Fax: +1 805 692 6787
Email: custerve@maplink.com
URL: http://www.maplink.com/

Maps Worldwide Ltd
Datum House 6 Lancaster Road, Melksham, Wilts, SN12 6SS UK
Tel: +44 1225 707 004
Fax: +44 1225 709 384
Email: customers@mapsworldwide.com
URL: http://www.mapsworldwide.com

OMNI Resources
1004 South Mebane Street, PO Box 2096, Burlington, NC 27216–2096, USA
Tel: +1 336 227 8300
Fax: +1 336 227 3748
Email: orders@omnimap.com
URL: http://www.omnimap.com/

Edward Stanford Ltd
12–14 Long Acre, London, WC2E 9LP, UK
Tel: +44 020 7836 1321
Fax: +44 020 7832 8928
Email: sales@stanfords.co.uk
URL: http://www.stanfords.co.uk/

(Continued)

(Continued)

Treaty Oak
PO Box 50295, Austin, TX 78763–0295, USA
Tel: +1 512 326 4141
Fax: +1 512 443 0973
Email: maps@treatyoak.com
URL: http://www.treatyoak.com/

World of Maps
1235 Wellington Street, Ottawa, Ontario, K1Y 3A3, Canada
Tel: +1 613 724 6776
Fax: +1 613 724 7776
Email: info@worldofmaps.com
URL: http://worldofmaps.com/

Increasingly of course, hard-copy printed mapping is being supplemented and often replaced by digital alternatives available over the internet. Accessing digital-map data presents different challenges. It may require GIS software, storage media, hardware and output devices. Some data are distributed with viewing and interactive mapping software, perhaps on CD-ROM, but increasingly data are being served on the internet. Even greater disparities of availability exist between the economically developed and less developed worlds, in part because of the economics of digital map production. Digital mapping is usually much more expensive to buy than printed maps and may only be available to use under strictly regulated licence conditions. In the UK, the Joint Information Services Committee has negotiated a number of deals with data providers to release digital data to universities and colleges. Students can register to use an increasing range of digital data in teaching, learning and research. Since 2000 selected Ordnance Survey data have been available via the DIGIMAP service from EDINA, and the scope of data available from this source has been expanded to encompass historical Ordnance Survey mapping, geological survey data and marine charting. Other notable online sources of data relating to Great Britain are available through CHEST (Combined Higher Education Software Team). In the USA, Federally produced digital map data are in the public domain and may be readily accessed over the internet. Map libraries in North America, western Europe and Australasia increasingly offer access points to the more useful sources of digital mapping. The internet also offers a plethora of copyright-free maps. The majority of mapping websites still serve static images, but notably rich online libraries include the University of Texas for contemporary mapping and the David Rumsey collection for scanned historical mapping. Increasingly, the web is also a valuable source of more interactive mapping, such as the now ubiquitous virtual globes such as Google Maps. It also serves as a delivery gateway to data warehouses storing digital map data that may be imported in GIS (see Chapter 25). Box 22.3 lists a number of the more important online sources of digital mapping.

Box 22.3   Online sources of mapping

Search engines
Oddens list
http://oddens.geog.uu.nl/index.html
Around 25,000 links to cartographic-related sites on the web. The richest and most current way in to spatial data sources.

Gigateway
http://www.gigateway.org.uk/default.html
Metadata and search engine for many available UK digital spatial data.

Gateway sites
Atlas of Cyberspace
http://www.geog.ucl.ac.uk/casa/martin/atlas/atlas.html
A rich and multimedia grouping of virtual geographies.

History of Cartography Gateway
http://www.maphistory.info/index.html
Over 5000 annotated links to historical sources.

Sources of online map data
University of Texas
http://www.lib.utexas.edu/maps/
A rich source of scanned conventionally published mapping. Huge collection of maps, mostly produced by the CIA, available as gifs, jpegs or pdf files plus links to other sites (including historical maps, city plans and cartographic reference sources).

United Nations Cartographic Section
http://www.un.org/Depts/Cartographic/map/profile/nonselfe.pdf

David Rumsey Historical Map Collection
http://www.davidrumsey.com/
The richest global collection of scans of old maps.

Ordnance Survey
http://www.ordnancesurvey.co.uk
The national mapping agency of the UK, with a wealth of information available about mapping products and some on-screen map downloads.

Digimap
http://digimap.edina.ac.uk/
The EDINA Digimap site makes some OS digital datasets accessible to the higher education community in the UK and includes current and historic topographic maps as well as geological and marine coverage.

Trails.com
http://www.trails.com/maps.aspx
Digital USGS mapping at 1:100 000, 1:25 000 and 1:24 000 scales and aerial coverage.

OpenStreetMap
http://www.openstreetmap.org
The most comprehensive global collaborative free map.

(Continued)

(Continued)

Mapquest
http://www.mapquest.co.uk
The world's busiest mapping site. Road maps and routefinding software. Worldwide coverage.

Virtual globes and mapping portals
Google Earth
Industry leading virtual globe.

Google Map
http://maps.google.com
Global coverage and an ever-expanding range of mapping, applications, mashup capability and interface design, including urban *Streetview* photographic coverage.

Multimap
http://www.multimap.com
Microsoft-owned map, aerial, birds eye and hybrid interfaces to location aware data.

Yahoo Maps
http://uk.maps.yahoo.com/

Placename finding aids
GEOnet Names Server
http://earth-info.nga.mil/gns/html/index.html
Access to the National Geospatial Intelligence Agency database of foreign place and geographic feature names. Approximately 20,000 of the database's 4 million features are updated monthly with names information approved by the US Board on Geographic Names (US BGN).

Census and administrative geographies
CASWEB
http://census.ac.uk/casweb/
The spatial interface to census mapping and data in the UK.

UK Borders
http://edina.ed.ac.uk/ukborders/
Boundary files for use with the census.

Neighbourhood Statistics
http://www.neighbourhood.statistics.gov.uk/dissemination/
Interactive mapping tools for UK census data.

Data archives
UK Data Archive
http://www.data-archive.ac.uk/home/
A specialist national resource containing the largest collection of accessible computer-readable data in the social sciences and humanities in the UK. Plus catalogue searching of other national archives for computer-readable data.

National Geophysical Data Center (NGDC)
http://www.ngdc.noaa.gov/
A wide range of science data services and information. Well documented databases from many sources, and value-added data services. NGDC acquires and exchanges global data through the World Data Center system and other international programs.

US Geological Survey
http://edcwww.cr.usgs.gov/webglis/
Provides data on climate, digital line graphs, elevation, geology, hydrology, land cover, maps, photography, satellite imagery – mainly for the USA but also with global coverage.

ESRI
http://www.esri.com
The software house responsible for industry-standard GIS data sources. Wide range of GIS data and digital map samples.

Air photographic data
Landmap
http://landmap.mimas.ac.uk/landmap/
Satellite images and Digital Elevation Models of the British Isles, plus aerial photography for selected British conurbations.

UK Perspectives
http://www.ukperspectives.com/
Aerial photographs for the UK.

Cities Revealed
http://www.citiesrevealed.com/
Many different types of aerial photographs for the UK and other countries.

Getmapping
http://www.getmapping.co.uk
Global and UK aerial photography, height data and mapping.

Land cover and thematic data
Multi-Agency Geographic Information System (MAGIC)
Web-based interactive environmental mapping and data downloads for the UK.

Cranfield University National Soils Research Institute
http://www.landis.org.uk/downloads/index.cfm
Interactive soils mapping of the UK.

British Geological Survey
http://www.bgs.ac.uk/
The national earth science agency of the UK – rich source of geoscientific data.

National Atlas of the United States
http://www.nationalatlas.gov/
The official interactive atlas of the USA, multi-thematic layered mapping and animations.

Climate Research Unit, University of East Anglia
http://www.cru.uea.ac.uk/
Provides climatic data for the British Isles, Europe, the USA and global data too.

HOW MAPS WORK

Once you appreciate the complex roles they play, it makes sense to try to understand how maps work. This involves understanding basic spatial properties, and appreciating how maps simplify and represent the constraints of symbolizing data, as well as appreciating the wider social contexts in which mapping operates.

Spatial properties

All maps are about places and represent distance, direction and location in a graphical medium. For the tool to work, these spatial properties have to be mapped out in a consistent way. The surface of the earth is not flat, so a mechanism is needed to translate the relative positions of places to the flat sheet of paper in a way that minimizes distortion of scale, area, direction and shape. The science of map projection regulates this process and for many years the mathematics and technology of projection dominated cartography. More recently, the politics of projection have been emphasized – for example, in the controversy over the use of the Peters projection (Monmonier, 1996). If you need to use a small-scale map of the world then take advice on the 'best' projection to use (American Cartographic Association, 1991).

Projection maps out the position of the lines of latitude and longitude that comprise the graticule. Absolute locations of places on the earth's surface may be defined in spherical co-ordinates, but the graticule does not intersect at right angles. Grids are a rectilinear net of lines that allow eastings and northings to be defined. They allow space to be structured and were first employed by ancient Greek cartographers. Grids may be arbitrary, like the alphanumeric references used by A–Z-style town atlases, or may be mathematically related to the projection, like the British National Grid.

Consistent representation of distance implies the use of a mathematical linear scale linking the map to the world, that may be expressed on the map as a representative fraction (1:50,000), as a bar scale or a verbal-scale statement (one and a quarter inches to one mile). Very few maps use scale consistently for every object; most will exaggerate the size of some features so they can be read as a symbol (for example, a road). Many small-scale maps of the world will also have to distort linear scale because of the projection used.

Generalization and classification

Scale regulates how much detail the map can show. Larger-scale maps depict more detail and with greater accuracy. As scales get smaller, so features must be more generalized (see Figure 22.4). There are a number of different ways of generalizing: simplifying, enlarging, displacing, merging and selecting, and so on. Classification can be another effective way to map out complexity. Alternatively, you may have to leave features off the map altogether. Matching the level of detail shown to the scale is an important aspect of design: a cluttered map will not work as well as one with the right balance between content and space. The process of selection might be seen as a technical issue, but in many maps is also a political outcome, with omissions or 'silences' reflecting cultural values (Harley, 2001: 84–107).

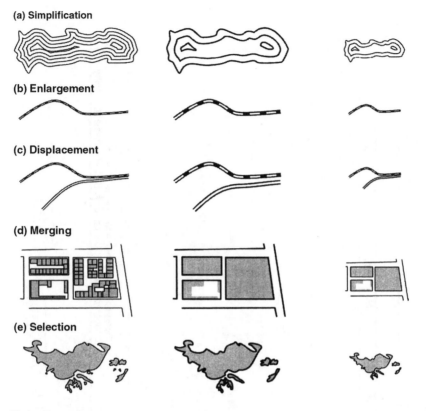

(a) Simplification

(b) Enlargement

(c) Displacement

(d) Merging

(e) Selection

Figure 22.4   Generalization

Symbolization

The cartographer also has to symbolize the world in a regulated graphic language in which text and the visual properties of symbols are combined. This combination often takes place in quite standardized ways – common elements of a map on a screen or sheet of paper may be identified (see Figure 22.5; Dent, 1999: 242). Objects in the map itself may be thought of as having different numerical qualities. These *measurement levels* are important for design. Nominal data show the presence or absence of information; ordinal data imply that a feature is larger or smaller (but do not indicate how much larger). Interval data involve ordering with known distances between observations – e.g. Fahrenheit measurement – whereas ratio measurement is an interval scale with a known starting point. Symbols and objects also have *dimensions*: points, lines, areas, volumes and duration. They may be distributed in discrete, sequential or continuous patterns.

On the map itself, the geometry and measurement level of symbols have *attributes* that allow information to be communicated. For example, a road may be red and

Figure 22.5 Map elements

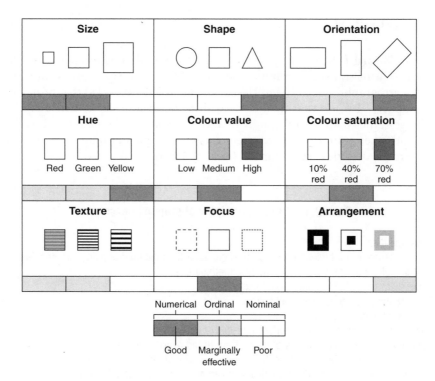

Figure 22.6   The graphic variables

it may have a label indicating that it is the A57 Snake Pass. The effective use of lettering on maps, and the rules governing how it should be used, are one of the most difficult areas of cartographic design. Name placement is a complex and often intuitive process.

The map designer can use only a limited number of graphic variables. Figure 22.6 illustrates how they might be used for point symbols and suggests there are rules governing inappropriate use. For example, shape should not be used to suggest variation in quantitative data (MacEachren, 1994). Symbols constructed with these variables may be iconic, geometric or abstract (MacEachren, 1995).

PRACTICAL SUGGESTIONS FOR DESIGN

Having obtained source material, how do you know whether it will make a 'good' map? Putting together all the elements in Figure 22.5 does not automatically result in a map that works well: these elements need to be combined in a mean-ingful aesthetically pleasing design. Can there be universal rules defining aesthetic quality? How might these be influenced by production technologies and are they transferable to different contexts?

Universal rules?

In 1983, Edward Tufte came up with a list of qualities that might define excellence in the design of statistical graphics (see Box 22.4). A survey carried out by the British Cartographic Society Map Design Group in 1991 revealed that professional cartographers also felt that quality resided in the overall perceptual qualities of the design rather than with the individual components of the map (see Box 22.5). Maps and other graphics seem to operate as wholes, greater than the sum of their parts. Artistic qualities are important (Keates, 1984).

Box 22.4   Tufte's principles of graphical excellence

Show the data.
Induce the reader to think about the substance rather than the methodology, graphic design, the technology of graphic production or something else.
Avoid distorting what the data have to say.
Present many numbers in a small space.
Make large datasets coherent.
Encourage the eye to compare different pieces of data.
Reveal the data at several levels of detail, from broad overview to the fine structure.
Serve a reasonably clear purpose: description, exploration, tabulation or decoration.
Be closely integrated with the statistical and verbal descriptions of a dataset.

Source: Tufte (1983: 13)

Box 22.5   Maps as communication graphics

Contrast between symbol and background and between symbols is vital.
The symbols themselves should be clearly legible and unambiguous.
The amount and nature of the data depicted should be appropriate to the main purpose of the map.
The overall appearance should be clear, simple and uncluttered.
The metrical attributes of the map should be both appropriate and clear.
The ordering of the data should be made clear by the hierarchical organization of the map image into recognizable visual levels.

Source: British Cartographic Society (1991)

There are two key areas involved. Our perceptual systems are programmed to respond to the *visual organization of images* (Dent *et al.*, 2008). A good map should be balanced, with spatial layout allocated according to the Golden section. (The Golden section refers to rectangles with sides at a ratio of 1:1:6 that offer the most pleasing appearance to the eye.) It should be organized so that the eye's area of maximum attention (just above the geometric centre) corresponds to the central focus of the map. The individual elements in the design ought to work together as an integrated unit. The second key aim should be to maintain a *clear hierarchy between different visual levels* in the map (Dent *et al.*, 2008). The 'figure' needs to stand out from the 'ground'. The most important objects should contrast most with their surroundings.

Production technology

By 2010, almost all maps produced by students in the UK were created using computer-based technologies. Choosing an appropriate type of software, and using it to best effect, is now probably the single most important impact on design. Be realistic and aware of some of the factors listed in Box 22.6. Five examples illustrate this process:

Box 22.6 What software should I use?

How much time do you have?
How ICT literate are you?
What packages are you already aware of?
How much support would you get for learning a new package?
How does the software link to other packages?
What file formats does it support, both for importing material and for pushing out completed graphics?
What output devices does the software talk to?
What kind of hardware environment do you intend to use: Mac or PC?
Do you intend to use someone else's base material and edit it up or design from scratch?
Do you have access to a fast scanner?
What role do you want the map to play?
What kind of use is the map intended for: presentation, analysis or exploration?
Is the map static or dynamic?
How complex is the information you want to show?
In which medium is the map going to be published: printed or electronic, black and white or colour?

---

1   Serving mapping on the web requires a number of different packages, in addition to the vehicle you use to design the map. The medium delivers maps to many users independent of platform, and maps are updatable. Production depends on the configuration of the site (whether processing is client or server based), the site format (whether maps are delivered in single pages, multiple pages, or a frames environment), the nature of the web interface (which browser and plug-ins are being used in web-applications), and the data type and the content interface (Cammack, 1999).

Be aware of the implications of using different file formats and the differences between serving maps in raster formats such as .jpeg, .tiff, or .png, as against vector formats like .sfw, .svg, or pdf. Seek advice from standard texts such as in the appendices in Kraak and Brown (2000: 177–209) or Dent *et al.* (2008). Remember that the web environment offers different constraints to the designer. In particular, remember that screen resolution, the aspect ratio of the monitor viewing the site and the size of the monitor reduce the available space to view the map, and that monitors vary in the ways they display colours (Dent *et al.*, 2008).

2   GIS software, such as MapInfo, ArcGIS or Idrisi, is designed for analysis but also offers a wide range of visualization options for map design. It is complex to learn, but if you know the package, have the data and want to display results of your analysis then it offers a realistic choice. If you need to produce many statistical maps for a case study, designed to a common template and from a wide range of variables, this automated kind of production will save time. Advice and practical examples about this kind of software are included in Brewer (2005).

3   Some mapping software will also allow you to create maps automatically but will not include the range of analytical tools you would expect in a full GIS. Mapviewer from Golden Software is a good example of this kind of program. It offers a compromise between analytical and design capability, and is relatively straightforward to use. It, too, supports a number of predefined thematic map types that automatically create maps from data held in spreadsheets. The package also automatically creates associated support information such as legends. The designer simply has to work out which map type is appropriate for the data (see Figure 22.7).

4   If you want real design quality, have complex ideas to map and the time to spend learning a sophisticated piece of software, then a professional drawing package such as CorelDRAW or Illustrator is the best option.

5   The majority of static maps displayed in undergraduate dissertations do not need this sophistication. A simple drawing package will suffice. It needs to have drawing tools and to allow you to import and edit graphics created elsewhere by adding symbols, text and marginal information.

The context

Even if design rules are followed and the appropriate technology used, maps still need to be matched to the medium in which they are to be published (MacEachren, 1994). Maps are not read in a cultural vacuum. They are interpreted in relation to what Denis Wood has termed the *paramap*, comprising other printed elements associated with the image (*the perimap*), and wider cultural referents that allow us to understand the medium (*the epimap*; Wood and Fels, 2008). Most maps produced by students are still designed for presentation in printed reports: descriptors for these might specify how a map should be designed. But increasingly, they may be used

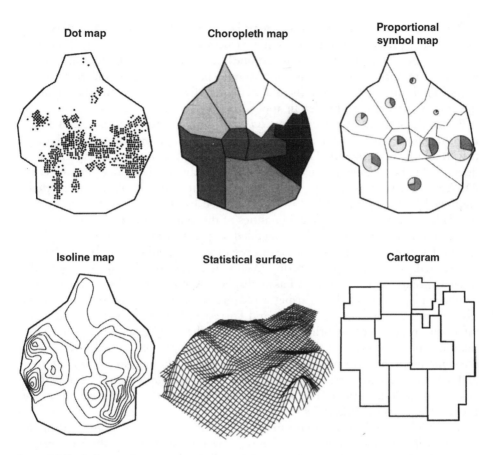

Figure 22.7   Different thematic map displays of the same data set

in PowerPoint displays, slides or overhead transparencies, on posters, displayed on computer monitors, delivered via the web or as animated, dynamic visualizations. The medium may support colour, or perhaps designs will be limited to grey scale. The mapping may be static or animated. Practical decisions that need to be taken are listed in Box 22.7 and many of these flow from wider cultural concerns.

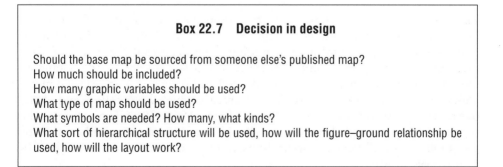

Box 22.7   Decision in design

Should the base map be sourced from someone else's published map?
How much should be included?
How many graphic variables should be used?
What type of map should be used?
What symbols are needed? How many, what kinds?
What sort of hierarchical structure will be used, how will the figure–ground relationship be used, how will the layout work?

With a coloured map, resolution may be inferior, but you can use two additional visual variables to improve visual structure. It makes no sense, though, to design in colour and print on a monochrome laser printer. Relating the map to other associated textual elements is also important. The more complex the medium, the more important the links. So, include mapping close to the words that relate to the graphic, rather than hidden in appendices.

Figure 22.8 shows a typical location map designed for incorporation into a report or dissertation and 'placing' the research by mapping a simple spatial context in which it is situated. The annotations draw attention to key aspects of the design. This book does not use colour, so the map has been designed in Core-lDRAW and saved as an .eps file for incorporation into book design software employed by the publishers. Most maps created by students can be exported as .wmf files that preserve the visual qualities of the image, such as line weight, area tones and lettering, and can be embedded as pictures in word-processed documents. The same map in a PowerPoint presentation would have to be much more generalized, and use larger lettering. Maps designed for display on paper also often translate badly to screen-based displays. On-screen colours are lit from behind and usually displayed using a red–green–blue colour model, in contrast to maps printed on paper (Gooding and Forrest, 1990). Screen resolution is still inferior to process colour-printed mapping, and the viewing area is much smaller than in printed versions of the same map. More complex multimedia versions of the same map will also have to reflect the design of the graphical user interface and incorporate marginalia concerned with screen navigation and manipulation. Display is no longer the only goal for design: functionality and navigation also become important (Miller, 1999).

Above all else when designing maps, seek out examples of graphical quality and consult the standard works by Cindy Brewer, Borden Dent, John Krygier and Denis Wood, Alan MacEachren and Edward Tufte. Use their advice, but be very aware of the context for which you are creating the map and what you want the map to say. Remember, someone will be reading the map, and reading it in the light of a very particular set of circumstances.

CONCLUSION

The following statement concluded a text about British mapping published just as the internet started to have a significant effect on the distribution of mapping, but well after digital map technologies had changed the world of visualization: 'Maps are important. Technology and society are rapidly changing the ways in which they are constructed, used and regarded, but as visual metaphors they will continue to provide statements which both reflect and shape our perception of the world' (Perkins and Parry, 1996: 380). The nature of the representation and its role may have altered, but graphicacy is still central for geographers. We all have a responsibility to make better maps and use them in a more critical way.

Figure 22.8   A location map placing a field course to Crete

Summary

- Mapping works as a social process.
- Maps are available from publishers/map sellers, libraries and online.
- Accessing digital maps requires GIS software, storage media, hardware and output devices.
- Maps work as representations, highlight spatial properties and provide a visual means of creating generalizations, classifications and symbolization.
- Good maps should be organized so that the eye's area of maximum attention corresponds to the central focus of the map, and there should be a hierarchy between different visual levels of the map.
- There is a need for more critical use and design of mapping.

Further reading

- The book by Dorling and Fairbairn (1997) is still probably the single most accessible overview aimed at the undergraduate student, as it introduces how maps work as images.

- More challenging chapters brought together in Harley (2001) or in Dodge *et al.* (2010) explore the many roles played by the medium and establish the changing social context of the history of cartography.

- Parry and Perkins (2000) is still the best printed source about published map availability, providing vital publication details and contacts, but is increasingly being superceded by web sources.

- The most useful introduction to the practical issues in map design and production are Brewer (2005), Dent *et al.* (2008), Krygier and Wood (2005) and MacEachren (1994).

- John Krygiers's blog Making Maps (http://makingmaps.net/) is a rich source of recent design ideas.

- The Cartographic Communication section of the Geographers Craft home page (http://www.colorado.edu/geography/gcraft/notes/cartocom/cartocom_f.html) also offers useful practical advice on a wide range of map-design issues.

Note: Full details of the above can be found in the references list below.

References

American Cartographic Association (1991) *Matching the Map Projection to the Need.* Falls Church, VA: American Congress of Surveying and Mapping.

Balchin, W.G.V. and Coleman, A.M. (1966) 'Graphicacy should be the fourth ace in the pack', *The Cartographer*, 3: 23–28.

Barr, R. (2001) 'Spatial data and intellectual property rights', in R.B. Parry and C.R. Perkins (eds) *The Map Library in the New Millennium.* London: Library Association, pp. 176–187.

Black, J. (1997) *Maps and Politics.* London: Reaktion.

Blaut, J.M. (1991) 'Natural mapping', *Transactions, Institute of British Geographers*, 16: 55–74.

Board, C. (1984) 'New insights in cartographic communication', *Cartographica*, 21: 1–138.

Borchert, J.R. (1987) 'Maps, geography and geographers', *The Professional Geographer*, 39: 387–389.

Brewer, C. (2005) *Designing Better Maps: a Guide for GIS Users.* Redlands: ESRI Press.

British Cartographic Society (2009) *Directory of UK Map Collections.* London: British Cartographic Society (available at http://www.cartography.org.uk/default.asp?contentID=705).

Cammack, R.G. (1999) 'New map design challenges: interactive map products for the WWW', in W.E. Cartwright, M.P. Peterson and G. Gartner (eds) *Multimedia Cartography.* Berlin: Springer-Verlag, pp. 154–172.

Cartwright, W., Gartner, G. and Lehn, A. (eds) (2009) *Cartography and Art*. Berlin: Springer.

Cosgrove, D. (2001) *Apollo's Eye*. Baltimore, MD: Johns Hopkins University Press.

Dent, B.D., Torguson, J. and Hodler, T. (2008) *Cartography: Thematic Map Design* (6th edn). New York: McGraw Hill.

Dodge, M. and Perkins, C. (2008) 'Reclaiming the map: British Geography and ambivalent cartographic practice', *Environment and Planning A*, 40: 1271–1276.

Dodge, M. Kitchin, R. and Perkins, C. (eds) (2009) *Rethinking Maps*. London: Routledge.

Dodge, M. Kitchin, R. and Perkins, C. (eds) (2010) *The Map Reader*. Chichester: Wiley.

Dorling, D. and Fairbairn, D. (1997) *Mapping: Ways of Representing the World*. Harlow: Longman.

Gooding, K. and Forrest, D. (1990) 'An examination of the difference between the interpretation of screen-based and printed maps', *The Cartographic Journal*, 27: 15–19.

Harley, J.B. (1989a) 'Deconstructing the map', *Cartographica*, 26: 1–20.

Harley, J.B. (1989b) 'Historical geography and the cartographic illusion', *Journal of Historical Geography*, 15: 80–91.

Harley, J.B. (2001) *The New Nature of Mapping*. Baltimore, MD: Johns Hopkins University Press.

Harley, J.B. and Woodward, D. (1987) *The History of Cartography*. Chicago, IL: University of Chicago Press.

International Cartographic Association (1995) *Achievements of the ICA 1991–1995*. Paris: Institut Géographique National.

Keates, J. (1984) 'The cartographic art', *Cartographica*, 21: 37–43.

Kraak, M.-J. and Brown, A. (2000) *Web Cartography*. London: Taylor & Francis.

Krygier, J. and Wood, D. (2005) *Making Maps: a Visual Guide to Map Design for GIS*. New York: Guilford Press.

Loiseaux, O. (ed.) (2000) *World Directory of Map Collections* (4th edn). Munich: Saur.

MacEachren, A.M. (1994) *Some Truth with Maps*. Washington, DC: Association of American Geographers.

MacEachren, A.M. (1995) *How Maps Work*. New York: Guilford Press.

MacEachren, A.M., Buttenfield, B.P., Campbell, J., Diabiase, D.W. and Monmonier, M. (1992) 'Visualization', in E. Abler, R.F. Abler, M.G. Marcus and J.M. Olson (eds) *Geography's Inner Worlds: Pervasive Themes in Contemporary American Geography*. New Brunswick, NJ: Rutgers University Press, pp. 99–137.

Miller, S. (1999) 'Design of multimedia mapping products', in W. E. Cartwright, M.P. Peterson and G. Gartner (eds) *Multimedia Cartography*. Berlin: Springer-Verlag, pp. 51–63.

Monmonier, M.S. (1996) *How to Lie with Maps* (2nd edn). Chicago, IL: University of Chicago Press.

Oddens, R. (2009) Oddens Bookmarks (available at http://oddens.geog.uu.nl/ index.html).

Parry, R.B. (1999) 'Finding out about maps', *Journal of Geography in Higher Education*, 23: 265–272.

Parry, R.B. and Perkins, C.R. (2000) *World Mapping Today* (2nd edn). London: Bowker Saur.

Parry, R.B. and Perkins, C.R. (2001) *The Map Library in the New Millennium*. London: Library Association.

Perkins, C.R. (2002) 'Progress in tactile mapping', *Progress in Human Geography*, 26(4): 521–530.

Perkins, C.R. and Parry, R.B. (1996) *Mapping the UK*. London: Bowker Saur.

Peterson, M.P. (2001) 'The development of map distribution through the Internet', *Proceedings of the 20th International Cartographic Conference*, 4: 2306–2312.

Pickles, J. (2004) *History of Spaces*. London: Routledge.

Robinson, A., Morrison, J.L., Muehrke, P.C., Kimerling, A.J. and Guptill, S.C. (1995) *Elements of Cartography* (6th edn). Chichester: Wiley.

Thrower, N. (1996) *Maps and Civilization: Cartography in Culture and Society* (2nd edn). Chicago, IL: University of Chicago Press.

Tufte, E.R. (1983) *The Visual Display of Quantitative Information*. Cheshire, CT: Graphics Press.

Turnbull, D. (1989) *Maps are Territories, Science is an Atlas*. Geelong: Deakin University Press.

Wood, D. (1992) *The Power of Maps*. London: Routledge.

Wood, D. and Fels, J. (2008) *The Natures of Maps*. Chicago: University of Chicago Press.

Using Statistics to Describe and Explore Data

Danny Dorling

Synopsis

In this chapter I argue that simple statistics are better than complex statistics in the study of geography, that statistics are of themselves dull and that the majority of university geography students and staff are ill-equipped to use them, let alone enliven them. Five rough and ready rules are suggested for when and where to use statistics in the study of geography. One origin of the word statistics is that it implies 'facts about the nation-state'. The development of both 'state-istics' and the academic subject of statistics is briefly summarized to provide a context for the argument presented here that geographers should concentrate on simple statistics. Finally, the chapter ends with three examples of how things may not be as they seem when viewed statistically.

The chapter is organized into the following sections:

- Introduction
- Five rules
- What are statistics?
- Why use simple statistics?
- Three statistical examples
- Conclusion – possible futures

INTRODUCTION

Statistics are duller than ditch water. In and of themselves they tend to be of interest only to people who are not very interesting. To me, it is only when statistics are set in a wider context that they begin to come to life. In geography this wider context obviously also varies widely. Suppose you are swamped by a series of floods and want to know how likely they are to happen again and how large they could be. Statistics provide you with techniques and tools to make good guesses – guesses that don't necessarily rely on you knowing much about floods. Suppose you are interested in poverty and how poverty rates compare between countries. Poverty rates are statistics. You need to understand the statistics if you are interested in the issue of poverty. You could, of course, ignore the statistical study of floods or poverty while still being interested in either subject. Were you to do this, however, you would be missing out on a great deal that has been learnt about these things. Statistics on floods might miss out the nature, causes and meaning of floods. Similarly, statistics on poverty may dehumanize suffering. But if you are concerned about getting wet or about why higher levels of poverty persist in some places and not others (and hence one key aspect of what reduces poverty – variation in, say, policies that influence poverty), you are unlikely to get far without the numbers and methods which make up statistics.

Statistics are dull because they are so general – because so many different things can be turned into a percentage; because similar techniques can be used to study so many different processes. The generality of statistics is also their main weakness as well as being a strength. Often issues and problems in geography are shoehorned into a question that can be addressed, and perhaps answered, statistically. A researcher turns an interesting research question into a series of dull hypotheses, the answer to none of which quite gets to addressing his or her original question. It is, for instance, much easier with classical statistical tests to suggest that certain things are (probably) not true – rather than to assert what is true. Statistics can also have the disadvantage that many students have built up a resistance to them, an inherent dislike of their use and feel a chill wind travel down their spine when the word 'statistics' is mentioned. Ask yourself this (assuming you are a student studying for a geography degree): what proportion of your fellow students are likely to be reading this book and, of those, what proportion would skip this chapter? Are you in a minority of a minority to have got this far? Perhaps I am wrong, but hopefully if you read a little further I will try to show you that statistics are not quite as painful as they may be thought to be – particularly if you only use them when you need to.

There is at least one other probable reason why you may have got this far through this chapter. You bought or read this book a year or so ago when you were first being introduced to methods in geography. You read about the interesting ones, the novel and the new, but you skipped this chapter. Later, in a seminar or tutorial, someone asks 'What statistics are you using?' for your research project. That feeling of panic returns. Or you referred to some statistic in an essay and your marker irritatingly scribbled 'Explain!' next to it. In this short chapter I am going to show you only one statistical method (calculating and interpreting a confidence limit). No statistical test and no actual statistics – of the '95 per cent of statistics are wrong' variety. Instead I'll follow this introduction with five simple rules for using statistics in geography and then talk a little about the origins of statistics, give some common misinterpretations of statistics and end with possibilities for the future use of statistics in geography.

FIVE RULES

Often there is little point in using statistics

All too often statistics are included in a study or student essay to try to show that the writer or student had been working hard. An essay on the geography of employment in Britain might begin: 'Thirty million people work in Britain today.' So what? Statistics in and of themselves are dull and pretty meaningless (unless dull things excite you!). 'There are thirty million people working in Britain and forty million jobs' would be a little more interesting – many people would have to have two jobs – but that is only interesting if it is new to you as a reader. 'There was a positive correlation between the number of people working in each county and the number of jobs' – hardly surprising, and what's your point? Perhaps what you were trying to say was:

> More people are working than ever before at more and more tasks in Britain and at even more in the USA. Work, rather than removing human drudgery, has added to it. As we work harder we have created more work. Huge numbers of people have to work at two or three jobs a week compared to our labours in the recent past. The proportion of both children and the 'retired' who work has risen. For the affluent, childcare and housework are increasingly subcontracted as work for others, as for many is cooking when we eat out. It is difficult to see a smooth end to the apparent unstoppable commodification of human time. Work turns time into money. We now have both more money and less time.

Only the very crudest of statistics were used in that argument. Things had simply gone up or down – were 'more' or 'less'. It would have helped the argument if I had listed some sources for this information. It is probably not all true and what is described is certainly not true for many people. However, the argument did not need numbers or tests to strengthen it. There is a time and place for statistics.

If you do use statistics make sure they can be understood

Most people, including a majority of human-geography lecturers and a fair share of physical-geography lecturers, are innumerate – as are most students of geography. By innumerate I do not mean innumerate in the sense of 'could not get an "A" grade at GCSE maths'. By innumerate I mean do not have an innate feel for numbers, mathematics and simple algebra in the same way that the minority of students who are dyslexic do not have an innate feel for words and their spelling. Because dyslexia is a minority condition it is given a name. Because innumeracy is the norm in America, Britain and most of the western world, we do not call it as such but save the word for people with almost no ability to handle numbers. If you are in the innumerate majority you may have thought you had some weakness and everyone else could understand maths and statistics and did not need to learn examples by rote as you did. You are wrong: you are normal. If you are in the numerate minority (that is, you could walk into an exam and get an 'A' in maths without trying) you have a problem – many people (including often those reading or marking your work) might not understand you because they do not think like you. Statistics in geography are primarily a form of communication – a way of trying to convey information convincingly. You will only convey information convincingly if, to begin with, the person you are conveying that information to can understand the language you are using.

Do not overuse statistics in your work or methods

How many facts should there be to the page? Occasionally a great many simple statistics all pointing to the same things or suggesting a trend is going one way or another using many examples can be a powerful tool in making an argument. Generally, however, I'd try to stick to two or three facts on a page unless you are writing a reference book. You are trying to build up a picture in the mind of the reader/marker – explaining an argument through an essay. Similarly, how many numbers should tables of statistics contain? Part of the history of geography has been that multiple regression was a favoured statistical technique for several decades. Multiple regression allows dozens of variables to be entered into a model to predict an outcome. Each variable can have numerous parameters associated with it – for example, its 'effect', 'significance', 'error' and so on. When computers first began to be used in geography, a huge amount of work was required to enter the data and variables to undertake this technique. One result was that huge tables of 'results' were often printed in papers which in fact only referred to a few of the numbers in the tables of sometimes hundreds of figures. Have a look at a journal like *Environment and Planning A* in the 1970s and 1980s for these giant tables. The tables give the impression of thoroughness, but if you want people to *read* the numbers in your tables I'd aim to have only a dozen numbers there. Finally, concerning tests and techniques, there should be little need to use more than one statistical test or technique on a set of data. If you are using more than one, the chances are that only one is appropriate.

If you find a complex statistic useful then explain it clearly

Sometimes, for some problems, only a complex statistical method or complex statistics get to the heart of the issue you are interested in – and it is the issue rather than the method that we are interested in. A typical case in geography might be that you have access to a whole series of river-water samples and know the level of a particular pollutant in each sample. You also know whether the sample was taken from water near the surface of the river or near its bed, whether from a pool in the river, a riffle, or a bend and what both the slope and velocity of the river were at each sample point. You want to know whether some rivers are more polluted than others or whether they just appear to be so because of the nature of the way the samples of water were collected from each river. Your data are arranged in at least two geographical levels: that of the river and that of geographical location within each river. Each individual sample point also has a number of characteristics you want to keep taking into account. There are a number of techniques you could use to study these data. They range from various types of ANOVA (analysis of variance) tests to MLM (multi-level models). The one you are likely to choose will depend most on what you have been taught and remembered, or perhaps on what you have read. My recommendation is that, when using a complex technique in a complex situation like this, explain every step simply – but above all understand why you are using these techniques. The above techniques are interesting for their abilities to identify 'interactions'. Within a particular set of rivers you might expect more pollution in riffles than in pools and vice versa in another set of rivers. If you are not interested in interactions, why are you using these complicated techniques? Furthermore, check first that there is not a very simple pattern in your data that simple statistics could undercover more persuasively. For example: 'Pollution levels are on average ten times higher in the pools of rivers than when measured at any other point.'

Recognize and harness the power of statistics in geography

Statistics have a political power across much of geography. In physical geography their use implies you are (or are becoming) a competent 'scientist'. Many scientific (and, in particular, medical) journals use statisticians to referee papers before publication as well as referees who understand the substantive subject of the paper. Statistics has become a language of scientific credibility – rather like Greek, and then Latin, was the language of religious credibility in the Christian church for most of its history. Languages are used both for their ability to communicate and to exclude the uninitiated. The power of statistics (in particular, statistical methods) in human geography has fallen in recent years. This fall was partly the result of researchers seeing through the way statistics have been used to exclude and also partly because fewer numerate human geographers choose to teach (or research) geography than once did. However, within human geography, statistics placed carefully can still add authority to an argument that, without them, appears little more than a considered rant. This is particularly true when you are arguing against the generally accepted case. As I argue (or rant!) below, statistics are increasingly used in general debates outside of geography. It is possible both to appreciate the weaknesses of statistics and to use them with effect. But it helps first to know where they have come from.

Before considering the origins of statistics, in the pursuit of honesty and to sum up, it is probably worth explaining the situation in which I find myself most commonly introducing geography students to statistics (I don't teach statistics in lectures as I can't find a way of doing so that would keep 150 undergraduates awake or at least attentive for longer than a few minutes).

Here's the scenario: 'So, you've thought of your research questions, done your survey/questionnaire/measuring, got your data, and are ready for some "statistics"…' It would have helped if the students had thought about what they were going to do with their data before it had been collected but, for many, even if they had done that they would not have known the possibilities: what are these statistics things and why do you need them? For most undergraduate students in geography, and much of the rest of the social or earth sciences, statistics are the things that change a 2.2 dissertation into a 2.1 dissertation – or at least the things they think might do that. They are the things 'taught' by the lecturer who didn't get to teach what he or she wanted to teach – or, even worse – by the lecturer who actually chose to teach statistics! They were the bane of your life while revising for 'A' levels. They are the things you dread in exams. You had to learn a few, learn to use a few techniques and learn some more general statistical methods. A dislike of statistics was the reason you studied geography rather than psychology, economics or biology. Students who can cope with statistics don't tend to opt for a geography degree. However, they find them difficult rather than repulsive. If you could not stand statistics you might have picked English, history, art or drama instead. This chapter is for these kinds of student. If you have read this far then it might well be for you too. Above I tried to explain why I think it is worth using some simple statistics. Below I define statistics more fully, give some reasons for not using complex statistics too often and then I end this short chapter by giving a few simple statistical problems.

WHAT ARE STATISTICS?

Statistics are many things and mean different things to different people. My favourite definition is that they are 'facts about the state' – meaning the nation-state people inhabit and usually involving aspects of their lives. This is a very human-geography definition but it fits the origins of statistics where the word began to be used in earnest two hundred years ago to describe the numerical portrayal of countries, which was beginning then. (For more on these, consult William Playfair's writing and a little of what has been written on their origins – see the Further Reading section at the end of the chapter.) The first modern census – collecting useful statistics about the entire population of a country – was undertaken in the USA in 1790. It is perhaps telling that the counting of people in this way as a new technique began in the New World. Statistics as a word has come, in recent times, to mean more often than not numerical facts about the state. Most statistics that people will encounter on an average day in the country that I am currently writing this chapter (Britain) are about that country (e.g. the rate of unemployment, crime statistics, recovery rates from different illnesses). The same is even more the case for the USA. Two centuries ago, as today, economists in various guises were and are responsible for producing the majority of these kinds of statistics. Their use in political arguments, however, has involved a far wider set of disciplines than economics (Dorling and Simpson, 1999, provide examples of these).

Of course, statistics has a second meaning far removed from simply being about facts. Just over one hundred years ago a small group of people who were interested in chance (probabilities) created the academic discipline of statistics. They were also, incidentally, quite interested in society, and many had views about, say, 'criminals' and 'races' which would be rightly seen by most today as abhorrent. This meaning of the word is a world away from the first definition. Most large universities have either a statistics department in them or statistics is part of their school of mathematics. Statistics in this sense has been dehumanized partly due to a wish to forget part of the history of statistics. You may have taken part of an 'A' level in these kind of statistics. Half my first university degree was spent in a statistics department. There, the working definition of statistics was very different from that used across the campus in geography. One of my first textbooks was entitled *Statistics: A Guide to the Unknown* (Tanur, 1989), the implication being that for those things we know about there is little need for statistics, but for anything we don't it has a purpose. The introductory examples I was taught as an undergraduate ranged from assessing the likelihood that Shakespeare was the author of each play attributed to him, to calculating how many whales were likely to be living in the seas around Japan in the future, given current trends (and not knowing how many whales were actually in the sea). The one area that examples were not often drawn from was the study of human societies, and so I often wondered whether this was because we know too much about people and how they organize their lives to make them a suitable subject for statistical study. Many human geographers would argue that people's lives are too intricate to summarize meaningfully statistically. But again, perhaps the lack of statisticians now studying societies is partly due to embarrassing links between the origins of the study of statistics and movements in the first few decades of the last century which sought to claim that some people were superior to others? These are elitist movements which remain strong but which operate far more covertly today. You are almost certainly reading this because you are a product of the continued success of such elitist movements. You were selected to be 'reading' for the degree that leads you to read this book because you passed a series of tests and went to a series of schools for those deemed to be 'more able' (including comprehensive schools in affluent areas). Most sensible social researchers know that you did not pass those tests because of some inherent position you hold on a supposed 'normal curve' of ability, but because of where and to whom you were born, the wealth and caring of your family. But you are also where you are because statistics were misinterpreted in the past to create an education system in Britain that remains highly segregated and shockingly elitist.

WHY USE SIMPLE STATISTICS?

I have not found that the use of complex statistics has greatly enhanced my understanding of people and how their lives are organized over space. Nor have I generally found that reading other people's accounts of such models has helped me much either. Perhaps it's just me, but I think it's worth saying.

From looking at complex models of who chooses how to vote for political parties, to why some people in some places are more likely to be ill than others – to me, the results of complex statistical tests and models depend in most cases more on how the tests and models are put together than on the data. This is a very personal and generalized view (which is why I have written it and why most of this chapter is in the first person). But it is half the explanation I have as to why I think simple statistics are usually better. When you have less control over the numbers and how you manipulate them it is harder to condition the answers you get to what you want to find.

A second reason for tending to use simple statistics is that you are more likely to have something to compare with. Take, for instance, researchers who study pollen in geography departments. These are people who are trying to construct past climate records from the historical record of pollen in a soil core. They tend to call themselves things like Quaternary scientists interested in palaeoclimatic reconstructions rather than people looking at specks of pollen in old mud. Note that how we choose to define ourselves is often as much to confuse as enlighten! What little I know of them and their use of some statistics suggests to me that they mostly tend to use a similar method of cluster analysis to study some aspects of their data. They do that because it worked pretty well for the first person who used it and who taught the second and, although it may not be perfect, it makes more sense if almost everyone uses a method he or she can understand rather than each invent his or her own. Cross-comparison and communication between studies are then possible. Statistics is a language that has different dialects in different places – even within different parts of one small academic discipline. A 'total fertility rate' and 'bank-full discharge' are two common simple geographical statistics but hardly anyone will know the meaning of both. If you do want to know the meaning of these terms, key them into a search engine on the internet – but be careful to confirm the definition you are given on more than one website. Anyone can write almost anything on the web! Wikipedia, much maligned by lecturers and the publishers of textbooks, is a far better source than most others that can be accessed for free. It is particularly good on statistics by the way (less so on opinion, in my opinion).

Even if you use a simple statistic in one part of geography, it is often peculiar to that part of geography. You are choosing to speak to a very small audience. There are, however, even simpler statistics. On the physical side of the subject these are your basic rates, weights and measures. On the human side simple statistics are the stuff of social summaries. They appear in everyday settings most commonly delivered through newspapers, radio, television or the web. These are the statistics most worth using and which put a geographer into a very good position to question the facts. Many, if not most people, accept facts at face value, which is part of the reason why they are such powerful tools to use in argument (see Dorling and Simpson, 1999, if you are interested in how such things are questioned). In short, simple statistics are more easily understood and more convincing. Within a subject such as geography in the first decade of the twenty-first century – where so few geographers feel (or are) numerate – the use of complex statistical techniques has to be questioned: whom are they being presented to? So, if you as a student find them hard to understand I would say to you, don't worry, and ask yourself why you were being asked to understand them in the first place! If you find a good reason for that, *then* try to understand them.

THREE STATISTICAL EXAMPLES

It is hoped this chapter so far has given you some food for thought about the use of statistics in geography today. Rather than list a series of techniques and refer you on to books about their use, I've tried to give you a more personal view about one of the supposedly most impersonal of subjects taught within geography. My view may be a little odd: I use numbers but believe they are deadly dull; I like both to undermine them and use them in the arguments I make. I think our legacy of complicated statistical techniques in geography is more of a historical accident of what was found to work at certain times than a particularly useful set of tools – at least where the study of the geography of society is concerned. I'd like to end this chapter, however, with three examples of how statisticians think differently – which, hopefully, show some things that can be learnt from statistics.

Example 1: a simple prisoner's dilemma

You are playing a quiz game on TV. In front of you are three doors. Behind one door is the prize and behind the other two doors nothing. You pick a door. The quiz host then picks another door. If you picked the right door the host picks a wrong door; if you picked a wrong door the host picks the right door. You then have a chance to change your mind over which door you enter. Which do you go through?

The answer is relatively simple, or it is if you find this kind of thing easy: you pick the door the host picked. Your chance of winning the prize is 2 out of 3. Why? Because 2 out of 3 times you will have initially picked the wrong door. Why does this matter to a geographer? Well, go back to where we started this chapter. There's been a large flood and you are interested in the probability of another large flood happening. The fact that you are suddenly interested in the probability of the flood is not independent of the flood occurring. In fact, the chance of the flood having occurred after the event was 100 per cent. You cannot say 'that was the 1 in 100 year flood', just that 'this is what it would be like' (or perhaps, better, might be like!). Probabilities are usually conditional. In geography we often make simple statistical mistakes as geographers learnt most of their statistics during the discipline's classical phase (when statisticians didn't worry about conditional probabilities so much).

Example 2: another conditional probability

You are worried you have a disease. One per cent of students have the disease. You take a test for the disease. For people who have the disease the test is accurate 95 per cent of the time. The test is positive for you. What is the chance you have the disease?

As you might have guessed, the answer is not 95 per cent. There are two possibilities: either you have the disease and tested positive or you didn't have the disease but the test gave you a false positive. Take 10,000 students. Of the 100 who had the disease, 95 would test positive. Of the 9,900 who did not have the disease, 495 (5 per cent) would test false positive. Thus the chance of you having the disease, having tested positive for it, is 95/(95 + 495) or roughly 1 in 6. You are unlikely to have the disease. Note that almost 6 per cent of students would think they had the disease had they all taken the test. Of course, if you think there was a particular reason for you to be worried and could quantify that, more conditional probabilities would be introduced!

Now, substitute for disease and students, heavy metal pollution and soil samples. You are working in the labs and have a test for heavy metal pollution in a sample of soil that is 95 per cent accurate. About 6 of your 100 soil samples appear to be polluted after you have tested them. Three of these are located in the same village – have you found a cluster? (Answer: no, probably not.)

Example 3: classical confidence limits

There is only one even mildly complicated statistic I now work out on a regular basis and even that has a more complicated interpretation than is usually taught to geography students. I often work out rates in areas – comparing how many things have happened in an area compared to how many you might have expected to have happened in that place given certain assumptions. Most commonly, this is how many people died in an area compared to how many you would have expected to die there given the ages and sexes of the people living there and national mortality rates. Given these things I might work out, for instance, that 120 people died in an area where you would expect 100 people to have died (this is called the indirect method of standardizing rates). Thus, it looks as if 20 per cent more people have died than could have been expected. However, it might just have been a bad year in that place. So with what confidence can I say that 20 per cent more people died there? The equations to approximate the confidence limits for this statistic are as follows:

Given a standardized rate of $100 * O/E$, where O = the number observed and E = the number expected:

Lower confidence limit (95%) = $100*O*(1 - 1/(9*O) - 1.96/(3*\sqrt{(O)}))^3/E$
Upper confidence limit (95%) = $100*(O + 1)*(1 - 1/(9*(O + 1)) + 1.96/(3*\sqrt{(O + 1)}))^3/E$

Thus, our standardized rate is $100*120/100 = 120$ with a lower confidence limit of approximately:

$$100*120*(1 - 1/(9*120) - 1.96/(3*\sqrt{(120)}))^3/100 = 120*(1 - 1/1080 - 1.96/32.86)^3 = 99.49$$

And an upper confidence limit of approximately:

$$100*(121*(1 - 1/(9*121)) + 1.96/(3*\sqrt{(121)}))^3/100 = 121*(1 - 1/1089 + 1.96/33)^3 = 143.49$$

So what does this say? That we can be 95 per cent sure that the real death rate in that area lies between 99 and 143? That we can't be 95 per cent sure that the death rate in that area is not average (100)? What it actually says it that given, say, lots of years of data, 95 per cent of the time the true death rate of that area will lie within these limits – so that if we were to work out confidence limits for the Standardized Mortality Ratio (SMR) of an area each year for 20 years, on average the true rate would lie within those limits for 19 of those years. This assumes there is an unchanging 'true' rate – that something about the area raises or lowers the mortality rates of the population living there.

Given this, these classical confidence rates do not tell us what we thought they told us. It is not the case that the true mortality rate of this area lies between 99 and 143 with 95 per cent certainty. This has not, however, stopped researchers, including myself, from labelling areas on maps as 'significant' if their lower confidence limits exclude 100, but strictly speaking we have little idea what level of significance applies! (For more on this, see Congdon, 2001.)

CONCLUSION – POSSIBLE FUTURES

When this chapter was first being written it was government policy to attempt to kick start a renaissance in quantitative methods in the social sciences. Evidence-based 'this and that' were all the rage, and research training money was supposed to be being redirected slightly towards the modellers and away from more qualitative studies. The intended effect is that more of your lecturers in the future should have more of a grounding in statistics. What effect is this likely to have? Very little, I suspect. Only a minority of us are confident with numbers by the time we (the third of children who go by the time they are aged 20) turn up at university. The contortions of logic required to understand that significance tests and confidence limits are not quite what they seem to be – after all, you have only just managed to learn them – are a further great disincentive to use complex statistics. This should not, however, put you off using simple facts and quoting basic trends in sustaining the arguments you make in studying geography. Just because the study of statistics is becoming ever more complex does not mean the practice of using statistics within geography should cease altogether – just try to recognize the breadth of meaning to the word 'statistics'. Writing a short chapter on statistics within geography is rather like trying to define 'geography'. It means different things to different people. What you have just read has been my take on the subject. If you find contortions of logic of interest, please use the Further Reading section below. If you have been concerned about your insecurities with statistics, don't be – you are normal – just try to use a few more simple facts to strengthen your arguments and try to feel less intimated about the complex methods.

Summary

- Often there is little point in using statistics.
- If you do use statistics make sure they can be understood.
- Do not overuse statistics in your work or methods.
- If you find a complex statistic useful, explain it clearly.
- Recognize and harness the power of statistics in geography.

Further reading

For a chapter such as this the most useful set of further readings is probably a collection of the books and papers which inspired the point of view I've put forward here. The more recent they are the easier they are to obtain.

- First, on the origins of statistics, why not go straight to the source? The links between geography and statistics which were there from the beginning are hard to miss, given Playfair's (1800) extremely descriptive title: *A Geographical, Historical and Political Description of the Empire of Germany, Holland, the Netherlands, Switzerland, Prussia, Italy, Sicily, Corsica and Sardinia: With a Gazetteer of Reference to the Principal Places in those Countries, Compiled and Translated from the German: To which are Added, Statistical Tables of all the States of Europe.* For students in the USA, a search for material on the 1790 census can begin with http://fisher.lib.virginia.edu/census/. If you are interested in worldwide statistics then take a look at www.worldmapper.org and the 400 or so tables of data in excel files available there.

- Two hundred years on and a hugely less ambitious work is the collection of papers that encouraged me to think that statistics are still useful in the study of society: Dorling and Simpson (1999). The book contains examples of the use of statistics in studies across many aspects of life, mostly in Britain and mostly where political debate is currently raging.

- For statisticians, most have their basic reference book. Mine (before I stopped using complicated statistics!) was Breslow and Day (1980). Books such as Breslow and Day are very detailed and assume a great deal of knowledge, but it is worth knowing where people using statistics refer to for their basic formulae.

- A much simpler, if now quite old, introduction to the study of statistics is still available in many university libraries and, if you are interested in learning more about statistics in general, I'd start with Tanur (1989).

- For what may be both the most up-to-date and the most complex book on statistics written by a research professor in geography, see Congdon (2001). The first few pages help explain why 95 per cent confidence limits are not what you thought they were!

- Finally, Plummer (2007) provides an impassioned argument in favour of a renewed statistical approach to the sub-discipline of economy geography.

Note: Full details of the above can be found in the references list below.

References

Breslow, N. and Day, N.E. (1980) *The Analysis of Case-control Studies*. Lyon: International Agency for Research on Cancer.

Congdon, P. (2001) *Bayesian Statistical Modelling*. Chichester: Wiley.

Dorling, D. and Simpson, S. (eds) (1999) *Statistics in Society: the Arithmetic of Politics*. London: Arnold.

Playfair, W. (1800) *A Geographical, Historical and Political Description of the Empire of Germany, Holland, the Netherlands, Switzerland, Prussia, Italy, Sicily, Corsica and Sardinia: With a Gazetteer of Reference to the Principal Places in those Countries, Compiled and Translated from the German: To which are Added, Statistical Tables of all the States of Europe*. London: J. Stockdale.

Plummer, P. (2007) 'Economic geography, by the numbers?', in A. Tickell, E. Sheppard, J. Peck and T. Barnes (eds) *Politics and Practice in Economy Geography*. London: Sage, pp. 199–209.

Tanur, J. (1989) *Statistics: A Guide to the Unknown* (3rd edn). Pacific Grove, CA: Wadsworth.

ACKNOWLEDGEMENTS

I am grateful to Chris Keylock for suggesting some of the examples used here.

24   An Introduction to Geostatistics

Adrian Chappell

Synopsis

Geographers and environmental scientists are interested in the variation of properties in space and the reasons for this variability. The investigation of spatial data and interpretation of arising trend or pattern is the aim of geostatistics. At its simplest, geostatistics is a suite of tools that provides a framework for that investigation. The framework is based on the notion that things closer together are more similar than things further apart and it provides the opportunity to account for the relative position of data in space. Geostatistics is different from 'classical' statistical approaches to handling data sampled in space, and consequently it provides the potential to explore further the geography of environmental variables. In this chapter, the key elements in the geostatistical approach are explained (the theory of regionalized variables; the use of the variogram to characterize spatial processes; and local estimation or prediction – often termed kriging). Examples are provided to illustrate the way in which information contained in the variogram (its structure and model parameters) may be used to interpret the processes and scale of spatial variation in geographical phenomena.

The chapter is organized into the following sections:

- Introduction: what is geostatistics?
- Regionalized variables
- Characterizing spatial processes using the variogram
- Local estimation or prediction
- Conclusion: how geostatistics can help you

INTRODUCTION: WHAT IS GEOSTATISTICS?

Geostatistics is a suite of tools for solving problems with spatial data. It is commonly applied to, but not limited to, the investigation of spatial variation of processes and landforms on or near the Earth's surface (for example, de Roo, 1991; McBratney *et al.*, 1991; Gallichand and Marcotte, 1993; Henebry, 1993; Odeh *et al.*, 1994; Webster, 2000; Chappell and Agnew, 2001; Heuvelink and Webster, 2001). Although many applications appear to be in the realm of the geographer or environmental scientist, many other subject disciplines have used geostatistics to investigate spatial processes. Webster and Oliver (2001: 6–8) trace across several disciplines, the history of developments in spatial variation that resulted in the theory responsible for geostatistics. Present-day practice is influenced most by Matheron's (1971) theoretical

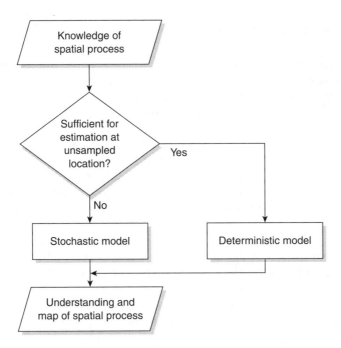

Figure 24.1   Flowchart of the decision-making process and consequence for model
development

work and by Krige's applications in mining (Krige, 1966). Consistent with the
highly applied nature of geostatistics, the intention here is to provide sufficient background knowledge to enable a basic geostatistical analysis to be
undertaken.

The key to this analysis are algorithms represented by flowcharts or tables of
procedures for decisions that are based on the aim, objectives, type of data available and the nature of the sampling etc. For example, the flowchart (Figure 24.1)
shows the highest-level decisions that must be made about the level of understanding for spatial processes, as discussed above. Since there is space here to outline
only three components of geostatistics, a considerable amount of information is
omitted. An attempt to direct the reader to some of the other important aspects of
geostatistics is provided in the annotated guide to Further Reading.

The flowchart serves to remind us that geostatistics is a response to a system
that is currently too complicated for us to make reasonable estimates about the
nature of a property found at an unsampled location. In other words, we do not
know a sufficient amount of information about the processes responsible for the
formation of the property in space (Isaaks and Srivastava, 1989). In practice,
there is rarely such a discrete separation between entirely stochastic and deterministic models and geostatistics is no exception as use is often made of deterministic
information. Consequently, the remainder of this chapter focuses on those three
key elements in the development of a stochastic, geostatistical model of spatial
processes.

REGIONALIZED VARIABLES

The investigation of the spatial variation of a property (for example, those properties related to soil) may be regarded as a response to classification, with the implicit assumption of homogeneity within classes. In turn, homogeneity was probably assumed historically because of the apparent complexity of variation, especially in soil. Intuitively, soil properties and many other properties that vary in space are controlled by spatial processes. Unfortunately, our understanding of these processes is often so limited that predictions or estimates at unsampled locations are often somewhat imprecise and highly inaccurate. Thus, our *deterministic* understanding may not sufficiently reduce uncertainty in the required estimate. Geostatistics recognizes the complexity in spatial processes and utilizes a *stochastic* approach for analysis which presumes that a value of a property may vary at a point in space, in other words there is uncertainty about the value. This is a major departure from the classical statistical approach to estimation and sampling (Chapter 17) which assumes that, for a given property a single value occurs at each point in space. This introduction provides an outline of the stochastic approach to spatial variation (Theory of Regionalized Variables) that supports the practice of geostatistics. It overlaps the more pragmatic considerations of the following section by also providing the underlying concept and assumptions necessary for characterizing spatial variation.

With a limited understanding of the factors controlling spatial processes there is uncertainty about what happens at unsampled locations. The classical statistical approach to assessing uncertainty (Chapter 17) is to assume that samples of a property are selected uniformly (random sampling, stratified sampling, etc.) and are independent. In this case, the statistical rules for sampling design (Cochran, 1977; Yates, 1981) apply, such as that for the variance of the mean. The geostatistical approach to estimation at unsampled locations is based on a probabilistic model that recognizes these uncertainties (Isaaks and Srivastava, 1989). In other words, at one point in space there is *not* just one value for a property but a whole set of values that could be drawn:

- This is regarded as a random variable for a single location and as a single realization of an ensemble of possible values in space.
- Many randomly located samples of a property in space are regarded as a set of random variables (or realizations) that constitute a random function, a random process or a stochastic process (Webster and Oliver, 2001; Chilès, and Delfiner, 1999).
- The actual values that form the realization of the random function are known as a regionalized variable.

Values of regionalized variables tend to be related; their variation appears to depend on the distance that separates them. It is desirable to describe the way in which they vary in space as this will better inform our estimations at unsampled locations. Regionalized Variable Theory (Matheron, 1971) is derived and explained elsewhere (cf. Webster and Oliver, 2001: 51). Here, the focus is on the underlying assumptions and their implications for conducting a geostatistical analysis. In practice, the assumptions of Regionalized Variable Theory are reduced to establishing that the means and the variances of differences between samples over small local neighbourhoods are very similar over the study area (called quasi-stationarity of the Intrinsic Hypothesis). In situations where this does not hold, the local mean values vary predictably or deterministically from one part of the region to another. This is evidence of trend or drift which violates the assumptions of geostatistics. It is common in elevation data (cf. Leenaers *et al.*, 1990; Chappell *et al.*, 1996) and an example of its treatment is provided in the next section.

An emerging development, perhaps parallel to geostatistics, is the proposal (Diggle *et al.*, 1998) that there are some undesirable consequences of the origins of geostatistics. In particular, geostatistical inference is often *ad hoc* in nature, with explicit stochastic models rarely declared and, consequently, little use is made of the likelihood-based methods of inference which are central to modern statistics. The reader can follow this discussion and consider the relative merits elsewhere (Diggle and Ribeiro, 2007).

CHARACTERIZING SPATIAL PROCESSES USING THE VARIOGRAM

Geostatistics provides a different way of thinking about variation in space. As described earlier, it makes use of autocorrelation or the common observation that samples of a process are spatially dependent. In other words, samples that are closer together are more similar than those further apart. This approach requires a tool that incorporates the separation distance and difference in magnitude between samples to characterize the variation in space (the *variogram*). This section provides an outline of the important considerations when computing a variogram and the significance and procedures for fitting models to the variogram. Examples are provided to illustrate the way in which information contained in the variogram (structure and model parameters) may be used to interpret the processes and scales of spatial variation for different properties. Improvements over conventional sampling strategies are illustrated with the combined use of the variogram and nested sampling.

A common approach to understanding the complexity of the natural environment was to simplify it and divide the variation on the basis of certain attributes. This was widespread in land resource surveys and stratification and classification was the norm in soil survey (Webster and Oliver, 1990). This approach aimed to use characteristic information (for example, the mean and the standard deviation of the samples) for any one class, to predict conditions elsewhere in space within the same class. The prediction was based on samples of the population and was dependent on a reduced within-class variance and increased between-class variance. The precision of prediction was limited by the appropriateness of the classification: variation within classes (in other words the continuous variation of a property) was ignored and local variation was not resolved. Regionalized Variable Theory enables us to take this behaviour into account when building a model of the spatial variation of a property (a variogram) necessary for estimation at unsampled locations.

The semi-variogram (or commonly the variogram) is a practical tool for estimating the expected squared difference between two values and their respective separation distance. The semi-variance is commonly computed as:

$$\hat{\gamma}(h) = \frac{1}{2m(h)} \sum_{i=1}^{m(h)} \{z(\mathbf{x}_i) - z(\mathbf{x}_i + \mathbf{h})\}^2 \tag{1}$$

where the z(**x**) and z(**x+h**) represent actual values of a property Z at places separated by the lag vector **h** for a set of data z($\mathbf{x}_i$), i=1, 2, ... and where m(**h**) is the number of pairs of data points. There are a number of important issues arising from the calculation of the variogram that are identified and revisited in the examples below:

- The variogram is the average of all pairs of semi-variances at each separation distance and consequently it is dependent on the amount of data and the consistency of those data. In other words, the variogram can be very erratic when few data are available (Webster and Oliver, 1992) and when those data include extreme values or outliers. Although Krige and Magri (1982) have shown outliers do not present a serious problem, Gilbert and Simpson (1985) suggest the conventional variogram estimator is not appropriate for this type of data.
- Note that **h** is a vector and this implies direction as well as distance. The importance of direction in the variogram will be illustrated below, but directional variograms will not be considered here in detail. Those readers interested in anisotropic variation (spatial variation that is different in different directions) such as that found in fluvial or aeolian geomorphology are encouraged to investigate directional variograms further (Oliver and Webster, 1986b; Webster and Oliver, 2001). The focus here will be the model of spatial variation for a property that does not specify a direction – instead, all directions are included. This is commonly termed an omni-directional variogram.

The semi-variances ordered according to their separation distances (lag) **h** are in practice grouped according to a user-defined lag interval, averaged for each interval and plotted against the lag interval. Experimental variograms for electrical conductivity (EC) surveys in soil forming a playa in Australia are shown in Figure 24.2. Measurements of EC were made approximately every 1 m over a 25 km² area to investigate the spatial variation of salinity in the soil. A separate, complex (nested, stratified and gridded) sampling strategy of 160 locations was also used to make the same measurements. The variogram of EC from the detailed survey is shown using a very small lag interval because of the large amount of data available. The variogram of EC using few sample locations in the same area follows a similar pattern. However, the latter is much more erratic and appears to under-represent the variation of samples spaced between 500 m and 1500 m apart (less so for lag spacings between approximately 1000 m and 1500 m). This is a consequence of the sample locations. In this case, the nested sampling strategy has not represented the spatial variation of the property as

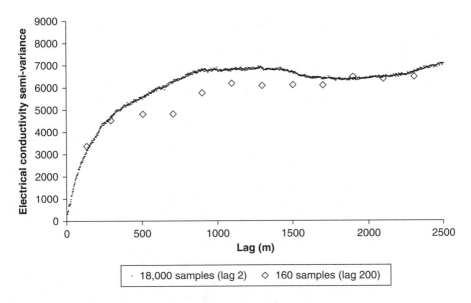

Figure 24.2   Experimental variograms of electrical conductivity measured using electro-magnetic (EM) induction in a 25 km² area of Diamantina Lakes National Park, western Queensland, Australia calculated for (a) 18,000 samples approximately every 1 m and (b) 160 samples using a nested stratified strategy

Source: Data provided by J. Leys

well as that of the detailed survey. However, with so few samples (ca. 160 samples) the nested sampling strategy has captured a remarkably large proportion of the variation at all spatial scales. This is a consequence of tailoring the nested sampling strategy to the geomorphic information of the spatial scale of variation, for example using catena or topography (cf. Oliver and Webster, 1986a; Chappell and Warren, 2003). Conventional sampling strategies using relatively few samples are often incapable of capturing all scales of variation in a single campaign and hence would perform poorly by comparison with the detailed survey. Furthermore, where little is known about the spatial variation of a property, conventional sampling is also often inadequate and a nested sampling strategy may be the only realistic method for capturing the scales of variation (Oliver and Webster, 1986a; Chappell *et al.*, 2003a; Webster *et al.*, 2006). Improvements to the sampling strategy can later be made by using the variogram and the map of spatial variation. In this case (Figure 24.2) improvements to the nested sampling strategy could be undertaken by inserting more samples with spacings from existing samples at the under-represented lag spacings (McBratney *et al.*, 1981). It is evident from this example that sampling and geostatistics are inextricably linked, and this area of research has developed greatly in recent years (Brus and de Gruijter, 1997; de Gruijter *et al.*, 2006).

In order to interpret the spatial variation summarized by the variogram, it is necessary to characterize the structure and provide parameters that represent it. For example, Figure 24.2 shows that, with few samples and a large lag interval, it is difficult to know with certainty that the semi-variance decreases to zero as the lag distance decreases. Even when many samples are available, the semi-variance does not quite reach the origin. The semi-variance at a very small lag spacing (tending to zero) is an important characteristic of the variogram structure. Furthermore, the experimental variograms in Figure 24.2 suggest that as lag spacing increases (samples are further apart) the semi-variance does not continue to increase. In general, the rate of change in semi-variance decreases with increasing separation until some point at which it appears to remain relatively consistent despite lag spacing increasing.

The experimental semi-variances are generally modelled mathematically to estimate the underlying regional variogram. The characteristics of the spatial variation can then be extracted and compared with other properties and those parameters of the variogram model can be used for kriging. Only models that meet several assumptions, that is, are 'conditional negative semi-definite' (CNSD), may be used for kriging (Oliver *et al.*, 1989a). For simplicity, groups of models known to be CNSD are used and linear additions of these models may be combined for more complex structures. In general, there are two families of functions that describe the simple forms of variograms: those that are bounded and those that are not (Figure 24.3).

The steepness of the initial slope of the variogram indicates the intensity of change with distance in a property and the rate of decrease in spatial dependence. Where the extent and intensity of sampling enables the scale of spatial dependence to be determined, the variogram will reach a maximum, called the sill variance, where it flattens (Figure 24.3a). The lag at which the sill is reached is the range, or limit of spatial dependence (Figure 24.3a). This bounded form of the variogram is generally interpreted as representing variation that consists of transition features such as different types of soil or lithology (Burgess and Webster, 1980; Oliver and Webster, 1987). It is commonly represented by the spherical model (Figure 24.3b) but the circular model (Figure 24.3c) often fits equally well. Experimental variograms with the form of the exponential model (Figure 24.3d) are expected where differences in soil type are the main contributions to soil variation and where the boundaries between types occur at random (Webster and Oliver, 2001). In an unbounded variogram, the decrease in spatial dependence may continue indefinitely (Figure 24.3). This unbound variation has been attributed to an increasing number of sources of variation as the area of interest increases (cf. Chappell *et al.*, 1996; Chappell, 1998). It is typically represented by a linear model which is a special case of the more general power model. If an experimental variogram is unbounded and concave upwards near the origin, this may indicate trend or drift which violates the assumptions of stationarity and must be removed before the analysis can proceed. This is discussed in more detail below using elevation data as an example. The mathematical functions of these and other commonly found models in environmental geostatistical applications are described in detail elsewhere (Oliver and Webster, 1986b; Webster and Oliver, 2001).

Experimental variograms often have a positive intercept on the ordinate called the nugget variance and they sometimes appear completely level, exhibiting pure nugget variance. Webster and Oliver (2001) suggest that, if the property of

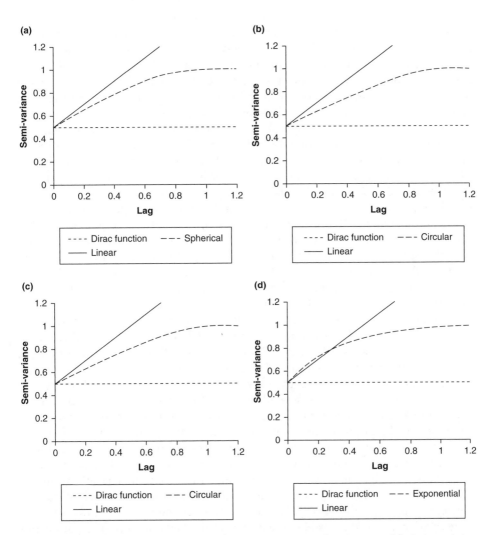

Figure 24.3 Theoretical models used for fitting to experimental variograms: (a) characteristic parameters of bounded models; an unbound linear model and dirac function (nugget model) are shown for comparison with (b) spherical; (c) circular and (d) exponential models

Source: Modified after Webster and Oliver (2002)

interest is continuous, then a variogram exhibiting a pure nugget variance has almost certainly failed to detect the spatially correlated variation because the sampling interval was greater than the scale of spatial variation. The nugget variance was used successfully as an indicator of sampling in rain gauge networks (Chappell and Agnew, 2001) and as a measure of noise in remote sensing images (Curran and Dungan, 1989; Chappell *et al.*, 2001). The other geostatistical parameters have been diagnostic of the structure of spatial variation in soil geomorphology (cf. Chappell *et al.*, 1996; Chappell and Oliver, 1997) and were essential for estimation at unsampled locations (kriging).

The presence of trend or drift violates the assumptions of stationarity in the data. This systematic trend may be removed prior to computing the experimental variogram. The trend is common in elevation data, for example over a hillslope or along a river where the difference between the elevation of locations is not wholly dependent on location but it is also determined by the systematic change in slope. An example of trend in the experimental variogram for elevation in a river channel is shown in Figure 24.4a. The trend may be removed by fitting a low-order polynomial (for example the polynomial $z = ax + by + c$ where z is the elevation and x and y are the location co-ordinates) to the co-ordinates of the data. The residuals of the polynomial are themselves random variables (Olea, 1975, 1977). The use of a polynomial assumes that it adequately explains the systematic variation of the slope. Some geostatisticians (cf. Chilès and Delfiner, 1999) argue that the separation between global systematic change and local variation is problematic and recommend the use of structural tools that are insensitive to drift and rely on a special theory (Intrinsic random function IRF-k; Knotters *et al.*, 1995). Recently, Lark and Webster (2006) proposed the use of residual maximum likelihood to estimate both the trend and the variogram of the residuals simultaneously. In many cases, the residuals of a polynomial fitted to data are adequate to identify the majority of the spatial variation (Figure 24.4a). The parameters of the model fitted to the omni-directional variogram for elevation above an arbitrary datum (Table 24.1) show that the range of spatial dependence is approximately 20 metres.

This variogram is effectively the average of the variation in all directions. It hides the variation in spatial dependence in different directions (Figure 24.4b) and the range parameters of the directional variograms illustrate the difference in magnitude of the distance of spatial dependence. Most of the variation in elevation has been captured by the sampling frequency since the nugget variance for all variograms are very small (Table 24.1). This is not surprising, since more than a thousand sample locations were used to sample the variation. The sill variance is also considerably larger than the nugget variance suggesting that much of the spatially correlated variation has also been captured (see the ratio of the uncorrelated nugget variance and the correlated variance in the last column). The parameters of these models may be used for kriging. However, the anisotropic variogram fitted to all of the directional experimental variograms (Figure 24.4b) is the best approximation of the variation in all directions. It may be used for kriging that takes account of different variation in different directions. A more detailed account of anisotropic modelling and the mechanism for anisotropic kriging is beyond the scope of this introduction and interested readers are directed to the core geostatistical textbooks described in Further Reading.

A procedure for characterizing spatial variation using the variogram is provided in Table 24.2. The aim is to model the spatial variation of a property (theoretical variogram) so that: (a) interpretations of the variogram can be made to characterize variation over space; and (b) estimates may be made at previously unsampled locations. The model is estimated by the experimental variogram and a function is fitted to it. The procedure illustrates the basic decisions and considerations required when calculating, modelling and interpreting the experimental variogram and the model parameters. The calculation of the variogram is relatively

Figure 24.4 Experimental variograms of elevation above an arbitrary datum in a 1 km² river channel in UK. An omni-directional experimental variogram of elevation shows evidence of trend and an omni-directional variogram of the same elevation data is shown with trend removed and an exponential model fitted (a). The experimental variograms calculated in four directions (0°, 45°, 90° and 135°) are fitted with an anisotropic model (b)

Source: Modified after Chappell *et al.* (2003b)

straightforward and can be done with a spreadsheet or computer program unless many data are being considered. Model fitting is more complicated than the variogram computation, and is best left to the bespoke geostatistical software packages

Table 24.1 Parameters of models (omni-directional and anisotropic) fitted to the elevation relative to an arbitrary datum in a river channel in UK

Type	Model Type	Range (m)	Variance Nugget (C_0)	Variance Sill (C)	$C_0/(C+C_0)$ (%)
Omni-directional	Exponential	19.83*	0.05	0.48	9.43
0°	Exponential	34.05*	0.22	0.45	32.84
45°	Spherical	15.68	0.14	0.36	28.00
90°	Exponential	71.96*	0.11	0.58	15.94
135°	Spherical	28.03	0.19	0.44	30.16
Anisotropic	Spherical	23.47	0.18	0.35	33.96

*The effective range is used to represent the range of spatial dependence for the exponential model

Table 24.2 Procedure for computing an experimental variogram, fitting a model and interpreting the parameters of that model

Calculate the variogram

Omni-directional variogram	Calculate the omni-directional variogram using separation distances between samples in all directions.
Trend	Is there any systematic change in the mean of the property over space? Is there evidence in the shape of the variogram for this non-stationary behaviour (trend or drift)?
Directional variograms	Calculate the variogram separately for at least four directions (0°, 45°, 90° and 135°).

Characterize the spatial variation

Bounds	Does the variogram appear bounded, i.e. does the semi-variance reach a maximum within the distance computed or appear as though it would reach a maximum if the lag distance was extended somewhat (bounded)? Alternatively, does it look as though it would increase without limit (unbounded)?
Anisotropy	Does the variogram have approximately the same form and values in all directions? If so, then accept it as isotropic and compute an average experimental variogram over all the directions. If not, then in what way do the directions differ?
Nugget	Does an imaginary line drawn through the experimental values when projected cut the ordinate at a positive value (not 0)? If so, this intercept is known as the nugget variance.

Model the experimental variogram

Choose	Choose from one or more simple models with approximately the most appropriate shape and with sufficient detail to honour the principal trends in the expected values that you wish to represent.
Fit	Fit each model in turn by weighted least squares, i.e. by minimizing the sums of squares suitably weighted between the expected and fitted values. Calculate some statistics for goodness of it, for example residual sum of squared (RSS) difference between expected and fitted values.
Judge	Plot fitted model on the same pair of axes as the experimental variogram and if all looks well choose from among them the smallest RSS.

Source: Modified from Webster and Oliver (2001)

(computer code and software availability is described in the annotated guide to Further Reading below). The variogram is a tool to summarize spatial variation that is effective for elucidating the processes controlling landforms and a model of underlying spatial process that is used for geostatistical estimation at unsampled locations using kriging.

LOCAL ESTIMATION OR PREDICTION

When you have sampled the values of a property at many locations, it should be possible to make estimates at unsampled locations using a model of how the property varies over space (variogram). Estimation is the task for which geostatistics was initially developed and it is generally called kriging (Krige, 1966). A brief history of estimation is provided by Cressie (1993). This section provides an outline of the important differences between estimation and interpolation. The latter may be conducted using an arbitrary mathematical function such as found in graphical software (for example, inverse-distance squared). This section will also describe the procedures and important considerations necessary for geostatistical estimation (kriging). The derivation of the kriging equations are omitted here in favour of a more pragmatic approach. Those interested in understanding the mechanisms by which the kriging conditions are conducted are referred to the further reading. Kriging estimates are often far smoother than the variability of the sample data. Although not considered here, stochastic simulation is an alternative geostatistical mapping approach which may be very useful to some readers (Goovaerts, 2000, 2001). It emphasizes the global attributes of a property and approximates sample statistics (e.g. histogram, variogram, sample values) and is very useful for uncertainty analyses in spatial models such as those found in Geographical Information Systems (GIS) analysis (Goovaerts, 1997).

Interpolation methods

Interpolation methods are deterministic and they accord with our understanding that the variation in the environment has physical causes (it is physically determined). However, the spatial processes are so complex that our current understanding renders mathematical functions (inverse-distance squared, minimum curvature, etc.) inadequate to describe any but the simplest components (Webster and Oliver, 2001). Most methods of interpolation make predictions at unsampled locations from nearby data using an arbitrarily selected mathematical function (Webster and Oliver, 1990). Burrough (1986) and Webster and Oliver (2001: 37) provide useful reviews of various methods for interpolation at unsampled locations. Many of these methods take an average of existing samples from a local neighbourhood. The way in which the nearby data are included is often based on a mathematical relationship between distance and the variation of the property in space. For example, interpolation using inverse-distance squared (IDS) takes existing data within a certain distance of the location to be predicted (Figure 24.5) and

Figure 24.5   The location of nine samples of aeolian transport (g cm⁻¹) from Australia and the
unsampled target location for estimation (+)

Table 24.3   Values of the aeolian transport data at each of the selected sample locations
(Figure 25.5), the distance from the target location and the weights given to each location using
homogeneity, inverse-distance squared and ordinary kriging derived from the variogram

Location	Values	Distance	Homogeneous	Inverse-distance squared*	Ordinary kriging
1	4.24	1344.34	0	0.001	0.003
2	0.03	1167.68	0	0.001	0.010
3	0.23	917.33	0	0.001	0.038
4	2.76	655.40	0	0.002	0.123
5	4.99	469.51	0	0.002	0.203
6	5.42	342.25	0	0.003	0.290
7	0.02	1021.20	0	0.001	0.036
8	208.24	330.73	0	0.003	0.291
9	0.02	1285.16	0	0.001	0.007
Estimated location	–	–	25.11	47.25	63.64

*Weights are not complete for each location since aggregation for all locations is included in the calculation

calculates the influence of each sample on the prediction using the IDS function
and increases the importance it will have on the average of all values (Table 24.3).
Thus, the prediction at the unsampled location is a weighted average according to the mathematical function. The weights shown are for the distance and

the weighted values are not shown because the target value is the average of the result. An illustration of a spatial process predicted by an arbitrarily selected IDS function for aeolian transport is shown in Figure 24.5.

Geostatistical estimation: kriging

Kriging is the method of geostatistical estimation at unsampled locations. Since kriging was first formulated there have been many developments to tackle an increasing complexity of problems in a variety of disciplines. However, ordinary kriging is the most common type of kriging in practice, and the discussion below is limited to this method. It is similar to methods of interpolation in its use of a weighted average of nearby samples. However, the main difference is that the weights given to nearby samples are derived from the variogram model rather than using an arbitrary mathematical function. Since the variogram model is fitted to the experimental variogram, which itself was derived from the existing samples as described above, the variogram provides an appropriate relationship between distance and the variation of the property in space.

In addition to a weighted average, there are several other properties incorporated into ordinary kriging. The weights of those nearby samples required for an estimate at an unsampled location are made to sum to 1 (Table 24.3). The expectation is that there is no error of the estimate made at the unsampled location, but that the estimate has variability. For example, if the estimation location happens to coincide with an existing sample the value at that location is allowed to exist and the variance at that location is reduced to zero. Unlike many other methods, ordinary kriging provides estimates that are an exact interpolation. The kriging procedure also has the ability to make estimates at points and over blocks of any size. The method by which the weights are computed is not straightforward, and those readers interested in a mathematical explanation of the procedures for ordinary kriging are directed to the core geostatistical textbooks (Goovaerts, 1997; Webster and Oliver, 2001).

A comparison between the weights generated by an arbitrary mathematical function and ordinary kriging is provided in Table 24.3. The value at the unsampled location interpolated by inverse-distance squared and that estimated by ordinary kriging is also shown in Table 24.3. The influence of spatial variation on the prediction at the unsampled location can be seen by comparing these values with that of a simple average of the available data which implicitly assumes homogeneity over space. Webster and Oliver (2001: 155) provide further examples of the weighting procedure and also illustrate the effect that the variogram model parameters has on the weights and the estimation procedure. In general, they suggest that the nearest four or five points might contribute 80 per cent of the total weight and the next nearest ten almost all of the remainder. Goovaerts (1997: 125) provides two-dimensional examples of ordinary kriging and shows developments that include simple kriging and kriging with a trend model.

The procedure for ordinary kriging is provided in Table 24.4 and is based on that of Webster and Oliver (2001). It can be used as the basis for providing the necessary information for ordinary kriging in most computer programs. It is especially appropriate for providing some of the parameters for the GSLIB kriging

Table 24.4   Procedure for ordinary kriging to provide the appropriate parameters for computer code such as GSLIB (OKB2DM)

Punctual or block kriging?	If the targets are points (for example small soil cores) then punctual kriging is required. Alternatively, small blocks may be specified in block kriging. More information on block kriging and the number of points used to discretize the blocks can be found in the core geostatistical textbooks.
Number of data for each kriging estimate?	Ordinary kriging computes a weighted average of the data. The weights are determined by the configuration of the data in relation to the target in combination with the variogram model. In general, the nearest 20 points to the target should be specified. If the data points are exceptionally scattered then a local estimate can be retained by reducing the number of included points.
Search radius?	The search radius should be the same size as the range of spatial dependence. In some cases this may need to be slightly larger in order that sufficient data are included in each kriging estimate.
Transformation?	Simple linear geostatistics requires that the data conforms approximately to a normal distribution. If the data are skewed and have been transformed the variogram model parameters and the kriging estimates should be of the transformed data. If you want estimates to be of the original (untransformed) data then you must transform the estimates back.
Variogram parameters?	You will need to include the number and type of each structure for your model of spatial dependence, the amount of nugget variance, the range, spatially dependent variance and any angle and ratio of anisotropy (as described in the previous section).
Kriging for mapping	Specify a finely spaced interval on a square grid that is no more than 2 mm on the final hard copy. Once the kriging estimates have been written to a computer file, pass those values to a graphics program for the final display of the results as isarithms (contours of the same level).

code (OKB2DM or COKB3DM; Deutsch and Journel, 1992). Examples of the parameter files are included in the GSLIB manual and information on obtaining the computer code is provided in the further reading annotations below. For comparison with the previous inverse-distance squared interpolation, the kriging equations were used to estimate aeolian transport at points every 200 m across the 5 km × 5 km study area in Australia. Contours were threaded through the kriging estimates with the same contour frequency as the previous interpolation (Figure 24.6). The results presented here are similar to those found by many other studies (for example Chappell, 1999). The IDS function produces concentric artifacts at unsampled locations. This is primarily because of the arbitrary distance over which nearby samples influence the interpolation. The pattern of transport produced by ordinary kriging estimates is consistent with the observed continuous spatial transport process. Kriging has been shown to be one of the most reliable two-dimensional spatial estimators (Laslett *et al.*, 1987; Laslett and McBratney, 1990; Laslett, 1994) and it is expected to produce more reliable estimates of sediment transport than simple methods of interpolation. However, kriging tends to smooth the variability in the data and other geostatistical techniques are required to maintain the variability in the original data (for example stochastic simulation, cf. Goovaerts, 1997). It is good practice in geostatistical analyses to validate the geostatistical model and kriging plan by cross-validation (Deutsch and Journel, 1992). Each sample location is estimated under conditions which closely resemble those used later for kriging: variogram model(s), the type of kriging and the

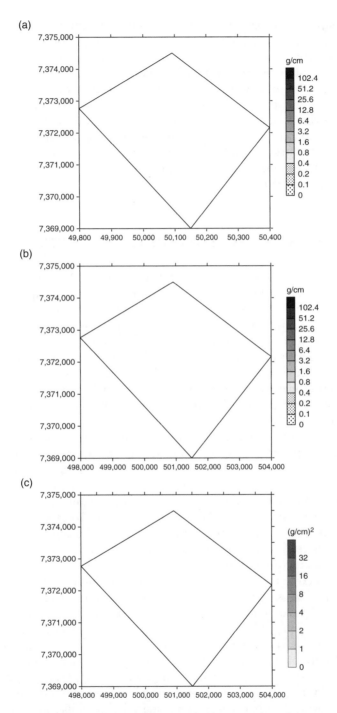

Figure 24.6   A Universal Trans-Mercator projection (in metres and north is oriented to the top) of aeolian transport (g cm⁻¹) from the south displayed using the same (doubling) isarithms for (a) inverse-distance squared (IDS) interpolation, (b) ordinary punctual kriging estimates and (c) sample locations and ordinary kriging estimation variance every 200 m over a 5 km × 5 km study area in Diamantina Lakes, western Queensland, Australia

Source: Modified after Chappell *et al.* (2003a)

search strategy. Actual data are removed (with replacement) one at a time and re-estimated using the remaining data. The estimation performance is commonly quantified using the root mean square error (RMSE) and mean absolute difference (MAE) between observed and estimated values.

An important by-product of the kriging estimation process is the estimation variance which may be considered to be an estimation reliability map (Figure 24.6c). The kriging and estimation variance weights obtained from the variogram depend on the configuration of the sampling points and not on the observed values (Burgess *et al.*, 1981). The estimation variances can be used to minimize the maximum kriging error and optimize future sampling campaigns (McBratney *et al.*, 1981).

CONCLUSION: HOW GEOSTATISTICS CAN HELP YOU

Geostatistics has a sound foundation in statistical theory (Regionalized Variables). The theory is underpinned by the assumptions that: (1) the spatial process is currently too complicated to make reasonable estimates about the nature of a property found at an unsampled location; (2) samples of the process adequately represent the model of spatial variation; (3) variation is spatially dependent. The explicit description of the theory provides the opportunity for its critical evaluation for the exploration of weaknesses and progressive developments. There are many developments in the literature for a plethora of applications over space, time and in space-time (Heuvelink and Webster, 2001) in Geography and Environmental Science.

The central tool of geostatistics is the variogram. This provides a summary for the variation of a property in space. Its structure may be used to characterize and interpret that variation in space. Repeated calculation of the variogram at different points in time enables a comparison of the spatial processes at different stages in time. More advanced geostatistics provide the opportunity to consider the variation of a property simultaneously in time and space. To estimate the model of spatial dependence, the mean and variance of a distribution are required to be similar everywhere. This constraint of second-order stationarity is often restrictive and may be relaxed to constraints within local areas. If this Intrinsic Hypothesis cannot be satisfied or there is evidence (in the variogram) or there is an intuitively reasonable basis, the systematic variation over space must be removed so that the residuals may be used in investigating the model of spatial variation.

The representation or visualization of the spatial dependence in a property usually requires some form of map. Kriging provides the ability to estimate values at unsampled locations with little compromise in the statistical theory, unlike other arbitrary mathematical interpolators. Despite this advantage kriging tends to smooth the variability in the data and other geostatistical techniques (for example stochastic simulation) provide the opportunity to maintain the variability in the original data (cf. Goovaerts, 1997). It is important to note that geostatistics (kriging especially) is unsuitable where there are abrupt spatial changes in properties and researchers have combined soil classification with kriging to utilize the advantages of both

approaches (Bregt *et al.*, 1987; Voltz and Webster, 1990; Heuvelink and Bierkens, 1992). Studies have improved kriging estimates by sampling more effectively
by using regional stratifications such as soil maps (Voltz and Webster, 1990; Yost
et al., 1993) or geomorphic information (McBratney *et al.*, 1991).

Summary

- Geostatistics is a suite of tools for solving problems with spatial data. It is commonly applied to, but not limited to, the investigation of spatial variation of processes and landforms on or near the Earth's surface.
- Geostatistics is a response to a system or process that is currently too complicated for us to make reasonable estimates about the nature of a property found at an unsampled location. It uses a stochastic approach for analysis which presumes that a value of a property may vary at a point in space; in other words there is uncertainty about the value.
- Geostatistics makes use of autocorrelation or the common observation that samples of a process are spatially dependent; in other words samples that are closer together are more similar than those further apart.
- One aim of geostatistics is to model the spatial variation of a property so that interpretations of the variogram can be made to characterize variation over space and estimates may be made at previously unsampled locations using kriging.

Further reading

Provided below are references to the core text books that have influenced my understanding of geostatistics. Although the diversity of texts is limited to my experience of the fields described, the annotations provide a brief outline suitable for those wishing to tackle geostatistics in a variety of fields. As geographers or environmental scientists, we should not limit ourselves to traditional subject boundaries and should embrace a holistic understanding of spatial analysis by investigating geostatistical applications in many disciplines.

- A shortage of geostatistics textbooks led several practitioners to publish comprehensive and detailed information on geostatistics (cf. Cressie, 1993; Goovaerts, 1997; Chilès and Delfiner, 1999). Although these books are required reading for the more adept geostatistician, most are not suitable as introductory undergraduate texts because of their in-depth treatment of the subject.

- Isaaks and Srivastava (1989) provide a book which progressively develops from simplistic to complicated geostatistics using a common database. However, it is difficult to develop from it a rationale for geostatistics. Webster and Oliver (2001) provided a book which is accessible to a new student of geostatistics and which gradually leads the inquisitive mind into greater detail and more opportunities for practice. Of particular importance for the novice practitioner is the section in the book on the reliability of the variogram which includes discussion on sample sizes. Their book provides plenty of applications and usefully an appendix for computer code in a programming language (Genstat 5 Committee, 1992). It is fast becoming one of the core introductory geostatistics textbooks for geographers and environmental scientists alike.

- The application of geostatistics to remote sensing data and its incorporation in Geographical Information Systems has made a considerable impact on land surface mapping (cf. Dancy *et al.*, 1986; Atkinson *et al.*, 1992) on the summary and interpretation (cf. Woodcock *et al.*, 1988; Curran and Dungan, 1989; Oliver *et al.*, 2000; Chappell *et al.*, 2001) and classification of digital imagery (cf. Atkinson and Lewis, 2000) and for reducing data redundancy (cf. Webster *et al.*, 1989; Atkinson *et al.* 1990; Atkinson and Curran, 1995). There are many texts on GIS that incorporate the use of remote sensing and geostatistics (cf. Burrough, 1986) but few which include discussion on the primary components of geostatistics outlined in the main text above that are available for use in remote sensing. Stein *et al*. (1999) gathered together remote sensing researchers who utilize specialist components of geostatistics. Consequently, their book provides a wealth of information on the application of geostatistics to remote sensing. It also provides the opportunity to appreciate the difference afforded by geo-statistics to the treatment of large quantities of data which contrasts markedly with limited samples of spatial processes considered in physical geography and environmental science.

- There are many more applications of geostatistics in physical rather than human geography. This is probably because data in the latter are often not in a form that is readily analysed in this way. However, Oliver *et al*. (1992) illustrate the use of geostatistics in human geography in their analysis of the spatial variation of child-hood leukemia. Their analysis was complicated by the form of the spatial data, but by overcoming it they were able to provide an explanation to the factors controlling leukemia that had not previously been considered. Similarly, Gething *et al*. (2006) used innovative space-time geostatistics to improve health management information systems in Africa. There appear to be many more geostatistical applications in soil science than in geomorphology. Oliver and Webster (1991) provide a brief history of geostatistics in soil science that shows that, in its earliest form, geostatistics had been considered as early as 1911. Their paper illustrates many of the areas in soil science that have improved with the use of geostatistics. There are many examples of the use of geostatistics in other fields of physical geography (cf. Oliver *et al*., 1989b; Leenaers *et al*., 1990; Chappell *et al*., 1996, 1998; Chappell, 1998; Chappell and Agnew, 2001).

- The paper by Sterk and Stein (1997) is interesting and has potentially wide applicability for use with many different environmental data. It illustrates one of the greatest problems with the application of geostatistics to environmental problems – the shortage of data in one or more domain. This is especially the case with climate data which often derive from few stations in space that monitor climatic conditions frequently over time (Chappell *et al*., 2003a). This is problematic for the reliable estimation of the variogram as discussed in the main text above. Sterk and Stein (1997) provide a method of overcoming the sampling difficulty by combining measurements made over time with variation in space. Although this may sound complicated, its application amounts to the careful design of sampling. Several authors have provided space-time theory and applications (Stein, 1998; Kyriakidis and Journel, 1999, 2001; Le and Zidek, 2006). One operational difficulty with space-time geostatistics is that computer code necessary for its implementation is not widely available. De Cesare *et al*. (2002) have recently provided code to tackle such applications and it is likely that the wide applicability of this theory will ensure its incorporation into major geostatistical software in the future.

- The proliferation of geostatistics was for a long time hindered by the availability of computer code suitable for its application. Deutsch and Journel (1992) provided a suite of programs that quickly became the standard probably because they enabled a diverse range of applications to be considered and because the programs offered the flexibility for modification for more specific solutions. The first edition of their book provides a comprehensive overview of the practical considerations for the application of geostatistical theory. The book has recently been revised to reflect a recent packaging of the programs for a more accessible operating environment. In recent years, other software has been developed so that now there is a wide range of geostatistical programs. A noteworthy recent addition is the open source software SGeMs (http://sgems.sourceforge.net) which provides many updates to the GSLib routines and integrates the output with 2D and 3D visualization. One of the best databases of available software and the most successful geostatistics email discussion list is 'AI-geostats' which can be found by searching the internet.

Note: For full details of the above can be found in the refrences list below.

References

Ahmed, S. and De Marsily, G. (1987) 'Comparison of geostatistical methods for estimating transmissivity using data on transmissivity and specific capacity', *Water Resources Research*, 23(9): 1717–1737.

Atkinson, P.M. and Curran, P.J. (1995) 'Defining an optimal size of support for remote sensing investigations', *I.E.E.E. Transactions on Geoscience and Remote Sensing*, 33: 1–9.

Atkinson, P.M. and Lewis, P. (2000) 'Geostatistical classification for remote sensing: an introduction', *Computers and Geosciences*, 26: 361–371.

Atkinson, P.M., Curran, P.J. and Webster, R. (1990) 'Sampling remotely sensed imagery for storage retrieval and reconstruction', *Professional Geographer*, 42: 345–352.

Atkinson, P.M., Curran, P.J. and Webster, R. (1992) 'Co-kriging with ground-based radiometry', *Remote Sensing Environment*, 41: 45–60.

Atkinson, P.M., Webster, R. and Curran, P.J. (1994) 'Cokriging with airborne MSS imagery', *Remote Sensing Environment*, 50: 335–345.

Bregt, A.K., Bouma, J. and Jellinek, M. (1987) 'Comparison of thematic maps derived from a soil map and from kriging of point data', *Geoderma*, 39: 281–291.

Brus, D.J. and de Gruijter, J.J. (1997) 'Random sampling or geostatistical modelling? Choosing between design-based and model-based sampling strategies for soil' (with Discussion), *Geoderma*, 80(1–2): 1–59.

Burgess, T.M. and Webster, R. (1980) 'Optimal interpolation and isarithmic mapping of soil properties. I: The semi-variogram and punctual kriging', *Journal of Soil Science*, 31: 315–331.

Burgess, T.M., Webster, R. and McBratney, A.B. (1981) 'Optimal interpolation and isarithmic mapping of soil properties. IV: Sampling strategy', *Journal of Soil Science*, 32: 643–659.

Burrough, P.A. (1986) *Principles of Geographical Information Systems for Land Resources Assessment*. Oxford: Clarendon Press.

Chappell, A. (1998) 'Mapping ^{137}Cs-derived net soil flux using remote sensing and geostatistics', *Journal of Arid Environments*, 39: 441–455.

Chappell, A. (1999) 'The limitations for measuring soil redistribution using ^{137}Cs in semi-arid environments', *Geomorphology*, 29: 135–152.

Chappell, A. and Agnew, C.T. (2001) 'Geostatistical analysis of west African Sahel Rainfall', in A. Conacher (ed.) *Land Degradation*. Dordrecht/Boston/London: Kluwer Academic.

Chappell, A. and Oliver, M.A. (1997) 'Geostatistical analysis of soil redistribution in SW Niger, West Africa', in E.Y. Baafi and N.A. Schofield (eds) *Quantitative Geology and Geostatistics. Volume 8/2* Dordrecht: Kluwer Academic, 961–972.

Chappell, A. and Warren, A. (2003) 'Spatial scales of ^{137}Cs-derived soil flux by wind in a 25 km^2 arable area of eastern England', *Catena,* 52(4): 209–234.

Chappell, A., Seaquist, J.W. and Eklundh, L.R. (2001) 'Improving geostatistical noise estimation in NOAA AVHRR NDVI images', *International Journal of Remote Sensing*, 22(6): 1067–1080.

Chappell, A., McTainsh, G., Leys, J. and Strong, C. (2003a) 'Using geostatistics to elucidate temporal change in the spatial variation of aeolian sediment transport', *Earth Surface Processes and Landforms*, 28: 567–585.

Chappell, A., Warren, A., Taylor, N. and Charlton, M. (1998) 'Soil flux in southwest Niger and its agricultural impact', *Land Degradation and Development*, 9: 295–310.

Chappell, A., Heritage, G.L. Fuller, I.C., Large, A.R.J. and Milan, D.J. (2003b) 'Geostatistical analysis of ground-survey elevation data to elucidate spatial and temporal river channel change', *Earth Surface Processes and Landforms*, 28(4): 349–370.

Chappell, A., Oliver, M.A., Warren, A., Agnew, C.T. and Charlton, M. (1996) 'Examining the factors controlling the spatial scale of variation in soil redistribution processes from south-west Niger', in M.G. Anderson and S.M. Brooks (eds) *Advances in Hillslope Processes*. Chichester: J. Wiley & Sons, pp. 429–449.

Chilès, J.P. and Delfiner, P. (1999) *Geostatistics: Modelling Spatial Uncertainty*. Chichester: J. Wiley & Sons.

Cochran, W.G. (1977) *Sampling Techniques* (3rd edn). New York: J. Wiley & Sons.

Cressie, N.A.C. (1993) *Statistics for Spatial Data*. New York: J. Wiley & Sons.

Curran, P.J. and Dungan, J.L. (1989) 'Estimation of signal-to-noise: a new procedure applied to AVIRIS data', *IEEE Transactions on Geoscience and Remote Sensing*, 27: 620–628.

Dancy, K.J., Webster, R. and Abel, N.O.J. (1986) 'Estimating and mapping grass cover and biomass from low-level photographic sampling', *International Journal of Remote Sensing*, 7(12): 1679–1704.

de Cesare, L., Myers D.E. and Posa, D. (2002) 'FORTRAN programs for space–time modeling', *Computers and Geoscience*, 28: 205–212.

de Gruijter, J., Brus, D.J., Bierkens, M.F.P. and Knotters, M. (2006) *Sampling for Natural Resource Monitoring*. New York: Springer.

de Roo, A.P.J. (1991) 'The use of ^{137}Cs as a tracer in an erosion study in south Limburg (The Netherlands) and the influence of Chernobyl fallout', *Hydrological Processes*, 5: 215–227.

Deutsch, C.V. and Journel, A.G. (1992) *GSLIB Geostatistical Software Library and User's Guide*. Oxford: Oxford University Press.

Diggle, P.J. and Ribeiro, P.J. (2007) *Model-based Geostatistics*. New York: Springer.

Diggle, P.J., Tawn, J.A. and Moyeed, R.A. (1998) 'Model-based geostatistics' (with discussion), *Journal of the Royal Statistical Society*: Series C (Applied Statistics), 47(3): 299–350.

Gallichand, J. and Marcotte, D. (1993) 'Mapping clay content for subsurface drainage in the Nuile delta', *Geoderma*, 58: 165–179.

Genstat 5 Committee (1992) *Genstat 5, Release 3, Reference Manual*. Oxford: Oxford University Press.

Gething, P.W., Noor, A.M., Gikandi, P.W., Ogara, E., Hay, S.I., Nixon, M.S., Snow, R.W. and Atkinson, P.M. (2006) 'Improving imperfect data from health management information systems in Africa using space-time geostatistics', *PLoS Medicine*, 3(6): e271. DOI: 10. 1371/journal.pmed.0030271.

Gilbert, R.O. and Simpson, J.C. (1985) 'Kriging for estimating spatial pattern of contaminants: potential and problems', *Environmental Monitoring and Assessment*, 5: 113–135.

Goovaerts, P. (1997) *Geostatistics for Natural Resources Evaluation*. Oxford: Oxford University Press.

Goovaerts, P. (2000) 'Estimation or simulation of soil properties? An optimization problem with conflicting criteria', *Geoderma*, 3–4: 165–186.

Goovaerts, P. (2001) 'Geostatistical modelling of uncertainty in soil science', *Geoderma*, 103: 3–26.

Henebry, G.M. (1993) 'Detecting change in grasslands using measures of spatial dependence with Landsat TM data', *Remote Sensing Environment*, 46: 223–234.

Heuvelink, G.B.M. and Bierkens, M.F.P. (1992) 'Combining soil maps with interpolations from point observations to predict quantitative soil properties', *Geoderma*, 55: 1–15.

Heuvelink, G.B.M. and Webster, R. (2001) 'Modelling soil variation: past, present, and future', *Geoderma*, 100: 269–301.

Isaaks, E.H. and Srivastava, M.R. (1989) *An Introduction to Applied Geostatistics*. Oxford: Oxford University Press.

Knotters, M., Brus, D.J. and Oude Voshaar, J.H. (1995) 'A comparison of kriging, co-kriging and kriging combined with regression for spatial interpolation of horizon depth with censored observations', *Geoderma*, 67: 227–246.

Krige, D.G. (1966) 'Two-dimensional weighted moving average trend surfaces for ore-evaluation', *Journal of the South African Institute of Mining and Metallurgy*, 66: 13–38.

Krige, D.G. and Magri, E.J. (1982) 'Studies of the effects of outliers and data transformation on variogram estimates for a base metal and a gold ore body', *Mathematical Geology*, 14(6): 547–555.

Kyriakidis, P.C. and Journel, A.G. (1999) 'Geostatistical space-time models', *Mathematical Geology*, 31(6): 651–684.

Kyriakidis, P.C. and Journel, A.G. (2001) 'Stochastic modeling of atmospheric pollution, a spatial time series framework. Part I: Methodology', *Atmospheric Environment*, 35(13): 2331–2337.

Lark, R.M. and Webster, R. (2006) 'Geostatistical mapping of geomorphic variables in the presence of trend', *Earth Surface Processes and Landforms*, 31: 862–874.

Laslett, G.M. (1994) 'Kriging and splines: an empirical comparison of their predictive performance in some applications', *Journal of the American Statistical Association*, 89: 391–409.

Laslett, G.M. and McBratney, A.B. (1990) 'Estimation and implications of instrumental drift, random measurement error and nugget variance of soil attributes-a case study for soil pH', *Journal of Soil Science*, 41: 451–471.

Laslett, G.M., McBratney, A.B., Pahl, P.J. and Hutchinson, M.F. (1987) 'Comparison of several spatial prediction methods for soil pH', *Journal of Soil Science*, 38: 325–341.

Le, N. and Zidek, J. (2006) *Statistical Analysis of Environmental Space-time Processes*. Amsterdam: Springer Verlag.

Leenaers, H., Okx, J.P. and Burrough, P.A. (1990) 'Employing elevation data for efficient mapping of soil pollution on floodplains', *Soil Use and Management*, 6(3): 105–114.

Margetson, D. (2000) 'Depth of understanding and excellence of practice: the question of wholeness and problem-based learning', *Journal of Evaluation in Clinical Practice*, 6(3): 293–303.

Matheron, M.A. (1971) *The Theory of Regionalized Variables and its Applications*. Fontainebleau: Cahiers du Centre de Morphologie Mathématique de Fountainebleau no. 5.

McBratney, A.B., Hart, G.A. and McGarry, D. (1991) 'The use of region partitioning to improve the representation of geostatistically mapped soil attributes', *Journal of Soil Science*, 42: 513–532.

McBratney, A.B, Webster, R. and Burgess, T.M. (1981) 'The design of optimal sampling schemes for local estimation and mapping of regionalized variables. I: Theory and method', *Computers and Geosciences*, 7: 331–334.

Odeh, I.O.A., McBratney, A.B. and Chittleborough, D.J. (1994) 'Spatial prediction of soil properties from landform attributes derived from a digital elevation model', *Geoderma*, 63: 197–214.

Olea, R.A. (1975) *Optimum Mapping Techniques Using Regionalized Variable Theory*. Series on Spatial Analysis, no. 2. Lawrence, KS: Kansas Geological Survey.

Olea, R.A. (1977) *Measuring Spatial Dependence with Semi-Variograms*. Series on Spatial Analysis no. 3. Lawrence, KS: Kansas Geological Survey.

Oliver, M.A. (1987) 'Geostatistics and its applications to soil science', *Soil Use and Management*, 3: 8–20.

Oliver, M.A. and Webster, R. (1986a) 'Combining nested and linear sampling for determining the scale and form of spatial variation of regionalized variables', *Geographical Analysis*, 18: 227–242.

Oliver, M.A. and Webster, R. (1986b) 'Semi-variograms for modelling the spatial pattern of landform and soil properties', *Earth Surface Processes and Landforms*, 11: 494–501.

Oliver, M.A. and Webster, R. (1987) 'The elucidation of soil pattern in the Wyre Forest of the West Midlands, England. II Spatial distribution', *Journal of Soil Science*, 38: 293–307.

Oliver, M.A. and Webster, R. (1991) 'How geostatistics can help you', *Soil Use and Management*, 7(4): 206–217.

Oliver, M., Webster, R. and Gerrard, J. (1989a) 'Geostatistics in physical geography. Part I: theory', *Transactions of the Institute of British Geographers*, 14: 259–269.

Oliver, M., Webster, R. and Gerrard, J. (1989b) 'Geostatistics in physical geography. Part II: applications', *Transactions of the Institute of British Geographers*, 14: 270–286.

Oliver, M.A., Webster, R. and Slocum, K. (2000) 'Filtering SPOT imagery by kriging analysis', *International Journal of Remote Sensing*, 21(4): 735–752.

Oliver, M.A., Muir, K.R., Webster, R., Parkes, S.E., Cameron, A.H., Stevens, M.C.G. and Mann, J.R. (1992) 'A geostatistical approach to the analysis of pattern in rare disease', *Journal of Public Health Medicine*, 14(3): 280–289.

Stein, A. (1998) 'Analysis of space–time variability in agriculture with geostatistics', *Statistica Nederlandica*, 52(1): 18–41.

Stein, A., van der Meer, F. and Gorte, B. (eds) (1999) *Spatial Statistics for Remote Sensing*. Dordrecht: Kluwer Academic.

Stein, M. (1999) *Interpolation of Spatial Data*. New York: Springer.

Sterk, G. and Stein, A. (1997) 'Mapping wind-blown mass transport modelling in space and time', *Soil Science Society of America Journal*, 61: 232–239.

Voltz, M. and Webster, R. (1990) 'A comparison of kriging, cubic splines and classification for predicting soil properties from sample information', *Journal of Soil Science*, 41: 473–490.

Webster, R. (2000) 'Is soil variation random?', *Geoderma*, 97: 149–163.

Webster, R. and Oliver, M.A. (1990) *Statistical Methods in Soil and Land Resource Survey*. Oxford: Oxford University Press.

Webster, R. and Oliver, M.A. (1992) 'Sample adequately to estimate variograms of soil properties', *Journal of Soil Science*, 43: 177–192.

Webster, R. and Oliver, M.A. (2001) *Geostatistics for Environmental Scientists*. Chichester: Wiley.

Webster, R., Curran, P.J. and Munden, J.W. (1989) 'Spatial correlation in reflected radiation from the ground and its implications for sampling and mapping by ground-based radiometry', *Remote Sensing Environment*, 29: 67–78.

Woodcock, C.E., Strahler, A.H. and Jupp, D.L.B. (1988) 'The use of variograms in remote sensing II: Real digital images', *Remote Sensing of Environment*, 25: 349–379.

Yates, F. (1981) *Sampling Methods for Censuses and Surveys* (4th edn). London: Griffin.

Yost, R.S., Loague, K. and Green, R. (1993) 'Reducing variance in soil organic carbon estimates: soil classification and geostatistical approaches', *Geoderma*, 57: 247–262.

25 Using Geographical Information Systems

Michael Batty

Synopsis

Geographical information systems (GIS) are organized collections of data-processing methods which act on spatial data to enable patterns in those data to be extracted, understood and visualized. In fact, GIS is often thought of as software which takes numerical spatial data which can be visualized in map form, stores it in the most efficient way in various computer environments, and allows an array of techniques of analysis to process the data and then map it. GIS, however, is broader than this, synonymous in some contexts with quantitative geography, while the term is increasingly being used to refer to 'geographic information science' which embraces a wide range of mathematical and statistical techniques. Here, first its origins will be noted and then GIS will be defined in terms of the way it represents geographical data. The various operations which enable spatial data to be analysed and visualized are referred to as 'functions' and this 'functionality' will be illustrated through ways in which GIS can measure, interrogate and manipulate maps as data. The power of visualizing data through GIS is then discussed. Finally the chapter presents issues concerning changes in the technology and its software which is moving the methods from the desktop to the internet and enabling ever user-friendly web applications, many of them using 'free' software such as Google Maps. This emphasizes the great diversity of applications which GIS offers and the impact it is having on geography, geographers, and society-at-large.

This chapter is organized into the following sections:

- What is GIS? Historical antecedents
- How a GIS is organized? Representing data
- Functionality in GIS: geometric operations, spatial queries, and map algebras
- GIS as visualization
- Technology for communicating geographic information
- Conclusions: applications

WHAT IS GIS? HISTORICAL ANTECEDENTS

GIS evolved slowly from several origins. As computers became ever more powerful, computer graphics came of age, and from early and somewhat painful beginnings with large and expensive map plotting devices, computer memories reached the point by the mid-1980s where maps could be displayed with ease on the screen. Computer cartography was complemented by the development of database theory, specifically for spatial data which involved linking geometry to geography, while quantitative models and spatial statistics gradually began to be used as key functions within evolving GIS software (Chrisman, 1988). Remote data acquisition through aerial and satellite sensing came to be linked with GIS, and today we stand at a threshold where the technology is spreading across the web and is being implemented in everything from hand-held devices and telephones to software for target marketing and climate modelling. The most recent of these technologies are satellite navigation systems, which by 2007 had penetrated 10 per cent of the European market and which have brought GIS-like technologies to the mass market.

It is important to separate technology from theory. At one level, GIS is simply visualizing map data in whatever context and in this sense, it is little different from much of the graphical computation that we see everyday on the desktop and the web. But GIS as geographic information science is of much wider import for geography and geographical method in that it contains many generic models that embody the most elaborate (and sometimes esoteric) of theoretical conceptions about how the geographical world works. Currently, GIS technology has advanced to the stage where the focus is no longer on graphical representation *per se* but on integrating visualization with method so that both quantitative and qualitative analysis might be enriched (Longley *et al.*, 2005). The next decade is likely to see some remarkable advances as theory and technology merge. Since the early 1990s when the first desktop GIS software systems emerged, the technology has crept out onto the internet but in doing so it is changing as it becomes 'commodified'. Moreover, there are now many open source ('free') GISs which one can use on the desktop but the fastest developments are GIS systems on the web evolving from free mapping software. But this is getting ahead of ourselves and before this future is charted, let us step back, first with an illustration of how a GIS is organized and then how it can be used to visualize and analyse spatial data.

HOW A GIS IS ORGANIZED? REPRESENTING DATA

Two kinds of data are required. As GIS always has the capability of displaying data in map form, there must be data about the way the map is configured, as boundaries and points for example; this is called *digital map data*. Usually there is more than this for the map has features or characteristics – called *attribute data* – and these data are associated or tagged to the map's configuration. A good example is a map of population by local authority area: the boundaries of the local authorities have to be input as digital map data and the population associated with each authority is attribute data. There is another twist to this for there are two different ways of representing the map as digital data. The simplest is to assume all areas are the same, as if you were representing the population in equal grid squares; this is called a *raster map*. Rasters are particularly useful for data produced routinely such as that from satellites where the easiest way to record it is using equal areas (like grid squares or pixels). However, more realistic configurations of maps are based on points and lines which are assembled into objects

Figure 25.1 Digital data types: raster and vector map representations

Source: Adapted from the University of Melbourne's GIS self-learning tool (http://www.sli.unimelb.edu.au/gisweb/)

such as polygons; this is called a *vector map* such as the map of local authority boundaries. Both raster and vector maps have attributes; the raster being associated with grid squares which become a computer screen, the vector with irregular areas which are defined by assemblages of points and lines. Raster data can be entered into a GIS directly as numbers from the keyboard (although it is normally entered as remotely scanned source data), whereas vector data is usually digitized using a mouse-like device called a puck, which is centred on each of the points and lines defining the map object in question. These distinctions are shown in Figure 25.1.

Much of GIS technology is concerned with visualizing these data in map form and many low level GIS functions (involving how closed polygons are drawn, for example) are buried away in the software, no longer of any real significance to geographical analysis. In fact there is little difference between raster and vector map data with many systems enabling users to integrate and move easily between each. However, GIS really comes into its own when more than one set of attributes is associated with a raster or vector map. The central organizing concept is based on treating different sets of attributes as *map layers*. It is useful to first think of geographical systems as being represented by a series of data layers which can then be translated into map layers, and this leads one to consider ways in which the layers might be related. For example, in a raster map, we might have layers that deal with topography, vegetation, geology, agricultural use and so on, where it is useful to relate these variables so that we might examine how they correspond or correlate with one another. GIS enables a first shot at such analysis by simply displaying the map layers and then by 'overlaying' them to see if there is any such commonality.

Another instructive example involves examining the relation between where people live and where they work. If these are mapped as separate layers, then comparing them will reveal that people do not usually live and work in the same place. If we do this for a large city, then it is likely that this will reveal the classic

Figure 25.2   A digital map base (bottom right) and three map layers of attributes in the desktop GIS MapInfo

pattern of people working at the centre and living at the edge. Illustrating the use of GIS in this chapter is the example of Greater London, first divided into its 33 boroughs. Figure 25.2 shows three different layers for the year 1991: total population, percentage of households who own their own homes and the percentage who rent from their local council. The maps showing these data have been created in the desktop software MapInfo, and the figure shows the typical desktop that a user would see. Apart from the ability of these software to display such data in many different map forms, the real power of GIS comes from being able to associate, combine, analyse and model these data, and this involves us in presenting the kind of tools that are available within GIS which, in turn, form the core of geographic information science.

FUNCTIONALITY IN GIS: GEOMETRIC OPERATIONS, SPATIAL QUERIES AND MAP ALGEBRAS

The most basic operations with a GIS involve measuring various geometric features of the digital map data. Once such data are within a GIS, several of these measuring functions are virtually automatic. For example, most GISs have rulers which enable you to measure straight line distances, functions to compute areas of polygons, and methods to count the density of points. There are many functions which derive from these, such as the ability to draw areas around points and lines – called buffers – which provide ways of computing 'nearness' which is a crude proxy for accessibility. If networks are represented within a GIS, then there is added functionality to find shortest routes. Map projections into many different coordinate systems are immediate, for all such functions are routine once the map's geometric data have been represented digitally.

The other set of routine functions within a GIS pertains not to geometry but to the attribute data themselves, and these involve various ways of interrogating them. Spatial queries of the simplest form are usually achieved by pointing at some area or point on the map and accessing its attributes directly. Much more sophisticated queries are possible, however, based on concatenating different requests. A typical one might be of the form: "Find all areas on the map which have a population density greater than 1,000 persons per square kilometre, within 5 kilometres of a main highway, and which are located on land with slopes less than 1 in 20." In fact, the current generation of satellite navigation devices available on mobile phones, PDAs, or in-car systems use such querying extensively where a user keys in location by name – city, street, etc. – or by post/zipcode – or simply navigates to the area by panning and zooming across the map.

Such queries can also be developed into ways of producing new data from the basic data layers and this introduces one of the most important functions of contemporary GIS software which is called 'map algebra' (Tomlin, 1990). The concept of data layers is useful here because it is a particularly simple way of thinking about how different attributes might be combined. Such combinations are a convenient way of representing new types of derived data. For example in our London example, we have population in 1991 – one of the maps in Figure 25.2 – and we can easily compute the area of each local authority as there is standard function in the GIS to do this. P_i is the population in each local authority i, L_i is the area of each authority, and we can then calculate the density $D_i = P_i/L_i$. To do this in GIS, we can invoke a standard calculator which enables us to plot the map of density by combining the data according to this formula. The result is shown in Figure 25.3 with the typical GIS dialogue boxes used to effect the calculation shown alongside. This is, in essence, map algebra, although in a sense, what we are doing is ordinary arithmetic on the data and presenting it through visualizing the result as a map. This is a powerful but simple way of thinking about geographical relationships.

A further example impresses the point. For London, we have population at two dates in time – 1981 and 1991 which we call $P_i(t)$ and $P_i(t + 1)$. We can easily compute the growth rate for each local authority area i as $\lambda_i = P_i(t + 1)/P_i(t)$ and using these rates, we can project the population forward one step at time as $P_i(t + 2) = \lambda_i P_i(t + 1)$, $P_i(t + 3) = \lambda_i P_i(t + 2)$ and so on. Those of you familiar with ordinary algebra will see that this kind of relation is recursive and that we can project the population forward any number of time steps with the assumption of constant growth rates, of course. The formula for this would be $P_i(t + n) = \lambda_i^n P_i(t)$ which illustrates the 'compounding' or exponential nature of the growth, like the way 'compound' interest is used in simple discounting

Figure 25.3 Computing a map of population density for London boroughs within the GIS MapInfo

operations. This is illustrated for London in Figure 25.4. There we show the two populations, the growth rates and the population 100 years on from 1981 which involves using the growth rate on each newly computed set of populations some ten times. In essence, what we have done is to show how a simple population forecasting tool can be built into a GIS by thinking of this as updating a map of population.

The idea of map overlay is even more extensive. One of the original areas where GIS was developed was for landscape planning where the typical problem was to find the suitability of land for different uses (McHarg, 1969). This invariably required different factors to be considered which could be represented as maps, each providing a different and often contradictory impression of where the most suitable land was located. Methods for reducing conflict between such factors are usually based on overlaying these maps and often weighting each factor differentially to provide some sort of combined suitability surface. This is the classic example of map algebra which is illustrated schematically in Figure 25.5 showing how maps are weighted and added. Of course the decision to weight and add in this fashion is not something that is intrinsic to GIS for it depends on the use to which the GIS is put by those involved in such problems (Heywood *et al.*, 2006).

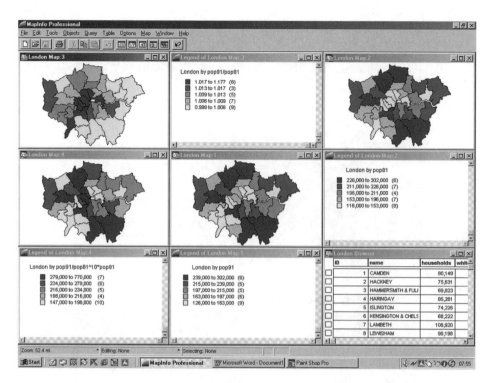

Figure 25.4   Projecting the population of London boroughs for 100 years from 1981. Top left: growth rates from 1981 to 1991, top right: population 1981; middle middle: population 1991, and middle left: population in 2081 using 1981–1991 growth rates; legends for these maps in other windows and a fragment of the data attributes table in the bottom right window. This panel of maps and legends illustrates that the software lets the user 'tile' the screen with the windows which have been created for the maps, data and legends

GIS AS VISUALIZATION

GIS is not computer cartography although it may be quite justifiable to use the cartographic functions in a GIS to produce maps if many variants of the map are required and if the overhead of just entering map data is justified. However, the visualization capabilities of most GIS are by no means restricted to presenting the two-dimensional map. There are extensive functions for presenting different types of 2-D map with a distinction here between vector and raster, vector maps being associated with thematic, perhaps more abstract mapping, and raster maps being more lifelike in appearance. Thematic maps with bars, pies, density dots and so on to represent various attributes are standard, while raster maps with various types of hill shading giving the appearance of 2½-D (2-D with the third dimension simply extruded) and oblique views are part of many packages. In fact, although GIS has not been traditionally used for presentation in map design, software is beginning to incorporate tools for good design while in the GI science domain, tools are being added to enable model and simulations to be generated (Maguire *et al.*, 2005).

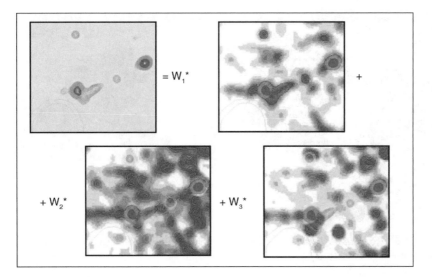

Figure 25.5   Map algebra: adding (+) and weighting (W_i) sustainability surfaces based on economic employment, property values and leisure potential in an area of West London

There are usually other graphical functions for visualizing data within most GIS. For example, there are drawing capabilities which enable maps and any other graphic to be constructed in the map window. There are usually some graphing facilities. For example, MapInfo is organized around four different graphical 'views': the map itself of course, but also a graphing routine which enables scatter graphs, pie charts and so on to be presented (of the attribute data) in non-spatial form. The other two views are of the data itself – the numeric tables and a layout window which allows the user to place maps, graphs, and tables together on a bigger canvas, a minimal form of presentational design.

The greatest advances in visualization, however, are coming with the extension of GIS to the third dimension. For a long time, GISs have dealt with landscape data, and various techniques of developing 3-D landscape models are included in some packages such as ArcGIS. However, the move to explicit 3-D within GIS indicates the weakness of conventional forms of 3-D representation. Computer-aided design (CAD) models, for example, do not have anything like the data functionality that GIS has but increasingly, those involved in 3-D modelling wish to attribute other data to the 3-D geometry, and then use query, overlay and spatial analysis techniques in their exploration of such models. An example of what is possible in current desktop GIS software is shown in Figure 25.6 where we illustrate a 3-D block model of part of inner London, constructed from population densities at small area enumeration district level and displayed within the desktop ArcGIS. This is typical of many low cost GISs in that it can be upgraded by purchasing plug-ins – add-on modules which extend its functionality. One of these – ArcScene – enables users to extrude 2-D areas into the third dimension, to then pan and zoom around such 3-D scenes, and to navigate through the scene, querying the data as one goes.

Figure 25.6   Representing attribute data in 3-D: population density in small census areas in the London Borough of Hackney

Figure 25.7   A portion of virtual London created in ArcGIS (left) and displayed in Google Earth (right)

This kind of tool is providing entirely new ways of visualizing cities (Batty and Hudson-Smith, 2005). 2-D and 3-D map systems, however, are fast moving from the desktop to the web. Tagging vector map data with a third dimension in terms of representing point and line data as x,y,z coordinates essential to its extrusion into the third dimension as in Figure 25.6, enable such data to be used in a variety of CAD and other software. These can now be imported quite easily into systems such as Google Earth, which is rapidly becoming (like its sister software Google Maps as we show below) non-proprietary CAD and GIS for the mass market. We have built a large 3-D model of Greater London – called *Virtual London* – so that the Greater London Authority and the boroughs can visualize the impact of high buildings, development proposals, air pollution and so on in user-friendly ways. The model was built in state-of-the-art desktop GIS software but is available for use in Google Earth, an example of which is shown in Figure 25.7. Into this web-based software can be embedded many other media, still pictures, panoramas, attribute data of a textual and numerical kind, links to other web sites and so on, while different layers can be added to the 3-D building content. We can intersect the 3-D content with surfaces, we can flood the model, we can query so that the user can pull out the data in terms of the location of a building or street in 3-D. There are many such applications, and viewing such data in these terms is likely to become widespread within the next decade.

TECHNOLOGY FOR COMMUNICATING GEOGRAPHIC INFORMATION

Although the technology and software of GIS do not dominate the geographical science that they support, there are remarkable changes still being worked out with respect to what this technology is able to offer. The ability to work graphically with data at immense speed, accuracy and the best visual quality is a remarkable-enough phenomenon in the light of how geographers visualized and explored data a generation or more ago. But the real impact is on how software is being distributed across networks – across the internet – and how many new users of geographic data and science are being drawn-in by GIS. Extending GIS into 3-D is only one cutting edge. Another is the ability to explore geographic data remotely using internet GIS which links digital to attribute data and serves it to users in remote locations. The functionality of such internet map servers is fast reaching that of desktop and workstation GIS but many of these applications are now being developed as part of enterprise computing, for large corporations who have distinct and customisable requirements. In contrast, GIS on the web is becoming more open-source like with various kinds of map and 3-D display device acting as free viewers of map data. To these, a generation of programmers are adding the sort of functionality that exists in mainstream GIS with these applications being referred to as 'maphacks' which originated in the games industry.

Two rather different, but equally instructive, examples of how GIS is being developed on the internet will be shown. First, an internet map server has been constructed at University College London to deliver information about town centres to those interested in their planning. A simple overlay method has been developed based on the logic illustrated in Figure 25.5 above, but instead of being on the desktop, this is available through web pages to as many users as the network can handle at remote locations. What has been done is to design a map overlay procedure which takes different indicators of town centre sustainability and enables users to combine these. There are many data layers which are based on very detailed employment, turnover, rents, floorspace, social composition, consumer profiles and so on which all have different implications for the economic sustainability of different areas. Users can select, combine and differentially weight these to produce new surfaces of sustainability. This supports the idea that a very wide range of users with different viewpoints need to discuss these issues, and this is helped by the delivery of information and the need to actively work with it in this fashion. A picture of the interface to this tool is illustrated in Figure 25.8.

Figure 25.8   Combining map layers, as in Figure 25.5, to form indices of urban sustainability in Greater London using internet GIS through a web browser

The second example involves the use of 'free' internet mapping software. A resource has been built based on fast map creator software that turns GIS files into layers that can be displayed in Google Maps. The application is to geo-demographics data in Greater London with attribute layers pertaining to education, deprivation, ethnicity, and so on, forming the data in the London Profiler (www.londonprofiler.org). It is possible to load and combine different map layers and even add other map layers from remote locations which pertain to London and are already displayed using Google Maps technology. An example is shown in Figure 25.9 where we have brought up the index of multiple deprivations coded to the lower-level super output area geography (on average about 7,500 persons per area) for Greater London and displayed on a ten-point scale. This gives an idea as to how GIS is being opened up to the wider world and how simple functionality can be added to the various online mapping systems that are now available. What this heralds is a revolution in GIS in particular and IT technologies in general with much of the basic technology now being widely available and enterprise bespoke applications being the focus of more specialist GIS software and systems.

The development of networked GIS is only one part of the story, however, as GIS is now being linked to other kinds of software at many different levels of sophistication. Spreadsheets such as Microsoft Excel let you import maps from MapInfo while ArcGIS files can be converted into just about any graphical format

Figure 25.9   Map layers on the web: the London profiler built on top of Google Maps

one could want and therefore be used in related software. Most GISs are now open to programs written in one of many high-level languages and we are moving to a world where any kind of software can be linked to any other within the limits of the application, of course. Many of the links are loose, although increasingly, software is merging across the desktop and the net and the days are probably numbered for the stand-alone all-purpose GIS. Software will increasingly be delivered as different modules to be plugged together and there will be a proliferation of systems applicable in different contexts. Convergence from the many to the few is not likely to be a feature of this domain of GIS for some time to come.

CONCLUSIONS: APPLICATIONS

There is virtually no geographical problem that is not touched by GIS, if not by the software or the system, then by the science. Whether or not GIS is applicable will depend on many things, not least the quality and extensiveness of the data. If all that is required is computer mapping, then GIS may not be worthwhile unless the maps are to be reworked many times. But, as in all problem-solving, the problem must be well-formulated for GIS to be of any use. As already implied, the diversity of application areas is staggering and all this Conclusion can do is to point to significant ones. Although GIS came from landscape planning and automated cartography, the biggest substantive areas currently lie in urban planning and in environmental analysis (Heywood *et al.*, 2006). GIS began at rather coarse scales, but it is gradually being applied at finer and finer scales and there is every prospect that entire new professional areas in real estate, architecture and urban design will embrace these tools during the next ten years.

Niche areas of GIS have also emerged, one of which is 'business geographics', called more broadly geo-demographics, with a focus on marketing and retailing, utilizing new sources of data on consumers. In terms of physical systems, apart from remote sensing where GIS and RS packages are often tightly linked, there has been less development with the exception, perhaps, of geology and terrain analysis (Bonham-Carter, 1994). Much of the current functionality deals with networks, spatial data structures, and specifically more human kinds of analysis than physical, although this is changing and GIS is making substantial inroads into ecology and related sciences. There are some areas that GIS has not touched and one is meteorology. This is largely because such areas have always had their own software and systems. In fact, where there have already been well-established models and analytical techniques, GIS has been less evident in terms of its applications. Transport is another case in point and this illustrates an important limitation of the approach. GIS is inevitably data-driven and this is often a precursor to modelling and simulation. In this sense, GIS is prior to prediction and design, and although many problems require large arrays of diverse data, those that are very focused like transportation planning or weather prediction tend to use GIS, if at all, purely as a display medium.

We also need to note the difference between routine and the more infrequent, more strategic uses of GIS. Routine usage is for rather low-level queries and for scheduling and much of the future use in hand-held systems, for example, is of this nature. To an extent, although GIS is used in these areas, much of its geographic software is purpose-built and often embodied directly into the control systems used in such ongoing tasks. The best examples are in the current generation of satellite navigation devices where GIS functionality is widely used, particularly in querying, but the data itself on street and road networks is also organized in GIS structures and available for many applications other than navigation *per se*. In the strategic context, the kinds of functions illustrated earlier for map algebra are those that are likely to be used. This emphasizes support for decision-making and planning in the widest sense which again is an area of increasing interest and application. There are very rapid developments in GIS at present, particularly with respect to open source and internet access. Within a decade, many of the techniques within GIS will have changed beyond recognition as GIS diversifies and fragments. What is certain, however, is that as most human activities now occur in digital environments, GIS-like functionalities, systems and software will play an ever more important part.

Summary

- Geographical information systems (GIS) are organized collections of data processing methods which act on spatial data to enable patterns in those data to be extracted, understood and visualized.
- For some, GIS covers a very broad area of research and research methods, synonymous in some contexts with quantitative geography, while the term is increasingly being used to refer to 'geographic information science' which embraces a wide range of mathematical and statistical techniques.

- Although GIS came originally from landscape planning and automated cartography, the biggest substantive areas currently lie in urban planning and in environmental analysis. GIS began at rather coarse scales but it is gradually being applied at finer and finer scales. There is every prospect that entire new professional areas in real estate, architecture and urban design will embrace these tools during the next ten years.
- There are differences between routine (e.g. satellite navigation) and the more infrequent, more strategic uses of GIS which support decision-making and planning.
- There are very rapid developments in GIS at present particularly with respect to open source and internet access. Within a decade, many of the techniques within GIS will have changed beyond recognition as GIS diversifies and fragments.

Further reading

- The textbook by Longley *et al.* (2005) provides a good introduction while the bigger two-volume edited reader by Longley *et al.* (1999, 2005) contains 72 articles on many different aspects of GIS.

- Niche areas are worth exploring: for example in Batty *et al.* (1999) explore how GIS might influence small scale site design.

- There are many new applications of GIS in Heywood *et al.* (2006) where there is an example of the *Virtual London* model (pp. 269–271).

- A good book stressing applications in physical geography is Burrough and McDonnell (1998).

- The flagship journal with the greatest technical flavour is the *International Journal of Geographical Information Science* (*IJGIS*), while there are a host of magazines, which appear at least monthly, targeted mainly at the industry but useful to see how the field is developing. The main one at present in the UK is the monthly *GEO: connexion* (http://www.geoconnexion.com/).

- There are several really good websites where you can learn about GIS. Visit ESRI's site and the Virtual Campus at http://training.esri.com/. The site at Edinburgh is a good teaching and learning resource http://www.geo.ed.ac.uk/home/gishome.html, and the SPLINT project at UCL, Nottingham and Leicester is providing many teaching materials (see http://www.spatial-literacy.org/ for material on geo-demographics). The site at UCL http://www.casa.ucl.ac.uk/ reports a number of GIS projects and the digital-urban blog is an enormous archive of visual material about maps and 3-D applications of GIS to cities (see http://www.digitalurban.blogspot.com/). Finally, UK Census data can be downloaded from Manchester University (http://www.mimas.ac.uk/) and boundary data from Edinburgh University (http://www.edina.ac.uk/) and read directly into various GIS packages and spreadsheets.

Note: Full details of the above can be found in the references list below.

References

Batty, M. and Hudson-Smith, A. (2005) 'Urban simulacra: from real to virtual cities, back and beyond', *Architectural Design*, 75(6): 42–47.

Batty, M., Dodge, M., Jiang, B. and Smith, A. (1999) 'Geographical Information Systems and urban design', in J. Stillwell, S. Geertman and S. Openshaw (eds) *Geographical Information and Planning*. Heidelberg: Springer, pp. 43–65.

Bonham-Carter, G.F. (1994) *Geographic Information Systems for Geoscientists*. Oxford: Pergamon.

Burrough, P.J. and McDonnell, R.A. (1998) *Principles of Geographical Information Systems*, Oxford: Oxford University Press.

Chrisman, N.R. (1988) 'The rise of software: a case study of the Harvard lab', *American Cartographer*, 15(3): 291–300.

Heywood, I., Cornelius, S. and Carver, S. (2006) *An Introduction to Geographical Information Systems* (3rd edn). Harlow: Pearson Educational.

Longley, P.A., Goodchild, M.F., Maguire, D.J. and Rhind, D.W. (2005) *Geographical Information Systems and Science* (2nd edn). New York: John Wiley & Sons.

Longley, P.A., Goodchild, M.F., Maguire, D.J. and Rhind, D.W. (eds) (1999, 2005) *Geographical Information Systems: Principles, Techniques, Management and Applications* (2 vols). New York: John Wiley & Sons.

Maguire, D.J., Batty, M. and Goodchild, M.F. (2005) (eds) *GIS, Spatial Analysis, and Modeling*. Redlands, CA: ESRI Press.

McHarg, I. (1969) *Design with Nature*. New York: Doubleday-Anchor.

Tomlin, C.D. (1990) *Geographic Information Systems and Cartographic Modeling*. Englewood Cliffs, NJ: Prentice Hall.

26 Statistical Analysis Using PASW (formerly SPSS)

John H. McKendrick

Synopsis

Predictive Analytics Software (PASW) is a user-friendly software system which provides a multifaceted means of working with and analysing numerical datasets. Prior to 2010 Predictive Analytics Software was known as SPSS (Statistical Package for the Social Sciences). Despite the name change, for some time to come, PASW is likely to continue to be referred to as SPSS by existing users, in older student guides and textbooks, and to appear as SPSS on some university machines. In order to assist understanding and to minimize confusion, the double abbreviation of PASW/SPSS will be used throughout this chapter. PASW/SPSS assists users working with quantitative data from the stage of data collection, through data access, data preparation, data management, data analysis and report writing. This chapter explores some of the common misconceptions about PASW/SPSS and explains its role in geography. It then describes the product range and provides ten steps to guide student geographers from the point of data collection to the point of data analysis.

The chapter is organized into the following sections:

- What PASW/SPSS is not
- What is PASW/SPSS, and who are SPSS Inc.?
- PASW/SPSS, the research process and the student geographer
- What exactly does PASW/SPSS offer the student geographer?
- Beyond university ... with PASW/SPSS

WHAT PASW/SPSS IS NOT

If all of the readers of this chapter were asked 'What are you studying at university?', what proportion of readers would be expected to answer 'essay-writing skills'? Not many, one would hope! It might reasonably be assumed that those who offered this response had misinterpreted the question: this answer does not appear to make sense. More typically, we would have anticipated answers such as 'geography', 'human geography', 'physical geography', or one of those three descriptors, and 'geology' or another of the many cognate disciplines alongside which students pursue geographical studies. Yet, to the chagrin of many a dissertation tutor, 'SPSS' – the same type of answer – is frequently offered in response to the question 'How do you intend to analyse your [quantitative] data?' PASW/SPSS is to data analysis what essay-writing skills is to doing geography: merely one possible means through which to ensure efficient and effective practice. Just as the student geographer could hone skills other than essay writing to improve geographical understanding (reading skills, observational skills, etc.), so he or she could utilize tools other than PASW/SPSS to analyse quantitative data. Just as essay-writing skills comprise many elements (paragraph construction, citation practices, etc.), so PASW/SPSS comprises many tools for working with quantitative data (frequency counts, cross-tabulations, etc.). In short, PASW/SPSS is not a method. Rather, Predictive Analytics Software/Statistical Package for the Social Sciences (PASW/SPSS) is a software package – a user-friendly, widely used and multifaceted means of working with numerical datasets.

Although familiar to student geographers, PASW/SPSS is not particularly popular among them and tends not to have a constant presence in undergraduate geography studies. These two 'negative' statements are inextricably linked. Students tend to be confronted with PASW/SPSS in limited duration, intensive, research methods modules that are set apart from their substantive geographical studies. This can render it more difficult to appreciate the wider significance of PASW/SPSS and tends to encourage learning that is narrowly focused on process rather than application (how, rather than why, to apply PASW/SPSS). Even when sequenced in a degree as a progressive programme of building competence in statistical data analysis, there is often a gap between advanced learning and preliminary studies, requiring students to spend time refamiliarizing themselves with basic statistical concepts and the PASW/SPSS software. Add to this the gradual shift away from quantitative research that is an inadvertent by-product of 'the cultural turn' of the 1990s, it should come as no great surprise to find that a special place is reserved for PASW/SPSS/statistical data analysis classes at the bottom of the league table of end-of-term student module evaluations!

Sounds familiar? Why, then, should you read on? There are four reasons why student geographers should persevere with PASW/SPSS. First, many of the research questions that concern geographers are amenable to, or are best answered by, extensive comparisons between places and people, and PASW/SPSS is an appropriate tool to use to obtain answers to these questions. For example, PASW/SPSS could be used to ascertain whether there are significant differences in rates of wave erosion at different points around Britain's coastline. Similarly, PASW/SPSS could be used to calculate the proportion of families that are headed by a lone parent, and to ascertain whether there are significant differences in their incidence across New York's neighbourhoods. Second, statutory bodies are extending the range of national survey datasets that are freely available to those involved in higher education. These datasets enable many geographical issues to be researched (see Chapter 5). Many of these datasets can be downloaded as PASW/SPSS data files, ready prepared for data analysis. Third, a wide range of organizations outside academia, many of which employ geography graduates, use PASW/SPSS. These issues – utility for academic geography studies, working with data and utility beyond academia – are considered, respectively, in the next but one, penultimate and concluding sections of this chapter. Beforehand, contextual information is provided on PASW/SPSS, the company and the product. This reflects the broad aim of this chapter: to provide a general introduction to PASW/SPSS. Specialist textbooks are necessary to provide a comprehensive guide to using PASW/SPSS and some recommended key texts are suggested in the Further Reading section at the end of this chapter. And the fourth reason for reading on? Well, it is highly probable you will encounter PASW/SPSS (again) in your degree studies: perhaps these few thousand words will leave you better motivated to make the most of this learning opportunity. Or, at least, you will be better placed to assess the relevance of PASW/SPSS ... and perhaps even inclined to be a tad more generous in your assessment of statistical laboratory classes!

WHAT IS PASW/SPSS, AND WHO ARE SPSS INC.?

PASW/SPSS is a range of products supplied by SPSS Inc. Founded in 1975 and based in Chicago, Illinois (USA), SPSS Inc. was acquired by the global IT giant IBM in October 2009. The first SPSS program, which provided the impetus for the subsequent establishment of SPSS Inc., was developed in 1968 by three graduate students from Stanford University. It has evolved ever since with a new version being released, on average, a little over every two years. Much has changed between the first and the current (eighteenth) version and significant milestones have been achieved in passing. In 1984, SPSS/PC+ became the first mainframe-capable statistical software package that could be used on personal computers, and in 1992 SPSS launched a product for use with the Microsoft Windows operating system. Although the old name for the software (the Statistics Package for the Social Sciences) was perhaps reassuring to many geographers, particularly those interested in human geography, PASW may at first appear less student-friendly. However, even SPSS Inc. was keen to stress that SPSS "had ... *evolved from its academic roots* to become a leading enterprise analytical solutions provider" (*emphasis added*). Perhaps then the incorporation of SPSS Inc. in the IBM Group in 2009 was a logical step in the evolution of the company and the product.

Thus, SPSS was never just a software package for academia and PASW/SPSS will surely remain a key resource for data analysis in the undergraduate geographer's statistical laboratory class. Indeed, the learning of statistics is not the only way in which PASW/SPSS is applied in academia: many lecturers will use the software to analyse data in their academic research and university administrators may be using PASW/SPSS to count you as part of their work in monitoring changes in the characteristics and performance of student populations, for example.

Thus, applications for the PASW/SPSS product range extend beyond simple data analysis. PASW/SPSS products are conceived as belonging to one of four families: the Statistics Family; the Data Collection Family; the Modelling Family; and the Deployment Family. Most readers of this chapter will encounter PASW/SPSS as PASW/SPSS Statistics Base (from the Statistics family). Other PASW/SPSS products are designed to enhance the effectiveness of data collection and application. And all of these products can be integrated to deliver a comprehensive support system from design through to final report or presentation.

The statistics product is an integrated system of modules, built upon the core package, formerly known as SPSS Statistics Base. PASW/SPSS Statistics Base is

7 Deployment
SmartViewer Web Server,
SmartScore

6 Reporting
SPSS Base, SPSS Tables
SPSS Report Writer, Sigma Plot
SPSS Maps

5 Data analysis
Predict numerical outcomes –
SPSS Base, SPSS Regression
Models, SPSS Advanced
Models, Amos

Identify groups –
SPSS Base, SPSS Regression Models,
SPSS Advanced Models,
SPSS Categories, Answer Tree,
SPSS Exact Tests

Forecast time-series data –
SPSS Trends

1 Planning
SamplePower, SPSS Conjoint

2 Data collection
SPSS Data Entry

3 Data access
SPSS Base

**4 Data management and
data preparation**
SPSS Base, SPSS Missing
Value Analysis

Figure 26.1   PASW/SPSS and the research process

a comprehensive software package with an easy-to-use graphical interface that assists users working with quantitative data from the stage of data collection, through data access, data preparation and data management to report writing (Figure 26.1). While most users eventually require supplementing the base package with specialist modules (e.g. PASW/SPSS Advanced Statistics or PASW/SPSS Regression), few users would routinely use all the PASW/SPSS products.

PASW/SPSS, THE RESEARCH PROCESS AND THE STUDENT GEOGRAPHER

Many human geographers would now argue against the depiction of the research process as a simple linear sequence of pre-defined steps from problem specification through to writing up results (see Chapter 1). In particular, those deploying multiple methods in multi-stage research (for example, using a questionnaire survey to obtain generalizable findings and following this up with key informant interviews to explore issues or processes in more detail) may find themselves writing up findings from introductory methods, while formulating

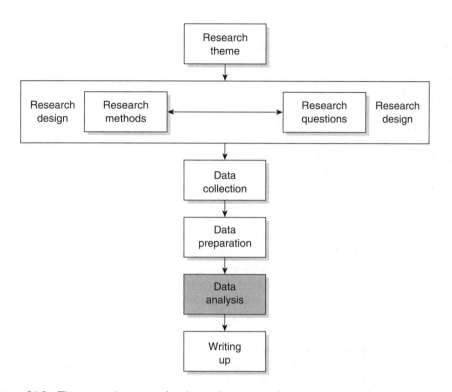

Figure 26.2   The research process for the student geographer using quantitative data

research questions or supplementing their original research themes for the next stage of their research. Those working within a 'grounded theory' tradition view research as an iterative process in which themes and theoretical understanding emerge from field research rather than theory providing an overarching framework at the outset. These 'fuzzy' models perhaps better reflect the reality of undertaking research, but they contrast with earlier models proposed by SPSS Inc. (Figure 26.1) and by the author of this chapter for student-led research with quantitative data (Figure 26.2).

In student-led research, 'PASW/SPSS' (or any other statistical software package that could be used to analyse quantitative data) tends to be the fifth stage in a six-step process (the shaded box in Figure 26.2). Having decided upon a research theme (which might involve reviewing literature and engaging with theory and/ or policy) and conceived of a research design (which involves choosing research methods and formulating specific research questions), the next task is to collect quantifiable data (which may, for example, be derived from systematic field measurement, content analysis of documents or survey questionnaires), prior to preparing data for analysis (which involves, for example, assigning numerical codes to categorical data). This – the data analysis stage – is the point at which student geographers tend to utilize PASW/SPSS products. Having completed the analysis, student geographers tend to move away from PASW/SPSS to write up their findings using word-processing software.

PASW/SPSS performs a critical role in the research process but it is far from all encompassing or all important. Effective quantitative data analysis requires clarity in the specification of the research problem (step 1), a research design that provides a meaningful sample and a dataset that is amenable to quantitative data analysis (step 2). Without these fundamental inputs – which are researcher driven – PASW/SPSS will be of little value. In short, competence in research design is as important as competence in using PASW/SPSS to analyse data. It is very easy to design a survey but very difficult to design a *useful* survey. Sufficient effort must be invested in what are typically pre-PASW/SPSS stages in Quantitative research.

It should be noted that this 'typical' depiction of the use of PASW/SPSS by student geographers does not preclude alternative models of application. Indeed, when Figure 26.1 (the SPSS Inc. model of how PASW/SPSS products could be used from the stage of data design to writing up) is compared to Figure 26.2 (the 'normative' model for quantitative research by student geographers), it could be concluded that the full potential of PASW/SPSS is not being realized. Student geographers tend not to use PASW/SPSS products to design their research. A simpler model than Figure 26.2 is required to depict the use of PASW/SPSS in analysing existing datasets. In such secondary analysis step 2 is simplified and steps 3 and 4 are superfluous as the researcher moves straight from question formulation to data analysis (having secured data access). Finally, steps 1–4 are often bypassed in introductory statistical laboratory classes, in which the learning objective is often to 'learn PASW/SPSS'. The challenge and joy of identifying themes to research, designing research methods and formulating questions, collecting data and preparing them for analysis are far removed (and may even seem far fetched!) from the PASW/SPSS laboratory class using pre-prepared data and schedules of analysis.

WHAT EXACTLY DOES PASW/SPSS OFFER THE STUDENT GEOGRAPHER?

The answer to the question posed in this section's title is unequivocally 'far too much for one short chapter!'. Another, equally valid, answer might be to note that PASW/SPSS offers the means to undertake the tasks outlined in Chapters 5 (making use of secondary data), 6 (conducting questionnaire surveys), 17 (sampling), 19 (exploring systems using mathematical models), 22 (basic numeracy skills), 23 (using statistics to describe and explore data) and 24 (introductory geostatistics). Perhaps the most satisfactory answer is to direct you to the annotated guide to Further Reading that is presented at the end of this chapter. In the mean time, here are ten steps to guide the student geographer from the point of data collection to the point of data analysis.

File type

Two files are automatically created when you open PASW/SPSS Base (hereafter, PASW/SPSS). The Data Editor file ('untitled data') is a spreadsheet-like grid of

Figure 26.3 PASW/SPSS files and the PASW/SPSS Applications menu

numbered rows and labelled columns. At the outset, the row and column headings are dimmed, as there are no data in the cells. It is into this file that you will enter your statistical data. The Output file ('!untitled output 1') is a blank file into which PASW/SPSS will enter the results of any analysis you undertake (e.g. if you command PASW/SPSS to undertake a frequency count). There is a third file type – a Syntax file ('untitled syntax 1') – which may be opened when you start PASW/SPSS, depending on how your department or school computing officer has chosen to set up PASW/SPSS. The Syntax file is a record of commands that you have either keyed in directly or have pasted in while working with PASW/SPSS's graphical user interface. Syntax files are extremely useful if you want to repeat an analysis or if you want to maintain a record of your analysis. For more information on using Syntax files, refer to one of Norusis's books (see Further Reading). Finally, the PASW/SPSS Application menu – the headings of File, Edit, Data, Transform, Statistics, Graphs, Utilities, Windows and Help, each of which has drop-down options – is listed at the top of the screen. The data file, output file and PASW/SPSS Application menu are illustrated in Figure 26.3.

Structure of a PASW/SPSS data file

Each row of the Data Editor file consists of cases and each column consists of variables. Thus, each questionnaire survey (or each field site) should have all of its responses (or field measurements) kept within the same row. Similarly, each question that is asked of all survey respondents (or each type of measurement that is taken for all field sites) should be kept within the same column. In effect, the PASW/SPSS data file is a structured, shorthand summary of your data in which you reduce a mountain of paperwork (questionnaire surveys) or collate several series of data (measurements for field sites) into one record.

Preparing the data file

Your first task is to translate your data into an annotated language of numbers, which you will always be able to translate back to its original form at any time. This is made easier if your survey (see Chapter 6) or field measurements (see Chapter 16) are well designed. For example, if you ask survey respondents to record their gender, you should be able simply to translate the responses into numeric data for the PASW/SPSS file: 'men' becomes '1', 'women' becomes '2' and 'rather not say' becomes '3'. You will always know what 1, 2 and 3 in the gender column of your data file represents. Or will you? It is surprisingly easy to forget what the numbers of your PASW/SPSS data file represent, particularly if you have a large dataset or if you return to analyse your data some time after you prepared the data file. One of the most useful functions of PASW/SPSS – and one that makes PASW/SPSS more user friendly than spreadsheets – is the function that allows the user to add descriptive labels to variables and values.

If you double-click your mouse on the dimmed header (var) at the top of the column of your data file, the 'Define Variable' dialogue box will open (top of Figure 26.4). PASW/SPSS gives you the option of changing the settings for variable type, missing values, labels and column format. You can also change the variable name as it appears at the top of the column on your PASW/SPSS data file. At this stage, those new to PASW/SPSS should make five changes.

First, change the variable name for the data file. You are permitted to use up to eight characters. Keep it simple and either use an obvious descriptor (e.g. 'SlopAng1' for the first slope angle that you measured) or use the ID number you used when you were collecting data for that variable (e.g. 'Q_02' for the second question of your survey). When you have accepted these changes, the dialogue box will close and your data file will now show the variable label at the top of the column. You should repeat this process for every variable in your data file.

Second, define the variable name and the value labels for any output files you will generate at the stage of data analysis. Click the 'Labels' button of the Define Variable dialogue box (top of Figure 26.4) to open the Define Labels dialogue box (bottom of Figure 26.4). You should enter a longhand description of your variable, which can be up to 256 characters in length (but keep it shorter!). You should ensure that you provide enough information to make sense of the abbreviated variable label that appears on the header row of your data file. For example, 'SlopAng1' might be described in this dialogue box as 'Slope Angle at Findhorn Creek'. For ordinal and categorical (nominal) data (refer back to Chapter 22), you should enter a descriptive label for each value. Enter the value (your numerical code) in the Value box (e.g. '1'), and then in the Value Label box enter your longhand

Figure 26.4   Defining variables and labels dialogue boxes

description of this value (e.g. 'men') before clicking on the Add box to register this addition. At this point, your value label and value will be added to the record box. You should repeat this process for every value of your variable (e.g. 'men', 'women' and 'rather not say' for gender) before clicking on the Continue box to confirm all of the additions you have made. These changes should make no difference to your data file: it seems like a lot of effort for not much reward at this stage! You should repeat this process for every variable in your data file.

Third, you should register any missing values of which you are aware – for example if a respondent failed to provide an answer to a survey question or if your data-recording instruments failed to provide a measure at a field site. For surveys, you may also wish to have 'don't know' or 'other' responses treated as missing values. You can specify discrete missing values or a range of missing values. Again, the value of recording missing values is not readily apparent – no changes are instantly evident to your data file. However, this is important preparatory work for the stage of data analysis.

Fourth, if some of your data files contain values that are alphabetic (words) rather than numeric (numbers), you will have to change the variable Type. This should be avoided if at all possible, as the value of PASW/SPSS (or any statistical data-analysis package) rests in its ability to work with numbers. However, you may find it useful to enter alphabetic responses into your data file before converting them into numerical codes (e.g. typing in job titles before translating these into social-class categories). In such instances, you should change the variable type to a String variable (the label PASW/SPSS uses for alphabetic data). PASW/SPSS automatically gives you the option of increasing the number of characters (in this instance letters) from its default setting of eight. You may also find it useful to increase the width of the Column Format for the purpose of displaying your data.

Two further points are worth noting. First, you must click the OK button (in the top right-hand corner of the Variables Label dialogue box) to confirm any of the changes referred to above. If you don't, PASW/SPSS will not register your changes. Second, it may be useful to prepare your data file, as outlined above, before entering your data. This may make it easier to identify data entry errors.

Importing data

It is possible to import data into PASW/SPSS from Microsoft Excel, Lotus, SYLK, dBASE or tab-delimited files. You should ensure that the file you intend to import is structured in the same way as a PASW/SPSS data file. The data you import become your PASW/SPSS data file. Data files can be imported by using the Open drop-down subcommand from the File menu of the PASW/SPSS Application menu (Figure 26.3). It is also possible to cut and paste data into a PASW/SPSS data file.

Entering data

You can, of course, enter your data directly into a PASW/SPSS data file. This would be the preferred option if you were working with data that have not yet been entered into a spreadsheet or database. Simply position the cursor in the first cell of the first column then enter your data (for one questionnaire survey or one field site) one variable at a time. You may find it quicker to move across columns by using the Tab button on your keyboard rather than by repositioning your cursor. As you type in your data it will appear in the bar above the row with the variable labels (Figure 26.3): it will only be added to your data file once you have pressed the Tab button or positioned the cursor in the next cell.

When you make a mistake (which you will!), simply position the cursor in the cell with the data you want to change and key in the correct value. One way to minimize these errors is to take regular breaks from data entry and to postpone data entry if you are becoming tired. Another useful aid is to switch on the Value Labels mode of data presentation. This is only helpful if you have defined your value labels prior to entering your data. When the Value Labels option is selected, your data file should comprise a list of value labels rather than a list of numerical code (e.g. 'Women' and 'Men' instead of '2's and '1's). When a number is shown in your data file, this indicates that either you have not defined the value label for this valid code or that you have entered the wrong code. Releasing the cursor when you highlight the Value Labels option from the Utilities button of the PASW/ SPSS Applications menu can activate the Value Labels mode of presentation.

Saving a file

It is good practice to save your PASW/SPSS files at frequent intervals. PASW/SPSS has been known to freeze on networked systems and university computing networks have been known to crash from time to time. One useful rule of thumb would be to save your data file each time you have completed entering row data (that of a questionnaire survey or field site). This takes a few seconds for large data files and is a useful way of recharging the batteries before starting another round of tedious data entry!

To save a PASW/SPSS file, select either the Save Data (for an existing file) or Save Data As (for a file you have not yet saved) drop-down subcommand from the File menu of the PASW/SPSS Applications Menu (Figure 26.3). It may be wise to use the PASW/SPSS file naming suffix conventions: 'YOURNAME.*sav*' for Data files, 'YOURNAME.*spo*' for Output files and 'YOURNAME.*sps*' for Syntax files.

Identifying data errors

Having survived the marathon that is data input, it may be disconcerting to learn that you are not yet ready for data analysis. Even if you were aware of some errors you made when you entered data and managed to correct them, it is highly likely other errors in data entry will have passed unnoticed. Thankfully, there are quicker ways to check for errors than having to confirm that you have entered the correct data value by returning to your original questionnaires or field records. Although if you do not, you may have to concede that your data file will contain a small proportion of errant answers which are valid values – for example for gender, typing '2' (for women) instead of '1' (for men).

The most straightforward way to identify data entry errors is to undertake a frequency count of all variables and to scrutinize the results for unexpected values. This is a three-step process. First, you need to open the Frequencies dialogue box (Figure 26.5). This can be found at the top of the Summarize submenu, which can be accessed through the 'Statistics' heading on the PASW/SPSS Applications menu (Figure 26.3). Second, you should request a frequency count for each of your variables. This, in turn, is a four-step process. First, select the variable at the top of your list (the box in the top left-hand corner of the Frequencies dialogue box). Second, press the Shift key and select the variable at the bottom of your list (you may have to use the arrows to reach the bottom of this list). Now all your variables are highlighted. Third, click on the arrow button to move all your variables from the untitled box in the left-hand corner to the box labelled Variable(s). Finally, click on the OK button in the top right-hand corner of this Frequencies dialogue box. At this point, your '!Untitled Output 1' file comes to the front of your screen. The third, and final, stage of the identifying data errors process is to review the results in the '!Untitled Output 1' file to identify variables for which the wrong codes have been recorded. Note the number of cases of each errant code (e.g. four instances of a code '12' for a gender variable, which should only have returned codes 1, 2 or 3).

Figure 26.5   Frequencies dialogue box

Figure 26.6   Data search dialogue box

Changing data errors

The next stage in data preparation is to find and correct the errors in the data file. To do this, you should have handy your desk notes from the 'identifying data errors' stage, the original data (questionnaire surveys or field records) and you should open your PASW/SPSS data file. Correcting data errors is a six-step process. First, select the variable (column) that contains the data error. Second, open the Search For Data dialogue box (Figure 26.6). This can be accessed through the Edit heading on the PASW/SPSS Applications menu (Figure 26.3). Third, in the Search for box, enter the number of the errant value you wish to correct. Click on the Search Forward button and this will take you to the first instance of the data error. Fourth, identify the case (the questionnaire or field site) with this data error by referring to the row number (or case ID if you added one to the data file). Fifth, return to your original data source and identify the correct data value. Finally, correct your entry in the data file. You should repeat this process until all errors are corrected for all variables. Once you think you have amended all the errors, it would be wise to repeat the process to confirm this.

Recoding existing variables

At this point you are almost ready to start data analysis. However, it may be prudent to manipulate your data to facilitate effective analysis. 'Reducing' or 'recoding' the number of values for any given variable is often essential if you are to undertake meaningful analysis. For example, if you surveyed 100 people from two neighbourhoods to ascertain their opinion in response to the question 'Local government should do more to encourage residential households to recycle household waste', you may find that 30 people responded 'strongly agree', 20 responded 'agree', 20 responded 'neither agree nor disagree', ten responded 'disagree' and 20 responded 'strongly disagree'. If you wished to determine whether attitudes varied according to the area of residence, it would be helpful to reduce the number of attitudes. You could recode to two categories of agree (merging 'strongly agree' and 'agree' to give 50 cases) and disagree (merging 'strongly disagree' and 'disagree' to give 30 cases). In this instance, you would temporarily set aside the 'neither agree nor disagree' responses and code them as 'missing'. It is, of course, essential that your recoded categories maintain some substantive value.

You should use PASW/SPSS to create a new variable by recoding an existing variable – rather than amending the original variable – as you may wish to return to the original variable at a later stage (e.g. to conduct analysis on those who 'neither agree nor disagree' or to recode the original variable in some other way). To recode using PASW/SPSS is a nine-step process. First, you need to open the Recode into Different Variables dialogue box (Figure 26.7). This can be found at the bottom of the Recode submenu, which can be accessed through the Transform heading on the PASW/SPSS Applications menu (Figure 26.3). Second, select the variable you wish to recode (from the box in the top left-hand corner). Third, click on the arrow button to move this variable from the untitled box in the left-hand corner to the box labelled Numeric Variable > Output Variable. Fourth, give the new variable (the Output Variable) a name and a label by completing the box in the top right-hand corner of the dialogue box. At this point, the Change button becomes operational (it is no longer dimmed). Fifth, click on the Change button to register the new variable name and label. At this point the new variable name is added to the Numeric Variable > Output Variable box. Sixth, click on the Old and New Values button to open its new dialogue box (Figure 26.8). Seventh, complete this box by adding new values (for the new variable) for each of the old values (of the original variable). For example, for the attitudinal variable referred to above, old value '1' and '2' may become new value '1', old value '4' and '5' may become new value '2' and all other values should be defined as 'system missing'. Eighth, when all transformations have been added to the Old > New window (on the right-hand side of Figure 26.8), click the Continue button to accept these changes. The Old and New Values dialogue box (Figure 26.8) closes and you are returned to the Numeric Variable > Output Variable dialogue box (Figure 26.7). Finally, click the OK button to create the new variable. At this point the dialogue box closes and a new variable (column) is automatically added to the right-hand side of your data file. But before you set to work on this file, you should add value labels and check for data errors.

When you have created new variables from your old variables, you are ready for data analysis!

Figure 26.7   Recode into different variables, main dialogue box

Figure 26.8   Recode into different variables: old and new values dialogue box

Step ten – don't expect things to run smoothly

They won't – trouble-shooting and problem solving are an integral part of the PASW/SPSS experience!

BEYOND UNIVERSITY ... WITH PASW/SPSS

This chapter has provided student geographers with a contextualized introduction to PASW/SPSS, the product and the company. In addition to outlining ten steps to guide you from data collection to data analysis, it has clarified the role of PASW/SPSS in the research process and has introduced the history of the organization whose fortunes have flourished since the first statistical data-analysis package for social scientists was designed in the late 1960s.

Just as PASW/SPSS, the organization and the product have grown and developed through time, so too can the utility of PASW/SPSS for student geographers. Initial encounters with PASW/SPSS – most likely in computing laboratory classes that support introductory data-analysis modules – soon become a hazy recollection of statistical concepts, instructions, numbers, error messages, frustrations, system crashes/freezes, complicated printouts and expletives. However, it is to PASW/SPSS that many students return in the hope of making sense of questionnaire survey responses and systematic field measurements (to name but a few applications) that have been collected in the name of the honours level dissertation. Without the back-up of a laboratory class tutor and in the absence of the step-by-step guidance of a class exercise, this often is as much a test of patience as it is a test of competence in statistical data analysis. A *truly* student-friendly PASW/SPSS textbook (see the Further Reading section below), a dissertation tutor with much time to spare, a philanthropic postgraduate student or one who enjoys problem solving and a shoulder to cry on/bottle to empty are prerequisites for success.

However, familiarity with PASW/SPSS – if not, competence in its application – has value beyond university studies. In the cut-throat world of graduate recruitment, 'PASW/SPSS' on a CV serves as a marker that the graduate has grappled with statistical data, perhaps through developing skills in data classification, interpretation, analysis and reporting. The 'numerate' graduate is a sought-after commodity and it is more convincing to provide evidence of personal competence (use of PASW/SPSS) than it is to claim by proxy (through a vague recognition of having completed a degree in a 'numerate' discipline such as geography). At the very least, it might be added to a list of other computing software packages (Microsoft Word, Adobe Acrobat, etc.) to demonstrate computer literacy.

Yet, PASW/SPSS may be of even more direct value to the geography graduate's CV and long-term career prospects. In addition to serving as a marker of numeracy and computer literacy, the ability to use PASW/SPSS is, in itself, a selling point to potential employers. The PASW/SPSS home page provides a synopsis of 'success stories' – that is, reviews of how 144 companies have applied PASW/SPSS to overcome a challenge. You might be destined for a 'PASW/SPSS future' with institutions such as Yorkshire Water (who identify areas at risk of flooding using PASW/SPSS Predictive Analysis), Lloyds TSB (which has a dedicated section within the main fraud department whose sole purpose is to undertake data analysis to reduce card fraud) or First Community Housing (a non-profit housing organization serving low-income families who use PASW/SPSS to improve programmes and services). Then again, you might catch the geo-research bug and your destiny may lie in the university classroom … teaching PASW/SPSS to the next generation of undergraduate geography students!

Summary

- PASW/SPSS is not a method: it is a range of products that facilitate the analysis of statistical data. SPSS Inc. is an IBM company, acquired in 2009.
- PASW/SPSS assists geography students to undertake extensive comparisons between people and places.
- The skills acquired in using PASW/SPSS are transferable and are being used by geography graduates in a variety of ways in their careers.

Further reading

At the time of writing all available further reading pertained only to SPSS. It is likely, however, that subsequent PASW editions of the following books will be available by the time you read this chapter.

- Norusis (2008a, 2008b, 2009) is the front-runner for the title of PASW/SPSS student's best friend! Marija Norusis was SPSS's first professional statistician and has written many guides to SPSS and statistics. *SPSS 17.0 Guide to Data Analysis (Prentice-Hall)* aims to both introduce the reader to data analysis and SPSS and comprises 24 chapters. It is written in straightforward language and provides chapter exercises, comprehensive examples of basic statistical techniques, solutions to selected exercises, step-by-step procedures and data files specific to chapter examples and exercises. This is an essential reference text. It is likely that your library bookshelves will contain a wide range of her earlier titles – be sure to use the guide that matches the edition of the software that you are using. Also worth consulting are Norusis' *SPSS 17.0 Statistical Procedures Companion* and the *SPSS 17.0 Advanced Statistical Procedures Companion*. Visit www.norusis.com for information.

- Field (2009) is fast becoming a serious contender to Norusis's title as student's best friend. A psychologist from the University of Sussex, his introductory text is written in a lively style and clearly aims to lighten the experience of learning statistics and SPSS. It might not be to everyone's liking, but it is certainly worth having a quick browse through the pages – it might just be the PASW/SPSS companion you have been waiting for.

- Kitchin and Tate (2000) (Chapters 3–5). Although this student textbook does not offer any specific advice on using SPSS, and although it is a little dated, it does offer hand-worked examples with clear step-by-step instructions of how to undertake most of the introductory statistical techniques for which geography students are likely to use SPSS. At the very least, these hand-worked examples will enable you to appreciate what SPSS is helping you to miss! More importantly, the hand-worked examples illustrate what SPSS is doing to your data to reach the answers you need.

- Rogerson (2006) is an introductory student-level textbook, now in its second edition, which is clearly written with examples to illustrate calculation procedures and how to interpret the results. At the end of each chapter there are instructions on how to apply statistical techniques using SPSS for Windows. It includes self-assessment exercises and downloadable datasets. Most examples are drawn from human geography, although climatological applications are also used.

- Fielding and Gilbert (2000) provide an extremely well-written and accessible introduction to social statistics and data analysis for social-science students. This book is much more of an introduction to elementary statistical applications than, for example, Rogerson (2006). It refers to SPSS throughout and uses screenshots to illustrate how to use SPSS. Students with an inclination towards physical geography may find it more difficult to relate to the examples and may find the range of applications with interval level data too limiting.

- You should also visit the home page of PASW/SPSS (www.spss.com or www.pasw.com). This provides comprehensive information on the product range and the company. It is an enlightening read for the student geographer who, in all probability, will bemoan the utility of PASW/SPSS at some point during his or her undergraduate studies. Stato-phobes may fail to perceive any value in PASW/SPSS, while stato-purists may perceive PASW/SPSS to be too limited. The home page provides a comprehensive list of company 'success stories' which may encourage all to think again – i.e. it details many examples of private- and public-sector organizations that have used PASW/SPSS to reach a 'solution' to an 'information problem'. Optimistic stato-phobes may find it reassuring that their undergraduate studies are not being wasted in statistical laboratory classes; pessimistic stato-phobes may find the prospect of post-university life more daunting than ever!

Note: Full details of the above can be found in the references list below.

References

Field, A. (2009) *Discovering Statistics Using SPSS* (3rd edn). London: Sage.

Fielding, J. and Gilbert, N. (2000) *Understanding Social Statistics.* London: Sage.

Kitchin, R. and Tate, N.J. (2000) *Conducting Research in Human Geography: Theory, Methodology and Practice.* Harlow: Prentice Hall.

Norusis, M. J. (2008) *SPSS 17.0 Guide to Data Analysis.* Harlow: Prentice-Hall.

Norusis, M. J. (2009a) *SPSS 17.0 Statistical Procedures Compendium.* Harlow: Prentice-Hall.

Norusis, M. J. (2009b) *SPSS 17.0 Advanced Statistical Procedures Compendium.* Harlow: Prentice-Hall.

Rogerson, P.A. (2006) *Statistical Methods for Geographers* (2nd edn). London: Sage.

27 Coding Transcripts and Diaries

Meghan Cope

Synopsis

Coding is the assigning of interpretive tags to text (or other material) based on categories or themes that are relevant to the research. This chapter reviews strategies surrounding the coding of qualitative text, including how to evaluate your sources, identify topics and refine research questions, construct and fine tune your coding structure, build themes and maintain an accounting of the process of coding. Coding itself is broken down to include steps such as identifying patterns and forming categories as well as the procedures of defining first-level descriptive codes, developing second-level analytical codes and coding along a particular theme or concept. Different types of materials are discussed, including historical sources such as diaries and other 'pre-existing' documents, as well as 'self-generated' materials such as transcripts of interviews, focus groups, life histories and respondent diaries.

The chapter is organized into the following sections:

- Introduction
- What is coding? A brief description and an example
- Preparing to code
- Coding
- Conclusion

INTRODUCTION

Text-based materials (such as diaries, letters, oral histories, transcripts of interviews or focus-group sessions) and other similar sources are rich in information (see Chapters 7–9 and 14) but present the researcher with unique challenges of interpretation and representation. In this chapter, I discuss a series of strategies for organizing data, developing an analytical structure using coding, identifying trends emerging from the text materials and building themes that connect your empirical findings with a broader literature. Although the process of interpreting and coding data does not follow a linear path, the strategies discussed here are presented in a roughly sequential order for clarity.

It is often the case that text materials can fill gaps in information that otherwise could not be known, and in that role they are extremely valuable. For example, less is known about people who have typically been left off the historical record or are located outside the gaze of the mainstream of research (e.g. women, ethnic minorities, poor people, children), and we also tend to know less about historical 'everyday' events and life conditions than about events that have large-scale or long-term significance. Many researchers in geography and other disciplines have been working in these areas of non-elites and, at more local, day-to-day levels, and are finding primary text materials to be very valuable in filling those gaps in knowledge. The question then arises of how to go about incorporating diaries, interview transcripts or other materials into a research project and how to make the most of the sources while avoiding some potential pitfalls.

One of the key debates about using these data sources surrounds the fact that, because these types of materials often represent only a small group of people's views and ideas, the data are highly specific to those individuals. Even when the *n* is large (i.e. you have more than 30 respondents or text documents), the open-ended nature of these data forms raises issues for interpretation and representation. Often, the traditional (quantitative) standards of 'good research', such as objectivity (being value-neutral) and the ability to generalize to larger populations, are not applicable for such materials. However, qualitative researchers have also developed standards for doing 'rigorous' research because the standards for quantitative inquiry do not necessarily transfer well (see Baxter and Eyles, 1997; Bailey *et al.*, 1999). For example, as Silverman (2006) points out, the meaning of 'validity' in quantitative research is not appropriate for ethnographic work, but there are other strategies such as 'triangulation' (confirming results by consulting multiple and varied sources) that are viable and should be part of the qualitative researcher's tool kit to insure *reliability* of results. As part of making sense of these subjective and often problematic forms of data, while sticking to principles of rigorous inquiry, many researchers use strategies of coding and theme building. There are many ways to go about this, but some common principles and practices are reviewed in this chapter as a way to introduce the process.

WHAT IS CODING? A BRIEF DESCRIPTION AND AN EXAMPLE

Coding is basically a way of evaluating and organizing data in an effort to understand meanings in a text. First, coding helps the researcher identify *categories* and *patterns*. For example, in studying transcribed interviews done with low-income mothers, a researcher might see a pattern of daily challenges poor mothers face (she misses the bus, her child is sick, her caregiver shows up late) and code these as 'daily challenges', keeping a memo on what has so far been included in that category, and perhaps constructing subcategories (those related to transportation, those related to childcare, etc.). By identifying categories and patterns, we can begin to make more sense of the data and start to ask new questions. Following on the example of low-income mothers, we might find several subcategories of daily challenges (housing, transportation, childcare) but then also several bigger-picture categories of challenges they face (children's future, discrimination, violence, personal and financial security, self-esteem). Coding for these can help identify how these challenges intersect and may produce new codes, new research questions and new understandings of meaning in the data.

However, as Strauss (1987) points out, the identification of patterns and the formation of categories are only the most basic level of coding. He suggests that those beginning analysis consider four types of themes in the data: *conditions*, *interaction among the actors*, *strategies and tactics* and *consequences*. Many of these are indicated to the researcher directly by the subject. 'Conditions' can be indicated by such phrases as 'because' or 'on account of' (Strauss, 1987: 28) or passages like 'when I was in that situation …'. For example, if a low-income mother referred to a time when she lost her housing and had to live with her sister, we would code it as a particular type of housing 'condition'. Similarly, 'interaction among the actors' means looking for how the informants engage with others, what they think of others, what others do to them. So we might be interested in how low-income mothers interact with their children, their partners, their case workers, their employers and so on. 'Strategies and tactics' refers to what people do in certain situations or how they handle particular events. For example, poor women with children often exchange favours with friends and neighbours as a strategy to survive on low wages and/or state benefits (Gilbert, 1998). Again, there will usually be subcategories of strategies that become relevant, and these are very likely to be tied together with both 'conditions' and 'interaction with others'. Finally, 'consequences' are often easy to identify because the informant makes the connection for us: a woman says, 'My child was sick so I missed work too many times and lost my job. Then I couldn't pay the rent so we got evicted and had to move in with my sister.'

These are the *consequences* of certain *conditions*, but also indicate *interactions* and *strategies*. The idea, then, is to use these special types of categories to start analysing the data and pulling out new themes.

Of course, the idea is not merely to code something as a 'strategy' but simultaneously to name that particular strategy or set of tactics that was used. In the above example, the strategy of 'depending on family' enabled the woman and her children to find a temporary place to live after being evicted. This strategy is a code that then becomes connected to other codes, such as the conditions codes for housing and the interactions codes for family members. Codes do not stand alone but are part of a web of interconnected themes and categories.

Coding enables the researcher to make new connections. For example, the connection between a child getting sick and the family losing housing may not be immediately apparent, but when it is considered in the context of the fragile balance low-income women must achieve the relationship becomes more clear. The researcher is then sensitized to such connections and seeks to identify other similar relationships within the data, and may even decide to follow up with the individual woman for more detail or ask about similar circumstances with other informants. This is why coding should not be seen as something that is done *after* data collection; rather, the practices of data collection and analysis can be seen as blending together, affecting each other, and, through their mutual impact, they help contribute to more rigourous conclusions.

PREPARING TO CODE

Evaluating your sources

Researchers are often confronted with two scenarios when working with text materials – either they are working with pre-existing documents that were created without specific reference to their project (such as diaries written in the past, other historical documents, secondary sources such as newspapers, oral histories and transcripts generated from others' research), or they have constructed a research project and carried out interviews, focus groups, respondent diaries, or participatory research which they will then code. In my own experience I have used many different kinds of text documents which fall into one or the other of these two categories. Pre-existing sources I have studied include diaries from women living in the late nineteenth century, transcripts of oral histories that were done by a local historian and deposited in the archives of a city I was researching, and various other historical materials such as personal letters, social-club newsletters, factory records and memoirs. In the self-generated category, I have used transcripts of interviews and focus groups I personally conducted, field notes of my own and my students' observations, as well as art materials, journals, photos, videos and maps that were generated by children in a participatory research project (see Cope, 2008, for an example).

Clearly, these two categories of documents (pre-existing and self-generated) require somewhat different approaches, particularly in the ways they connect to your research questions. Using pre-existing material, especially historical data, tends to be an even more inductive process than using self-generated text – that is, the researcher's initial approach must be one of broad evaluation to see what trends come out of the material. In this approach the research questions must be flexible and open to change, depending on what is contained in the documents. Imagine, for example, that you wanted (as I did, when I was a Master's student) to learn about migration and citizenship from diaries kept by white women moving to the western USA in the 1860s – it makes sense to keep an open mind while reading through these before deciding on your specific research questions.

On the other hand, having the opportunity to generate original text documents from interactive research you yourself are conducting means you can begin by linking questions for your respondents directly to your research interests. As an example of tailoring your data collection, imagine you were interested in young people's daily negotiations of space as they pass through school, work, home and public space. You might ask them to keep diaries of their daily activities and draw maps of their patterns of movement and then accompany them on a typical neighbourhood walk to listen to their own explanations of what challenges they face, how they experience mobility, and sources of friction in their negotiations of public space (see Chapter 14). In both these scenarios (using pre-existing texts or generating new ones), the development of strong research questions is essential. Ideally, the research questions reflect some element of what we already know (from related literature and theory) *and* incorporate initial findings or hypotheses of the empirical component of the research.

Whether using pre-existing sources or self-generated texts, researchers doing qualitative work spend a lot of time reading and thinking about their material. By approaching the data with an open mind, researchers allow the data to 'speak' to them. This is important because even in research with self-generated materials it may require several readings for the full diversity of topics and meanings to begin to reveal themselves. When allowing data to 'speak' researchers need to consider their own listening (reading) biases and decide whether to include an emerging theme or not. For example, if a diary goes into great detail on family relationships but the researcher is primarily interested in travel, a decision needs to be made on whether to stick to the original topic (travel) or shift the theme to family relationships, or perhaps combine elements of both themes and, say, view travel through the lens of family relationships.

Identifying topics and forming research questions

Reading and rereading transcripts, diaries and other textual material allows the researcher to start identifying topics that are recurrent or demonstrate important insights. These may come out of a query that is generated by theoretical literature (e.g. 'Does this theory hold true in that geographical/historical context?'), or they may emerge from the data themselves (e.g. 'Many women diarists of the American West mentioned feelings of depression when confronted with the bleak prairie landscape'), or the researcher may identify a theme with both theory and data in mind (e.g. 'Do current theories of migration apply to marginalized people's experiences of relocation as domestic workers?').

Some initial familiarity with both the relevant literature and the empirical data to be used is desirable for the formation of a few key research questions (see Chapter 1). The successful creation of original academic work often begins by bringing existing literature and theory together with an empirical context or dataset in a way that has not been done before. From that point new insights start to flow as you become more familiar with the data, are engaged in existing scholarly literature and become increasingly convinced that your project has something to offer. (If things do not progress in this manner, it may be time to re-evaluate!)

Here is an example from my own experience of the development of a key topic and set of research questions. My theoretical readings in contemporary economic geography demonstrated that so-called 'contingent' labour – that which is short-term, part-time, contract or seasonal work – was perceived to be on the rise in the 1980s and 1990s as the USA and other countries moved from a primarily Fordist industrial period of capitalism to a more flexible, post-industrial economy. But empirically, I was interested in the economic conditions of women's everyday lives in Lawrence, Massachusetts (a 'textile town') during the Great Depression. In reading transcripts from oral histories that were completed by a local historian, involving men and women who had been working in Lawrence textile mills in the 1920s and 1930s, I had seen that the women's jobs were often paid by piecework, while men's jobs carried straight salaries. This sounded like 'contingent' work to me and, indeed, on further digging I found that women were often employed by the mills on a seasonal, short-term or part-time basis, while men had more stability in their mill jobs. So my research questions became: how were women's and men's jobs constructed differently through the gender division of labour in the mills to produce both fixed and flexible components to the labour force? What impacts did this have on people's everyday lives and their household strategies? And, theoretically, what were the implications for our economic theories of industrial capitalism if there were strong elements of contingent labour within what has been perceived as highly standardized labour relations of the 1930s? This is an example of linking the theoretical and the empirical in one's research questions, and also of remaining open-minded in both data mining and in literature reviews.

CODING

Once you are familiar with the relevant literature, have begun reading through your pre-existing texts or are on your way towards generating data through interactive research, you probably have built some initial themes that both interest you and appear in the data – these are the beginnings of coding, even if you have not identified them as such. There are many approaches to taking the next steps toward formalizing this process but I will outline some general steps here to provide some basic guidelines.

Building and refining the coding structure

The process of developing the coding structure for your project is one that is inevitably circular, sporadic and, frankly, messy. Some scholars have tried to standardize the coding process (see, for example, Strauss, 1987; Strauss and Corbin, 1990) with some degree of success, but even they acknowledge that it is not a clear, linear process for which you can follow step-by-step instructions and at some point say that you are 'done'. Rather, coding involves reading and rereading, thinking and rethinking, and developing codes that are tentative and temporary along the way, even during an on-going research project. However, coding is also really rewarding in that it enables the researcher to know his or her data intimately and see patterns and themes emerging in a way that would not be possible otherwise. For a lively account of some of the challenges and rewards of coding, see Crang (2001).

The most common way to construct the first set of codes is to start by reading through your first text document, marking important sections, phrases or individual words and assigning those a code. After reading through all your materials with a critical eye, you should have a list of codes you think are important, along with your notes about them (keeping notes or 'memos' on your coding process is very valuable). If you are using a qualitative software package, the codes can be entered and organized in various ways and their notes are linked to them automatically (see Chapter 28). Recall that the initial set of codes will be changing: there will be some things you find you don't use and others that should be added. Strauss calls this process *open coding*. He says open coding 'is unrestricted coding of the data. This open coding is done by scrutinizing the fieldnote, interview, or other document very closely: line by line or even word by word. The aim is to produce concepts that seem to fit the data' (1987: 28). The purpose of this stage is to 'open up' the data, fracturing them along the way if necessary, and breaking the data down so that conceptual implications can emerge in the later steps.

Strauss (1987) also recommends two other kinds of coding. One is called *axial coding* because it proceeds along an *axis* or key category. Axial coding can be part of the open-coding process but it allows the researcher to follow a particular category for a while as a way of testing its relevance. To return to the above example of interviews with low-income mothers, imagine that as the open coding progresses the axis of 'dependence on friends/family' catches our interest. We may follow that particular category in a bout of axial coding by focusing on different ways the respondents depended on their friends and family, soon returning to the freer open coding. Strauss's third type of coding is called *selective coding*. This is a more systematic approach to coding that is done when a central or 'core' category is identified and followed. For example, after some open coding and axial coding of the interviews from low-income mothers, we might decide the core theme emerging is the struggle for a better future for their children. We say it is 'core' because we have found that most of what the mothers talk about is related in some way to this theme. From that point on other themes become secondary and the main lens through which the data are viewed is based on the core category (i.e. we are being 'selective'). While there are many approaches to coding, Strauss's characterization of different types of codes and approaches to coding can be a helpful starting point in organizing what is a somewhat chaotic process.

Another way to think of different stages of coding is to consider the first level as 'descriptive' and the next level as 'analytic'. Descriptive codes contain mainly what we call *in vivo* codes – that is, they appear in the text and we use respondents' own words as codes. Analytic codes emerge from a second level of coding that comes after much reflection on descriptive codes and a return to the theoretical literature.

Here are two examples from former research projects I completed. The first is drawn from the historical work I mentioned above and the coding procedure is described step-by-step. The second example is from some later research and appears in Box 27.1. It shows how an interview question was coded in a project that looked at the impacts of welfare reform on low-income families, in part by interviewing the directors of social service organizations in Buffalo, New York in the late 1990s.

In the first example, the following passage is from an oral-history interview recorded and transcribed in the 1980s (by the local historian) from a woman who was recalling her working conditions in the 1930s in the textile mills of Lawrence, Massachusetts. She said:

> I was on piecework and I found my work very tedious and hard and you had to keep running and working all the time. See, the difference between my husband's salary and mine was that he was a day worker, like he got paid by the day and I didn't. I got paid piecework for the amount of work I could produce in a day. So, at the end of the week I was never sure if my pay would be larger or smaller according to the amount of work I did.
> (Lucienne Adams, *Immigrant City Archives*)

In the first-level (descriptive) coding, this passage was marked for the following codes (mostly *in vivo*): work, salary, piecework, uncertainty of income and difficulty of work. I found these themes coming up again and again in reading through over 50 oral histories and innumerable other documents and, because I was reading about similar themes in the scholarly geography journals regarding women, work, household relations and company production strategies, I felt they were codes that would lead to some interesting insights. So when I went back to the passage later in the analysis and did a second-level coding, I added analytic codes for this passage including 'gender divisions of labour', 'women's contingent work' and 'household relations'. These later codes had more connection with and significance to the broader literature and my research questions than the simple descriptive codes I had initially assigned and, indeed, they were indicative of some developing themes in the research project.

Box 27.1   Coding an interview

Text	*In vivo* codes/description	Analytic codes
Q: What do you expect to see happen as people hit [welfare] time limits or are cut off from programmes?		
A: Yeah, the safety net thing. I see them coming into that. I suspect, and the big picture here if you look at the [welfare] population approximately a third of the individuals can probably get work on their own or with a little bit of help, another third need some fairly intensive services, basically all the services that the employment and training and other human services community can provide them, and then I believe – and this is my own opinion – that there is probably another third or so that really are going to have a very very hard time of things after the five-year limit. They may be in a situation where they're really not going to be able to work. There are some people who have serious	safety net who can work who needs services who is unable to work	issue of 'creaming' – early success of those ready to work role of job training and support services 'hard to reach' population

(Continued)

<table>
<tr><td>

(Continued)

problems if you want to call
them problems, if you want to
label them as such, that are
going to prevent them from
working.
</td><td>

respondent is
sensitive to 'labelling'
</td></tr>
</table>

Source: Interview conducted in July 1997 by Meghan Cope with a staff member of a social service organization in Buffalo, NY

Theme building

The coding process is fluid and dynamic but it is not the end-product of analysis. As codes become more complex and more connected to the project's theoretical framework, they start building into *themes* that can then serve as the main topics for the final product (paper, report, dissertation). Connections between simple codes – for example, my finding 'uncertainty of income' co-occurring with 'women's work' repeatedly in multiple oral histories and other sources – can generate new paths of exploration, both into the data and back into the framing literature. Ultimately, this co-occurrence of codes turned into a major section of results from this project: women's work was purposely constructed as contingent (part time, seasonal, uncertain) within the mills in part to keep costs down (contingent workers typically earn less) and in part to take advantage of the fact that women moved in and out of the waged workforce as they reared children or attended to other family and household obligations. This finding also fed back into my critique of the theories of economic geography at the time that were – in some cases – so focused on men's labour that the historical contingencies of women's labour were overlooked.

The process of theme building is central to qualitative, interpretive work because it allows for the organization of information into trends, categories and common elements that are theoretically important. Themes may be based on similarities within the data or, conversely, on differences that appear and are interesting for some reason. The important thing to realize about theme building is that it is an ongoing process throughout the qualitative research project: themes may be identified before, during or after the data collection and analysis stages. Indeed, many of the best research projects are quite fluid in that they shift focus in response to the data and findings that emerge along the way.

One of the best approaches to theme building is to read *across* the materials being used rather than *solely* within them (Jackson, 2001). That is, after having spent time coding and interpreting individual documents, sit down with several or all your materials, including your notes and memos on the coding process, and work with a particular topic or code while drawing from multiple texts. This process aids in seeing trends that manifest themselves in many different ways. For example, poor mothers may use different vocabularies to talk about their struggles to maintain stable environments for their children but the theme is apparent when reading across several transcripts at a time.

Looping back to your research questions

Once you have built some themes based on interesting coding insights, revisit your research questions to evaluate and perhaps refine them. It may be that what you had hoped to find really is not apparent in the data sources you are using but that another unexpected (and – one hopes – equally compelling) theme has emerged and you will adjust the direction of your project accordingly. This can happen both when using pre-existing documents and in cases in which the researcher structures the questions for the respondents. In the former, you may discover, for example, that a theme you saw emerging in the first document or two you looked at was not evident in any other materials – somewhat of a dead end. In the latter case, you might begin a set of interviews with one idea of what is happening but find that an initial idea is not borne out by respondents but a new (and even better?) theme emerges unexpectedly.

For example, the project outlined in Box 27.1 started out by looking at the social networks poor people use to obtain jobs and asking directors of social-service organizations about these networks. It soon became apparent that many poor people couldn't use social networks to obtain employment because they didn't know anyone who was working (or working happily), so I had to abandon that line of inquiry. However, through discussions with the social service directors a new theme emerged: their organizations' job-training programmes were often reinforcing the existing racial and gender divisions of labour in the local workforce, and this became a new focus for the project.

Issues to consider

In the coding process there are always additional issues and challenges to consider. First, it may be helpful (or even necessary, as in multi-investigator projects) to have multiple coders. If you are working alone it may be constructive to give your code book and coded materials to a colleague or mentor to see if you are missing important themes and to test the strength of your interpretations. In group projects, the code book is ideally developed with everyone who will be involved in the coding; if that is not possible, at least have good notes attached to each code and discuss what different codes represent so there is minimal ambiguity and unevenness.

Second, you may need to eliminate some codes as the project progresses. It is very easy to get caught up in designing codes for finer and finer resolutions of a theme, but having an unwieldy code book just means some codes will not be put to use very often and it may make sense to combine or consolidate them. This is another instance in which having a colleague or your tutor to serve as a sounding board is valuable.

Third, if you are working on a project using self-constructed data materials (that is, your respondents are available for further interaction), you may want to review your interpretations with some or all of your respondents as a way to check that your findings reflect what they intended. This practice, called 'member checking' is indicative of a larger shift in qualitative research in geography that has been growing in popularity over the past decade in which researchers assume a much deeper involvement with community members, such as through participatory action research (Kindon *et al.*, 2007). Indeed, some researchers have co-authored books and journal articles with the people they worked with in an effort to challenge typically power structures and issues of who 'speaks' for whom (see, for example, Pratt and the Philippine Women Centre, 1999; Sangtin Writers and Nagar, 2006) (also see Chapter 11). In the context of member checking, some questions to think about include: how do you deal with respondents who contradict their earlier statements? How much of your project do you want to reveal to respondents (especially that which may be critical or unflattering)? At what point (if at all) will you stop involving your subjects in the research process? What is your philosophy on the power relations set up between researcher and 'subjects'? How much time are you willing/able to devote to collaboration with your respondents?

For an excellent review of some other issues to address in the interpretation of qualitative data, see Jackson (2001), especially his 'checklist' in the article's conclusion. Jackson points out, for example, the need to consider the silences, hesitations, uses of humour or irony and other non-verbal cues that may not be conveyed in text transcriptions of interviews or focus groups but that are evident in the tape recordings. This is one reason it is helpful to have notes from the interviewers and/or focus group facilitators that are attached to each transcript (something that is particularly easy when using a computer analysis package). Silverman's (2006) chapter on transcripts is very useful in demonstrating how notes on gestures, facial expressions and other cues can be incorporated into a text transcript in order at least partially to convey these important non-verbal elements of interviews and focus groups.

Additionally, Jackson (2001) suggests that, at some point in the interpretation, researchers should consider what is *absent* from their respondents' accounts and think about how these might be important. In his work on masculinity, for example, he found that certain themes were rarely or never addressed by the men in his focus groups, including fatherhood, race, friendship and all things domestic.

The way respondents frame the *position* of the researcher is also important. Participants in Jackson's projects' focus groups displayed several different types of attitude towards the university researchers in his project. He also found evidence of respondents moderating what they said in his presence and in that of his co-investigators (particularly his female colleague). Indeed, feminist researchers have explored the issue of positionality (see also Chapters 8 and 12) and the relations between researcher and researched, but these issues are also a matter for all researchers and should be addressed seriously (see Valentine, 2002; Cope, 2003; Johnson, 2008).

All these issues are important to consider, depending on your types of sources, the scope of the project, the type of data collection that was used (semi-structured interviews vs. life histories vs. activity mapping), who is involved in the research and who the informants are, and whether your materials were pre-existing or self-generated. The final suggestion, then, is to keep a detailed account of which techniques you have used, what problems you have encountered and how you have dealt with issues such as those discussed in this section. Qualitative research has often been critiqued for its 'hidden' methodologies: researchers need to be much more open about their procedures of data collection, coding and interpretation and should always include an explicit discussion of their methods in any presentation, whether written or oral. Coding is one way of ensuring a more systematic methodology, but the coding process also needs to be disclosed fully to readers in appropriate ways.

CONCLUSION

Coding should be seen as an active, thoughtful process that generates themes and elicits meanings, thereby enabling the researcher to produce representations of the data that are lively, valid and suggestive of some broader connections to the scholarly literature. By approaching coding as both systematic and flexible, it becomes an enlightening, fruitful and revealing process that allows for final products that are rich with meaning and true to the initial respondents or 'subjects'. The methods presented in this chapter attempt to balance the need for a systematic approach with the advantages of remaining flexible and open to emerging themes and multiple interpretations. Coding should not be seen as tedious and boring but rather as a type of detective work – we are trying to solve mysteries using varied clues and we are open to surprises in the data that generate those 'aha!' moments of investigation and inquiry.

Summary

- Coding enables qualitative researchers to make sense of subjective data in a rigorous way.
- It is a way of evaluating and organizing data in order to understand meanings in the text.
- Coding helps researchers identify categories, patterns, themes and connections in the data.
- When preparing to code, evaluate your sources, read and re-read your data.
- Building and refining code structures is a dynamic process which can be messy.
- There are different levels of coding: descriptive and analytic.
- When coding, revisit your research questions, ask others to look at your codes, show your interpretations to informants and consider what is absent.
- Researchers need to be more open about their procedures for coding.

Further reading

There are a lot of social-science books and articles about how to code qualitative data, but I have found the following particularly useful:

- Jackson (2001) is a quick and easily comprehended chapter that raises many important issues for qualitative research, specifically the interpretation of results. Jackson's conclusion 'checklist' is especially valuable.

- Now in its third edition, Silverman (2006) is a truly comprehensive treatment of how to go about coding and analysing all kinds of verbal and text-based data, with exhaustive step-by-step instructions and plenty of real-life research examples – this is the book!

- Flowerdew and Martin (2005). This is the definitive guide for geography students with a wide range of coverage and suggestions for related sources.

- Hay (2010) is an excellent collection of 'how-to' chapters on a range of methods.

Note: Full details of the above can be found in the references list below.

References

Bailey, C., White, C. and Pain, R. (1999) 'Evaluating qualitative research: dealing with the tension between "science" and "creativity"', *Area*, 31: 169–178.

Baxter, J. and Eyles, J. (1997) 'Evaluating qualitative research in social geography: establishing rigour in interview analysis', *Transactions, Institute of British Geographers*, 22: 505–525.

Cope, M. (2003) 'Feminist epistemology in geography', in P. Moss (ed.) *Feminist Geography in Practice*. Oxford: Blackwell, pp. 43–56.

Cope, M. (2008) 'Patchwork neighborhood: children's real and imagined geographies in Buffalo, NY', *Environment and Planning A*, 40: 2845–2863.

Crang, M. (2001) 'Filed work: making sense of group interviews', in M. Limb and C. Dwyer (eds) *Qualitative Methodologies for Geographers*. London: Oxford University Press, pp. 215–233.

Flowerdew, R. and Martin, D. (eds) (2005) *Methods in Human Geography: A Guide for Students Doing Research Projects* (2nd edn). Boston, MA: Addison-Wesley.

Gilbert, M. (1998) 'Race, space and power: the survival strategies of working poor women', *Annals of the Association of American Geographers*, 88: 595–621.

Hay, I. (2010) *Qualitative Research Methods for Human Geographers* (3rd edn). South Melbourne: Oxford University Press.

Immigrant City Archives, 6 Essex Street, Lawrence, MA 01840.

Jackson, P. (2001) 'Making sense of qualitative data', in M. Limb and C. Dwyer (eds) *Qualitative Methodologies for Geographers*. London: Oxford University Press, pp. 199–214.

Johnson, L. (2008) 'Re-placing gender? Reflections on fifteen years of gender, place and culture', *Gender, Place and Culture*, 15(6): 561–574.

Kindon, S., Pain, R. and Kesby, M. (2007) (eds) *Participatory Action Research Approaches and Methods: Connecting People, Participation and Place*. London: Routledge.

Limb, M. and Dwyer, C. (eds) (2001) *Qualitative Methodologies for Geographers*. London: Oxford University Press.

McKay, D. (2002) 'Negotiating positionings: exchanging life stories in research interviews', in P. Moss (ed.) *Feminist Geography in Practice*. Oxford: Blackwell, pp. 187–199.

Pratt, G. with the Philippine Women Centre (1999) 'Is this Canada? Domestic workers' experiences in Vancouver, BC', in J. Henshall Momsen (ed.) *Gender, Migration and Domestic Service*. London: Routledge, pp. 23–42.

Sangtin Writers and Nagar, R. (2006) *Playing with Fire: Feminist Thought and Activism through Seven Lives in India*. Minneapolis, MN: University of Minnesota Press.

Silverman, D. (2006) *Interpreting Qualitative Data: Methods for Analysing Talk, Text, and Interaction* (3rd edn). Thousand Oaks, CA: Sage.

Strauss, A. (1987) *Qualitative Analysis for Social Scientists*. Cambridge: Cambridge University Press.

Strauss, A. and Corbin, J. (1990) *Basics of Qualitative Research: Grounded Theory Procedures and Techniques*. London: Sage.

Valentine, G. (2002) 'People like us: negotiating sameness and difference in the research process', in P. Moss (ed.) *Feminist Geography in Practice*. Oxford: Blackwell, pp. 116–126.

Computer Assisted Qualitative Data Analysis

Bettina van Hoven

Synopsis

CAQDAS is the acronym for Computer Assisted Qualitative Data Analysis Software. It comprises a variety of programs with different capabilities to assist the analysis of qualitative data. Simple programs can be used for text searches, while more sophisticated ones have functions especially for theory building. This chapter examines the ways in which you might use CAQDAS.

The chapter is organized into the following sections:

- Introduction
- Computer software in qualitative research
- Key functions of CAQDAS
- Further considerations for choosing a program
- An example of CAQDAS in practice: experiences with QSR N4
- Some advantages and concerns about CAQDAS
- Conclusion

INTRODUCTION

Qualitative research techniques, such as interviewing and participant observation (see Chapters 8 and 9), are commonly used by human geographers. They generate rich data in the form of interview transcripts and diaries. Traditionally, geographers have analysed this material by coding it manually by pen and paper (see Chapter 27) but this approach has a number of limitations. Richards and Richards (1987) note that key worries in the use and analysis of qualitative material include the volume of data generated; the complexity of data analysis; and the notion that clerical tasks required for data preparation and management may prevent thorough data analysis. As a result in recent years, computer packages have increasingly been adopted for qualitative data analysis because they are regarded as useful tools for handling and coding large amounts of written data and because they facilitate the in-depth examination of relations between and within these data.

This chapter discusses how Computer Assisted Qualitative Data Analysis Software (hereafter CAQDAS) can be used in qualitative research. Rather than dealing with the specifics of software on the market, it focuses on the fit between different types of software and the needs of research projects. The key functions of packages are outlined, such as 'text retrieval/text management', 'coding and retrieving' and 'theory building'. In addition, the specific considerations you need to take into account before choosing software are summarized. The chapter then illustrates the use of computer software by describing my own experiences of using the computer program QSR N4 in qualitative research. Finally, the chapter draws attention to concerns regarding the use of CAQDAS and its impact on the way in which qualitative research is done.

COMPUTER SOFTWARE IN QUALITATIVE RESEARCH

Most students at undergraduate level are familiar with preparing reports using a PC or Mac. Indeed, a number of activities that are part of qualitative research may have already been completed using word-processing programs such as transcribing interview recordings or making fieldnotes (using a laptop) and editing these. Equally, data display in the form of matrices or networks and graphic mapping may be familiar using 'traditional' graphics programmes. Box 28.1, compiled by Miles and Huberman (1994), gives an overview of these and other uses of computer software in qualitative research. However, for points 4, 8, 11 and 12, simple word-processing packages are not sufficient; rather, for speed and efficiency more specialized software is required. It is these issues this chapter addresses.

Box 28.1　Uses of computer software in qualitative studies

1　Making notes in the field.
2　Writing up or transcribing fieldnotes.
3　Editing: correcting, extending, or revising fieldnotes.
4　Coding: attaching key words or tags to segments of text to permit later retrieval.
5　Storage: keeping text in an organized database.
6　Search and retrieval: locating relevant segments of text and making them available for inspection.
7　Data 'linking': connecting relevant data segments to each other, forming categories, clusters or networks of information.
8　Memoing: writing reflective commentaries on some aspects of the data as a basis for deeper analysis.
9　Content analysis: counting frequencies, sequence or locations of words and phrases.
10　Data display: placing selected or reduced data in a condensed, organized format, such as a matrix or network, for inspection.
11　Conclusion-drawing and verification: aiding the analysis to interpret displayed data and to test or confirm findings.
12　Theory-building: developing systematic, conceptually coherent explanations of findings; testing hypotheses.
13　Graphic mapping: creating diagrams that depict findings or theories.
14　Preparing interim and final reports.

Source: Miles and Huberman (1994: 44)

Several computer packages to aid qualitative data analysis have entered the market since the introduction of commercial programs in the 1980s. Every program has its specific focus and, of course, strengths and weaknesses. An assessment of computer packages aimed at the analysis of unstructured non-numerical data can be found in Tesch (1990), Prein *et al.* (1995) or Weitzman and Miles (1995). In these publications, a review of software can be found as well as their technical requirements. Keep in mind, though, that programs are constantly being improved and updated. Some have become very alike in terms of their purpose and capabilities. As a result the reviews provided by these authors (and anything else you find to read on this subject) may, at least in part, be slightly out of date. Nevertheless, they remain an invaluable source and starting point when investigating the scope and functions of programs.

KEY FUNCTIONS OF CAQDAS

Three types of CAQDAS can be distinguished according to their key functions: (1) text retrievers/textbase managers; (2) coding and text retrieval; and (3) theory-builders (Fielding, 1994; Weitzman and Miles, 1995). It is useful briefly to outline what programs in these categories do as it helps to narrow down the multitude of software packages according to the analytic needs of the project. Although various programs are available for each category, I include only some. The programs chosen here have been on the market for some time and have benefited from continuous improvements through interactions between developers and users (see Richards, 1997). An extensive overview is provided in Table 28.1. The examples suggested below are also available as demo versions (distributed by Scolari) for those who wish to examine the programs first:

1   *Text-retrieving* programs, in brief, find words or combinations of words in so-called string searches and put these into new files. The degree of sophistication of these programs varies. Some can, for example, assist in content analysis with additional functions such as counting word occurrences in the source text or creating word lists. Textbase managers are more advanced, in particular regarding their capability to organize data and their more sophisticated text search-and-retrieve functions (Fielding, 1994; Weitzman and Miles, 1995). One example of such a program is WinMAX with which data can also be coded and combined with statistical packages.
2   *Code-and-retrieve* packages can do what qualitative researchers do with cut-and-paste techniques using paper and scissors or coloured pencils. They are used to mark text segments for coding, cut and sort text segments, and find, retrieve and report on data. However, as Weitzman and Miles (1995: 18) claim, this software is a 'quantum leap forward' compared with manual practice regarding its working speed as well as being more systematic and thorough. Some packages also have a memo function. In 'the Ethnograph', for example, these memos can be hyperlinked to text segments.

Table 28.1   Overview of CAQDAS

Text retrievers/ textbase managers		Coding and text-retrieval		Theory-builders	
PC Mac		**PC Mac**		**PC Mac**	
✓	AskSam (2)	✓	Ethno (1, 4)	✓	AQUAD (1, 2, 3)
✓ ✓	Metamorph (2)	✓	The Ethnograph (1, 2, 3, 4)	✓	ATLAS/ti (1, 2, 3)
✓	Orbis (2)	✓ ✓	FolioVIEWS (1, 2)	✓ ✓	HyperRESEARCH (1, 2, 3)
✓ ✓	Sonar (1, 2)	✓	FuzzyStat (1)	✓	Hypersoft (1, 3)
✓ ✓	Tabletop (2)	✓	GATOR (1)	✓ ✓	QSR N4
✓	TextCollector (2)	✓	HyperFocus (1)	✓	Qualog (1)
✓	WordCruncher (2)	✓	HyperQual (1, 2, 3, 4)	✓	QCA (2)
✓	ZynINDEX (2)	✓	KWALITAN (1, 2, 3)		
		✓	Martin (1, 2, 3)		
		✓	MAX International (1, 2, 3)		
		✓	WINMAX (1)		
		✓	QUALPRO (1, 2, 3, 4)		
		✓ ✓	SQL Text Retrieval (1)		
		✓	Textbase Alpha (1, 3, 4)		

Notes: (1) Grbich (1999); (2) Weitzman and Miles (1995); (3) Kelle (1995); (4) Tesch (1990). There are several other programs on the market that have not been reviewed (see, for example, http://caqdas.soc.surrey.ac.uk/packages.htm)

3   In addition to the aforementioned functions of CAQDAS, *theory-builders* have the capacity to test hunches, ideas and hypotheses that are pre-existing or emerge from data in the project. They allow the researcher to make connections not only between code and data but also between code and code. Relationships can often be displayed in a hierarchical index system or a graphic network. Examples of such software are QSR N4 and ATLAS/ti, respectively.

QSR N4 consists of two components: the Document System and the Index System. The first holds the source data, such as texts, video fragments and/or pictures. The second stores codes, either unstructured or arranged in a hierarchical form (tree), and text searches. It also contains memos and queries about data. A code network (i.e. the tree structure), can be displayed, and hyperlinks from nodes to source documents or memos established. In the example described below, these functions and the use of QSR N4 are illustrated in greater detail.

FURTHER CONSIDERATIONS FOR CHOOSING A PROGRAM

The choice of a particular type of software depends on such issues as the methodological orientation of the researcher (see also Further Reading at the end of this chapter), the type of data, whether a project is conducted individually or in a research team, or if expectations from CAQDAS will increase over time. There are five key considerations for choosing qualitative software that are outlined below: (1) data entry/storage; (2) coding; (3) memoing/annotation; (4) data linking; and

(5) search and retrieval (see also Fielding, 1995; Miles and Weitzman, 1996). Again, I refer to Weitzman and Miles (1995) for an evaluation of these functions per package in the form of a table.

Some programs, particularly older versions (which are both still useful and used often), require *data entry* in ASCII format. Text needs to be saved following formatting rules such as a limited number of characters per line, single spacing or special characters to indicate the beginning of a new text segment. Once saved, the files can be imported and stored as internal documents or left on the original PC/Mac as external documents. *Coding* can be done in several ways, of which one or more are supported by the package. Options vary from on-screen to off-screen coding or both, from coding one or several chunks of text to coding 'nested' or overlapping text. Some software permits hierarchical coding and the renaming and reorganizing of codes. Last but not least, a number of packages have hyperlinks from code to source text and are able to display relationships between codes graphically (see above). When aiming to build theory from project data, for example using a 'grounded theory' approach as described by Glaser and Strauss (1967), a function to store definitions of, and thoughts about, data and codes in *memos* is important. The capability of programs varies as the length of memos may be restricted or reference to source data is included while the actual (hyper-)link is absent. A part of theory building is the constant comparison of data (per category, case or person, for example) (see also Glaser and Strauss, 1967). In order to do this, quick *retrieval* of text segments is desirable. The time with which project data is searched can vary significantly. Furthermore, *searches* can be simple text searches, such as for specific words or phrases using Boolean functions ('and', 'or', 'but not'), or searches for codes or a combination of codes. Again, the level of sophistication for such searches varies. Some programs allow searches that include overlaps – 'A' in 'B' or 'A' near 'B', for instance. The results of these searches can be displayed per find, per source document or per category.

With regards to the requirements of a project, it is worth adding that some software is designed to accommodate multiple researchers, in particular those with more elaborate functions. Furthermore, software that is capable of performing all the above operations is more likely to 'grow' with increasing research demands.

AN EXAMPLE OF CAQDAS IN PRACTICE: EXPERIENCES WITH QSR N4

Having briefly outlined the types of software available and the specific considerations for choosing a particular package in a theoretical way, the remainder of this chapter explores the nuts and bolts of CAQDAS. QSR N4 is used to illustrate the discussion. It combines several functions mentioned above. It can be used for relatively simple projects in the first and second year of the course but it is also able to accommodate more demanding projects, such as an undergraduate dissertation or perhaps a thesis (it also has a very useful help network). There is a classic version (QSR N4) and a more sophisticated 'sister version' (NVivo) for

more experienced researchers. The developers of the program have an extensive
website (www.qsrinternational.com) with demos, downloadable handbooks and
lists of literature both by the developers and users. In addition, the University of
Surrey set up the CAQDAS networking project (caqdas.soc.surrey.ac.uk) offering
workshops, downloadable demos and literature as well as discussion forums for
users. Last but not least, Kitchin and Tate (2000) give a quick overview of how
data can be entered into and analysed using QSR N4. I recommend using the
latter in addition to this chapter.

In the following I will illustrate a number of the issues drawing on a project I conducted for my PhD between 1996 and 1999 (for a full account, see Hoven-Iganski,
2000). I had decided to use a modified grounded theory approach for my data analysis which had consequences for my data collection and analysis. I adhered largely
to analytic steps suggested by Glaser and Strauss (1967) and Strauss and Corbin
(1990) including coding, constant comparison and theory building. These steps are
also supported by QSR N4 as this methodological framework is its 'bumper sticker'
(Crang *et al.*, 1997: 776).

Context of the study 'Made in the GDR'

The project 'Made in the GDR – the changing geographies of women in the
post-Socialist rural society in Mecklenburg-Westpommerania' draws on data
from 83 key informant interviews (e.g. politicians, civil servants at ministries or
communities, chairpersons of NGOs), 12 group interviews (two interviews per
'case village') and letters from 39 women throughout the wider study area. The
interviews and correspondence were initiated using a topic guide and developed
further based on the respondents' comments. The study had a strong exploratory
character with the general aim to identify processes that help explain the formation of gendered identities and exclusionary practices in rural East Germany
since unification. Although I took photographs in and around the case villages,
I did not include them as data for analysis. Instead, they were used to illustrate
the story or when they were pertinent to the argument. Regrettably, it was only
upon my return from the last phase in the field, and when I became more familiar
with the scope of QSR N4, that I realized I could have included video recorded
'data' in my project (see Chapter 9 for discussion of video material).

Moving from paper piles to PC files

As part of my chosen methodology, I collected data in various stages, allowing
time to transcribe these data before the next period of data collection. In a first
stage, I corresponded with women in the study area in order to get a very general impression of the changes in East Germany and of the issues that occupied
them most. I had aimed to analyse the material by hand (see Chapter 27), and
started with the coding using coloured pencils and post-it notes. Although this
helped with the development of a topic guide used for interviews as themes

emerged from the data, I soon began to feel overwhelmed. I had generated so many codes in a line-by-line analysis that they had become illegible on paper and difficult to administrate. I worried that I would only be able to investigate fully some aspects while, unwillingly, ignoring others. In addition, the mere thought of the amount of data I would generate (95 x 1.5 hours of tape recordings) was almost paralysing. I had images of piles of colourful paper with post-its sticking out like a hedgehog, and of paper stacks being attacked by my cat in an idle moment. In other words, I did not feel very organized and not able to present a neatly finished project that would survive the critical scrutiny of my peers.

Preparing the data for QSR N4

More or less at this point I became aware of QSR N4. Learning the package at a basic level was quick. I completed the tutorial provided on the demo, did a one-day workshop through the CAQDAS networking project at the University of Surrey and was ready to go. Although QSR N4 is capable of combining the functions outlined above, it requires data entry in ASCII format. I used the text font 'Courier 10' – i.e. limiting the number of characters per line – and saved the file as 'Text with line breaks' – i.e. without any formatting. I also created a header for each document (i.e. interview or letter) and made sure to enter an asterisk (*) at the start of each speaker in a group interview, or every question or answer in individual interviews. In so doing, I produced text blocks the program recognized when searching and retrieving coded text later. Once all data were saved in this format, I imported them into the QSR N4 project.

Coding and retrieving

Two of my initial objectives were to assess the general implications of unification for women in rural areas; and to explore the dynamics behind the formulation of norms and meanings in the GDR. During my literature review, the reading of media reports and, of course, my correspondence, I identified several themes such as employment, unemployment, modernization, freedom of choice and travel. I expected these themes to emerge from 'my' data as well. Initially I had begun to 'make sense' of these, summarizing them per respondent or per theme. The 'expanded sourcebook' on qualitative data analysis by Miles and Huberman (1994) was an invaluable starting point for a research novice like me in that it illustrated various ways of reducing data. As a consequence, once I shifted to using QSR N4, I had already generated a number of codes from the initial coding of the correspondence and my literature review. These codes were entered first and stored as 'free nodes' in the Index System, meaning that I had not ordered them in any hierarchical way. I used these nodes for subsequent line-by-line coding but generated a multitude of other nodes, 205 in total. In particular regarding the second research objective, which proved to be under-researched in the literature, the majority of codes were generated from the interviews.

At first all codes were 'open codes' (refer to Strauss and Corbin, 1990) – i.e. describing what 'my' data were about. They were relatively broad containers for the contents of one or more lines, often using the respondents' own words as code labels. For example, while reading the transcripts I noted that many respondents described experiences in the 'working collectives' – a small task-orientated unit at the workplace. Until then I had not been familiar with this theme but decided to explore this phenomenon further by including this issue in my interview topic guide. While coding a text chunk 'working collective', I also coded the same segment in many other ways, such as 'GDR', 'agricultural co-operative' (i.e. work-place), 'family', 'identity', 'democracy', 'socialist services' and others. QSR N4 stored the codes in the Index System while hyperlinking each code to the text segment, leaving my transcript uncluttered, unlike its paper counterpart before.

As a story began to evolve I saw connections between respondents and themes. I felt it would be worth exploring differences and similarities between respondents with different age groups, occupational status, gender, 'function' (i.e. key inform-ant versus focus groups) and case villages. I was already well into my coding and had not yet coded for 'type of data'. Had I worked on paper, re-reading every page of every transcript would have been required in order to assign the above labels to the relevant text chunks. In QSR N4, however, this was less time-consuming and more thorough using a text search. For example, I entered '*130' (i.e. respondent Veronica) upon which QSR N4 retrieved everything Veronica had said. This text was stored in a separate file which I could then code '65+', 'pensioner', 'female', 'focus group' and 'Wilmersdorf'. I repeated this procedure for each respondent.

Constant comparison and theory building

Having coded 'my' data for content and 'type of data', I proceeded to compare codes. With the simple search mentioned above, coded segments can be extracted by respondent or theme.

I now wanted to combine different features to explore relationships between codes or between codes and type of data. In so doing, I would investigate hunches and ideas and test initial hypotheses. For example, I had the impression that the key informants suggested there was a difference between the experiences of men and women in the working collectives. Thinking about the narrations by women in the focus groups or accounts in my correspondence, I recalled very positive experiences of women. I therefore combined different codes to check out if this 'hunch' could be verified from my data. First, I retrieved everything women said about 'working collective'. Later, I created a file that also displayed other codes I had assigned to the same text chunk. Investigating this new file, further questions arose wherefore I needed to double check in which context something was said. As QSR N4 hyperlinks code to the source text, I could view the code in context by means of a simple mouse click. By so doing, my coding refined and my theory about the meaning of working collectives for gendered identities developed. This was also a part of exploring the dynamics behind the formulation of norms and meanings in the GDR, my second research objective.

What I did was to 'decontextualize' data (i.e. rip bite-size chunks out of their original context) and 'recontextualize' them in a new context (see the example in Box 28.2). Tesch (1990) called this having a 'playful relationship' with data. According to the literature, I should have played more – i.e. continue this process until a state of 'saturation'. However, going back to the field collecting more data until absolutely nothing new would emerge could not be achieved due to the financial and time constraints of the project.

Box 28.2 Decontexualizing and recontextualizing

Within a two-hour group interview during my study, respondents discussed their experiences in the GDR and those since unification. The following conversation took place sometime during the interview:

014: Well, twice a week, the butcher came to the village [in the GDR]. It wasn't like everyone had a car, like they do today. We would have to go all the way into town and especially for the pensioners that would have been too much … Today, the collective spirit is missing. It's missing, it used to be …
016: It's not there anymore.
014: We are … it's *missing* these days, we … everyone just thinks of their own job … one only thinks of oneself these days. We used to have a collective spirit.

The respondents were discussing their memories of the working collective. When I searched the other interview transcripts for this theme, I found that 39 documents contained the code 'collective'. I wanted to investigate more closely what insight I could gain from the women's experience of 'collective spirit' in relation to norms and meanings in the former GDR. Therefore, I displayed only those text segments that were coded 'collective'. This extraction of text segments from the interviews is called decontextualization. Other segments were, for example:

013: Women used to be able to discuss their problems and worries in the collective … Today they don't know where to turn and they don't dare talk about the problems.
011: It was different then, you were in a collective and somebody was asked to do this or to organize that. Like when a trip to the theatre was organized or so, you know? It was always somebody's turn and everyone participated in the organization at some point in time. Today, people find this very difficult.
120: The collective was promoted, you probably heard that already. One could get an award, a prize, and we always tried to be supportive of each other. That is what people miss so much today, the collective spirit. We were there for each other, we supported each other.

By bringing these text segments together, I recontextualized them. There were several indications that the collective was indeed an important part of the formulation of norms and meanings in the former GDR. Many women had fond memories of the collective because they experienced it as supportive and motivating. As the collective was organized according to a political plan and was also a significant part of the state's policy to form socialist identities, this finding suggested that the collective was an important means to achieve this.

In Chapter 27 of this book, Meghan Cope argues that qualitative researchers need to be more transparent about how they code and analyse their material. One advantage of using QSR N4 is that it helps to create a so-called audit trail for external scrutiny (Stanjek, in Gildchrist, 1992). This is done using memos alongside my coding. Such memos can have different functions, such as keeping track of organizational issues, developing and defining coding and preserving steps in analytical thinking. I clarified why I coded a text segment in a certain way, why I related it to another code and why I felt it illustrated a theory particularly well. These thoughts were stored in easily accessible 'containers' (memos).

SOME ADVANTAGES AND CONCERNS ABOUT CAQDAS

I described several advantages of using CAQDAS in my experiential account above. QSR N4 largely worked for me but it is also important to be aware of the limitations of CAQDAS. Table 28.2 gives an overview of advantages and concerns described in the literature. The list of concerns in this table is considerable and does not yet include the critique that the use of CAQDAS distances the researcher from the 'real world' of ethnography (Coffey and Atkinson, 1996; Coffey *et al.*, 1996; Hinchcliffe *et al.*, 1997).[1] The lesson drawn from the table is perhaps that a researcher will never be able to press a button and get a ready-made dissertation or project; nor is the software some kind of 'Frankenstein monster' (Lee and Fielding, 1991: 8). There is a danger that researchers may get lost or trapped in the data – particularly the logic flow of different software – and lose sight of what they are trying to write or achieve. In particular, researchers may struggle to reconcile their approach to qualitative data analysis (e.g. grounded theory) with the software (e.g. QSR N4 works on a tree-branch structure). One way to avoid this is to use CAQDAS in a critical and reflective way, seeking feedback

Table 28.2   Advantages and concerns about CAQDAS

Advantages	Concerns
Managing of large quantities of data	Obsession with volume
Convenient coding and retrieving	Mechanistic data analysis/taken-for-granted mode of data handling
Comprehensive and accurate text searches	Exclusion of non-text data
Quick identification of deviant cases	Over-emphasis on 'grounded theory'
More time to explore 'thick data' as clerical tasks become easier	Loss of overview
Playful relationship with data-enhanced creativity	The machine takes over–alienation from data
	Makes qualitative research look more scientific
	Limitations for connecting with geographical data such as GIS-type systems

Sources: Tesch (1990); Seidel (1991); Richards and Richards (1992); Dembrowski and Hanmer-Lloyd (1995); Lonkila (1995); Coffey *et al.* (1996); Fielding and Lee (1996); Kelle (1996); Crang *et al.* (1997); Tak *et al.* (1999)

from peers and supervisors as some kind of 'rain check'. A project does not have to be analysed in a standardized way – i.e. using a grounded theory approach. The suggestions in the Further Reading section below describe other techniques. In sum, it is advisable to know about qualitative data analysis and its underlying methodologies before beginning to use CAQDAS, and to adopt an approach to your analysis and software package that is compatible.

CONCLUSION

In this chapter I have illustrated some uses of CAQDAS in qualitative research. Drawing on previous publications, I have outlined different types and key functions of CAQDAS and illustrated these issues using an example from my own research where I adopted a grounded theory approach to my data analysis. Overall, my opinion of the software is positive as I found the use of CAQDAS helpful, in particular for ordering and analysing data in an efficient and thorough way. However, I have also drawn attention to the dangers of seeing qualitative analysis software as a seductive technology whose benefits may be overestimated/emphasized. In particular, I have highlighted the danger of losing sight of your own methodological questions and analytical approach to the data. As such its important to retain a healthy scepticism about the value of CAQDAS and not to forget the importance of reading about approaches to qualitative data and thinking about your material (sometimes with pen and paper!). After all, the software only processes the data; it does not in itself understand or analyse the material, and it will not write your dissertation for you. At the end of the day a computer will not deliver anything that it is not commanded to!

Summary

- When using computers in qualitative research, the use of more specialized software is recommended for activities such as coding and searching text, finding relations and keeping an organized database.
- In general, CAQDAS has three key functions: *text retrieval*, *coding* and *theory building*. The choice of software depends on the methodological orientation of the researcher.
- More sophisticated programs usually combine several functions and accommodate different kinds of data, such as texts, statistics, pictures and video segments.
- Concerns about CAQDAS include the danger of mechanistic data analysis, the detachment from one's data and a false sense of science.

NOTE

1 It must be noted that further issues have transpired from more recent discussions about CAQDAS which should also be integrated in Table 28.2. Key concerns address the expansion from the current focus on textual data to the possibilities of hypertext and hypermedia, and the convergence towards one

prevailing method of analysing these. I refer the reader to more comprehensive accounts – for example, in Lee and Fielding (1995), Coffey and Atkinson (1996: Ch. 7), Coffey *et al.* (1996), or Hinchcliffe *et al.* (1997).

Further reading

There is a growing body of literature about how to use CAQDAS. Key references include the following:

- Weitzman and Miles (1995) should be consulted for an evaluation of what is on the market.

- Theoretical and practical questions about how to use CAQDAS, illustrating pros and cons, are explored in Fielding and Lee (1991), Burgess (1995), Kelle (1995) and Bazeley (2007).

- Possible ways of accommodating approaches to qualitative research other than those described in this chapter – such as narrative studies, ethnography, interpretive/ hermeneutic analysis, critical theory, collaborative or action research, or content analysis – can be found in Tesch (1991), Kelle (1996) or Miles and Weitzman (1996).

- An invaluable internet source is the site maintained by the CAQDAS networking project (caqdas.soc.surrey. ac.uk).

- Last but not least, it is advisable to immerse yourself in the methodological literature as well. For the grounded theory approach, Strauss and Corbin (1990) provide an accessible account of 'how to . . .'

Note: Full details of the above can be found in the references list below.

References

Bazeley, P. (2007) *Qualitative Data Analysis with NVivo* (2nd edn). London: Sage.

Burgess, R.G. (1995) *Computing and Qualitative Research. Volume 5: Studies in Qualitative Sociology*. Middlesex, UK: Jai Press.

Coffey, A., Holbrook, B. and Atkinson, P. (1996) 'Qualitative data analysis: technologies and representations', Sociological Research Online, 1 (available at http://www.socresonline.org.uk/socresonline/1/1/4.html).

Crang, M.A., Hudson, A.C., Reimer, S.M. and Hinchcliffe, S.J. (1997) 'Software for qualitative research. 1. Prospectus and overview', *Environment and Planning A*, 29: 771–787.

Dembrowski, S. and Hanmer-Lloyd, S. (1995) 'Computer applications – a new road to qualitative data analysis?', *European Journal of Marketing*, 29: 50–62.

Fielding, N. (1994) 'Getting into computer-aided qualitative data analysis', ESRC Data Archive Bulletin, September (available at http://caqdas.soc.surrey.ac.uk/getting.htm).

Fielding, N. (1995) 'Choosing the right qualitative software package', ESRC Data Archive Bulletin, 58 (available at http://caqdas.soc.surrey.ac.uk/choose.htm).

Fielding, N.G. and Lee, R.M. (1991) *Using Computers in Qualitative Research*. London: Sage.

Gildchrist, V.J. (1992) 'Key informant interviews', in B.F. Crabtree and W.L. Miller (eds) *Doing Qualitative Research*. London: Sage, pp. 70–89.

Glaser, B. and Strauss, A.L. (1967) *The Discovery of Grounded Theory*. New York: Aldine.

Grbich, C. (1999) *Qualitative Research in Health: An Introduction*. London: Sage.

Hinchcliffe, S.J., Crang, M.A., Reimer, S.M. and Hudson, A.C. (1997) 'Software for qualitative research. 2: Some thoughts on "aiding" analysis', *Environment and Planning A*, 29: 1109–1124.

Hoven-Iganski, B. van (2000) *Made in the GDR: The Changing Geographies of Women in the Post-socialist Rural Society in Mecklenburg-Westpomerania*. Utrecht: KNAG/NGS.

Kelle, U. (1995) 'An overview of computer-aided methods in qualitative research', in U. Kelle (ed.) *Computer-aided Qualitative Analysis: Theory, Methods and Practice*. London: Sage, pp. 1–18.

Kelle, U. (1996) 'Computer-assisted qualitative data analysis in Germany', *Current Sociology*, 44: 225–241.

Kitchin, R. and Tate, N.J. (2000) *Conducting Research into Human Geography: Theory, Methodology and Practice*. Harlow: Pearson Education.

Lee, R.M. and Fielding, N.G. (1991) 'Computing for qualitative research: options, problems and potential', in N.G. Fielding and R.M. Lee (eds) *Using Computers in Qualitative Research*. London: Sage, pp. 1–13.

Lee, R.M. and Fielding, N.G. (1995) 'User's experiences of qualitative data analysis software', in U. Kelle (ed.) *Computer-aided Qualitative Analysis. Theory, Methods and Practice*. London: Sage, pp. 29–40.

Lonkila, M. (1995) 'Grounded theory as an emerging paradigm for computer-assisted qualitative data analysis', in U. Kelle (ed.) *Computer-aided Qualitative Analysis. Theory, Methods and Practice*. London: Sage, pp. 41–51.

Miles, M.B. and Huberman, A.M. (1994) *Qualitative Data Analysis. A Sourcebook of New Methods*. Thousand Oaks, CA: Sage.

Miles, M.B. and Weitzman, E.A. (1996) 'The state of qualitative data analysis software: what do we need?', *Current Sociology*, 44: 206–224.

Prein, G., Kelle, U. and Bird, K. (1995) 'An overview of software', in U. Kelle (ed.) *Computer-aided Qualitative Analysis: Theory, Methods and Practice*. London: Sage, pp. 190–210.

Richards, L. (1997) 'User's mistakes as developer's challenge: designing the new NUD.IST', *Qualitative Health Research*, 7: 425–434.

Richards, L. and Richards, T. (1987) 'Qualitative data analysis: can computers do it?', *Australian and New Zealand Journal of Sociology*, 23: 23–35.

Richards, L. and Richards, T. (1992) 'Hard results from soft data? Issues in qualitative computing', paper presented at the British Sociological Association Annual Conference, Manchester.

Richards, T. and Richards, L. (1993) 'Qualitative computing: promises, problems, and implications for research process', paper presented at the British Sociological Association Annual Conference, Colchester.

Seidel, J. (1991) 'Method and madness in the application of computer technology to qualitative data analysis', in N.G. Fielding and M.R. Lee (eds) *Using Computers in Qualitative Research*. London: Sage, pp. 107–116.

Strauss, A.L. and Corbin, J. (1990) *Basics of Qualitative Research: Grounded Theory Procedures and Techniques*. Newbury Park, CA: Sage.

Tak, S.H., Nield, M. and Becker, H. (1999) 'Using a computer software program for qualitative analyses. Part 2. Advantages and disadvantages', *Western Journal of Nursing Research*, 21: 436–439.

Tesch, R. (ed.) (1990) *Qualitative Research. Analysis Types and Software Tools*. Basingstoke: Falmer Press.

Tesch, R. (1991) 'Software for qualitative researchers: analysis needs and program capabilities', in N.G. Fielding and R.M. Lee (eds) *Using Computers in Qualitative Research*. London: Sage, pp. 14–15.

Weitzman, E.A. and Miles, M.B. (1995) *Computer Programs for Qualitative Data Analysis: A Software Sourcebook*. London: Sage.

29 Analysing Historical and Archival Sources

Iain S. Black

Synopsis

Students using historical and archive sources should clearly identify whether their research project is 'problem orientated' or 'source orientated'. Source-orientated research relies primarily upon the detailed examination of a single source, such as a diary. Problem-orientated research involves defining a research question through conceptual and theoretical reasoning before the initial engagement with archives and historical sources. This chapter considers three groups of key historical sources in detail: official and private documentary sources, visual evidence and literary sources. Each group is reviewed to assess their potential and problems in historical geographical research.

The chapter is organized into the following sections:

- Introduction
- The nature of historical evidence
- Documentary sources
- Visual texts
- Literary sources
- Conclusion

INTRODUCTION

Geographers have long been interested in the nature of past societies, economies, cultures and environments. This chapter introduces you to the nature of source material available to reconstruct past geographies and reviews some of the key ways in which historical research is undertaken. It begins by considering the nature of historical evidence, indicating the need to critically evaluate the creation of particular historical sources as part of any well-grounded research project. Following this, three groups of key sources are identified for more detailed scrutiny: official and private documentary sources, visual evidence and literary sources. Each group is reviewed to assess their potential and problems in historical geographical research.

Students using historical and archive sources should clearly identify whether their research project is *problem-orientated* or *source-orientated* (see Baker, 1997). *Source-orientated* research relies primarily upon the detailed examination of a single source, such as a diary, a set of business records or a census enumerators' book. The strength of this approach is that it allows the researcher to build up a detailed picture, or 'thick description', of a particular aspect of the past. The weakness is that such research is often led by the source material, decontextualized from wider questions, and can present problems of establishing a broader interpretive framework. *Problem-orientated* research involves defining a research question through conceptual and theoretical reasoning before the initial engagement with archives and historical sources – in short, establishing what questions to ask of the data (see Kearns, 2007). Once research questions are established it is often possible to identify a series of pertinent sources, which can be linked for consistency and greater understanding. Before commencing the research process, however, a clear understanding of the nature of the source material is necessary.

THE NATURE OF HISTORICAL EVIDENCE

Historical geographers engage with a wide range of evidence, both verbal and non-verbal, in their research. Harley (1982: 261) remarks that evidence giving a window on the past includes 'words written on paper or parchment; words or symbols carved on stone or wood; maps, paintings and photographs; monuments and landscape evidences; artefacts dug up from the ground; and the living documents of oral history'. But it is important to realize, along with Harley, that evidence is not a pre-given entity waiting to be discovered and marshalled into service to answer questions of historical research. The past can never be recovered as a solid whole, as history is always fragmented. Historical research can only deal with residues or, to use a geological metaphor, sediments left over as a result of past action (Dirks, 2002). White, in *Machina Ex Deo* (1968: 4), claims that 'history does not exist; all that exists is debris – scattered, mutilated, very fragmentary – left by vanished ages'. This 'debris' – the sources – is the stuff of history.

We can never simply approach sources directly in the course of historical research, for three main reasons. First, the sources themselves have already been constructed by those who recorded the information and, subsequently, those who have selected what to preserve and what to discard. Second, an archive is itself socially constructed and dynamic, wherein professionals catalogue and order data according to the changing needs and values of their institutional paymasters (Kurtz, 2001). Third, the construction of history depends on a process of retrieval, selection, contextualization and ascription of meaning. This is always more than simply putting 'the facts' into a chronological sequence. It involves establishing relationships of cause and effect. As Harley (1982: 272, emphasis in original) notes, 'as scholars *we* communicate with the sources and simultaneously try to recreate the way in which *they* formed part of a communication system involving people in the past'. Historical facts are, therefore, constructed by historical geographers through questioning based on an a priori understanding of what is significant. Le Goff (1992: 113) quotes the historian Lucien Febvre thus: 'to elaborate a fact is to construct it. It is, so to speak, to answer a question. And if there is no question, there is nothing at all'. Thus, without questions or hypotheses, 'facts' cannot be generated or placed in context. According to Max Weber: 'any attempt to understand [historical] reality without subjective hypotheses will end in nothing but a jumble of existential judgements on countless isolated events' (quoted in Le Goff, 1992: 113–114).

Historical sources, then, must be critically evaluated as part of any research project in historical geography. If we accept that sources are made and not simply given, a series of key questions about the nature of any particular source can be raised. First, there is a need to establish the authenticity of a source. Is it genuine? In most cases it is important to consult material in its original form rather than any subsequent transformations of the data, though evidence edited by scholars as part of a critical evaluation of a source can be very useful. Second, any source must be assessed for accuracy. How close is the source to the events or phenomena it is recording? How accurately was the information recorded? This highlights the need to cross-check with other sources to establish whether the information is consistent with established facts. Third, it is important to understand what the original purpose for gathering the information was. How might this have influenced what was collected? Historical data are always filtered so it is necessary to establish for whom and by whom information was recorded and for what purpose. Further, this highlights the need to consider the 'silences' in the historical record whereby the actions of the powerful and the unusual have often taken precedence over the everyday life of the mass of the population (Duncan, 2001). Fourth, how has the process of archiving the information imposed a classification and order upon historical events? Archiving can be thought of as a further filter between the 'raw' information and the historical researcher. It should be clear from this that sources themselves convey residues of power structures, and this reinforces the importance of scepticism in historical research. According to Thompson (1978: 220–221), 'historical evidence is there, in its primary form, not to disclose its own meaning, but to be interrogated by minds trained in a discipline of attentive disbelief'.

A common distinction made by historical researchers is that between *primary* and *secondary* data. Primary data are generally taken to mean raw data in an unprinted or unpublished form, usually located in a record office or archive. Secondary data are generally understood to have been transformed from their raw state, often collated, classified and/or tabulated for publication. The question of superiority between these types of data has often been raised (see Dennis, 1987, 1988; Harvey, 1988) but, as Butlin (1993: 73) notes: 'there is nothing particularly sacrosanct about the nature of "primary" as opposed to "secondary" data or information, for both need careful scrutiny and verification'. In what follows, three examples of different groups of sources are presented to indicate the range of possible approaches to understanding the geographies of the past.

DOCUMENTARY SOURCES

Documentary evidence – in essence, the written word – can be usefully subdivided into *official* and *private* sources. *Official data* sources comprise those records produced by and for the national and local state. Such bodies had a constant need to generate information in the course of government, relating in particular to assessing population, taxation, land ownership and monitoring the state of agriculture, industry, the courts and colonial development. In Britain the legacy (contained in British parliamentary papers, reports of royal commissions and committees of the House of Commons and Lords) is vast. Examples of historical work based on such data include analysing agricultural enclosure in Britain (Turner, 1984), mapping agricultural returns (Coppock, 1984) and reconstructing the transition from domestic to factory production in the West Yorkshire woollen industry (Gregory, 1982).

Private archives are a source of growing importance for historical geographers too (Hall, 1982). Traditionally, a very high proportion of data used in historical geographical research was drawn from the official record. However, the development of new areas of research, including the construction of biographies of explorers and travellers or the analysis of private institutions, has demanded sources of greater specificity than those which official archives typically provide. Indeed, official archives, concerned mainly with aggregate information, often have relatively little to say about individuals. Even when they do appear (such as witnesses in select committee reports, for example) they usually comprise only the prominent and powerful (for a fascinating example of reversing this dominant tapestry see Burton, 2003).

Private archive sources, such as diaries or letters, can help to overcome these problems of aggregation and the recording of 'official' rather than personal views. But private archives also need care in their use. Precisely because they comprise records created by an individual or an individual institution, they possess a unique quality requiring comparison and evaluation with data from other sources. The following examples illustrate the use of documentary sources drawn from both private and public collections.

Geographies of people in the past

A concern to identify and characterize the nature of people in the past has a long tradition in historical geography. Such analyses range from the reconstruction of individual biographies, through the reconstitution of family and household units, to more general explorations of urban or rural socioeconomic structures based on large-scale official survey data contained in national censuses. The researching and writing of historical biographies are typically based upon records of population, family papers, educational and business records, and contemporary newspapers or other accounts. Clearly, the better known the individual in question, the greater likelihood that substantial bodies of evidence on his or her life will be extant. But historical geography is not only concerned with the rich and powerful; it is also possible to reconstruct something of the lifeworlds of a whole range of people from all groups in past societies.

The start and end points for reconstructing a biography of a particular individual are conventionally the dates of birth and death, though often it is useful to locate these within a broader family history. In England and Wales systematic, though by no means comprehensive, information on the lives of individuals begins in 1538 when the recording of christenings, weddings and burials by parish began. Used as proxies for births, marriages and deaths, these parish registers provide a valuable source for the identification of the key life events of individuals in the past, despite clear problems in their use, including 'variation in accuracy through time, the exclusion of data for the non-established churches, clandestine marriages, and birth-baptism shortfalls through very early infant mortality' (Butlin, 1993: 77). From 1837 the researching of individuals becomes more straightforward as registration of births, marriages and deaths became a legal requirement.

The following example of Onesiphorus Tyndall, an eighteenth-century banker and merchant in Bristol, is based on a biographical essay commissioned for the *Oxford Dictionary of National Biography* (Black, 2004). It indicates the range of sources available that provide details of people's past lives and how these can be interpreted. It is not possible to establish Tyndall's date of birth, but it is known he was baptized on 28 May 1689. This datum was located by a search through the parish registers of the City of Bristol. No family trees or other papers can be found, but something of his family background was traced in a nineteenth-century history of banking in Bristol. Details on his early life are opaque but it is possible to infer from the banking history that Tyndall was apprenticed to his father in the trades of grocery and dry salting. On 6 November 1717 Tyndall married Elizabeth Cowles, also of Bristol, and together they had two sons and two daughters. The date of marriage was traced from marriage licence records, while details on Tyndall's own family were reconstructed from the relevant parish registers. Tyndall was notable for his role in the formation of the Old Bank, Bristol's first bona fide banking house, which commenced business on 1 August 1750. Details on his banking and business career were found in contemporary publications on Bristol's commercial life. In addition to banking, Tyndall had a long and varied association with the activities of the port of Bristol, including the West Indies trade in sugar and tobacco. He was still listed as trading to Africa in 1755, two years before his death, in a manuscript collection on West Indian traders in Bristol City Library. On 4 June 1757 *Felix Farley's Bristol Journal* carried an obituary of Tyndall giving the details of the date of his death (30 May 1757) and burial (3 June 1757). The latter date was cross-checked in the burial registers of Christchurch Parish Church. For historians there is also life after death, a principal source for which is an individual's will. Tyndall's will, proved in the Prerogative Court of Canterbury, enabled a detailed estimate of his wealth at death and identification of his beneficiaries. Combining a range of such detailed biographies, together with wider documentary evidence and published sources, could enable a 'thick description' of a city's civic and business elite and the multiple ways they negotiated their power (see Billinge, 1982).

However, to increase the level of aggregation significantly, to look at the changing composition of the family unit for example or the demographic response of local communities to the external shocks of food shortage or disease, it is necessary to move beyond the individual. Before the nineteenth century, data contained in the parish registers can be aggregated by calculating totals of births, marriages and deaths for a specified geographical area over a given time period. By tabulating the frequency of such events, either monthly or annually, patterns of population growth or decline can be identified. A particular concern has been the specification of years of 'crisis mortality', where an unusually large number of deaths have occurred. To try to explain such events, attempts are made to correlate population dynamics with wider socioeconomic fluctuations, such as the state of harvests or the incidence of major diseases such as plague.

Family reconstitution, by contrast, involves gathering data on vital demographic events for individual families, in which the surname is the link. As Morgan (1979: 74) notes, this involves:

> recording the names of all those baptized and tracing them among the marriages 15 to 50 years later and among the burials up to 105 years later ... [building] a figure of the actual number of people alive at any one time, the size and extent of the families, at what age people got married, had children and died.

The most ambitious and comprehensive use of data in parish registers, involving the calculation of population size and age structure and the use of backward projections to derive population trends over time where data are lacking, is Wrigley and Schofield's (1981) study of English population between 1541 and 1871. This work is a major starting point for any study of the historical geography of population prior to the census.

The nature and use of the census is given comprehensive treatment in Chapter 5. It is sufficient to note here two key forms of census data of especial significance to historical geographers. First are the census enumerators' books, which are available for consultation at ten-year intervals between 1841 and 1901 and to which a hundred-year confidentiality rule applies. The books provide a comprehensive set of data, in a standardized format, for each individual in the relevant enumeration district, including the following variables: relation to head of household, marital condition, age, sex, occupation and birthplace. This basic set of data can then be used to study the demographic, social, economic and migration structure of places ranging in scale from the individual household, through streets and districts, to large urban areas. A useful review and bibliography of work based on this source has been produced by Mills and Pearce (1989). Second are the census abstracts, which consolidate information derived from the original enumerators' books and are available from 1801 in published parliamentary papers and online. They contain tables of information on population, age, sex, marital status, birthplace, occupation and housing for a range of geographical areas. From 1841 it is possible to distinguish four spatial scales of information: county, registration district, subdistrict and parish. A principal use of the abstracts has been the historical analysis of patterns of migration and the changing industrial structure of cities and regions (Lawton, 1982).

Historical geographies of money

The records of business institutions have considerable value in addressing important research questions in historical geography. Black (1995, 1996) has used data contained in financial records, for example, to research spatial patterns and processes of financial integration in early industrial England, developing a critical perspective upon regional dynamics by reconstructing flows of money, short-term credit and commercial information never recorded in official surveys. Where sufficient data exist to undertake quantitative and cartographic analysis, research based on such private archives can provide an important corrective to some of the aggregate conclusions drawn from official data sources. The following examples, drawn from work in bank archives, indicate the potential of these sources for reconstructing past monetary geographies.

Consider Figure 29.1, which shows the spatial structure of correspondence between the private bank of Peacock & Co. of Newark, Nottinghamshire, and its business and private customers. The data are drawn from letter books surviving for the period between 1809 and 1813. Though the data are relatively simple, the patterns revealed through mapping the distribution of correspondence are full of interest in the context of debates over the nature and extent of regional economic integration in the early industrial economy of England and Wales. The map shows a focused intraregional circuit of commercial information, centred on the bank's sister branch at Sleaford, and the proximate commercial centres of Nottingham and Lincoln. But more striking is the overwhelming dominance of London in the bank's business correspondence in the early nineteenth century. Given there was no formal institutional link to the capital, this pattern demands explanation. A careful reading of the content of the letters mapped in Figure 29.1 indicates that the bank had an agency link to a London private bank, which handled its business affairs on an interregional scale. Already the consideration of this small provincial firm begins to provide suggestive evidence of widespread and interlocking patterns in the circulation of commercial information on an intraregional and provincial-metropolitan level.

Moving from letter books to historical documents that recorded financial transactions, such as ledgers, bill books or remittance books, it is possible to build on the initial mapping of correspondence to reconstruct geographies of past capital flows. Figure 29.2 shows the value of bills of exchange received for discount at the Newark bank between 1807 and 1809. The bill was the key financial instrument used in the provision and circulation of short-term paper credit in the early industrial English economy (Black, 1996). The Newark bank's bill book, on which Figure 29.2 is based, includes valuable data on the places from which bills were received and their value. Again the patterns revealed are suggestive. Two features are immediately apparent: first, the high value of bills drawn and cashed locally; and, second, the flow of bills from key centres of early industrialization. The first pattern is expected as the bank provided ready cash to local businessmen by buying their bills drawn in trade. The second pattern is more surprising and demands explanation.

What is revealed by a close analysis of the data is that bills drawn in trade in the industrial districts of Manchester, Leeds, Sheffield and Birmingham found their way to this small country bank in Newark for discount (sale). The Newark district was predominantly agricultural in nature at this time and had surplus capital for investment. The industrial districts of the north and midlands were capital hungry. What we have evidence for, then, is a distinctive geography of capital flow as bills circulated from the industrializing regions of the country to regions of capital surplus in the east and the south. But a final question arises: how did manufacturers

Figure 29.1   Peacock & Co., Newark, Nottinghamshire. Correspondence with banks and private customers, 1809–1913

in the north and midlands know that any particular bank in the east or south had surplus funds for investment? The answer cannot be found in the data themselves, but only by linking the primary evidence shown in Figure 29.2 to a broader under-standing of the role of the London money market in the early nineteenth century, where the London agents and bill brokers specialized in matching up bills seek-ing discount with provincial banks seeking outlets for their surplus capital. Thus,

Figure 29.2   Peacock & Co., Newark, Nottinghamshire. Value of bills received for discount, 1807–1809

evidence drawn from an intensive study of financial documents can lead to wider conclusions about the role of London in a period of English economic history often exclusively seen as dominated by the industrialization of the provinces.

VISUAL TEXTS

Although the term 'archives' commonly conjures up an image of reams of documents in labelled boxes and filing cabinets, geographers also draw upon a wide range of visual evidence in their reconstructions of the past, notably in seeking to analyse and interpret the production and representation of landscape (Cosgrove and Daniels, 1988; Cosgrove, 1993; Daniels, 1993; Domosh, 2001). Paintings, sketches, engravings and architectural drawings are all valuable in the essentially archaeological practice of recovering the symbolic geography of past landscapes, be they rural or urban. Indeed, in cases where the landscape in question has been substantially modified or even erased by subsequent human agency, such visual evidence may provide the only representations extant for such a reconstruction to take place.

Though a wide range of approaches have been adopted in the historical study of landscape, two methodologies have gained prominence: one relying upon textual metaphor and the other drawing upon the techniques of iconography. The metaphor of seeing the landscape as a text, which draws upon the influential work of the cultural anthropologist Clifford Geertz, suggests landscapes can be read as a social document, using the methods and techniques of literary theory (Barnes and Duncan, 1992). With reference to specific landscape forms, textual metaphors can be pursued to illuminate the crucial relationships between landscape and ideology, helping to identify how landscapes can transform ideologies into a concrete, visible form. Therefore, the role of the critical reader of landscapes should be, in part, to penetrate what Barthes called their 'layers' of ideological sediment (Duncan and Duncan, 1988). The second approach, that of iconography, is drawn explicitly from studies in art history. Cosgrove and Daniels (1988: 1) define iconography as, broadly, 'the theoretical and historical study of symbolic imagery'. Although the use of a formal iconographical approach to interpreting images reaches back to the Renaissance, in its modern form the concept was deployed by the discipline of art history to:

> probe meaning in a work of art by setting it in its historical context and, in particular, to analyse the ideas implicated in its imagery ... consciously [seeking] to conceptualise pictures as encoded texts to be deciphered by those cognisant of the culture as a whole in which they were produced. (Cosgrove and Daniels, 1988: 2)

This broad conceptualization of iconography was refined and codified into a systematic approach to the study of art by Erwin Panofsky, whereby iconography in its narrower sense 'was the identification of conventional, consciously inscribed symbols', while a more interpretative conception of iconography (termed iconology) sought a deeper level of meaning in art by reconstructing the cultural, political and ideological context(s) surrounding its production and which the art communicated consciously or unconsciously (Cosgrove and Daniels, 1988: 2). These principles of iconographical study have become widely used in a variety of attempts by geographers to interpret symbolic landscapes in a range of different settings.

An important methodological statement on the interpretation of the built environment was set out by Domosh in 1989. In that paper she sought to develop a multi-layered understanding of the design and construction of the New York World Building by linking this important early skyscraper to 'its socio-economic and aesthetic contexts, and the actors who directly produced and/or created that artifact' (Domosh, 1989: 347). Beginning with a single building, the argument is constructed to counterpose a series of inter-related 'layers' of explanation, encompassing the functional, symbolic and ideological qualities of this monumental corporate headquarters. After setting the new building within the social, economic and spatial context of later nineteenth-century New York, Domosh proceeds to link the aesthetic form and social meaning of the building to the individual aspirations of its owner (the newspaper magnate Pulitzer) and his place within the highly competitive elite of the city. Paralleling this discourse of architectural aesthetics was the changing nature of technology involved in skyscraper construction, which allowed height to be used as a competitive strategy in the buildings of new corporate capital. Domosh clearly demonstrates throughout how the 'text' surrounding the creation of this particular landscape artefact is informed by a continuous dialogue with the 'context' of the history of the city in which it was built and the social and economic conditions of late nineteenth-century American capitalism. In later work, Domosh goes on to apply this methodological framework, broadly conceived, in more detailed studies of New York's skyscrapers and in a more fully developed historical and comparative analysis of the landscapes of New York and Boston in the nineteenth century (Domosh, 1992, 1996).

Dennis, in a perceptive essay on the methodologies of historical geography and their application to the modern urban landscape, refers to Domosh's study of the New York World Building as a 'multi-layered approach to landscape interpretation' (Dennis, 2001: 24). However, he suggests her approach could have been strengthened by making it multi-*method* too. Rather than relying principally on secondary sources, chiefly newspapers, magazines and 'existing accounts of ... land values and skyscraper technology that were not specific to the building' (2001: 24), Dennis argues that a wider range of sources and methods would have deepened her analysis. In particular, he refers to the important work of Holdsworth who, in seeking to interpret past built environments, draws upon 'corporate archives, fire insurance plans, assessment records, property transfer records, mortgage records, and the manuscript census to try to make visible, to bring out of concealment, what is not visible in today's landscape' (Dennis, 2001: 24). Indeed, Holdsworth's essay is an important reminder of the need to combine analyses of visual evidence with a wider set of documentary sources to develop a fuller understanding of the creation and functioning of built environments in the past (Holdsworth, 1997). The following example considers a way of 'reading' a monumental commercial building within a past landscape that has all but disappeared from view: the central financial district of the City of London in the early Victorian period (see Black, 2000).

Money, power and landscape

Figure 29.3 shows the headquarters built by the celebrated early-Victorian architect C.R. Cockerell for the London & Westminster Bank in 1838. The site, in Lothbury, was at the heart of the City's central financial district, just opposite the northeast corner of the Bank of England. Careful observation of this section of streetscape suggests a number of key questions which can be used to structure an analysis of its meaning. Compared with its immediate neighbours the new headquarters was clearly dominant in both size and style, dwarfing the domestic

Figure 29.3   The London & Westminster Bank, Lothbury, 1838

townscape of the surrounding Georgian urban fabric. Even the business premises of the neighbouring private bank of Jones, Loyd & Co., modelled on a Georgian town mansion, seem to belong to another age. A closer look at the London & Westminster's façade reveals additional features requiring interpretation: How was this break from domesticity underlined by a more formal, public quality of architecture? Why was there much greater attention paid to the visibility of the entrance doorway? What messages can be found in the deliberate use of ornamentation and statuary? To move beyond architectural description, however, requires a consideration of the place and purpose of the London & Westminster Bank in the financial world of the metropolis in the 1830s.

Prior to 1833 joint-stock banking was prohibited in the capital, with the sole exception of the Bank of England. The London private bankers were legally restricted to private partnerships of a maximum of six persons. All this was to change in 1833 when, following decades of financial instability, the state finally moved to remove the Bank of England's monopoly of joint-stock banking. Joint-stock organization was a form of corporate structure similar to today's public companies, with capital owned by shareholders and the day-to-day business of the company entrusted to a board of directors. The London & Westminster Bank was the first new joint-stock bank to be established in London, bringing with it a greatly enhanced capacity to expand its capital base through share subscription. Already, therefore, we can begin to see an *economic* rationale for the design of its new headquarters. But the story cannot be reduced to one of simple economic power. In 1830 the City was still a closed and private world where the wealth of banks was materially and symbolically embodied in the private bankers themselves. They had nothing but contempt of the *arriviste* status of the new joint-stock banking companies. The Bank of England too was hostile, jealous of the erosion of its long-standing monopoly privileges. A closer evaluation of the London & Westminster's new headquarters, in terms of style and location, therefore requires cognizance of the deep-seated differences in business culture associated with the private and joint-stock forms of banking capital.

Figure 29.4 shows an engraving of the new headquarters in 1847. The solidity, purpose and strength of the design were noted by contemporaries as befitting the office of a capital-rich banking institution providing a home for the savings of

Figure 29.4   Engraving of the London & Westminster Bank, Lothbury, 1847

the growing metropolitan middle class. Decoration and ornament were kept to a minimum, though the first-floor windows possessed side panels between the rusticated piers available for embellishment. Here *caducei* and *fasces* were sculptured, in alternate disposition. The *caduceus* was a staff entwined with two serpents and bearing a pair of wings at the top which, according to Greek mythology, was carried by Hermes, messenger of the gods and representative of the divinity of commerce. *Fasces*, a Roman symbol, displayed a bundle of rods with a projecting axe blade. Traditionally carried before a Roman magistrate, fasces were a symbol of authority. At the extremities of the front were bold piers, each displaying a statue of a seated female figure. These figures were emblematic of the commercial interests of the bank, with the statue at the eastern end carrying a shield decorated with the arms of the City of London, while that at the western end displayed the arms of the City of Westminster. In alluding to the two cities that lay at the heart of the expanding metropolis, they also represented the two distinct banking traditions which the new institution was seeking to incorporate. The 'City' symbolized mercantile interests and the importance of trade finance, while 'Westminster' referred to the long tradition of private deposit banking that the new institution was opening up to a wider middle-class market. The scale and boldness of the design were unprecedented in the previously closed world of metropolitan banking, save for the Bank of England. The new and more public form of money represented by the London & Westminster relied upon such distinctive architectural references to underline its legitimacy and authority.

    This breakthrough to a wider middle-class market was reinforced by the scale and design of the principal banking hall shown in Figure 29.5. Gone was the simplicity of the bankers' shop, coupled with the Georgian parlour for the discussion

Figure 29.5   Principal banking hall of the London & Westminster Bank, Lothbury, 1845

of private financial affairs. The new banking hall provided an air of confidence to the setting of daily formalized money exchanges, where the separation of clerks and customers shown clearly by the counter underscored the mixing of public and private space so important for reputable money transactions. It was precisely these qualities of internal space that caught the attention of contemporary critics, who compared it with the baths of Imperial Rome where rounded forms of arch and dome enclosed a space in which public association reinforced moral order. The town banking hall announced a new era in the scale and purpose of banking, adding to the already-established traditions of discreet personal banking for the aristocracy and gentry, and the specialized business of the City mercantile houses, the popularization of the banking habit among the metropolitan middle class.

This example emphasizes a number of key methodological principles: first, the importance of maintaining a close relationship between visual evidence and other archival and historical sources, taking Geertz's notion of 'thick description' seriously to outline the detailed historical and geographical contexts surrounding an interpretation of particular landscape forms; second, to give due weight to the importance of politics and ideology embedded within visual evidence by providing a close reading of the signs and symbols encoded within particular landscapes; third, to ensure a rigorous study of the aesthetics of landscape by drawing upon the methods and techniques of iconographical analysis; and, fourth, recognizing the importance of linking the 'deep descriptions' of particular building projects, or other refashioning of landscape, to wider social, cultural and economic processes, maintaining a conscious dialogue between 'text' and 'context' in the social production of specific landscape forms (see Black, 2003 and Chapter 30).

LITERARY SOURCES

Literature produced in past periods and places constitutes the final group of sources to be considered. Such work, especially novels and poems, has long provided important evidence for the reconstruction and interpretation of past geographies, though its potential as a source demands critical examination (for examples, see Pocock, 1981; Lucas, 1988; Daniels and Rycroft, 1993; Wiley, 1998). Moretti (1998: 3), in a general survey of the geography of the European novel, suggests that space 'is not an inert container, is not a box where cultural history "happens", but an active force, that pervades the literary field and shapes it in depth'. Further, he sees the geography of the novel as comprising two distinct possibilities:

> it may indicate the study *of space in literature*; or else, *of literature in space.* In the first case, the dominant is a fictional one: Balzac's *version* of Paris, the Africa of colonial romances, Austen's redrawing of Britain. In the second case, it is real historical space: provincial libraries of Victorian Britain, or the European diffusion of *Don Quixote.* (Moretti, 1998: 3, emphases in original)

In historical geography the traditional approach to literature as a source has been rather more prosaic. A classic example is Darby's (1948) paper on the regional geography of Thomas Hardy's Wessex. Darby was concerned to correlate the 'fictional' places in Hardy's work with 'real' places in late nineteenth-century Dorset, to reconstruct something of its topography and landscape character. For him, novels were considered as a reliable source for analysis akin to many other forms of geographical data. While such topographical details can indeed be usefully combined with other types of evidence, such as maps, drawings or directories, for example, this is an unduly limited conception of literature as a historical source. One of literature's key characteristics that drew humanistic geographers to its analysis in the 1970s was its power to render palpable the essence of place in a way that many scholarly works of geography singularly failed to do (Tuan, 1978). It was argued that descriptions of people and places in the past, captured in literary form, could provide a valuable way of approaching the *experience* of the past; an experience otherwise impossible to capture by conventional ethnographic means, of course.

Although there are no simple methodological rules for the use of literature as a historical source, it is possible to set out a series of critical points to be borne in mind when approaching novels and poems written in the past. Sharp (2000) outlines some key dimensions of literature that require critical scrutiny. First, literature cannot be simply appropriated and used directly as a source, decontextualized from the conditions of its own production. She notes (2000: 328) how many geographers 'display a naïveté about the form of literary writing: it is seen as unproblematic and self evident in its immediate beauty'. This cautions against the use of literature as simply a storehouse of descriptions which can be culled as exemplifications of other empirical knowledges validated in traditional ways from more orthodox sources. Literary sources themselves require a scrupulous examination of the 'positionality' of their authors, their readers and the historical context in which they were produced. Second, literature is too often presumed to possess a universal quality, which masks differences between individuals and social groups. The meanings and values ascribed to a piece of literature as historical evidence cannot be presumed to be valid for all people in all times and places. Such a qualification seems obvious, but the fact remains that much literature is seen to deal with the general essence of the human condition. Third, geographers have largely ignored the status of literature as a form of writing, seeing it merely as another form of 'data'. Thus, according to Sharp (2000: 329), 'geographers are neglecting the most vital part of literature, which is the challenging potential of this form of expression compared to orthodox modes of geographical writing'.

This latter point has important methodological implications for the use of, say, novels as historical sources. Crang (1998) indicates, through his discussion of a range of authors writing on 'modern' cities, how their textual strategies help to shape their interpretation of modern urban life, beyond any simple attempt to capture the empirical detail of, for example, consumer cultures in nineteenth-century Paris. Thus, 'far from treating literary works as things that simply portray or describe the city, a source of data, we must look at how they construct the city in different ways' (Crang, 1998: 55). In this sense, the use of literary sources in historical geography goes beyond a search for 'evidence' to confirm or deny existing knowledges and moves towards considering how the production of the texts themselves may tell us something significant about the experience of places in the past.

CONCLUSION

This chapter has sketched out some broad parameters for the conduct of historical analysis in geography. It emphasizes the importance of defining research questions and their relationship to sources well in advance of actual engagement in the archive, as well as making a close assessment of the nature, potential and problems of any particular source material consulted. That said, archives are full of surprises and it would be a poor historical geographer who sought to apply a rigid theoretical framework or set of research questions to a particular set of sources. The relationship among theories, concepts and evidence in historical research should be open and recursive, and the researcher should remain flexible in his or her pursuit of answers to questions about the past. Following trails in archives can and often does lead to the discovery of unexpected evidence which questions or contradicts what might have been expected. The range of potential sources for the study of past geographies is undoubtedly vast and it has only been possible here to indicate three key groups of evidence for more detailed scrutiny. However, the separation of these sources into 'types' – *visual*, *literary* and *documentary* – is also potentially misleading. A hallmark of good historical scholarship is the ability to identify and link a series of sources, not only to deepen the analysis of interesting research questions but also to cross-check and validate different forms of historical knowledge. In short, it is in the *combination* of sources and methods that much of the most exciting historical geography is done. Finally, a word about presenting the results of an analysis of historical and archive sources. As Baker (1997: 238) has noted, many of the problems faced here are generic to all geographers, where writing involves constructing 'a series of balances between fact and interpretation, between the particular and the general, between the empirical and the theoretical, and between the objective and the subjective'. An additional problem for the historical geographer is, of course, that the past has already happened. What remains of it for interpretation is necessarily partial. Therefore, a key quality of successful writing in historical geography must be its ability to persuade others not only that it is soundly based upon reliable sources and methods but also that it accords with accepted facts and, above all, deals convincingly with the gaps in both evidence and interpretation that inevitably remain.

Summary

- Students using historical sources need to identify whether their research is *problem-orientated* or *source-orientated*.
- It is important to understand how and why particular sources were created and to establish their authenticity, reliability and partiality.
- Historical researchers make a distinction between primary and secondary data.
- Documentary sources include official/private archives, biographical and population data and financial records.
- As well as documentary evidence, visual evidence and literary sources can be used by historical geographers to reconstruct the past.

Further reading

- Baker (1997) provides a clear and concise introduction to the process of researching historical data in geography.

- Harley (1982) explains the need critically to evaluate sources and discusses questions to ask on the reliability of evidence in historical geography.

- Thompson (1978) is still one of the best general accounts of the need to think critically about the relationships between data and theory in historical work.

- Butlin (1993) discusses the range of documentary sources that have been used in historical geography.

- On visual evidence as a geographical source, see Burke (2001); on the built environment, see Domosh (1989) and Black (2003); and on literature, see Sharp (2000).

- Gagen *et al.* (2007) and the collection of essays on 'Practicing Historical Geography' edited by Schein (2001) in Historical Geography 29, both contain many excellent discussions of themes raised in this chapter.

Note: Full details of the above can be found in the references list below.

References

Baker, A.R.H. (1997) 'The dead don't answer questionnaires: researching and writing historical geography', *Journal of Geography in Higher Education*, 21: 231–243.

Barnes, T. and Duncan, J. (eds) (1992) *Writing Worlds: Discourse, Text and Metaphor in the Representation of Landscape*. London: Routledge.

Billinge, M.D. (1982) 'Reconstructing societies in the past: the collective biography of local communities', in A.R.H. Baker and M. Billinge (eds) *Period and Place: Research Methods in Historical Geography*. Cambridge: Cambridge University Press, pp. 19–32.

Black, I.S. (1995) 'Money, information and space: banking in early-nineteenth-century England and Wales', *Journal of Historical Geography*, 21: 398–412.

Black, I.S. (1996) 'The London agency system in English banking, 1780–1825', *London Journal*, 21: 112–130.

Black, I.S. (2000) 'Spaces of capital: bank office building in the City of London, 1830–1870', *Journal of Historical Geography*, 26: 351–375.

Black, I.S. (2003) '(Re)reading architectural landscapes', in P. Richards and I. Robertson (eds) *Studying Cultural Landscapes*. London: Arnold, pp.19–46.

Black, I.S. (2004) 'Onesiphorus Tyndall of Bristol, banker and merchant', *Oxford Dictionary of National Biography*. Oxford: Oxford University Press.

Burke, P. (2001) *Eyewitnessing: The Uses of Images as Historical Evidence*. London: Reaktion Books.

Burton, A. (2003) *Dwelling in the Archive: Women Writing House, Home, and History in Late Colonial India*. Oxford: Oxford University Press.

Butlin, R.A. (1993) *Historical Geography: Through the Gates of Space and Time*. London: Arnold.

Coppock, J.T. (1984) 'Mapping the agricultural returns: a neglected tool of historical geography', in M. Reed (ed.) *Discovering Past Landscapes*. London: Croom Helm, pp. 8–55.

Cosgrove, D.E. (1993) *The Palladian Landscape: Geographical Change and its Cultural Representations in Sixteenth-Century Italy*. Leicester: Leicester University Press.

Cosgrove, D.E. and Daniels, S. (eds) (1988) *The Iconography of Landscape: Essays on the Symbolic Representation, Design and Use of Past Environments*. Cambridge: Cambridge University Press.

Crang, M. (1998) *Cultural Geography*. London: Routledge.

Daniels, S. (1993) *Fields of Vision: Landscape Imagery and National Identity in England and the United States*. Cambridge: Polity Press.

Daniels, S. and Rycroft, S. (1993) 'Mapping the modern city: Alan Sillitoe's Nottingham novels', *Transactions, Institute of British Geographers*, 18: 460–480.

Darby, H.C. (1948) 'The regional geography of Thomas Hardy's Wessex', *Geographical Review*, 38: 426–443.

Dennis, R. (1987) 'Faith in the city', *Journal of Historical Geography*, 13: 210–216.

Dennis, R. (1988) 'Comment in reply: by the waters of Babylon', *Journal of Historical Geography*, 14: 307–308.

Dennis, R. (2001) 'Reconciling geographies, representing modernities', in I.S. Black and R.A. Butlin (eds) *Place, Culture and Identity: Essays in Historical Geography in Honour of Alan R.H. Baker*. Quebec: Les Presses de l'Université Laval, pp. 17–43.

Dirks, N.B. (2002) 'Annals of the archive: ethnographic notes on the sources of history', in Axel, B.K. (ed.) *From the Margins: Historical Anthropology and its Futures*. Durham & London: Duke University Press, pp. 47–65.

Domosh, M. (1989) 'A method for interpreting landscape: a case study of the New York World Building', *Area*, 21: 347–355.

Domosh, M. (1992) 'Corporate cultures and the modern landscape of New York City', in K. Anderson and F. Gale (eds) *Inventing Places: Studies in Cultural Geography*. Melbourne: Longman Cheshire, pp. 72–86.

Domosh, M. (1996) *Invented Cities: The Creation of Landscape in Nineteenth-Century New York and Boston*. New Haven, CT, and London: Yale University Press.

Domosh, M. (2001) 'Visual texts in historical geography', *Historical Geography*, 29: 68.

Duncan, J. (2001) 'Notes on emancipatory collaborative historical research', *Historical Geography*, 29: 65–67.

Duncan, J. and Duncan, N. (1988) '(Re)reading the landscape', *Environment and Planning D: Society and Space*, 6: 117–126.

Gagen, E., Lorimer, H. and Vasudevan, A. (eds) (2007) *Practising the Archive: Reflections on Method and Practice in Historical Geography*. London: Historical Geography Research Group.

Gregory, D.J. (1982) *Regional Transformation and Industrial Revolution: A Geography of the West Yorkshire Woollen Industry*. London: Macmillan.

Hall, C. (1982) 'Private archives as sources for historical geography', in A.R.H. Baker and M. Billinge (eds) *Period and Place: Research Methods in Historical Geography*. Cambridge: Cambridge University Press, pp. 274–280.

Harley, J.B. (1982) 'Historical geography and its evidence: reflections on modelling sources', in A.R.H. Baker and M. Billinge (eds) *Period and Place: Research Methods in Historical Geography*. Cambridge: Cambridge University Press, pp. 261–273.

Harvey, D.W. (1988) 'The production of value in historical geography', *Journal of Historical Geography*, 14: 305–308.

Holdsworth, D.W. (1997) 'Landscape and archives as texts', in P. Groth and T. Bressi (eds) *Understanding Ordinary Landscapes*. New Haven, CT, and London: Yale University Press, pp. 44–55.

Kearns, G. (2007) 'Taking theory for a walk in Ireland', in Gagen, E., Lorimer, H. and Vasudevan, A. (eds) *Practising the Archive: Reflections on Method and Practice in Historical Geography*. London: Historical Geography Research Group, pp. 9–22.

Kurtz, M. (2001) 'Situating practices: the archive and the file cabinet', *Historical Geography*, 29: 26–37.

Lawton, R. (1982) 'Questions of scale in the study of population in nineteenth-century Britain', in A.R.H. Baker and M. Billinge (eds) *Period and Place: Research Methods in Historical Geography*. Cambridge: Cambridge University Press, pp. 99–113.

Le Goff, J. (1992) *History and Memory*. New York: Columbia University Press.

Lucas, J. (1988) 'Places and dwellings: Wordsworth, Clare and the anti-picturesque', in D.E. Cosgrove and S. Daniels (eds) *The Iconography of Landscape: Essays on the Symbolic Representation, Design and Use of Past Environments*. Cambridge: Cambridge University Press, pp. 83–97.

Mills, D. and Pearce, C. (1989) *People and Places in the Victorian Census. A Review and Bibliography of Publications Based Substantially on the Manuscript Census Enumerators' Books*. Cheltenham: Historical Geography Research Group.

Moretti, F. (1998) *Atlas of the European Novel 1800–1900*. London: Verso.

Morgan, M. (1979) *Historical Sources in Geography*. London: Butterworth.

Pocock, D. (ed.) (1981) *Humanistic Geography and Literature*. London: Croom Helm.

Schein, R. (ed.) (2001) 'Special issue: practicing historical geography', *Historical Geography*, 29: 7–78.

Sharp, J.P. (2000) 'Towards a critical analysis of fictive geographies', *Area*, 32: 327–334.

Thompson, E.P. (1978) *The Poverty of Theory and Other Essays*. London: Merlin Press.

Tuan, Y.-F. (1978) 'Literature and geography: implications for geographical research', in D. Ley and M. Samuels (eds) *Humanistic Geography: Prospects and Problems*. London: Croom Helm, pp. 194–206.

Turner, M. (1984) *Enclosures in Britain 1750–1830*. London: Macmillan.

White, L. (1968) *Machina Ex Deo: Essays in the Dynamism of Western Culture*. Cambridge, MA: MIT Press.

Wiley, M. (1998) *Romantic Geography: Wordsworth and Anglo-European Spaces*. Basingstoke: Macmillan.

Wrigley, E.A. and Schofield, R.S. (1981) *The Population History of England, 1541–1871: A Reconstruction*. London: Edward Arnold.

ACKNOWLEDGEMENT

I would like to thank David Beckingham for his very useful comments and suggestions in revising the first edition of this essay.

30 Analysing Cultural Texts

Marcus A. Doel

Synopsis

Culture may be broadly defined as any way of life that can be differentiated from other ways of life, and texts as anything that signifies something for someone or other. So, far from being highly prized artistic and literary artifacts, most cultural texts appear mundane and are scattered among the debris of everyday life (e.g. graffiti, statutes, advertisements, photographs). As such, this chapter sets out to challenge the widely held view that 'culture' is a narrow concern of literary and artistic elites, and to show that 'culture' belongs as much to the 'low life' as it does to the 'high life', and to 'us' as much as to 'them'. At the same time it also challenges the bookish conception of 'texts' by showing not only how everything can come to signify something for someone or other, but also how texts give expression to social antagonisms.

The chapter is organized into the following sections:

- Hello culture!
- Culture is not what you think!
- Cultural texts are not what you think!
- Forget reading!
- Columbus's egg: 'mind that child!'

HELLO CULTURE!

Over the past couple of decades, more and more of human geography has taken a cultural turn. Or so the cliché goes (see, for example, Cook *et al.*, 2000; Anderson *et al.*, 2002). Virtually every part of the discipline – from economic geography to medical geography, and from political geography to rural geography – has become increasingly enculturated so that it is now not uncommon for geographers to study things like subcultures, cultures of nature and cultural capital (those cultural traits within any particular social setting that confer prestige on those individuals, groups and classes who possess them and shame on those who do not – Bourdieu, 1984). Meanwhile, cultural geography itself has changed from a relatively self-contained subdiscipline into one that embraces concerns that span the sciences, social sciences and the arts and humanities. As well as cultural geographies of nations, landscapes and built environments, there are also cultural geographies of money, domestic appliances and exotic fruit (Cloke *et al.*, 2005). Basically, there are cultures of everything just as there are histories and geographies of everything. Consequently, learning how to analyse cultural texts is fast becoming a vital skill for every human geographer. So in this chapter I want to help you set out in the right direction. I will do this by focusing on four key issues. First, I want to alert you to the fact that 'culture' and 'cultural texts' are almost certainly not what you think. Second, I want to encourage you to read – slowly, very slowly. As Dr Seuss famously put it: 'Go slowly. This book is dangerous.' Third, I want to demonstrate that reading cultural texts is nevertheless an easy feat to pull off – once you know the trick. Finally, I want to offer you an initial checklist for the effective analysis of cultural texts.

Now, while it would be easy to make this introduction difficult – by filling it with technical jargon and edifying quotations – I want to keep things as simple as possible. I just want to provide you with the *flavour* of textual analysis and trust that as you progress through your studies you will acquire the taste for more. My message is simple. The worst that can happen is not that you read badly. The worst would be for you to prefer not to read at all.

CULTURE IS NOT WHAT YOU THINK!

I suspect that the phrase 'cultural texts' will leave you cold. The word 'culture' still tends to conjure up images of refinement, elitism and exoticism. Little wonder, then, that culture is often assumed to pale into insignificance when compared to more pressing concerns like economic crises, regional conflicts and global warming. For those who are lucky enough to escape the numerous horrors that haunt the modern world, culture can be a sign of distinction, a weapon of class war and a way of distancing oneself from the trials and tribulations of everyday life. Whatever form it takes, culture is usually associated with the learned, the leisured and the tourist. Culture is what you experience in galleries, museums, theatres and heritage sites, not what you find in alleyways, squats, prisons and shopping centres. So, when you are offered the prospect of 'analysing cultural texts', I would not be at all surprised if you understood this to be an invitation to immerse yourself in myths, legends, scripture, poetry and literature in order to distil 'the customs, civilization, and achievements of a particular time or people', as the *Oxford English Dictionary* puts it. If you are anything like me – as I sit here on a cheap sofa, watching an inane TV sitcom while guzzling supermarket wine and listening to Gatecrasher Red – culture is probably not something for the likes of you. At a stretch, you might feel vaguely at home in so-called 'popular' culture, but even this is usually experienced as something beyond the reach of everyday life. It is something for the weekend, when one is released from the banality of everyday life: working, studying, shopping, sleeping, etc. Whether it is 'high' or 'low,' 'popular' or 'unpopular,' culture seems to exist in a parallel universe. Needless to say, many applied geographers are dismayed by the apparent irrelevance of cultural geography, and especially by the 'navel-gazing' analysis of cultural texts.

Yet culture is indeed to be found in alleyways, squats, prisons and shopping centres. One may not find much 'high' culture among the dustbins, tatty furniture, slops and carrier bags, but one will certainly encounter evidence of the customs, civilization and achievements of a particular time or people (see, for example, Buckley,

1996; Bell and Valentine, 1997; Gregson and Crewe, 1997). Graffiti, for example, make for interesting reading (Cresswell, 1996; @149 St; Banksy, 2006). For some, graffiti are merely signs of mindless vandalism: annoying, disrespectful and per-haps even threatening. For others, graffiti demonstrate an attempt by marginalized and alienated people to reclaim the streets: kids, gangs and wanna-be artists. For still others, graffiti are anonymous messages hurled into the world: 'MAD 4 AEP', 'Eat yourself fitter', 'Do you want to play toilet tennis?'. So, when one is invited to analyse cultural texts, one should read graffiti, dustbins and carrier bags with the same rigour as one would read literature, legislation and gene sequences. Indeed, I suspect you are already well versed in reading the signs of everyday life: words, pictures, gestures, facial expressions, ambiences, street furniture, footprints, etc. You are already analysts of cultural texts.

As a point of departure, then, do not begin your analysis with a prejudice about what a geographer like you *should* and *should not* set out to read. The types of text you will need to take up will depend upon the particular culture you are inter-ested in: British culture, urban culture, political culture, drug culture, neoliberal culture, pop culture, youth culture, etc. I could go on, but I suspect the message is crystal clear. Be guided by your interests rather than by presuppositions about what counts – and does not count – as culture. I am interested in consumer cul-ture, and my favourite cultural text is a two-page colour chart produced by Dulux for one of its range of paints. On the left-hand page there are 42 colours – with names like 'sheer amethyst', 'blue topaz' and 'rose lacquer' – arranged into three seemingly arbitrary groups. For over an hour my students failed dismally to work out what – if anything – these colours meant. When we could no longer stand the silence, I revealed the six-word solution that appears on the right-hand page of the colour chart: the first group of colours are 'Urban Discovery,' the second 'African Discovery' and the third 'Oriental Discovery'. For the rest of the session they were able to draw out a seemingly endless series of interpretations about the geographical imagination of western consumers that ranged from ideologies of home to the colonial foundations of the discourse of 'discovery'. For example, the Urban colour scheme is predominantly made up of airy and watery blue-greys: cold, muted and vapid. It is the perfect colour scheme for a superficial, artificial and post-industrial world of anonymity, alienation and interchangeability. When one notes the names of these blue-greys – such as madison mauve™, city limit™, manhattan view™, plaza™, dot com, loft™, café latte™, platinum and brushed steel™ – one can discern a highly gendered and class-specific lifestyle: single professional men immersed in the fast and furious world of twenty-first-century business relish their cosmopolitan comforts (cf. Baudrillard, 1996, 1998).

Meanwhile, the African colour scheme is dominated by earthy and fiery colours: warm, grounded and vibrant. While the Urban is alien, Africa is other: exotic, untimely and tied to the earth. It is rendered through white heat™, turmeric™, fired ochre, bazaar™, ancient earth, raw umber and beaten bronze™. Finally, the pastels of the Oriental colour scheme are mainly creams and greens. Yet rather than the *coldness* of the Urban (modernity) or the *heat* of Africa (tradition), they evoke a *calming* influence (nature): silken trail™, sea grass™, palm wood™, bamboo screen™, chi™, lotus blossom™, gold veil™ and eastern gold™. So, while the urban north/west and the African south pit the elements of air and water against those of fire and earth, the calming spirit of nature wafts in from the Oriental east. Such is the ideologically loaded world according to Dulux (cf. Cook and Crang, 1996). I hope this brief analysis has given you a flavour of how a few *divisions* – such as Urban/African/Oriental, North/South/East/West, modernity/ tradition/nature, worldly/spiritual, cold/hot/calm – can be arranged to organize a whole range of disparate materials into a functioning whole. So, try not to get bogged down in the details (specific content). Instead, seek out the divisions that structure those details (general form). While these divisions are sometimes hard to pin down and extract, more often than not they are in plain view – and therefore all too easily overlooked.

CULTURAL TEXTS ARE NOT WHAT YOU THINK!

Three things should now be apparent. First, culture is everywhere. Second, we all live among innumerable cultures that may or may not have clear boundaries, stable identities and coherent practices. Third, texts are not simply collections of words fixed onto paper. From now on, think of a text in terms of its original Latin root: *texere* – weave. A text is a tissue of signs. It is anything with a signifying structure. It is anything that leads one into decoding, exegesis, interpretation and translation. Quite simply, a text is anything that *refers* meaning – and thereby the reader – elsewhere: to other texts, languages, codes, situations, contexts, expectations, habits, etc. Needless to say, this process of referral can be prolonged infinitely and extended over innumerable domains, although in practice we tend to arrest this structure of dissemination in order to get things done (see Exercise 30.1).

Exercise 30.1 Signifying nothing – a flavour of moral geography

Do you believe me when I claim: (1) that a text necessarily refers you elsewhere; (2) that this process of referral is interminable; (3) that it will lead you all over everywhere; and (4) that sense, meaning and action are an interruption rather than a culmination of reading a text? No? Then consider this piece of graffito: ⊗. Once you have read it, read it again: slowly; very slowly. Now, try to answer the following two questions. What does ⊗ refer to? What else does ⊗ refer to? If you need some help, here is my starter for ten. First and foremost, ⊗ refers to 'no smoking'. Second, it refers to a place and a time for 'no smoking' (although the extent and duration of which remain unspecified). Third, it refers itself to you and to me: as readers of signs, users of space, smokers, non-smokers, smoking non-smokers (i.e. passive smokers and free riders) and non-smoking smokers (i.e. 'good' smokers who keep their cigarettes unlit). Fourth, it refers above all to actions: to what people do. Specifically, they do not smoke. But ⊗ is much more than a simple statement of fact: there is no smoking here. It prescribes a required practice: one must not smoke. Fifth, it thereby refers to sets of expectations, rules and sanctions that are meant to govern the relationship between people and place; to those people who devise, impose and enforce them; and to the sources of trust, expertise and authority that legitimate them – such as law, politics, commerce, culture, the media and health. Sixth, if the sign ⊗ can manage to elicit practices of 'no smoking', it actually helps to create a non-smoking space. Seventh, it is not entirely clear what 'no smoking' actually entails. Does ⊗ object to smoke, smoking or both? One can hold a lit cigarette without smoking it – as I often do when walking through 'no smoking' spaces en route from one smoking place to another – and one can smoke things other than cigarettes. Eighth, does the prohibition against smoke and smoking occur at the sign, around the sign or from where one reads the sign? Ninth, does ⊗ still function as a sign of prohibition when there is no one there to refrain from smoking, when it appears upside down, when it is obscured by foliage or when it is used as an example of dissemination in a text-book? Finally, although ⊗ seems to refer to 'no smoking' in general, it actually refers to particular cases. In short, we will never be finished with the interpretation of enigmatic signs. Nevertheless, we can gain a great insight into how they are socially and spatially structured along the way. So let me restate my original question. Can you think of anything that does not require interpretation and that does not lead into a spatial analysis?

Food, gadgets and clothes signify no less than written words, spoken language and The Human Genome (Barthes, 1993). Consequently, geographers can take virtually any artifact and attempt to draw out the meanings, values, dispositions, desires, knowledge, power relations and practices that are encoded into it. Consequently, what matters is not whether the artifact has a prominent place and obvious significance for the culture in question – such as Acts of Parliament, monumental architecture and famous works of art – but whether the artifact will enable you to access how that culture exists, experiences and acts in the world. Think of *the* international language of commerce: is it English, the US dollar or the computer spreadsheet? Or think of *the* technology that holds our world together: is it the transportation system, synchronized clocks or the humble screw? To put it another way, what would have the greatest impact on our world: the disappearance of Acts of Parliament, monumental architecture and famous works of art or the absence of money, spreadsheets, clocks and screws?

Since everything can be decoded, everything is a text. Everything opens out on to a social world. But a text is not simply something that purposefully carries a message or a meaning. Rather, a text is anything that *responds* to the call of an interpretative gesture. It is anything that leads someone to suppose – either implicitly or explicitly – that it might mean something or other. For example, take a landscape. On the one hand, this landscape can be read for the material traces of the various ways of life that have been encoded into it. So a landscape can be treated as the accumulated expression of innumerable modifications over time by a host of human and non-human agents: florae, faunae, technologies, political regimes, earth-surface processes, social formations, etc. Many people call this the 'Real'. No doubt you have gazed out from hilltops, studied maps and processed data on computers in order to make sense of a landscape and extract the truth about its history and geography. On the other hand, a landscape may also function as an enigmatic blank on to which meaning and significance can be projected by different groups of people with very different kinds of interest. As with astrology, numerology and conspiracy theories, values can be read *into* things. In this way, a landscape can come to signify all kinds of things for all manner of reasons: nature, perfection, order, beauty, desolation, belonging, the sacred, wealth, alienation, eternity, woman, the future, etc. Many people call this the 'Imaginary' or the 'Symbolic'. Needless to say, while these kinds of representation usually achieve a certain consistency through habitual associations – often to the point of seeming natural, commonsensical and self-evident to the groups whose interests they serve – they are nevertheless socially constructed and inevitably contested. For what appears to be natural and self-evident invariably turns out to be the product of a long and drawn-out social struggle (Schivelbusch, 1993; Perec, 2008). Like 'heterosexuality', 'lifelong learning' and 'working for a living', things are invariably naturalized rather than natural. In other words, what counts as natural self-evidence is not so much a quality inherent in the world as an outcome of specific social settings. It is a social construct. The city, for example, can be represented as anything whatsoever: monumental, alien, desolate, erotic, chaotic, unruly, sacred, fun, a desert, a jungle, sociable, fleeting, eternal, fearful, artificial, second nature, a wilderness, an ocean ... But these are not just representations since each elicits a certain set of *practices* that impact upon the city, transforming both its built environment and its social life (Clarke, 1997; Pile and Thrift, 2000). By systematically studying these material traces and representations – by reading the real and imagined landscape – you will be able to gain an insight into both the practices and the values that have shaped the relationship between people and place (de Certeau, 1984; Lefebvre, 1991). That will provide you with a basis for thinking about the social and spatial struggles that are articulated through a clash of representational practices: through a clash of different ways of enabling the world to make sense (e.g. painting versus photography, Creationism as opposed to evolutionism or humanism contra structuralism) (see Exercise 30.2).

Exercise 30.2   Signifying self-evidence in social space

Look around the room or space you are in. Note the people, the creatures, the plants, the objects, the ambience, the textures, the sensations and the dimensions. What do they tell you about who you are, your relationship to others and your place in society? What do they say about your social, cultural, political, economic, racial and sexual status? Take any object whatsoever. Why is it there? Where did it come from? What connections does it establish between you and the outside world? Now think of the room you are in as an ensemble. What kinds of activities does it enable? And what kinds of activities does it restrict or even preclude? What is the focal point of the room – and what does that tell you about the way of life to which you have become subject? Finally, think of three activities and three objects that would be completely out of place in this room. Now imagine the way of life – the 'spatial practices' – that would make them as self-evidently in place as that focal point around which your entire existence is presently arrayed.

Most people use the word 'discourse' when they want to consider both the representations *and* practices of a particular social group. A discourse is a specific constellation of knowledge and practice through which a way of life is given material expression. It engenders a discourse-specific (i.e. partial and relative) incarnation of the world that tends to become both naturalized and taken for granted. When writers draw attention to these material and immaterial constellations of knowledge and practice, they usually do so in terms of the social and spatial power struggle between 'dominant discourses' on the one hand and 'discourses of resistance' on the other. Discourse analysis discloses how this constellation of knowledge and power is structured, and situates it within its appropriate social, cultural and geo-historical context. For instance, think of the discursive conflict between adults and children, the rich and the poor, the colonizers and the colonized, and environmentalists and capitalists. Each draws upon a very different geographical imagination, and each deploys a very different repertoire of spatial practices that jostle for supremacy. A wonderful example of such a conflict is provided by Allen and Pryke's (1994) consideration of how foreign-exchange dealers, security guards, caterers and cleaners inhabit 'the floors of finance' in the City of London. Although they all more or less occupy the same physical *place*, they each live and work in very different social *spaces*. For example, while the foreign-exchange dealers are hooked into the hyperactive 'global space of flows' through their computer screens, telephones and social networks, the cleaners are expected to disappear without trace within the 'local space of place' after attending to the mundane materials that make up the floors of finance: glass, plastic, marble, paper, etc.

In summary, a geographical analysis of cultural texts and competing discourses needs to follow as rigorously as possible the spatial, temporal and social traces of both real and imagined signifying structures: representations and practices. The expertise you will need to draw upon will inevitably spin out, not only across the whole of human geography but also across many other disciplines. While this profusion is undoubtedly daunting, you should take comfort from the fact that the ability to make sense of the world is more likely to come from lifelong learning than divine inspiration. Perhaps the best advice is to move slowly, to remain alert for possible connections and to resist the trap of uncritically accepting common sense. So, you should try to enter into a photograph, artifact or document in the same way that you would enter into a film or a novel: they lead to worlds within worlds; social spaces within social spaces. But rather than passively submit to the line that has been laid out for you to follow, you should attempt to explore and survey systematically this world within a world for yourself. Take a landscape painting, an Act of Parliament or an advertisement for new housing. Or else a bridge, a nature reserve or a display of lucky thimbles:

- How, why and for whom has it been constructed?
- What are the materials, practices and power relationships that are assumed by it and sustained through it?
- What codes, values, dispositions, habits, stereotypes and associations does it draw upon?
- What kind of personal and group identities does it promote? And how do they relate to other identities?
- What does it mean? What are its main structuring devices: oppositions, divisions, metaphors, illustrations, exemplars, etc.? And how do they over-determine and constrain the choice and arrangement of content?
- More importantly, what kind of work does it do? And who benefits?
- What has been included, excluded, empowered and repressed?
- How might it be modified, transformed or deconstructed? How could this social space be inhabited differently?

- Finally, since nothing *ever* comes alone, what wider assemblages does it fit into and resonate with? Are these assemblages synergetic or contradictory? How does it relate to the assemblages that interest urban geographers, historical geographers, medical geographers, etc.?
- And since the work of contextualization and re-contextualization can be carried out indefinitely, there is never anything 'final' about following the manifold traces of textual analysis.

A good place to start is with a map – not least because, like a photograph or a measurement, most people tend to assume that a well-prepared map tells the truth. However, as with everything else, maps are made to serve particular purposes in specific social, cultural, economic and political settings. They edit, transform and remake the world in a way that suits the interests to hand (Wood, 1993; Pinder, 1996; Olsson, 2007; Wood and Fels, 2008).

FORGET READING!

So far I have tried to clarify how you should *approach* cultural texts: slowly, attentively, broadly, openly and with an eye towards the struggles between competing discourses for representational and practical supremacy over social space. However, I have been asked by the editors of this book to do something strange – perhaps even pointless. Since this is a 'how to' book I have been asked to *instruct* you in – of all things – *reading*. But how could you have got to Chapter 30 without already knowing how to read? What concerns readers like you is not learning how to read but *whether* to read, how *much* to read and *what* to read. And you don't need me to tell you the answers to these dilemmas because you have heard them repeated ad nauseam by teachers and tutors: 'You must read ... as much as possible ... of what is on your reading lists'. And the difficulties you will face will probably have less to do with reading per se and more to do with the nature of the writing. Academic texts are notoriously dry, long-winded, boring, jargon-ridden, humourless, irrelevant, pompous, obtuse and turgid.

Now, while I get endless queries from students about how things should be *written* and *presented* – structure, flow, balance, objectivity, referencing, quoting, exemplification, contextualization and especially length – I cannot recall anyone *ever* asking me how to read. This is odd: not only because many students experience an entirely understandable difficulty in reading 'academic' texts but also because reading in general is actually a demanding and skilled activity that requires training. Indeed, you should study techniques of reading in the same way as you would study any other qualitative or quantitative research technique. And believe it or not, there are countless ways in which to read – few of which boil down to reading word after word for page after page! I will *mention* just a few: hermeneutics, semiotics, psychoanalysis, structuralism, deconstruction, discourse analysis, postcolonialism, Marxism, feminism and reader response. To see them in action, see journals such as *Environment and Planning D* (a.k.a. *Society and Space*), *Gender, Place and Culture* and *Space and Culture*. For a more formal encounter, select a few introductory texts from your library's 'cultural studies' and 'literary theory' sections or see, for example, Rose (2007). For the flavour of things, try some of Icon Books' comic-book series designed 'for beginners'. I especially enjoyed *Introducing Baudrillard, Introducing Cultural Studies, Introducing Derrida, Introducing Postmodernism* and *Introducing Semiotics*. For now, however, I just want to give you a feel for the power of innovative reading strategies (see Exercise 30.3).

Exercise 30.3   A speedy literature review

My second-year tutees have been struggling to perform an onerous but nevertheless important task: a so-called 'literature review'. Basically, rather than write an essay on something or other out there in the world, they need to write about how geographers have written about something or other. Fifteen journal articles, several books and four weeks later the reviews are in. Although they have read a vast amount they still find it hard to get a grip on what academics think about things. Imagine their dismay when I suggested this short-cut after the event. Take any textbook – perhaps this one – and go and get a very old textbook on the same subject from the library (say something from the 1980s). Open their indexes and make three lists of words and phrases: (1) those that appear in both; (2) those that only appear in the old textbook; and (3) those that only appear in the new textbook. What are the dominant themes within *each* of the lists? Now, compare what has been *preserved* within the subject (list 1), what has been *purged* from the subject (list 2) and what has been *added* to the subject (list 3). What does this tell you about the changing interests of geography and geographers? Do these changes amount to *progress*? If you are really pressed for time, try the same procedure with the contents pages instead. This will give you a sense of how the subject is structured and of the relative importance of various issues and debates.

Note: When faced with the inconvenience of long indexes, feel free to restrict your textual analysis to a workable range of letters: say M–R. For those with a *quantitative* bent, try drawing *inferences* about the *population* of the entire index from this sample, write down how confident you are about your inferences and then compare your expected results with what is actually the case. This should alert you to the fact that quantitative analysis is a particular way of *thinking* rather than a fixation on numbers per se.

COLUMBUS'S EGG: 'MIND THAT CHILD!'

Let me bring this chapter to a close with an anecdote and a checklist. Apart from a few advertising slogans, the only phrase that I can remember from my childhood is: 'Mind that child!'. It was written on the back of the ice-cream vans that toured the streets where I lived. Although there was nothing difficult about the words, I always felt unnerved by them. On every occasion I encountered the phrase, the anticipated pleasure of eating ice cream was tainted by anxiety over these enigmatic and sinister words. Now, believe it or not, it was only a couple of months ago that this childhood association between ice-cream vans and foreboding dissipated. I was about to overtake an ice-cream van when I noticed those words: 'Mind that child!'. Suddenly, I realized that the words were addressed to *adults* rather than children. They simply warned drivers not to run over children distracted by the pleasures of ice cream rather than to warn *children* about the dangers of 'that!' – something so terrible that adults dare not even write its name! I had assumed that the statement 'Child: mind *that*!' was meant to warn me of evil or distracted car drivers who routinely ran people over (this was the moral panic par excellence of my infant years: soon to be followed by the horrors of sexually transmitted diseases, heroin abuse and glue-sniffing), but it always struck me as odd that one actually needed to be *in* the road to read the warning – as if the whole thing were a terrible ruse to expedite the killing of children. Years later I would be reminded of these deadly ice-cream vans when I read about the Nazis' highly effective discourse of dissimulation: the slogan 'Arbeit macht frei' (Work makes free) appeared above the gates to the Auschwitz-Birkenau death camp; its gas chambers were presented as if they were showers; and the conversion of vans into mobile gas chambers used at Chelmno and elsewhere were simply referred to as 'special vehicles' and often made to resemble furniture-removal vans (Lanzmann, 1985). The comeuppance is clear. When you start to analyse a cultural text, do not take its message at face value or assume that it was meant for people like you. It almost certainly wasn't. Once you have worked out for whom the text was produced you will have gone a long way to reading it effectively. So here is a final checklist to get you going. Investigate *who* produced the text, *why* they produced it, *how* they produced it and for *whom* they produced it (cf. du Gay *et al.*, 1997). Sometimes all of this will have been consciously intended. On other occasions, however, it will have been unconscious and unintended. Investigate the *form, content* and *assumptions* of the texts in question, paying just as much attention to what is *absent* from the text as to what is present. Set the text in appropriate *contexts* and compare and contrast it with other relevant material: texts, lifestyles, belief systems, practices, artifacts, etc. Investigate how they have been used and abused as mechanisms for articulating *power* and *resistance* by a range of people and groups. But above all else, investigate what *work* these texts do: how do they impact upon and affect society and space? Once you have done that you can begin to approach the analysis of cultural texts according to a wide variety of criteria that may interest you: power, knowledge, desire, truth, fidelity, meaning, motivation, intention, exclusion, class, race, gender, sexuality, etc. And if anyone mocks your scholarly and geographical interest in *reading* cultural texts, remember the story of Columbus's egg. In reply to a suggestion that *anyone* could have discovered America, Columbus challenged the guests at a banquet held in his honour to make an egg stand on end. When all had failed, Columbus did so by flattening one end with a sharp tap on the table, thus demonstrating that while others might follow, he had discovered the way. It is true that reading, like egg-standing, is an easy task to pull off – but only when you have learnt the trick.

Summary

- Cultural texts are not what you think: everything has a signifying structure, everything can be decoded, everything is a text.
- 'Culture' and 'textuality' have become key terms in geography.
- There are many different kinds of textual analysis (e.g. discourse analysis, semiotics and deconstruction).
- It is important to distinguish between the producers and consumers of cultural texts and the form and content of cultural texts.
- Power relations are encrypted into cultural texts.

- Cultural texts matter because they shape and inform social practices and social spaces.
- Effective reading is invariably more difficult than one would think.
- Take time to read – and to learn how to read.

Further reading

This guide includes an eclectic selection of examples of different forms of cultural analysis:

- @149 St. New York City Cyber Bench (http://www.at149st.com) (accessed January 2010). This website archives the history of the graffiti art form developed on New York City's subways. It includes a wide range of artists, crews, images, texts and links.

- Clarke (1997) is a collection of essays that explores how cities have been depicted in film and how cities themselves have been affected by the cinematic form.

- Cook *et al.* (2000) addresses the nature, significance and impact of the so-called 'cultural turn' in contemporary human geography which leads into a full-scale exploration of the relationship between culture and space.

- Cresswell (1996). From graffito to peace protests, this is a wonderful analysis of how power, transgression and resistance get played out in the geography of everyday life.

- Du Gay *et al.* (1997). Everything you ever wanted to know about the way in which the 'turn to culture' impinges on our lives using the Sony Walkman as an example.

- Rybczynski (2000), in attempting to identify the most significant invention of the past millennium, uncovers the mind-boggling secret history of the screwdriver and screw. This is a wonderful example of why one should never take anything for granted.

- Schivelbusch (1993) presents an astonishing world of social conflict and cultural struggle that surrounded the introduction of pepper, coffee, chocolate, tobacco and opiates into Europe. It is a remarkable study of the modernization and industrialization of culture.

Note: Full details of the above can be found in the references listed below.

References

@149 St. New York City Cyber Bench (www.at149st.com) (accessed January 2010).
Allen, J. and Pryke, M. (1994) 'The production of service space', *Environment and Planning D: Society and Space*, 12: 453–476.
Anderson, K., Domosh, M., Pile, S. and Thrift, N. (eds) (2002) *Handbook of Cultural Geography*. London: Sage.
Banksy (2006) *Banksy: Wall and Piece*. London: Century.
Barthes, R. (1993) *Mythologies*. London: Vintage.
Baudrillard, J. (1996) *The System of Objects*. London: Verso.
Baudrillard, J. (1998) *The Consumer Society: Myths and Structures*. London: Sage.
Bell, D. and Valentine, G. (1997) *Consuming Geographies: We Are Where We Eat*. London: Routledge.
Bourdieu, P. (1984) *Distinction: A Social Critique of the Judgement of Taste*. London: Routledge.
Buckley, S. (1996) 'A guided tour of the kitchen: seven Japanese domestic tales', *Environment and Planning D: Society and Space*, 14: 441–461.
Clarke, D. (ed.) (1997) *The Cinematic City*. London: Routledge.
Cloke, P., Crang, P. and Goodwin, M. (eds) (2005) *Introducing Human Geographies* (2nd edn). London: Arnold.

Cook, I. and Crang, P. (1996) 'The world on a plate: culinary culture, displacement and geographical knowledges', *Journal of Material Culture*, 1: 131–153.

Cook, I., Crouch, D., Naylor, S. and Ryan, J. (eds) (2000) Cultural *Turns/Geographical Turns: Perspectives on Cultural Geography*. Harlow: Prentice Hall.

Cresswell, T. (1996) *In Place/Out of Place*. Minneapolis, MN: University of Minnesota Press.

de Certeau, M. (1984) *The Practice of Everyday Life*. Berkeley, CA: University of California Press.

du Gay, P., Hall, S., Jones, L., Mackay, H. and Negus, H. (1997) *Doing Cultural Studies: The Story of the Sony Walkman*. Milton Keynes: Open University Press.

Gregson, N. and Crewe, L. (1997) 'The bargain, the knowledge and the spectacle: making sense of consumption in the space of the car-boot sale', *Environment and Planning D: Society and Space*, 15: 87–112.

Lanzmann, C. (1985) *Shoah: An Oral History of the Holocaust*. New York: Pantheon.

Lefebvre, H. (1991) *The Production of Space*. Oxford: Blackwell.

Olsson, G. (2007) *Abysmal: A Critique of Cartographic Reason*. Chicago: Chicago University Press.

Perec, G. (2008) *Species of Spaces and Other Pieces*. London: Penguin.

Pile, S. and Thrift, N. (2000) *City A–Z*. London: Routledge.

Pinder, D. (1996) 'Subverting cartography: the situationists and maps of the city', *Environment and Planning A*, 28: 405–427.

Rose, G. (2007) *Visual Methodologies: An Introduction to the Interpretation of Visual Materials* (2nd edn), London, Sage.

Rybczynski, W. (2000) *One Good Turn: A Natural History of the Screwdriver and the Screw*. New York: Simon & Schuster.

Schivelbusch, W. (1993) *Tastes of Paradise: A Social History of Spices, Stimulants, and Intoxicants*. New York: Vintage.

Wood, D. (1993) *The Power of Maps*. London: Routledge.

Wood, D. and Fels, J. (2008) *The Nature of Maps: Cartographic Constructions of the Natural World*. Chicago: Chicago University Press.

31 | Writing Essays, Reports and Dissertations

Michael Bradford

Synopsis

The essay, report and dissertation are three different forms of presenting (and assessing) an argument with evidence. The essay is discursive but short compared to the dissertation, which is a major piece of work usually based on original empirical research. The report is concisely and precisely written. Unlike the other ways of writing, it is often used outside academia.

The chapter is organized into the sections listed below. The sections are each written in the style of the form of writing they describe:

- Introducing the three forms of writing and the common skills required
- Guidelines on the essay, written as an essay
- Guidelines on the report carried out by a team, written in report style
- Guidelines on a dissertation: a model with various deviations

INTRODUCTION

This chapter discusses three major forms of presenting written work when studying geography at university: the essay, report and dissertation. Students are expected to learn some common and some different things from completing these three types of assessment or, in the current educational jargon, particular 'learning outcomes' are associated with them. The learning outcomes may also vary for each of the three, depending on the exact requirements. Some of these learning outcomes are discussed in this introduction together with the variations in the forms before the three are separately explored. The discussions on the essay and the report are presented as an essay and a report, respectively.

Essays vary in length. Some are completed to time within an unseen exam and fast thinking is necessary. Some comprise about 1,500 words and are submitted as part of tutorial work, which may or may not count towards the overall assessment. Others are extended essays of about 3,000 words, which are part of the coursework requirement of the degree programme. These last two types allow more considered thought. Essays are discursive while reports are written in a precise and concise style, with paragraphs numbered for easy reference. They have an executive summary that comprises the main findings of the report, which can vary in length. The dissertation or project is the longest of these forms varying from perhaps 6,000 to 15,000 words at undergraduate level. It is divided into chapters. It is usually research-orientated, based either on both primary and secondary evidence or on secondary evidence alone. In some universities, it can be entirely based on a review of literature and not involve the collection of primary data or the analysis of primary and secondary evidence.

Essays and dissertations are very much academic modes of writing, while reports are used widely outside academia – for example, in government and business. Gaining experience in writing reports is therefore very valuable for later life. The skills associated with all three, however, are very valuable both during and after university. All involve skills of organizing material and presenting it in a structured way. All involve developing and sustaining an argument and providing evidence to support that argument. All involve considerable preparation in terms of thinking and reading. The dissertation particularly requires pre-planning. It is a large piece of work and occupies a long period of time, so organizational and time management skills are very important. Networking skills can also be very useful when collecting evidence, whether through interviews, data gathering or searches of archives. Using these skills helps find the right people to answer questions or to give access to information. So, many types of skills are developed and assessed through these pieces of work.

All three will mostly be carried out as individual tasks but the extended essay, the report and the project can be completed in a team. Here, the guidance on writing a report is given for a team. The process of working as a team is very significant with organizational skills and time management becoming even more important to a successful outcome. In most situations, however, only the end-product will be assessed, and the process by which it was achieved will be assessed indirectly through the end-product. Sometimes the process is directly assessed – for example, the important stage of choosing a topic for a dissertation is often separately assessed by a written proposal. In this chapter, only the presentation of the end-products is considered.

Whichever of the three is being completed, the first thing to do is to study what is expected. There are separate guidelines available in most departments on dissertations and often, too, on essays. Good guidelines will include the marking schemes involved and the grade-related criteria that indicate what is expected to obtain certain levels of marks (Clark and Wareham, 2003). The structure of the argument used is likely to be one criterion on which all three forms are judged. It will be reflected in the grade-related criteria through a gradation of preceding adjectives – for instance, from basic to excellent. Certain aspects of the grade-related criteria may appear at only the top levels – for example, originality and evidence of synthesis. It is important to study any guidelines and to relate work to the relevant criteria. Within these guidelines will be tips on time management that are crucial to all three forms. They are particularly relevant to team reports and the dissertation. Many advise students to work backwards from deadline dates in establishing a timetable of work (Kneale, 2003). Having studied the guidelines and the intended learning outcomes, students are ready to approach the tasks.

THE ESSAY

Discuss the major challenges of writing an essay

University Challenge has been the British television audience's main view of universities for decades. The questions posed by Bamber Gascoigne and Jeremy Paxman may have changed, but the format remains the same. Over an even longer period of time, the essay has formed one of the challenges for students at university. Here, too, the questions may have changed, but the format of the essay has remained essentially the same. It is still the main mode of communication and assessment at university for geography students, even though its relative significance has declined. Since the importance of the essay has declined even more within schools, writing an essay, in itself, has become a major challenge to many geography students at university.

Going to university presents many challenges, from the more general ones of managing finance and time, to developing specific study skills such as writing essays. The extent of the challenge depends on students' previous experience and their ability to transfer it into a new context, and the standard to which they aspire. Experience varies in that, for example, some may have been trained at school or college to introduce and conclude an essay. Others may have been encouraged to be more instrumental, given the shortage of time within exams, and to score points without the overview provided by introductions and conclusions. Here they would be dispensing with the two ends of a traditional view of the structure of an essay:

- Say what you are going to say (introduction).
- Say it.
- Say what you have said (conclusion).

Many, however, would argue that an introduction catches the attention and sets the essay in perspective as well as indicating the structure of its argument, while a good conclusion is more than the sum of the preceding paragraphs, synthesizing points, developing implications and perhaps ending with related questions. At university, such introductions and conclusions are expected. Indeed, their quality may contribute very significantly to the difference between marks awarded. Understanding what is expected by the general and specific task is then fundamental to producing a high-quality essay. Preparing and gathering evidence for the argument to be used are crucial early stages that inform the writing and editing of the essay. The final stage is learning from the process and the assessment so that other essay-based tasks might be completed more effectively. These various stages will be discussed below in order to help students achieve the highest possible marks and, more generally, to help them learn from the overall process which underlies many assessment tasks.

The first challenge is understanding the task set, whether the essay is unseen within an exam, a tutorial essay or an extended essay submitted as a piece of coursework required for a course unit. Many essays will consist of one or more 'command' words such as 'discuss' or 'evaluate' or 'describe and explain', together with the subject such as 'the challenge of essay writing' or (evaluate) 'the success of regeneration schemes in British cities'. Understanding the 'command' word can be quite challenging, because academics who set the essays may have somewhat different ideas about what they expect by the word. 'Discuss', for example – a very common command word – means to some 'presenting a balanced argument', while to others, it requires 'presenting an argued case, preferring some viewpoints to others'. Expectations may also vary with the level of the task. 'Evaluate' is quite a high-level task because it requires deciding the criteria against which something (e.g. regeneration schemes) will be judged, as well as critically reviewing evidence that will permit judgements to be made. Matching students' expectations to those of staff is more generally one of the major challenges of university. Some degree programmes will articulate expectations very transparently and have guides to command words within handbooks.

What expectations are associated with the subject of the essay title? Discussing the title 'the success of regeneration schemes in British cities' will illustrate the required thought processes. 'British cities' sets the geographical context, but the timing of the schemes has been left open for the writer unless, of course, the course unit is about very recent change. If recent change is concerned, then the regeneration schemes might include urban development corporations, City Challenge, the Single Regeneration Budget and the New Deal for Community. If a wider time perspective is required, or could be argued, post-war comprehensive redevelopment might also be included. 'Success' is the key word because it begs the questions 'for whom?' and 'on what terms?' Success could be evaluated narrowly against the aims of the schemes – did they achieve their objectives? It could be viewed more widely against more general criteria, such as 'did they reduce social exclusion?' So, schemes might have been successful for the eventual residents but not for the displaced original residents. The challenge of the question lies not just in knowing the details of the schemes but also in understanding the more general economic, social, political and environmental contexts on which regeneration may be judged. A good answer might argue that some schemes have been successful in regenerating the built environment, but not in improving the economic opportunities and social inclusion of the original residents.

The task is not always fully set by staff. For an extended essay, students may be asked to decide the topic and/or the exact title. Even some unseen exam essays include 'write an essay on (global warming)'. To produce a good answer to this very open task the student has to decide and execute the task – e.g. 'evaluate the evidence for global warming' or 'discuss the degree to which global warming has been produced by recent human activities'. The setting of the task, as well as the degree to which it is achieved, becomes part of the assessment.

So, understanding and making the task explicit are the crucial first steps – defining the terms (e.g. regeneration schemes, global warming), the spatial and temporal contexts or scales that set the limits to the discussion and interpreting key words (e.g. the meaning of 'success').

Once the task has been understood, the next step is to construct a possible argument or approach so that students know which 'literature' to consult or upon which revised areas to draw. The process is made easier by having already clarified the task. 'Evaluating success' in the regeneration essay has suggested that there are many possible criteria to be considered, and these may be viewed from different perspectives – for example, have the schemes helped the city to position itself economically in a globally competitive way? Have they aided social inclusion? Have old buildings been refurbished for new functions in a way sensitive to the needs of conservation (Cochrane, 2007; Jones and Evans, 2008)? Just reading about or recalling the details of the regeneration schemes will not help to develop an argument for such an essay. Once the key areas of literature have been identified, there is often a further challenge when completing non-exam work: that of getting access to the appropriate journals and books and selecting appropriate web-based material. Another very significant, associated challenge is approaching this material in a critical way so that the argument developed is soundly based (see, for example, Cochrane, 2007).

The next challenge is central to the task: writing the essay. Writing an essay in an exam requires a linear development of the argument from introduction to analysis and conclusions, having established a structure through an initial short plan. Some regard the analytical part as the meat within the sandwich of the introduction and conclusion. When a word-processor may be used the development does not have to be linear. Some would write the meat of the analysis first and only then complete the introduction. The roles of the introduction are to catch the attention, set the topic in perspective and outline the points of the argument that need to be considered and thus the structure of the essay. Such an outline is called 'signposting'(MacMillan and Weyers, 2007). It is not necessary to outline the argument here, though some people advocate this. An outline at this stage does not have any evidence to support it.

The analytical meat addresses each point of the argument, ensuring that each paragraph contributes to answering the question and that the points occur in a logical order. Perhaps the two most often seen negative comments in the margin of essays are 'irrelevant' and 'order'. Editing on the word-processor allows the paragraphs to be reordered and the argument within paragraphs to be sharpened once drafted. In an extended essay, subheadings may be used to help the reader (and often the writer) to follow the structure.

When making the major points of an argument, it is important to present supporting evidence and to use examples to illustrate the points being made. The challenge is to demonstrate the breadth and depth of understanding possessed through the examples without allowing the argument to be dominated and obscured by the examples. If the example is dominating the essay and the reader is being left to infer the argument, it is time for a major rewrite. Developing and sustaining the argument are the major goal. Examples are used to help attain that goal.

The conclusion not only draws the points of the argument together but it also develops their implications and may raise further questions. For example, it might summarize the arguments on the degree of success of regeneration schemes against their objectives and against wider societal aims, and consider the degree to which there have been connections made between other schemes within the city, particularly those that are spatially contiguous. The possibility of schemes remaining 'islands of regeneration' raises the question whether spatially targeted regeneration alone can regenerate cities that have experienced major deindustrialization.

After proofreading and ensuring that all the references are included at the end of the essay, the final stage of the essay task is for the student to stand back and consider its evaluation. Have the aims been achieved? What are the positive points? What could have been improved? The submission of a brief evaluation helps the reader in his or her feedback but more importantly it helps students to be more generally reflective and to learn from their experience. This final challenge is only completed when feedback from the essay is received. Some tutors will give comments without a mark and expect the student to read the comments, compare them with grade-related criteria and arrive at his or her own assessment of the mark before learning and discussing the mark given by the tutor.

The overall challenge, of course, is learning from the process of completing the essay: learning about the subject matter, for example, points that might have been missed on urban regeneration; learning about interpreting command words and key words; learning about ways of introducing an essay, structuring an argument, using evidence and examples, and drawing together conclusions; writing references in the bibliography in a consistent and accepted fashion; and learning how to draw a line under the completion of the task. It is too easy with modern technology for a perfectionist to spend too long attempting to sharpen the argument or to achieve the best order. Too much time spent on one essay or, more generally, on one assessment task, may limit the performance achieved on other tasks. It may be galling to a geographer to conclude that the management of time, rather than space, is one of the most important skills necessary to succeed at *University Challenge*.

References

Cochraine, A. (2007) *Understanding Urban Policy: A Critical Approach.* Oxford: Blackwell.
Jones, P. and Evans, J. (2008) *Urban Regeneration in the UK.* London: Sage.
MacMillan, K. and Weyers, J. (2007) *How to Write Essays and Assignments.* Harlow: Pearson Education.

Self-evaluation of this essay

I think I have achieved my aim of writing an essay about writing an essay in a way that has conveyed the essential parts of the process. Perhaps I have dwelt too much on command words relative to key words. I would have liked to have given more examples of the process, especially one from physical geography, but I have concentrated on one, urban regeneration, to improve clarity, given the limitations of space. I have attempted to place the essay in the wider perspective of the challenge of university and linked it to *University Challenge* to catch attention. Perhaps, though, the eye-catcher is too stretched.

WRITING A REPORT AS A TEAM

Executive Summary (see section 7 of the report)

This report aims to outline the purpose and structure of reports. A report has to be concise and very well-structured. This report helps you to prepare, plan, write and edit a report as a team. The writing process involves adopting an appropriate structure, a clear style and language, and an attractive presentation. There is a recommendation to write the report from the inside out, starting with the analysis and ending with the conclusion, the introduction and finally the executive summary.

I Introduction

1.1 This report aims to outline the purpose and structure of reports and to present guidance on the ways they are produced by a team.
1.2 A report is the conventional method of presenting precise information. It may be used to convey an assessment of any situation or the results of an investigation.
1.3 A report is concise and succinct. It has clearly stated aims. It is tightly focused on the subject of investigation. A really effective report will also be compelling and stimulating to read.
1.4 This report is structured to give guidance on working in a team, to prepare, plan and write a report.

2 The importance of the team

2.1 This particular report is a team responsibility. As a team, you should be supportive of each other and be able to give constructive criticism on the project and each other's work.
2.2 You need to recognize and make best use of the talents of the team and to work in an organized way. The successful outcome of good teamwork will be more than the sum of the parts of the individuals' efforts.

3 Producing the report

Three stages in the process of producing a report have been identified in order to help you tackle key issues and understand the task set. These three stages are:

(a) preparation;
(b) planning; and
(c) writing and editing.

4 Preparation

4.1 Careful preparation is vital. It allows you to make the best use of the time available to the team. During this period you should collectively decide the aims (why), the process (how) and the contents (what) of your writing.
4.2 Establish the broad aims of the project and a draft plan of the report first. This enables you to identify the type of reading that each member of the team needs to undertake. Your aims and argument may subsequently be more precisely set.

4.3 When you have completed your individual reading you must meet as a team to discuss the key issues of your report.

4.4 You must always remember that report writing is a continuous process of decision-making. Your initial decisions will include:

    (a) the identification of the specific topic;
    (b) the aims of your report;
    (c) the precise message your team wishes to convey;
    (d) the appropriate structure and format; and
    (e) the correct vocabulary, style and tone.

4.5 When you have completed your discussion, note the decisions you have agreed as guidelines for your work.

5 Planning

5.1 Planning saves time and promotes clarity in collecting the information you require, organizing the material and writing the report.

5.2 Writing a report encompasses a number of autonomous activities. You will find it easier if you break the whole process down into a number of tasks:

    (a) collection and analysis of evidence can be organized according to the source or the subsection of the report; and
    (b) the writing process can be organized into writing subsections and presentation of evidence into graphs, tables, maps and quotations.

5.3 Planning and deciding on the method of work will eliminate much reproduction of effort. It will save time and it will also help you create a logical structure for the report. Good organization is the key to success.

5.4 The following sequence may help you to plan your report:

    (a) decide upon a working conclusion and the order of presenting your findings to create a preliminary structure;
    (b) identify the sources of evidence (data and/or literature);
    (c) decide what is the most appropriate and relevant evidence to collect and how to analyse and present it;
    (d) decide on the tasks to be completed and their order of priority;
    (e) draw up a realistic timetable for the completion of each task, including writing the draft of the report; and
    (f) allocate the tasks among the team members.

6 The writing process

6.1 There are three main factors to consider at this stage to give your report a sound framework, clear style and an attractive appearance:

    (a) structure;
    (b) style and language; and
    (c) presentation.

6.2 Structure

6.2.1 A clear structure with headings and subheadings helps the reader to digest the report. It also helps you write and organize your material logically.

6.2.2 A sensible structure is:

(a) title page;
(b) contents;
(c) executive summary;
(d) introduction;
(e) sources and methods of analysis;
(f) analysis and interpretation;
(g) conclusion;
(h) appendices;
(i) reference list.

6.2.3 The following order for writing is suggested:

(a) *Analysis and interpretation.* To help you present your findings have your material sorted, selected and arranged in note form. This section includes: (i) the results of your analysis; and (ii) your interpretation of them. You should help the reader by ending each part with its own conclusion.

(b) *Methods.* In this section you should discuss: (i) the sources of evidence; (ii) the collection and analysis of evidence; and (iii) the limitations of the sources and methods of collection and analysis.

(c) *Conclusions.* This section is a summary of all the major findings made throughout the report. No new evidence should appear here. The conclusion considers the evidence presented in the main body of the text, draws out the implications and brings it to one overall conclusion or an ordered set of final conclusions or recommendations.

(d) *Introduction.* After having written your findings and conclusions you now know clearly what you want to introduce. The introduction is where you indicate the purpose and structure of your report.

(e) *Appendices.* These are set aside for supplementary evidence not essential to the main findings, but which provide useful backup support for your main arguments.

(f) *Contents.* All the sections of the report should be listed in sequence with page references.

(g) *References.* This section covers the books, articles and web-sites that have been used in your research. It must include every reference mentioned in the text and be presented correctly.

(h) *Title page.* The title should indicate the central theme of the report. The page must also include the authors' names and the date of completion of the report.

(i) *Executive summary.* This is a very important part of the report and should be the last thing you write. Refer to the guide on executive summaries in section 7 of this report.

6.3 Style and language

6.3.1   Even though written by a team, the report should read as though one person has written it. So careful editing is needed.

6.3.2   Focus on the specific purpose of the report. Every part of the report should relate to it and this will help to keep the report concise and coherent.

6.3.3   Clarity and accuracy are vitally important so always be precise. You must know precisely what you want to say in each paragraph and sentence.

6.3.4   Each paragraph should contain one significant point so that, if the paragraph is referred to, its subject matter would be unambiguous.

6.3.5   Your sentences must be grammatically correct and well punctuated. Words must be spelt accurately.

6.3.6   Other important things to remember:

    (a) Keep sentences short and simple. Long complex sentences slow the reader down, confuse and impede understanding. The same applies to paragraphs; and

    (b) spell check and proofread the final document so that the reader is not deflected by poor spelling.

6.4 Presentation

6.4.1   The presentation of the report requires the same level of care that went into composing the text.

6.4.2   Adequate headings and numbering make it easier for the reader to comprehend what you are saying.

6.4.3   Numbering allows the writer and reader to refer to specific points. Using more than a three-number sequence can be unwieldy (e.g. 6.4.1.1) so use a letter or bullet point (e.g. 6.4.1. a).

    (a) The presentation of statistics is often more informative and eye-catching if they are shown visually – for example, by using graphs, pie charts or histograms. Remember that you should always discuss any tables or diagrams you use in the text.

    (b) Layout is important. This is the relation between print and space on the page. A crowded page with dense blocks of print and little space looks unattractive and is off-putting. Always ensure that there are:

      (i) adequate margins;

     (ii) either double or 1.5 spaced lines; and

    (iii) headings that stand out clearly from the page.

7 Writing the executive summary

7.1   An executive summary is defined as 'the main points of a report'. The purpose is to provide the briefest possible statement of the contents of the report, any significant findings, conclusions and recommendations.

7.2 It must cover all the essential points. It must be fully comprehensible when read independently of the full document. It is not a list of extracts, highlights or notes on the original.

7.3 The executive summary must:

    (a) introduce the subject of the full report, its aims, methods, findings and/or recommendations;

    (b) help the reader to determine whether the report is of any interest; and

    (c) save time for busy executives/officials in that it may be used by the reader to make initial decisions, so ensure that it fulfils its purpose.

7.4 The authors can use the executive summary as a rigorous check on the success of the full report – i.e. that the full report is a clear, concise statement that meets its aims.

7.5 The method of constructing an executive summary is to:

    (a) read the whole document;

    (b) isolate and summarize its central theme;

    (c) read each section and eliminate all repetition, lists, examples and detailed description;

    (d) identify and summarize the main statement of each section;

    (e) combine (b) and (d) into a set of major points. This may be written as a continuous narrative because your aim is to convey the overall impression of the full document in as clear and brief a way as possible, but this too may have numbered points; and

    (f) read through your summary to check that it gives a fair impression of the original while ensuring that it will make sense to the reader as a separate document from the full report.

(I would like to acknowledge the original work of Peter Shirlow and Laura Smethurst in writing this report. This is an amended version.)

THE DISSERTATION

There are many ways in which to present a dissertation but a model will be discussed with some deviations from it so that students can select the most appropriate form for their particular topic. In this model it is assumed that some original gathering of evidence is expected although, as noted in the introduction to this chapter, sometimes only secondary data are needed or, in the extreme case, only literature is reviewed. An outline of the structure of the dissertation is discussed first, then the deviations and, finally, the order in which it is often written.

Most dissertations will begin with a preface in which acknowledgements are made to those who have helped, with perhaps some comments on the process, particularly if the topic has changed in a major way.

Introduction

The first chapter introduces the general topic or problem and sets it in perspective within geography. It might include the reasons for the author's interest and the overall aims and objectives of the dissertation. It should outline the structure of the dissertation as a guide for the reader as to how those aims will be achieved, and objectives completed.

Review of the literature

The second chapter reviews the literature that establishes a context for the topic. It might identify a gap in the literature or a debate that needs to be addressed. It comprises a critical review that argues towards the specific research questions considered in the dissertation, which may well be stated at the end of this chapter. They may consist of questions or hypotheses, the latter being used more often in physical geography. It is through the analysis of these questions (objectives) that the aims of the dissertation are achieved. Their identification is, therefore, crucial to the success of the dissertation.

Methodology

The third chapter discusses the methodology used to address the specific research questions posed. It explains the research design and justifies the methods used to collect and analyse evidence. It can also include a justification of the choice of area in which the study has been carried out. The precise format will depend upon the approach to research used – whether, for example, it is extensive (seeking associations within representative samples) or intensive (seeking 'causes' and insights from a few in-depth investigations; Sayer, 1992). If it is the former, then the type of sampling used and the response rate should be discussed. If it is the latter then, for example, the original cues that were used in the first interview might be discussed, along with the way interviewees were chosen and the interviews changed.

Analytical chapters

The next few chapters form the main analytical sections of the dissertations where the results are presented and discussed. Each chapter, for example, may address a particular research question posed. The chapters are closely argued with evidence presented in maps, graphs, diagrams, tables, photographs and/or quotations to support the argument. The discussion will also feed back to the literature where it is appropriate to do so.

Conclusion

The conclusions to these chapters are brought together in the final chapter. It should include the limitations of the work, whether based on the methods used to collect or analyse the evidence. The overall conclusion is not just a summary of the individual conclusions to the research questions. It attempts to synthesize the findings (that is make more of them than just the sum of the parts) and to present their implications. It relates the findings back to the literature reviewed in Chapter 2. Finally, it poses further questions that arise from the research.

The references to all the work mentioned in the text follow. These are usually presented according to the Harvard convention, in alphabetical order, as in this book.

The appendices complete the dissertation. They may include such details as a questionnaire, an example of a transcribed interview or the description of a standard analytical method or laboratory technique used. It is too easy where there are word limits to put lots of extra work in the appendices. This should be avoided. If it is important to the argument, the material should appear in the body of the dissertation. Material that is irrelevant to the research questions should not appear anywhere.

In many physical geography dissertations, the structure of the analytical chapters is further divided. It is customary to present the results of the research separately from their interpretation (Parsons and Knight, 1995). The results are considered to be objective and therefore are presented in the form of maps, tables, graphs and diagrams. The interpretations of the results follow in a separate chapter. They relate back to the literature and are considered to be those of the author.

In most research in human geography (Flowerdew and Martin, 2005), the 'results' and their interpretation are presented together because they are not seen to be so clearly separate. An extensive approach to human geography research will identify the research questions, decide the methods and choose appropriate analytical techniques to address the research questions before the collection of evidence is begun. The three sets of decisions are made together so that, for example, questions in a questionnaire are asked in ways that allow the appropriate form of analysis to answer the research questions. It is not appropriate, for example, to collect answers in yes/no form if the research questions require degrees of agreement or disagreement. Although the decisions are reached in an iterative way, the research is carried out and appears in a consecutive manner – i.e. research questions, methods and data collection, and then analysis. Here, a standard data-collection technique will be used such as a questionnaire. The same one will be given to the whole sample and the analysis carried out after all the questionnaires are completed. In an intensive approach to research, on the other hand, the analysis is often carried out at the same time as the evidence is collected. An in-depth, semi-structured or unstructured interview may raise a number of questions or lines of inquiry that were not anticipated but are pertinent to the research. Further interviews will be amended to include them. Further research questions may also emerge. It is still possible to write-up the dissertation in a similar order (methods and collection of evidence and then analysis) but some would prefer to discuss the methods alongside the analysis, reflecting the rather different approach to research.

These are the major deviations from the model presented here and they involve different approaches to research (see Chapter 1 in Kitchen and Tate, 2000). There are some other variations that depend upon the topic. In some dissertations, the topic and research questions are decided and then a study area is chosen in which the evidence will be collected. In this case, the study area is like a laboratory. Its choice may be due to accessibility as well as appropriateness. In other dissertations, the study area is part and parcel of the topic. It is one of the reasons for studying the topic. In this case, the study area may well be discussed as part of the literature review rather than within a chapter on the methodology.

It is worth discussing the literature review in a little more detail to give some idea of its nature. As with all the chapters, it is helpful to emphasize the structure of the review using subheadings. Sub-subheadings may also be used, but it is advisable not to divide the text any further because it becomes too complicated to read. In a dissertation on 'Selling Birmingham: consensus or conflict?' a student began his review by setting the intellectual context. He discussed the re-emergence of 'civic boosterism' as the economy moved from being characterized by Fordism to flexible accumulation. He discussed 'the rise of the entrepreneurial city' and the 'selling of the city'. Later, he moved on to discuss the ways cities compete through 'flagship developments' and 'spectacle', as events such as major games and carnivals. He ended by discussing 'whose city?' it was and some of the 'false promises' that have been made in regenerating and selling cities. The phrases were subheadings within the review. Each was reviewed in a critical way.

Two excerpts from the dissertation demonstrate how quotations may be used within a review and within the dissertation, how the Harvard system of referencing works and how an argument may be sustained:

> Indeed Harvey (1989: p. 3) suggests that as a result of these economic changes, there has been an important shift in the way that cities in advanced capitalist societies are governed:
>
>> In recent years, urban governance has become increasingly preoccupied with the exploration of new ways in which to foster and encourage local development and employment growth. Such an *entrepreneurial* stance contrasts with the *managerial* practices of earlier decades, which primarily focused on the local provision of services, facilities and benefits to urban populations.
>
> Indeed, Colenutt (1993) asserts that 'place marketing' is now firmly embedded in the vocabulary of urban regeneration in Britain, used as a technique by city leaders to create or illustrate 'the niche in the world of inter-urban competition which their city occupies' (p. 187).
>
> (From Mole, J. (1998) *Selling Birmingham: Consensus or Conflict?* University of Manchester.)

Note how the first large quotation (from Harvey) is indented, as would be the case with a quotation that was introduced as evidence from an interview in an analytical chapter, whereas the shorter quotation (from Colenutt) is within the text. In both cases, the page number of the quotation is given. Figures and tables should be integrated into the text in an analogous way to quotations.

Finally, it is worth considering the order in which a dissertation may be written. Many will write it inside out, as in the report above, completing the methods and analytical chapters first, followed by the final version of the review of the literature, the conclusions and, lastly, the introduction, when what is to be introduced is known. The final touches ensure that the chapters lead-on from one another with links forward in the conclusions or links back in the introductions. Each introduction to a chapter sets out the structure of the chapter, 'signposting' as with the essay, while each conclusion brings together the main points of the chapter. Just as in the essay, each paragraph within a chapter should relate to the general argument. The final act will probably be the completion of an abstract, often no more than 200 words, which summarizes the aims and findings of the dissertation.

With all three forms of presentation the process of writing will yield or clarify ideas. It is a creative process that should excite the author as well as the reader.

Summary

- Understanding the assessment task is crucial.
- Planning the organization of the tasks to be tackled when writing essays, reports and dissertations is one key to successful completion.
- Developing a clear, structured argument that is supported with appropriate evidence is essential for all three.
- It is important to recognize the different formats, styles and audiences for the three forms.
- For all three, catch the attention in an introduction and make the conclusion more than a summary.

Further reading

There are very few articles or books written about how to write as a geographer or about different styles of writing for different forms of assessment/audiences. However, the following books are useful places to start:

- Kneale (2003) provides a practical guide to a whole range of study skills.

- Parsons and Knight (1995) and Flowerdew and Martin (2005) focus specifically on how to do a dissertation in geography, from the design to the writing-up stages.

- MacMillan and Weyers (2007) have a very useful generic guide to writing dissertations and reports in addition to their book on writing essays.

Note: Full details of the above can be found in the references list below.

References

Clark, G. and Wareham, T. (2003) *Geography @ University*. London: Sage.
Colenutt, B. (1993) 'After the urban development corporations: development elites or people-based regeneration?', in R. Imrie and H. Thomas (eds) *British Urban Policy and the Urban Development Corporations*. London: Paul Chapman, p. 175–185.

Flowerdew, R. and Martin, D. (eds) (2005) *Methods in Human Geography: A Guide for Students Doing a Research Project* (2nd edn). Harlow: Pearson Education.

Harvey, D. (1989) 'From managerialism to entrepreneurialism: the transformation in urban governance in late capitalism', *Geografiska Annaler*, 71: 3–17.

Kneale, P. (2003) *Study Skills for Geography Students: A Practical Guide* (2nd edn). London: Arnold.

MacMillan, K. and Weyers, J. (2007) *How to Write Dissertations and Project Reports*. Harlow: Pearson Education.

Parsons, T. and Knight, P.G. (1995) *How To Do Your Dissertation in Geography and Related Disciplines*. London: Chapman & Hall.

Sayer, A. (1992) *Method in Social Science: A Realist Approach* (2nd edn). London: Routledge.

32 Understanding Assessment

Robin A. Kearns

Synopsis

Assessment is a process by which performance is judged and through which, ideally, learning is both reflected and enabled. Assessment criteria are measures used by educators to judge the quality of students' work. There are two main approaches: norm-based and criterion-based assessment. In the former, work is assessed relative to that submitted by others working at the same level. In the latter, work is evaluated relative to standards (rather than a set of students). This chapter explains what academics are looking for when they assess your work.

The chapter is organized into the following sections:

- Introduction
- Why is work assessed?
- How is work assessed?
- Student and university perspectives on assessment
- Conclusion

INTRODUCTION

Earlier chapters in this book have surveyed issues associated with preparing for, and undertaking, geographical research, then analysing the information yielded by these activities. With the previous chapter's emphasis on presenting your findings, the emphasis shifted to the *product* of these labours. It is the purpose of this, the final chapter, to consider a frequently overlooked stage in the research process: the assessment of your written work.

To provide an understanding of the criteria used by university lecturers, this chapter discusses three sets of ideas: why work is assessed; how work is assessed (i.e. what criteria are commonly applied in assessment and why they are used); and how assessment is perceived and received. The central argument of the chapter is that whether assessment is undertaken by yourself, your peers or your teachers, it can and should be a positive process through which your learning as a geographer can be enhanced. In the chapter I pay particular attention to two different approaches to grading students' work: norm-based and criterion-based assessment. In the first of these approaches, teachers assess your work in relation to that submitted by others in your class. This can be contrasted with the second approach, criterion-based assessment, which involves evaluating how you perform according to sets of measures (rather than a set of students). I conclude that most assessment in universities uses a combination of these approaches. Further, if each party in the academic teaching relationship (i.e. students and lecturers) can come to appreciate the other's motivations and perspectives better, the assessment process will contribute to a more fruitful educational experience for all involved.

WHY IS WORK ASSESSED?

The radical educator and philosopher Ivan Illich (1971) believed all students should receive a university degree simply for participating in courses. While there may be a few lecturers sympathetic to this idea, even fewer are likely publicly to proclaim this view. This is because universities maintain their prestige through the common requirement that certain courses of study must be completed to a satisfactory standard and within an established time frame for a degree to be granted. Written work is assessed in order to determine whether students meet expected standards of scholarship. Therefore, while you may choose not to adopt one or more of the research methods outlined in earlier chapters of this book, there is no avoiding the experience of assessment. Indeed, assessment now occurs across a range of scales in the university system: lecturers themselves are assessed for promotion on their teaching and research, and universities at large are assessed and ranked on their research output.

Beyond the university, the criteria applied in assessing written work vary according to context. A poet, for instance, might assess her work as successful because it gets published and people attend her public readings. Alternatively, an editor might apply an economic criterion and assess his newspaper as successful if it sells well. In universities, the research work produced by students is assessed for a range of purposes. Gibbs (1999: 153) lists six purposes:

1 Capturing student attention and effort.
2 Generating appropriate learning activity.
3 Providing feedback to the student.
4 Developing in students the ability to monitor their own learning standards.
5 Allocating marks (to distinguish between students or degree classification).
6 Ensuring accountability (to demonstrate to outsiders that standards are satisfactory).

You might note that purposes 1–4 relate to student learning goals, whereas purposes 5 and 6 relate more to the functioning of the university as an institution. The question of why work is assessed leads us towards the closely related question of *how* assessment occurs. For, as Gibbs (1999) points out, larger classes and increasing marking loads have shifted the emphasis in assessment back towards tests and computer-based assessment (i.e. the fifth purpose above), whereas it is purposes 1–4 that actually work best to encourage and support student learning.

We can think more broadly than the Gibbs' (1999) categorization, however, and follow Neil *et al.* (1999) in recognizing that there are crucially different perspectives on assessment according to whether your institutional administrator is that of a student, teacher, institution or member of the public. To help you gain a rounded view of assessment at the outset, Box 32.1 lists some features of each of these perspectives. As the contents of Box 32.1 portray, there is clearly a range of 'stakeholders' who each serve to benefit from the existence of assessment. To develop an understanding of assessment criteria we must move from the 'why' question to examine the issue of 'how' more closely.

Box 32.1 Purposes of assessment – from student, teacher, institution and community perspectives

Students' needs
- To know how you are doing in general ('Am I on track?').
- To know whether you are reaching a required standard ('Will I pass/graduate?').
- To have something to show to others ('Please employ me/give me a scholarship').

Teachers' needs
- To know whether students are achieving the intended learning outcomes particularly at the subject level ('Am I getting through?' 'Are they doing the work?').
- To prove to others that they are effective teachers ('Please give me tenure/promote me').
- To certify that students can proceed to subjects for which theirs is a prerequisite ('A student who passes my subject should be able to cope with yours').

Institutions' needs
- To know whether students are achieving the intended learning outcomes ('Are our courses effective?' 'Are our staff effective?').
- To prove to others that graduates have achieved what the institution claims they have ('Please continue to fund/support us').
- To certify students can proceed to employment, professional practice or further study ('Our graduates meet your requirements').
- To know whether to accept applicants into programmes of study ('Do you have the required prerequisite knowledge?' 'What is your level of achievement compared to other applicants?').

Community needs
- To know whether the institutions and the teachers are effective ('We will continue to fund/ support you').
- To know whether individual graduates are employable, capable of practising, etc. ('You may work here, teach here, etc.').

Source: Nightingale *et al.* (1996: 8)

HOW IS WORK ASSESSED?

Within the university environment, standards are maintained through agreed-upon methods of assessment being used to judge the quality of students' work. Most universities have policies that outline agreed-upon methods or guidelines for assessing students' work. At the crudest level, an essay or research report is regarded as either passing or failing to meet the accepted standards of the university. While students are concerned in the first instance with simply 'passing', anyone with an ounce of ambition is likely to be concerned with *how well* he or she passes. This degree of passing a minimum set of standards is usually conveyed either by a numeric grade (expressed as a percentage of full marks) or, more commonly, as a letter grade. In some countries passing grades range from A to C with a D or E being a fail. When part-letter grades are employed in a grading scheme, brackets of numerical percentage marks can be summarized with, for instance, 90–100 per cent being otherwise expressed as an A+. In Britain, passing grades tend to be expressed as 'first', 'upper second (2i)', 'lower second (2ii)' and third class.

The next question is: what do these grades represent? While officially they symbolize an assessment made by the university itself, in reality it is the individual judgements of lecturers or tutors that eventually contribute to an official and over-all course grade issued in the name of the university. The question of 'judgement' deserves comment for, while a mathematical answer might be easily deemed to be objectively right or wrong, a geographical research report is not so easily categorized. Rather, what underlies the lecturer's assessment is usually the result of a *combination* of the application of identifiable assessment criteria and the harder to define issue of professional judgement. A measure of each of these processes (application of criteria and intuitive judgement) is commonly at work in grading systems. An example of a commonly used differentiation of grades is listed in Box 32.2.

Box 32.2   A letter-based grade distribution

A+ *Outstanding* – fulfils all the general assessment criteria to an unusually high standard.
A   *Excellent* – fulfils all the general assessment criteria to a very high standard consistently.
A–   *On the verge of excellence* – fulfils all the general assessment criteria to a high standard intermittently.
B+   *Very good* – fulfils most of the general assessment criteria to a very good standard.
B   *Good* – fulfils most of the general assessment criteria to a good standard.
B–   *Very reasonable* – fulfils some of the general assessment criteria to a good standard intermittently.
C+   *Reasonable* – fulfils some of the general assessment criteria to a competent standard.
C   *Fair* – fulfils some of the general assessment criteria to an adequate standard.
C–   *Marginally passable* – fulfils some of the general assessment criteria to an adequate standard.
D+   *Inadequate* – fails to fulfil enough of the general assessment criteria to a competent standard.
Grades below D+ indicate greater degrees of inadequacy.

Source: Department of Geography, University of Otago

The verbal descriptions of difference between the letter-based grades contained in the example in Box 32.2 are helpful in guiding teachers in applying grades and students in interpreting their meaning. However, it is also immediately apparent that there is subjective meaning attached to words used in the grade distribution such as 'good' and 'reasonable'. What do these words *really* mean? Educators who have been assessing student work for some time will often speak of being able to 'know an A-grade report' after only reading the first few pages of student work. This aspect of assessment may, on face value, seem a case of jumping to conclusions. However, the insight that comes with experience means a lecturer can be applying criteria subconsciously as she reads a research report. But regardless of whether these criteria are being painstakingly weighed up one by one, or are being intuitively considered as a whole set, what are the actual criteria applied by markers?

To answer the foregoing question, it will be useful first to distinguish between two distinctive ways of assessing student work: *norm-based* and *criterion-based* assessment. Explaining these might be helpful to you in recognizing the ways in which you are being, or will be, assessed by your lecturers or tutors.

Norm-based assessment '… uses the achievement of a group of students to set the standards for specific grades or simply passing or failing. The best X per cent of students get the best result and the worst X per cent fail' (Nightingale *et al.*, 1996: 9). Another term for this type of assessment is 'grading on the curve'. In other words, regardless of the individual achievements of students, a set proportion end up attaining a particular allocation of grades (e.g. 20 per cent get 'A' grades in any one year). This approach assesses your work by comparison with the work of others in your class.

By contrast, *criterion-based assessment* establishes criteria for specific grades or for passing and failing. As a result of these standards being applied, '… a student who meets the criteria gets the specified result' (Nightingale, 1996: 9). This direct link between fulfilment of criteria and attainment of a result signals some of this approach's immediate appeal: it appears more ethically sound to give a student a grade based on his or her performance according to a standard rather than according to how an entire class has performed. Whereas norm-based assessments judge student work by comparison with the work of others, criterion-based assessment is made with pre-established standards. What, commonly, are such standards? We will come to this, but for now let us consider advantages of criterion-based assessment (see Box 32.3). Box 32.3 clearly signals advantages of criterion-based assessment for both lecturers and students. However, notwithstanding these advantages, there are also drawbacks in using assessment criteria. Some of these disadvantages are listed in Box 32.4. The foremost of these is surely the first one listed: that students can be tacitly encouraged towards conformity, perhaps discouraging risk-taking and innovative styles of communication.

Box 32.3   Advantages of the criterion-based approach to assessment

For students
- You can compete with your own previous performance rather than with your peers.
- Your grades are based on your performance, rather than that of a cohort of your peers.
- You have a clear understanding of the standards required for a given outcome.
- Your outcomes are based on demonstrated competence rather than arbitrary standards.

(Continued)

(Continued)

- You can exercise greater judgement and choice regarding the outcome you target.
- There is greater transparency in the assessment process.

For lecturers
- It encourages the setting of clear goals and objectives.
- It encourages planning coherent courses that will achieve such goals and objectives.
- It encourages greater transparency in the assessment process.

Sources: Adapted from Abbiss and Hay (1992); Neil *et al.* (1999)

Box 32.4   Disadvantages of criterion-based assessment

For students
- You are encouraged to conform, perhaps discouraging risk-taking and innovative styles of communication.
- You can work on satisfying individual criteria of a project but lose sight of the overall objectives of the assignment.

For lecturers
- Little room for professional judgement.
- Inflexibility (it is hard to take account of the circumstances of individual students).
- Establishing valid and weighted marking systems is a very challenging task.
- Even with criteria, different markers can assess criteria in different ways, but still agree on a final mark.

For both lecturers and students
- This approach can lead to excessive regulation of learning that can lower morale.

Sources: Adapted from Hay (1995); Neil *et al.* (1999)

In summary, conformity, inflexibility and regulation describe key disadvantages of this approach, as noted by commentators. As Hay (1995) suggests, the criterion-based approach must be treated with caution because there are no true standards, in the strictly psychometric sense, to separate out those who have mastered some skill from those who have not. In other words, we are brought back to the issue of professional judgement and the ability of an experienced marker to spot an 'A' or, alternatively, a fail essay. So exactly what, to the experienced eye, constitutes an 'A' report in geography? Box 32.5 offers three examples drawn from communications with experienced university teachers. As the quotes imply, any university teacher might be expected to provide clearly stated and understood criteria for the assessment of your work. However, *how* that assessment is reached may have much to do with how familiar the lecturer is with the academic context in which the assignment is set. It may also relate to the degree to which the assessor is willing to reward creativity and risk-taking.

Box 32.5   Descriptions of an 'A' report in geography

'I think there is a real quality to an 'A' piece of work that is hard to decompose into different parts, but which involves a combination of insight, energy, inherent authority and rationality of argument and focus … It is the voice of an articulate, autonomous, independent mind engaged in deep thought … To borrow from an Incredible String Band song title, an 'A' answer belongs to someone who knows all the notes but also understands the song'. (P. Forer)

'… a student must show three things: a very good understanding of the issues, significant effort (in terms of research, structure and presentation for example), and some spark of originality or critical insight.' (R. Law)

'… an 'A' answer demonstrates flair, originality, insight, extension of ideas, is well structured and competently written, statements are substantiated, examples used to illustrate points, well-referenced, and goes beyond the requirements of the set question.' (J. Mansvelt)

Source: Personal communications, 2001

Few universities use entirely norm-based or criterion-based assessment practices. Indeed, most use both. As a very structured and transparent approach, criterion-based assessment is increasingly favoured for the extent that it both spells out expectations to you, the student, and requires us, as lecturers, to be more accountable in our teaching (and specifically, grading) practices. Given the likelihood you will encounter this form of assessment in your courses, there is merit in briefly reviewing some common components of assessment criteria. Scott *et al.* (1978) outline four key dimensions, or 'competencies', they see as central undercurrents to the assessment of all university-level work (see Box 32.6).

Box 32.6   Four key competencies underlying assessment criteria

1 *Process competence*: acquisition and exercise of specified skills in a subject.
2 *Content competence*: the acquisition of specified content on a particular subject.
3 *Affective competence*: evidence of affective (i.e. attitudinal or value-based) response.
4 *Skills competence*: acquisition and exercise of specified practical skills.

Source: Scott *et al.* (1978)

To focus on the first competency in Box 32.6, a key element of the *process* of completing a research report is demonstrating communication skills. There are six fundamental criteria for demonstrating communication competence that are applicable across all subjects and disciplines (Neil *et al.*, 1999). These criteria can be divided into aspects of *text* (the material you are communicating) and *context* (the setting in which you are communicating) (see Box 32.7).

Box 32.7   Aspects of communication competence

Textual features
1   Spelling, paragraph and sentence construction, punctuation and grammar
2   Word choice

Contextual features
3   Purpose
4   Language
5   Information sequencing
6   Ideas

Source: Neil *et al.* (1999), adapted from Anon (1998)

With respect to assessing the text, communication can be greatly improved by minimizing mistakes in the writing. Yet it is only rarely that university teachers have the time to correct every page of students' written work meticulously. It is therefore a good idea to ask a friend, family member or fellow student to 'proof' or check through your final draft before submitting it. Alternatively, it can be a good idea to seek out the services offered by your university's 'Writing Centre' or 'Student Learning Centre', as they are sometimes called. There is a great deal of practical assistance offered by such centres, for writing is as much an acquired skill as driving a car. Other sources of advice on writing include classic books such as *The Elements of Style* (Strunk and White, 1979), and more recent guides such as by Bonnett (2001). Ultimately, the way you write will serve as the gateway to your argument and ideas. Poor writing will turn a reader off and distract an assessor from your ideas through encountering grammatical and other errors. Compelling writing, conversely, can entice a reader into greater curiosity even about a topic he or she is not completely interested in. A good 'rule of thumb' is that you should be aiming for work that you would be proud to show a potential employer as typical of the quality and standards you are capable of attaining and maintaining (Neil *et al.*, 1999).

Based on the criteria for communication competencies listed in Box 32.7, a piece of work meeting a high and 'proficient' standard would typically contain the following features, in an approximate order of importance:

- A thoughtful progression of ideas.
- A clearly stated purpose.
- Clearly sequenced information.
- A lively and effective choice of words.
- An appropriate choice of language for the audience and type of communication.
- Rare spelling or grammatical errors.

For each of the key competencies identified in Box 32.6 (process, content, affective and skills) there are criteria that are equally applicable whether you are writing up research in human or physical geography (see Box 32.8). To elaborate, the first criterion in Box 32.8 deals with the *content* of your report. You might ask yourself: how appropriate are the materials I have drawn on and what is their scholarly quality? A good piece of work situates its argument or findings within a context built on the work of others. Usually, such work has been published in scholarly journals. Remember that there is a hierarchy of status among published sources. Referring to articles in *National Geographic*, for instance, tends not to be as highly regarded and considered as appropriate as work published in journals such as the *Transactions of the Institute of British Geographers*. Similarly, there is a hierarchy of credibility among websites. A government site, for instance is, self-evidently, a more credible source than Wikipedia (see also Chapter 5).

Box 32.8   Typical assessment criteria for a written research report in geography

1   **Content** (i.e. the nature and scope of the materials used):

   • Quality, relevance and depth of information and references.

2   **Process** (i.e. your comprehension and how you have made use of your materials):

   • Definition of the topic and/or problem in the broader context.
   • Analysis of key issues.
   • Logical sequence of argument.
   • Constructive discussion and consideration of contrary arguments.
   • Adequacy of supporting argument for recommendations and conclusions.

3   **Affect** (i.e. the originality of your insights and your handling of value positions):

   • Demonstration of original and independent thinking.
   • Presentation and examination of personal views and conclusions (supported by literature and logical argument).

4   **Skills** (i.e. your proficiency in the requirements for clear presentation):

   • Layout of the paper/essay.
   • Proficiency in language and communication.
   • Graphic and cartographic proficiency.
   • Accuracy, completeness and consistency of the way references and sources are cited.

Source: Adapted from Neil *et al.* (1999)

The second criterion in Box 32.8 relates to *process*: how you use the material you have assembled. The potter's craft is in working with clay to form an object of use and elegance. So, too, the writer's task is to work with the materials gathered and to construct a relevant and appealing report. Here the challenge is to explain findings clearly and reflect on them in light of the wider literature (Hay, 1999). As Holloway and Valentine (2001) point out, findings alone cannot carry a research project. Instead you need to create an argument. This involves integrating your empirical findings with theoretical material to show the reader their importance, demonstrating how your findings relate to previous work and indicating what new insights your work adds to your area of scholarship (Bonnett, 2001).

The third area, *affect*, relates literally to how effectively you are able to affect the reader. All research is value laden and it is up to you to find ways to signal 'where you are coming from'. Thinking critically and knowing where you stand on issues will help you signal this perspective in your writing. This, in turn, will assist your readers to appreciate what you see and why. While there are generally agreed-upon structures for research reports you should work within (discussed in Chapter 31), you should also strive to be creative within the acceptable format.

The fourth area of assessment criteria deals with the practical skills demonstrated in the course of presenting your work. Questions such as the neatness and accuracy with which work is presented are considered important skills. The layout and organization of your work are also important, and a good essay or report will always have clearly demarcated beginnings, 'bodies' and conclusions (see Fitzgerald, 1994). Precision in referencing is one skill considered to be crucial (Mills, 1994).

As the quotations included in Box 32.5 imply, there are some well-accepted qualities that distinguish a top from an average answer or piece of work. We can summarize by saying that key criteria for a first-class written response include: clear reference to relevant literature and examples; evidence of independent thought; a demonstrated ability to distinguish between ideas and arguments; and a high standard of writing, making it interesting to read. The difference between these criteria and those applicable to the next highest level of assessment ('upper second' in the British system) is subtle yet clear. The University of Sheffield criteria, for instance, refer to this level of achievement as signalled by '... reference to *particular* authors ... a *reasonable* breadth of knowledge, but lacks real original thought ... has an ability to write *good* English' (emphasis added). Note that the highlighted terms signal a qualified level of excellence (e.g. the use of 'good English' rather than 'well written and interesting to read').

One point worth making is that different sets of expectations tend to apply to assessment situations according to the level of constraint imposed on the exercise. For instance, students often ask whether they are required to provide citation details in a written examination. Most lecturers will accept a name and perhaps a date of publication as sufficient and certainly not expect the attention to detail required in a formal research report. In summary, therefore, some assessment criteria (e.g. writing and referencing standards) are often relaxed when applied by markers of invigilated examinations. In other words, evidence of scholarship is always sought, but the time and resource constraints of some assessment settings mean that less attention can reasonably be devoted to issues of presentation and detail.

STUDENT AND UNIVERSITY PERSPECTIVES ON ASSESSMENT

Despite both students and university lecturers being part of the same overall enterprise of tertiary education in geography, and despite all lecturers having once been students themselves, the two groups inhabit distinct and differently experienced 'worlds' or 'subcultures' within the university. How is assessment seen from a student perspective? You, the reader, may be in the best position to answer this question because I, the writer, have not been a student in many years. However, in talking with students I have increased my appreciation for your perspective.

The experience of being assessed is a cumulative learning process. As one student told me, it is wise to '... remember the comments made by a marker and take these into account recognizing that assignments are linked in your course of study'. In other words, you are likely to have multiple assignments from any one lecturer in any course or set of courses, so it is important to get to know their particular expectations, as well as the general expectations of your department or university. Research shows that students often fail to transfer the feedback on one essay or report to the demands of the next one very well. Too often there is an assumption that a disappointing result has been caused by not getting the answer right rather than a more general failure to understand how to write up a piece of research (Nightingale, 1996).

Students also complain that at some universities there can be a poorly defined upper limit to what grade you can achieve. In some places, this upper limit means that 'A+' grades are simply not given or only very rarely awarded. While it is generally recognized that it is harder to get 'A' grades in the humanities and social sciences than in the natural or physical sciences, there can also be an institutional reluctance to award grades signalling excellence. In the words of one graduate student working as a teaching assistant in Canada, 'professors say you can't give grades above 90 per cent'. This 'no-go zone' within a norm-based grading scale is one way that some universities, or schools within universities, maintain their prestige.

In the past, university-level assessment has involved ranking students according to the knowledge they have gained in a programme of learning. The methods of assessing this learning were seen to allow students to demonstrate their knowledge in ways that could easily be measured so as to ease comparison between students. Student achievements were almost always measured in quantitative terms raising questions such as 'how much do they know'. While this approach persists in universities in many forms, there are pressures for change. According to Nightingale *et al.* (1996: 14), these pressures are:

a) '*A desire to broaden university education and develop (and hence assess) a broader range of student abilities.*'

Research into student learning has shown that assessment can be powerful in determining what students perceive to be the 'real' curriculum in a subject area, as opposed to the 'espoused' curriculum. By this, the authors mean that how a course is assessed will, inevitably, affect what you as students consider are the important components of it. If a number of lectures are delivered in a course, but there is no assessment of what you have learnt from them, these topics tend to be disregarded as priority areas. Creative ways to involve students in demonstrating what they have learnt across the curriculum are therefore an important challenge for instructors. This recognition has, in turn, led to the development and use of a much broader range of assessment methods. Group assessment, for instance, tends to be less liked by students but, as Knight (2004) shows, can foster a wider range of skills.

b) *'A belief that university education should lead to students having the capacity to apply independent judgement and an ability to evaluate their own performance.'*

In universities, lecturers are encouraged to monitor their own performance (e.g. via annual performance reviews) as well as that of their peers (e.g. through the refereeing of research papers). This recognition has led some university-based teachers to advocate approaches to assessment that help students develop their own capacity for self-evaluation (e.g. Thompson *et al.*, 2005). The purpose in this approach is to develop in students the recognition that they have the capacity and can be trusted to evaluate their own learning so that they will graduate as professionals capable of this useful career skill. Despite the competitive atmosphere that often prevails at universities, one path towards evaluating your own learning is to participate in the informal evaluation of others' work. For example, you might consider trying out the idea of student writing groups in which members give each other feedback on drafts prior to submission (Hay and Delaney, 1994). Peer-assessment can be useful and does not necessarily involve assigning a grade. Indeed, its purpose can be formative rather than summative. Here the goal is to assess and assist towards improvement rather than offering a decisive judgement (see Box 32.9).

Box 32.9   Summative compared to Formative Assessment

Summative	**Formative**
(to decide)	(to improve)
end point	continuous
mainly numbers	mainly words
primarily judgemental	primarily helping to improve
e.g. final exam	e.g. pre-submission critique

Source: Bradford and O'Connell (1998)

---

c) *'A desire to use assessment in ways that might support learning processes themselves.'*

Some assessment methods can too easily lead students into a mere 'surface learning', a '... rote learning of isolated facts and formulae – quickly acquired to meet exam pressures and just as easily forgotten' (Nightingale *et al.*, 1996: 6). Recent developments in assessment attempt to go beyond this 'cramming/mental regurgitation' sequence that too easily typifies how students respond to traditional examinations. A more favourable situation is surely when you find a range of assessment tasks (including writing research reports) to be significant learning experiences in themselves that reinforce positive attitudes to learning.

In summary, it is worth bearing in mind that, while there have been huge strides towards taking teaching more seriously in universities, the emphasis given to teaching compared to research in the training of lecturers is still minimal (Gibbs, 1999). You may well, therefore, encounter teachers experimenting with assessment methods and who may be as keen to learn from the experience as you are. You should offer your lecturers feedback on their assessment practices through routine avenues such as evaluation questionnaires. You would also be wise to make use of information, guidelines and criteria provided to you in written and verbal form at the beginning of any course or assignment. You should also acquaint yourself with any relevant university policies on matters such as late submission and plagiarism. The latter term warrants some explanation. Plagiarism is the use of another person or organization's ideas or words without giving the source appropriate recognition (i.e. through standard referencing). It is, put simply, adopting, adapting or copying material without due attribution. Plagiarism is widely regarded as one of the more serious breaches of academic conduct and pleading ignorance is an unacceptable defence when policies have been spelt out in programme handbooks or course outlines.

Students can reasonably expect that they will be given both criteria to guide the writing of a research report and feedback after work is assessed. In the words of one student, it is simply unfair to be told that, after a research project is returned, '… to get an "A" you should have done this'. You are entitled to request clarification if the guidance you are provided seems insufficient. For instance, it is only reasonable to expect a course outline that supplies details of the goals, the assessment requirements and the criteria that will be used for the grading of your work. Just as I found it helpful to talk with students about assessment to gain the foregoing insights, so too it may be helpful to talk to your tutor or lecturers about assessment. Ask them how they apply assessment criteria and, if not immediately evident, what criteria they apply. Communication about this and other matters is surely the key to geographical education being a learning experience for all.

CONCLUSION

This chapter has attempted to 'unmask' some of what underlies the assessment of research in human and physical geography. It has discussed why work is assessed and how it is assessed, arguing that most assessment involves a mix of norm-based and criterion-based approaches. In conclusion, we can revisit the responsibilities of both parties in the learning dynamic that you are part of as university geography students. For us as lecturers, it is essential to improving student learning that we establish assessment criteria, communicate these clearly to students and take them into close account in marking student work. For you, as students, it is important that you identify and familiarize yourselves with what assessment criteria have been set, and reflect back on them when considering the feedback you receive after submitting work. By critically examining the feedback you receive, and thinking about its implications for your next piece of work, you will come to understand more deeply what is involved in producing advanced academic work within the discipline of geography. In this way, you will be mirroring the process that career academics participate in whenever they submit work to the peer-review process; for in the journey towards publication responding to reviewers' feedback is a critical opportunity to improve our writing and arguments, as well as participate in a broader community of scholarship.

Summary

- Within universities, standards are maintained through agreed-upon methods of assessment used to judge the quality of students' work.
- There is a range of 'stakeholders' who each serve to benefit from the existence of assessment.
- It is useful to distinguish between two distinctive ways of assessing student work: norm-based and criterion-based assessment.
- Most universities use both approaches to assessment.
- Competence in communication is central to assessment across all subjects.
- Criteria for communication can be divided into aspects of text (the material you are communicating) and context (the setting in which you are communicating).
- Common criteria used in assessment of geographical work relate to skills, content, process and affect.
- The experience of being assessed is a cumulative process.
- The way criteria are applied varies according to the situation in which work is produced (e.g. an examination essay vs. a non-invigilated essay).
- Communication with teachers about how they apply assessment criteria and what criteria they apply is important.

Further reading

While there is not a great deal written about assessment criteria per se, what is to be recommended is that you read some of the geographical education literature that provides guidance on aspects of constructing written reports. Five useful readings are as follows:

- Neil *et al.* (1999) discuss assessment criteria and present sets of criteria you might wish to take into consideration not only for your writing but also for the maps and graphics you include in research reports.

- Mills (1994). Correctly referencing the material that provides the context for your work is an important skill geography students are assessed on. This article is a helpful guide for citing sources in your academic writing.

- Hay (1999) provides helpful guidance on report-writing from a geographer who has contributed a great deal to thinking through the process of assessment.

- Having an argument is a key criterion in the assessment of geographic research reports. Both Valentine and Holloway (2001) and Bonnett (2001) provide very readable guides to the nature of an academic argument and how to create one in the course of writing up your research.

Note: Full details of the above can be found in the references list below.

References

Abbiss, J. and Hay, I. (1992) 'Criterion-based assessment and the teaching of geography in New Zealand universities', *New Zealand Journal of Geography*, 94: 2–5.
Bean, J.C. (1996) *Engaging Ideas: The Professor's Guide to Integrating Writing, Critical Thinking, and Active Learning in the Classroom.* San Francisco, CA: Jossey-Bass.

Bonnett, A. (2001) *How to Argue*. Harlow: Pearson Education.

Bradford, M. and O'Connell, C. (1998) *Assessment in Geography. Geography Discipline Network*. Cheltenham: Cheltenham and Gloucester College of Higher Education.

Fitzgerald, M. (1994) 'Why write essays?', *Journal of Geography in Higher Education*, 18: 379–384.

Gibbs, G. (1999) 'Improving teaching, learning and assessment', *Journal of Geography in Higher Education*, 23: 147–155.

Hay, I. (1995) 'Communicating geographies: development and application of a communication instruction manual in the geography discipline of an Australian university', *Journal of Geography in Higher Education*, 19: 159–175.

Hay, I. (1999) 'Writing research reports in geography and the environmental sciences', *Journal of Geography in Higher Education*, 23: 125–135.

Hay, I. and Delaney, E. (1994) 'Who teaches, learns. Writing groups in geographical education', *Journal of Geography in Higher Education*, 18: 217–234.

Holloway, S.L. and Valentine, G. (2001) 'Making an argument: writing up human geography projects', *Journal of Geography in Higher Education*, 25: 127–132.

Illich, I. (1971) *Deschooling Society*. New York: Harper & Row.

Knight, J. (2004) 'Comparison of student perception and performance in individual and group assessments in practical classes', *Journal of Geography in Higher Education*, 28: 63–81.

Mills, C. (1994) 'Acknowledging sources in written assignments', *Journal of Geography in Higher Education*, 18: 263–268.

Neil, D.T., Wadley, D.A. and Phinn, S.R. (1999) 'A generic framework for criterion-referenced assessment of undergraduate essays', *Journal of Geography in Higher Education*, 23: 303–325.

Nightingale, P. (1996) 'Communicating', in P. Nightingale *et al.* (eds) *Assessing Learning in Universities*. Sydney: Professional Development Centre, University of New South Wales, pp. 204–220.

Nightingale, P. and O'Neil, M. (1994) *Achieving Quality Learning in Higher Education*. London: Kogan Page.

Nightingale, P., Te Wiata, I., Toohey, S., Ryan, G., Hughes, C. and Magin, D. (eds) (1996) *Assessing Learning in Universities*. Sydney: Professional Development Centre, University of New South Wales.

Scott, E., Berkeley, G.F., Howell, M.A., Schunter, L.T., Walker, R.F. and Winkel, L. (1978) *A Review of School-based Assessment in Queensland Schools*. Brisbane: Brisbane Board of Secondary School Studies.

Strunk, W. and White, E.B. (1979) *The Elements of Style* (3rd edn). New York: Macmillan.

Thompson, G., Pilgrim, A. and Oliver, K. (2005) 'Self-assessment and reflective learning for first-year university geography students: a simple guide or simply misguided?', *Journal of Geography in Higher Education*, 29: 403–420.

Glossary

Abstraction The effect of moving from the 'real world' through collection and manipulation of data to generalizations and predictions.

Active sensors These emit electromagnetic radiation and record the amount of radiation scattered back from the Earth's surface.

ANOVA Analysis of Variance is a method whereby variation within the data being studied is partitioned into components that correspond to different potential sources that can explain that variation.

ASCII American Standard Code for Information Exchange is the 'translation' of characters into numbers that the computer will 'understand'. When an ASCII format is required, this normally means a plain text without formatting, i.e. no bold or underlined texts, no bullets or borders.

Attribute data Information about a geometric feature on a map that identifies what the feature represents.

Axial coding Coding along an 'axis' or theme; researchers may suspend 'open coding' in order to pursue a particular theme.

Bias The extent to which sample values deviate systematically from the population value precision (the size of the deviations between repeated estimates of a given statistic). Good sampling therefore aims to minimize bias and maximize precision.

Bivariate graph A graph which illustrates a single cause–effect relationship (e.g. scatterplot) between the response variable (data being examined) and the explanatory variable (possible cause).

CAQDAS The acronym for Computer Assisted Qualitative Data Analysis Software. It comprises a variety of programs with different capabilities to assist the analysis of qualitative data. Simple programs can be used for text searches, whilst more sophisticated ones have functions especially for theory building.

Closure The requirement to simplify a problem to obtain a solution by removing uncontrolled influences 'outside' the system under consideration.

Conditioning plot (coplot) A plot in which the range of values for the explanatory variable is split into segments, and then data for each segment are plotted on a separate scatterplot.

Confidence interval The probability that the sample mean lies within a given standard error of the true population mean.

Confidence limits Limits which describe an interval that contains plausible values for a statistic and which that statistic may lie within a certain proportion of the time given repeated sampling.

Cross-cultural research Researching cultures other than one's own, which may be spatially distant or closer to home. This requires sensitivity to cultural similarities and differences, unequal power relations, fieldwork ethics, the practicalities and politics of language use, the position of the researcher and care in writing up the research.

Dataframe Spreadsheets of data organized into rows and columns. Rows represent an attribute of the data 'case'; columns represent values of the attribute between all cases.

Data types Continuous data are measured on a continuous scale and bear a relational element; categorical data values signify membership of a group.

Dendroclimatology Analysis of annual variations in tree-ring widths.

Diaspora The dispersal or scattering of a population. It can also refer as a noun to dispersed or scattered populations such as the Black diaspora or Jewish diaspora. In more theoretical terms, the idea of diaspora, or diasporic communities (see also transnational communities) challenges notions of fixed connections between cultures, identity and place.

Digital data Data about the way the map is configured, such as boundaries and points.

Discretization In field studies, the ability to isolate a particular field site from 'outside' influences which may have an uncontrolled effect on observations and measurement.

Ecological fallacy Inferring the characteristics of individuals from information about the aggregate populations of which they are part. A common problem in the use of secondary data.

EDA Exploratory Data Analysis – graphical representation of data to help identify structure and pattern, often as a precursor to statistical inference or modelling.

Electromagnetic radiation A system of oscillating electric and magnetic fields that carry energy from one point to another, and used to measure the properties of features in an image.

Equifinality (or convergence)   When a similar end point in a system (e.g. a land-form) arises from the operation of similar processes, but from different initial conditions.

Essentialism/essentialist/essentialize   The assumption that particular character-istics, such as culture, gender or 'race', are determined by a single unchanging 'essence', often assumed to be a biological or 'natural' difference. The term is often used to criticize the stereotypes or the over-simplifications such views produce.

Ethnocentric/ethnocentrism   Prioritizing one's own world-view and experiences, one's own culture as the 'norm' against which others are measured for their strange-ness, lack of development, difference, or exoticism. It usually implies taking western experiences as the norm.

Factors   A categorical explanatory variable (e.g. gender, location).

Feedback   The positive or negative effects of process and process-form linkages in physical systems.

Field test-bed   Field site chosen to collect data with the purpose of answering a given research hypothesis, or in choosing between competing hypotheses.

Generalization   A set of processes to reduce complexity on smaller-scale maps.

Geodemographics   The analysis of data on the social, economic and population characteristics of areas as part of an exercise leading to classification of areas in different categories. Often used as a basic step towards targeted marketing.

Geographical information systems (GIS)   Organized collections of data-processing methods which act on spatial data to enable patterns in those data to be understood and visualized.

Geostatistics   A suite of tools for solving problems with spatial data. The term is commonly applied (but not limited) to the investigation of spatial variation of processes and landforms on or near the Earth's surface.

Globalization   A highly contested term which expresses the ways economic, tech-nical, political, social and cultural processes connect people and places at a global scale, albeit in highly uneven and unequal ways. While these processes have been occurring for several hundred years, it is generally argued that current processes have intensified the scale and extent of these globalized connections.

Graphicacy   Skills in reading and constructing graphic modes of communication such as maps, diagrams and pictures.

Graphic variables   The visual attributes of a symbol that may be changed by a map designer.

Human Development Index   An index designed for use in the international comparison of standards of life, based on a limited set of variables (usually life expectancy at birth, adult literacy and gross national product per capita).

Hybrid/hybridity   Hybrids result from the mixing of two or more distinct things. In relation to cultures, the terms emphasize the positive and productive results of mixing different cultures, producing new and distinctive cultural forms, often out of the experiences of diaspora populations or transnational communities. It is used to reject the negative ideas of cultural assimilation or pollution.

Hyperlink   This is an invisible link that connects you to text segments in the same or other, pictures or HTML pages. Through (double) clicking on the link, you can access the text, picture or Webpage.

Hypothesis   A testable research statement or question for investigation, often linking observation to an assumed cause or set of causes.

Image analysis   Either manual or digital analysis including filtering, enhancement, combination of images, and classification of data.

Indices of dissimilarity and segregation   Statistical measures of differences and similarities in population distributions. The index of dissimilarity measures the difference between two distributions of population sub-groups. The index of segregation measures the difference between the distribution of a sub-group and the total population of which it is part. The scale for both indices runs from 0 (no difference) to 100 (total difference).

Inductive   Generating understandings or theories from the data themselves ('from the ground up'), as opposed to 'deductive' research in which a theory is already adopted and a hypothesis is tested.

Inferential statistics   Procedures applied to sample data in order to make generalizations, validated by probability statements, about the entire population from which the sample was drawn.

Instrumental record   Direct measurements of environmental variables such as temperature and precipitation.

Interval data   A measurement level indicating order and precise distances between observations.

***In vivo* codes**   Codes that use words or phrases that appear in the text.

Kriging   A method of geostatistical estimation at unsampled locations, in contrast to arbitrary mathematical functions used for interpolation.

Layers (overlays)   Refer to either data or to maps where one set of attributes is compared with another to examine strength of associations.

Map projection   A mechanism for systematic rendering of geographic co-ordinates of features on the earth's surface onto a flat medium.

Microdata   Data on individuals, as opposed to 'traditional' census data which relate to the aggregate populations of areas.

MLM   Multi-level modelling: a method from studying variation within data where that data are known to be organized within a particular hierarchical or clustered structure.

Model   A simplification of reality used for conceptual, analytical or statistic purposes. Models of systems and system operation are used to obtain closure, and hence predict future outcomes in otherwise intractable situations.

Multiple regression   A statistical technique to analyse the possible relationships between several predictor variables and one possible response variable.

Multispectral images   Images with relatively high information content (i.e. using many wavebands) which can be represented in numerous colour combinations.

Multivariate analysis   Analysis designed to examine multiple influences on a given data set or pattern.

Natural kinds   An ancient concept relating to the 'real' essence of things, allowing causal mechanisms, powers, or processes to be identified, as well as natural categories (e.g. species, landforms) that are independent of the context of enquiry.

Nominal data   A measurement level indicating the presence or absence of information.

Non-stationarity   The presence in a series of observations of trend or cyclicity which makes the identification of representative conditions (e.g. mean/variance) time-dependent.

NUD.IST   Stands for Non-numerical Unstructured Data Indexing Searching and Theorizing. It is a rather sophisticated yet user-friendly computer package for qualitative data analysis. Its tools include functions for coding texts, exploring documents and ideas about them and developing theories.

Nugget variance   Positive intercept on the ordinate of the semi-variogram indicating the unexplained variance.

Numerical models Models which use a conceptual model to define links between fundamental physical, chemical and occasionally biological principles, which are then represented mathematically in computer code.

Open coding Also called 'free coding'. Refers to the first stage of coding when the text is read through thoroughly and anything of interest is coded as the researcher goes along; used to 'open up' the text.

Ordinal data A measurement level indicating the relative significance of a feature, but not its precise dimensions.

Orientalism/orientalist Drawing on the work of Edward Said, 'orientalism' describes and critiques the set of ideas common among Europeans in their depictions of 'the Orient' during the era of imperialism. These emphasized both the apparent attractions and exotic nature of such people, places and cultures, and their supposed danger and lack of civilization, helping to justify European colonial endeavours. The terms have subsequently been applied more widely to all colonial situations and also to contemporary representations of 'other' cultures, peoples and places.

Other Often used in cultural analysis with inverted commas, and sometimes an initial capital letter, e.g. the 'Other', or 'other' cultures, to imply that such cultures, social groups or societies may not be as different as is implied by cultural and social norms.

Palaeoclimatic reconstruction The reconstruction of environmental conditions over timescales from hundreds to millions of years based upon direct and proxy evidence.

Palaeoclimatic signal and noise These arise in interpreting proxy environmental records where the effects of climate (the signal) have to be separated from the effects of all other non-climatic influences (the noise).

Parameterization Use of a numerical quantity in a model to represent the effect of a more complex process or property, or set of process interactions.

Partial plot A graphical plot between two variables, after accounting for the modelled effects of other possible explanatory variables.

Participatory research Involves working in a collaborative fashion to develop alternative ways to generate data based on joint agenda setting, analysis and control of outcomes.

Passive sensors Receive electromagnetic radiation from an external source (primarily reflected sunlight).

Philosophy of science Studies of science traditionally focusing on instructing practising scientists on the manner in which scientific research is appropriately conducted; but more recently concerned with how practising scientists pursue their research.

Positionality Recognizing and trying to understand the implications of the social position of the researcher with respect to the subjects, particularly with regard to power relations or cultural differences that may influence the process of the research and its interpretation. For example, how we are positioned in relation to various contexts of power (including gender, class, 'race', sexuality, job status, etc.) affects the way we understand the world. Likewise, the information given by informants to a researcher may depend on how the researcher is viewed in that particular context (threatening, insignificant, powerful).

Precision The size of the deviations between repeated estimates of a given statistic.

Process-colour printing Output derived by overprinting selected values of additive primary colours.

Proxy data sources Historical records or measurable properties of biological, chemical or physical systems, that provide quantitative information about past environmental conditions. Biotic proxies are based on the composition of plant and animal groups and/or measures of their growth rates; geological proxies quantify changes (physical or chemical) in Earth's materials that have accumulated through time, most commonly sediments in oceans or lakes, or ice in polar or alpine glaciers; historical proxies are records of events such as droughts, floods or harvest yields.

Range The limit of spatial dependence in the semi-variogram, that is, the lag at which the sill is reached.

Raster map Particularly useful for data produced routinely from satellites where values are recorded for equal areas associated with grid squares or pixels on a computer screen. Raster data can usually be entered into a GIS directly as numbers from the keyboard.

Ratio data A measurement level indicating precise distances between observations and with a non-arbitrary starting point.

Reflexivity Critical and conscious introspection and analytical scrutiny of oneself as a researcher. Reflexivity is not simply 'navel gazing'; it is examining our own practice in order to gain new insights into research.

Regionalized variable The outcome of many randomly located samples of a property in space, assumed to be the statistical realizations of a set of random variables, or a stochastic process.

Remote sensing   The collection of images of parts of the Earth's surface using specialized instruments, commonly aerial cameras and satellite sensors.

Sampling   The acquisition of information about a relatively small part of a larger group (population), usually with the aim of making inferential generalizations about the larger group.

Scale-linkage   Phenomenon whereby things of one size and/or time-span are composed of objects and time periods that are smaller and, in turn, are themselves formative components of larger assemblages and/or time periods.

Selective coding   Selecting the 'core category' or primary theme of the research and relating other categories/themes to it, then coding along that core category.

Semi-variogram   A graphical tool to summarize spatial variation, and also a model of underlying spatial process used for geostatistical estimation at unsampled locations.

Significance levels   The probability (set by the researcher as part of the testing procedure) of incorrectly rejecting a true hypothesis.

Sill variance   Maximum value in a semi-variogram determining the scale of spatial dependence. The steepness of the initial slope of the semi-variogram indicates the intensity of change in a property with distance and the rate of decrease in spatial dependence.

Snowballing   A technique used by researchers whereby one contact, or participant, is used to help to recruit another, who in turn puts the researcher in touch with another. The number of participants soon increases rapidly or 'snowballs'.

Spatial resolution   The size of features discernible from remotely sensed data.

SPSS   A widely used and very comprehensive statistical package in the social sciences. It stands for Statistical Product and Service Solutions. It contains a data editor to enter and handle data as well as extensive graphic possibilities for the display of data in tables and graphics.

Standard error   The standard deviation of a sampling distribution.

Structured approach   A sequential process to site selection and the subsequent measurement programme designed to yield robust, reproducible and general conclusions from field study.

System   A set of objects, together with the relationships between the objects, and between their attributes.

Townsend index   A composite index attempting to measure affluence and deprivation. The index was devised for use in the UK in health studies, but its principles have been adapted for use elsewhere and on other topics.

Transition   May refer to any number of processes of societal change (from dictatorship to democracy, for example) but most often used to refer to the contested and problematic transformation of post-communist societies and economies.

Transnational community   Social and cultural relations that transcend and escape the bounded spaces of the nation-state, possibly as a result of migration and diaspora or because the social group does not fit existing national boundaries (such as the Kurdish population).

Triangulation   Using multiple data sources and/or research methods to strengthen your results. For example, interviewing different populations around the same issue (multiple sources) or combining focus groups, surveys and published data such as the census (multiple sources and methods).

Uneven and unequal social relations/power relations   The idea that groups of people and places are differently positioned in social, economic, political or cultural terms. Social relations are uneven (across space and time) and they both reflect and shape unequal social and power relations between people and places (for example along lines of class and gender, or between developed and less developed countries).

Validation   The process by which a model is compared with reality.

Vector map   A more realistic configuration of maps, based on points and lines which are assembled into objects such as polygons. Vector data is usually digitized using a mouse-like device called a puck, centred on each of the points and lines defining the map object in question.

Verification   The process by which a numerical model is checked to make sure that it is solving the governing equations correctly.

Web 2.0   A term describing trends in the use of *World Wide Web* technology which focus on increased information sharing and collaboration between users. The term includes *social-networking sites*, *video sharing sites*, *wikis*, *blogs* and *folksonomies* (also known as collaborative tagging or social indexing). Web 3.0, or the semantic web, is a newly emerging concept relating to new ways of organizing information based on machine-based comprehension.

Index